Proceedings

Today, a steadily growing store of information is called for in order to understand the increasingly complex technologies used in modern automobiles. Functions, modes of operation, components and systems are rapidly evolving, while at the same time the latest expertise is disseminated directly from conferences, congresses and symposia to the professional world in ever-faster cycles. This series of proceedings offers rapid access to this information, gathering the specific knowledge needed to keep up with cutting-edge advances in automotive technologies, employing the same systematic approach used at conferences and congresses and presenting it in print (available at Springer.com) and electronic (at SpringerLink and Springer Professional) formats.

The series addresses the needs of automotive engineers, motor design engineers and students looking for the latest expertise in connection with key questions in their field, while professors and instructors working in the areas of automotive and motor design engineering will also find summaries of industry events they weren't able to attend. The proceedings also offer valuable answers to the topical questions that concern assessors, researchers and developmental engineers in the automotive and supplier industry, as well as service providers.

More information about this series at http://www.springer.com/series/13360

Peter E. Pfeffer
Editor

10th International Munich Chassis Symposium 2019

chassis.tech plus

Springer Vieweg

Editor

Prof. Dr. Peter E. Pfeffer
Munich University of Applied Sciences
Munich, Germany

ISSN 2198-7432 ISSN 2198-7440 (electronic)
Proceedings
ISBN 978-3-658-26434-5 ISBN 978-3-658-26435-2 (eBook)
https://doi.org/10.1007/978-3-658-26435-2
Springer Heidelberg Dordrecht London New York

Springer Vieweg

Printed on acid-free paper

This Springer Vieweg imprint is published by the registered company Springer Fachmedien Wiesbaden GmbH part of Springer Nature.
The registered company address is: Abraham-Lincoln-Str. 46, 65189 Wiesbaden, Germany

WELCOME

The increasing automation of driving functions and the electrification of powertrains present new challenges for the chassis with regard to complexity, redundancy, data security, and installation space. At the same time, the mobility of the future will also require entirely new vehicle concepts, particularly in urban areas. The intelligent chassis must be connected, electrified, and automated in order to be best prepared for this future. Will the chassis remain a strategic key area of expertise in 2025 or will it be just another component? This is one of the questions that will play an important role at the anniversary event of the International Munich Chassis Symposium chassis.tech plus. 2019 will mark the 20th anniversary of chassis.tech, which is being held for the tenth time in its present form as chassis.tech plus and will bring together leading experts from all over the world for an exchange of ideas and experience on the chassis, steering systems, brakes, and tires/wheels.

After a short review of the past and a look into the future compiled by Prof. Dr. Ulrich Seiffert, renowned keynote speakers such as Friedrich Eichler (Volkswagen), Atsushi Tsubouchi (Honda), and former Formula 1 world champion Nico Rosberg (TRE) will provide an overview of the latest approaches. Speakers from industry and research will discuss current developments in the chassis, steering systems, brakes, and tires/wheels in four parallel strands of lectures. The program is rounded off by two plenary sessions at the beginning and end of the conference.

We look forward to welcoming you to the Bayerischer Hof in the center of Munich and hope you have an exciting conference.

Prof. Dr. Peter E. Pfeffer
Munich University of Applied Sciences
Scientific Director of the Symposium

INDEX

CHASSIS.TECH PLUS SECTION

KEYNOTE LECTURES I

KEYNOTE LECTURES II

KEYNOTE LECTURES III

PARALLEL STRAND I

INNOVATIVE CHASSIS SOLUTIONS

RISK ANALYSES AND VIRTUAL METHODS

PARALLEL STRAND II

AI IN THE CHASSIS

LIGHTWEIGHT DESIGN AND DRIVEABILITY

CHASSIS.TECH SECTION

CHARACTERIZATION OF ADAS DRIVING PROPERTIES

DEVELOPMENT METHODS

RIDE COMFORT

STEERING.TECH SECTION

DEVELOPMENT METHODOLOGY AND AI

NEW STEERING TECHNOLOGIES

HUMAN-MACHINE INTERFACE

BRAKE.TECH SECTION

SYSTEM REQUIREMENTS AND BRAKE FEEL

INNOVATIVE BRAKE COMPONENTS

SYSTEMS FOR ELECTRIC / CONNECTED VEHICLES

TIRE.WHEEL.TECH SECTION

WHEEL TECHNOLOGIES AND REQUIREMENTS

TIRE TESTS AND SIMULATION

VEHICLE DYNAMICS AND TIRE DEVELOPMENT

SPEAKERS

Claudio Annicchiarico
Meccanica 42 Srl, Italy

Patricio Barbale
IHS Global Srl, Italy

David Benz
Automotive Engineering,
Munich University
of Applied Sciences

Yves Billet
Dr. Ing. h.c. F. Porsche AG

Dr. Alexander Brandt
Robert Bosch GmbH

Jochen Buck
Daimler AG

Ralf Bux
ENTEC-STRACON GMBH

Francesco Calabrese
Fraunhofer Institute for
Industrial Mathematics (ITWM)

Xabier Carrera Akutain PhD
Toyota Motor Europe, Belgium

Pierre Chandezon
Michelin Technology Centre
France

Fan Chang
AUDI AG

Alessandro Contini
Hyundai Motor Europe
Technical Center GmbH

Luca Dusini
Maserati S.p.A., Italy

Dr. Philipp Ehring
branmatt || legal

Friedrich Eichler
Volkswagen AG

Rudolf Einsiedel
Cordenka GmbH & Co. KG

Dirk Ferge
JTEKT Europe

Dr. Sierk Fiebig
Volkswagen AG

Dr. Sebastian Gramstat
AUDI AG

Bruno Guimard
Michelin Technology Centre
France

Wenrui Han
Chair of Automobile Engineering,
TU Dresden

Matthieu Hascoet
Tallano Technologie, France

Johann Haselberger
Research ADAS/AD,
Kempten University
of Applied Sciences

Michael Haßenberg
DMecS GmbH & Co. KG

Dr. Manuel Höfer
Dr. Ing. h.c. F. Porsche AG

Dr. Jens Hoffmann
Continental Teves AG & Co. oHG

Dr. Thomas Kersten
Volkswagen AG

Guru Bhargava Khandavalli
Institute for Automotive Engineering (ika), RWTH Aachen University

Dr. Florian Klinger
Institute of Mechanics and Mechatronics, TU Vienna, Austria

Rob Kraaijeveld
FEV Europe GmbH

Frédéric Etienne Kracht
Department of Mechanical and Process Engineering, University of Duisburg-Essen

Dr. Manfred Kraus
Schaeffler Paravan Technologie GmbH & Co. KG

Patrick Krupka
Volkswagen AG

Peter Kuhn
SGL Technologies GmbH

Samuel Lago Places
Groupe PSA

Sebastian Lefèvre
Dr. Ing. h.c. F. Porsche AG

Dr. Günter Leister
Daimler AG

Yuran Liang
BMW Group

Christoph Loy
SGL Technologies GmbH

Prof. Dr. Giampiero Mastinu
Department of Mechanical Engineering, Politecnico di Milano, Italy

Diego Minen
VI-grade s.r.l.

Bertram Möller
Nexteer Automotive Germany GmbH

Stefano Murgia
FCA Italy S.p.A.

Marcin Nakielski
Nexteer Automotive Poland Sp. z o.o.

Dominik Nees
Schaeffler Technologies AG & Co. KG

Dr. Sergey Orlov
Hella GmbH & Co. KGaA

Dr. Marcus Perner
IAV GmbH

Prof. Dr. Peter E. Pfeffer
Automotive Engineering, Munich University of Applied Sciences

Nico Rosberg
TRE GmbH, F1 World Champion 2016

Alessandro Salgarello
Hyundai Motor Europe Technical Center GmbH

Christina Sax
Automotive Engineering,
Munich University
of Applied Sciences

Dr. Frank Schmidt
Bosch Engineering GmbH

Michele Sigilló
SIGICONTROL s.r.o.,
Czech Republic

Dr. Christian Strümpler
Joyson Safety Systems
Aschaffenburg GmbH

Guido Tosolin
Applus IDIADA Group, Spain

Atsushi Tsubouchi
Honda R&D Co., Ltd.
Automobile R&D Center, Japan

Sebastian Wagner
BMW Group

Felix Wenzel
Department of
Automotive Engineering,
TU Ilmenau

Matthias Wilmes
FEV Europe GmbH

CHASSIS.TECH PLUS SECTION

KEYNOTE LECTURES I

Vehicle development – how good are the forecasts? A look back and into the future

Prof. Dr. Peter E. Pfeffer, Automotive Engineering,
Munich University of Applied Sciences;
Prof. Dr. U. Seiffert, WiTech Engineering GmbH

This manuscript is not available according to publishing restriction. Thank you for your understanding.

How 'Automated Driving' changes the requirements for classic chassis development – a journey through time

Friedrich Eichler, Dr. S. Busch, S. Kreutz, Dr. J. Reichel,
K. Schebsdat, Volkswagen AG

This manuscript is not available according to publishing restriction. Thank you for your understanding.

Chassis for future mobility: challenges and solutions

Nico Rosberg, TRE GmbH, F1 World Champion 2016

This manuscript is not available according to publishing restriction. Thank you for your understanding.

KEYNOTE LECTURES II

The chassis market and technology trends – understanding the impact of automotive megatrends on the chassis sector

Patricio Barbale, IHS Global Srl, Italy

IHS Markit
The Chassis market and technology trends
Understanding the impact of automotive megatrends on the chassis sector
10th International Munich Chassis Symposium
Munich, 25th June 2019
Patricio Barbale

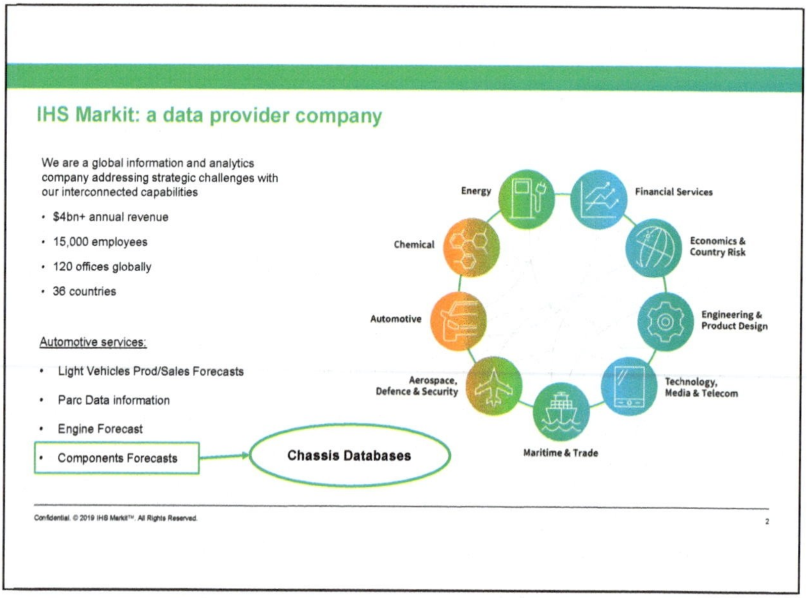
IHS Markit: a data provider company
We are a global information and analytics company addressing strategic challenges with our interconnected capabilities
• $4bn+ annual revenue
• 15,000 employees
• 120 offices globally
• 36 countries
Automotive services:
• Light Vehicles Prod/Sales Forecasts
• Parc Data information
• Engine Forecast
• Components Forecasts
Chassis Databases
Energy
Financial Services
Chemical
Economics & Country Risk
Automotive
Engineering & Product Design
Aerospace, Defence & Security
Technology, Media & Telecom
Maritime & Trade
Confidential. © 2019 IHS Markit™. All Rights Reserved.
2

Table of Contents

 3

Alternative Propulsion Vehicles

The powertrain revolution towards Electrification

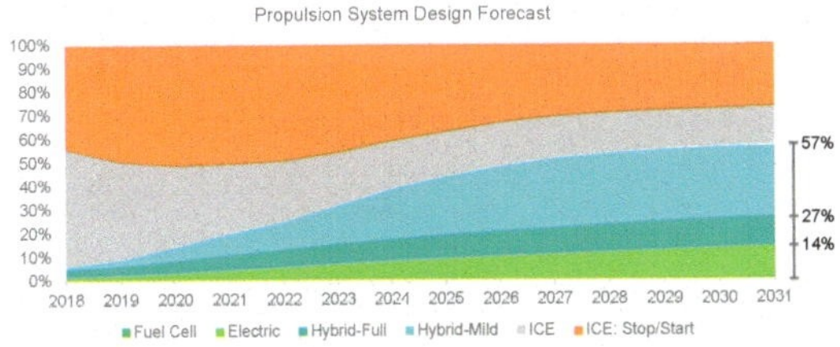

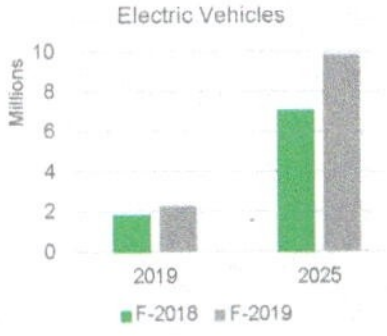

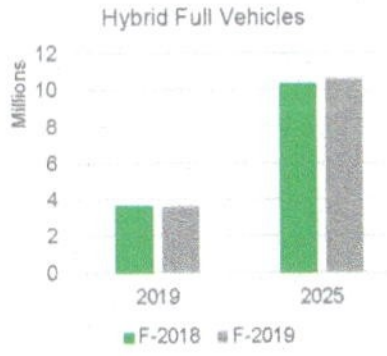

- Our monthly revision of engine database highlights a positive trend for Electric vehicles. Big efforts of Carmakers in this direction;
- Significant increase of EVs by 2031 (mostly in EU/CH), but not a revolution;
- 30% of Hybrid-Mild vehicles: what does that mean for Chassis?

 4

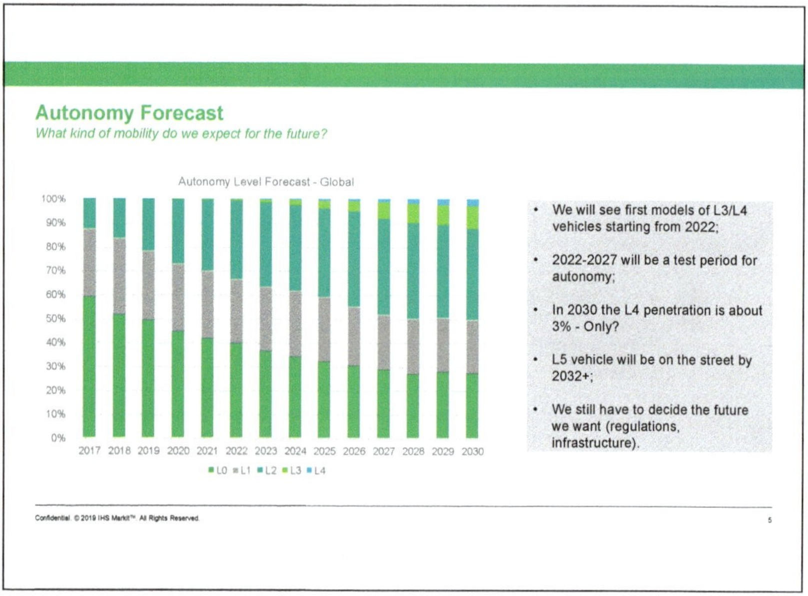
Autonomy Forecast
What kind of mobility do we expect for the future?
Autonomy Level Forecast - Global
100%
90%
80%
70%
60%
50%
40%
30%
20%
10%
0%
2017 2018 2019 2020 2021 2022 2023 2024 2025 2026 2027 2028 2029 2030
L0 L1 L2 L3 L4
We will see first models of L3/L4 vehicles starting from 2022;
2022-2027 will be a test period for autonomy;
In 2030 the L4 penetration is about 3% - Only?
L5 vehicle will be on the street by 2032+;
We still have to decide the future we want (regulations, infrastructure).
Confidential. © 2019 IHS Markit™. All Rights Reserved.
5

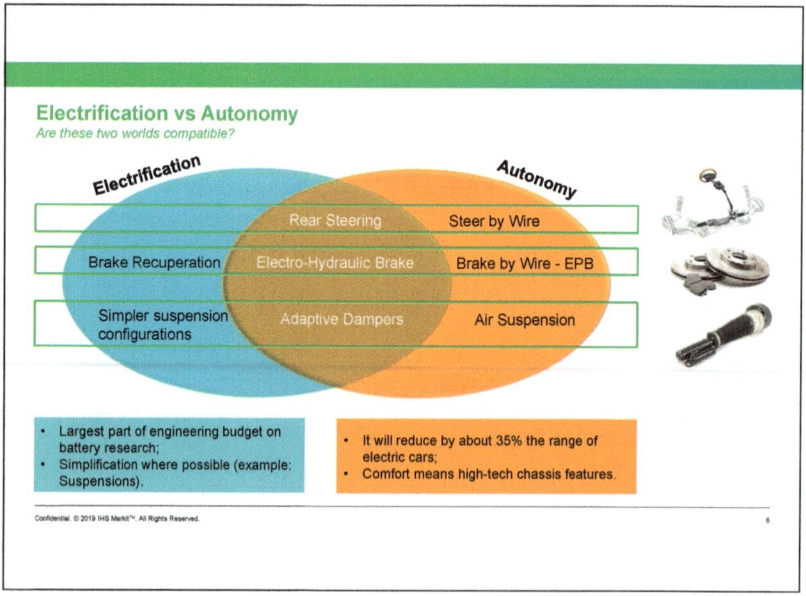
Electrification vs Autonomy
Are these two worlds compatible?
Electrification
Autonomy
Rear Steering
Steer by Wire
Brake Recuperation
Electro-Hydraulic Brake
Brake by Wire - EPB
Simpler suspension configurations
Adaptive Dampers
Air Suspension
Largest part of engineering budget on battery research;
Simplification where possible (example: Suspensions).
It will reduce by about 35% the range of electric cars;
Comfort means high-tech chassis features.
Confidential. © 2019 IHS Markit™. All Rights Reserved.
6

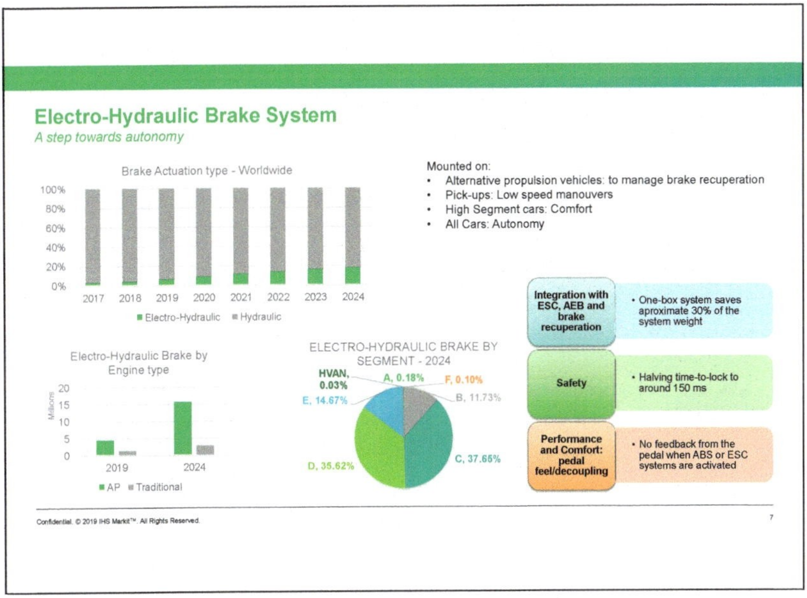
Electro-Hydraulic Brake System
A step towards autonomy
Brake Actuation type - Worldwide
100%
80%
60%
40%
20%
0%
2017 2018 2019 2020 2021 2022 2023 2024
Electro-Hydraulic Hydraulic
Mounted on:
• Alternative propulsion vehicles: to manage brake recuperation
• Pick-ups: Low speed manouvers
• High Segment cars: Comfort
• All Cars: Autonomy
Electro-Hydraulic Brake by Engine type
Millions
20
15
10
5
0
2019
2024
AP Traditional
ELECTRO-HYDRAULIC BRAKE BY SEGMENT - 2024
HVAN, 0.03%
A, 0.18%
F, 0.10%
B, 11.73%
E, 14.67%
C, 37.65%
D, 35.62%
Integration with ESC, AEB and brake recuperation
• One-box system saves aproximate 30% of the system weight
Safety
• Halving time-to-lock to around 150 ms
Performance and Comfort: pedal feel/decoupling
• No feedback from the pedal when ABS or ESC systems are activated
Confidential. © 2019 IHS Markit™. All Rights Reserved.
7

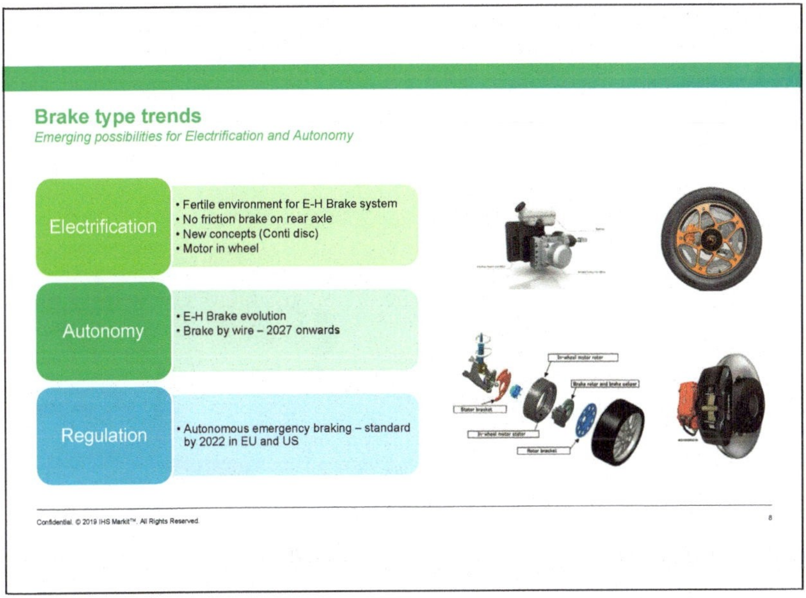
Brake type trends
Emerging possibilities for Electrification and Autonomy
Electrification
• Fertile environment for E-H Brake system
• No friction brake on rear axle
• New concepts (Conti disc)
• Motor in wheel
Autonomy
• E-H Brake evolution
• Brake by wire – 2027 onwards
Regulation
• Autonomous emergency braking – standard by 2022 in EU and US
Confidential. © 2019 IHS Markit™. All Rights Reserved.
8

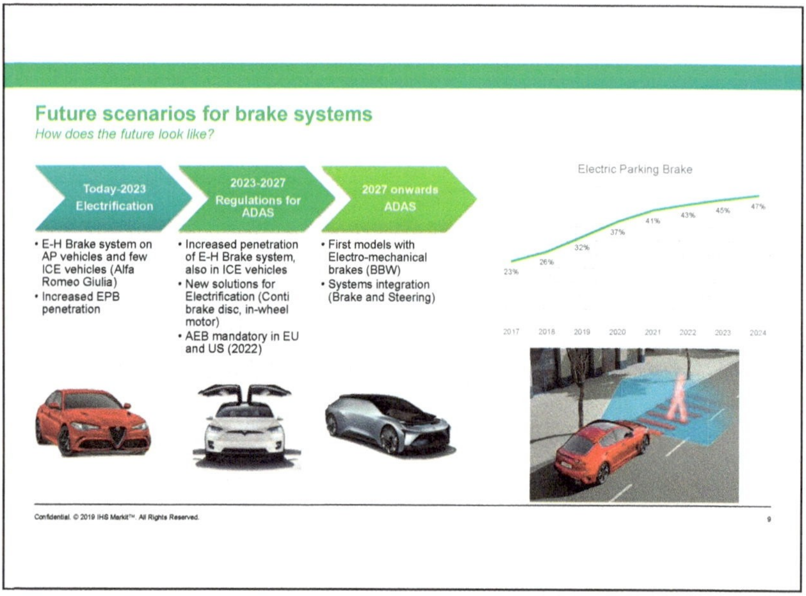
Future scenarios for brake systems
How does the future look like?
Today-2023
Electrification
2023-2027
Regulations for ADAS
2027 onwards
ADAS
• E-H Brake system on AP vehicles and few ICE vehicles (Alfa Romeo Giulia)
• Increased EPB penetration
• Increased penetration of E-H Brake system, also in ICE vehicles
• New solutions for Electrification (Conti brake disc, in-wheel motor)
• AEB mandatory in EU and US (2022)
• First models with Electro-mechanical brakes (BBW)
• Systems integration (Brake and Steering)
Electric Parking Brake
23%
26%
32%
37%
41%
43%
45%
47%
2017
2018
2019
2020
2021
2022
2023
2024
Confidential. © 2019 IHS Markit™. All Rights Reserved.
9

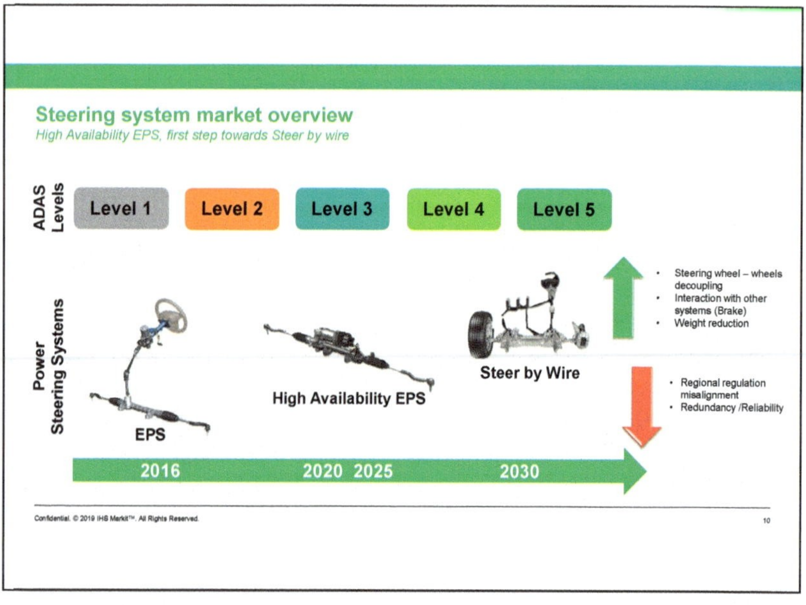
Steering system market overview
High Availability EPS, first step towards Steer by wire
ADAS Levels
Level 1
Level 2
Level 3
Level 4
Level 5
Power Steering Systems
EPS
High Availability EPS
Steer by Wire
• Steering wheel – wheels decoupling
• Interaction with other systems (Brake)
• Weight reduction
• Regional regulation misalignment
• Redundancy /Reliability
2016
2020 2025
2030
Confidential. © 2019 IHS Markit™. All Rights Reserved.
10

Steering technologies for Electrification and Autonomy

When will SBW technology be available?

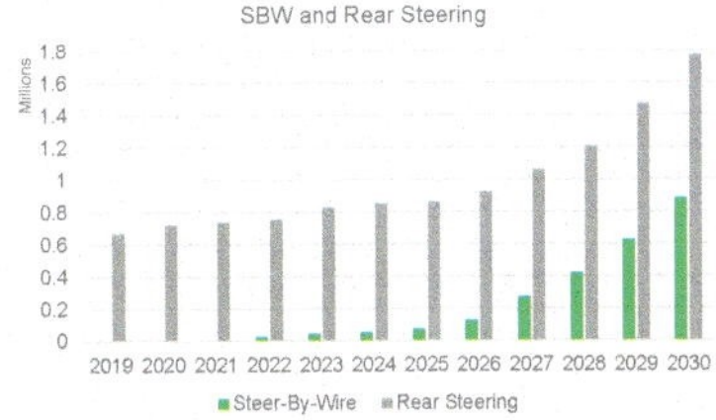

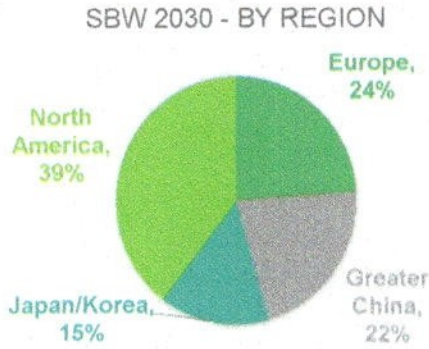

Rear Steering:

- Electrification: Longer wheel bases (due to batteries) could increase the need for rear steering technology;
- Autonomy: L4/L5 cars will need SBW system. Rear steering is a good redundancy system for SBW.

Steer by Wire:

- SBW technology will be fundamental for L4/L5 vehicles;
- Our forecast is conservative. 1 Million cars with SBW in 2030 is not so much, could be more and we expect to be more, but for the moment we don't want to over-estimate it;
- Regionally speaking, North America and Europe will lead the SBW market as per their technology advancement and customer requests.

 11

The future of Shock Absorber and Suspension

What kind of suspensions we will need in the future?

Electrification

- Cost saving: simplification of suspensions (more torsional beams)
- No need of a particular suspension configuration compared with ICE cars

Autonomy

- Comfort: air suspensions
- Control: adaptive dampers
- Preview systems: cameras, lidars, etc.

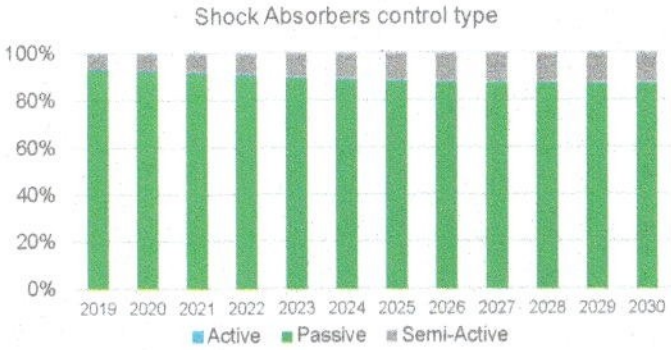

More than 55M semi-active shocks in 2030

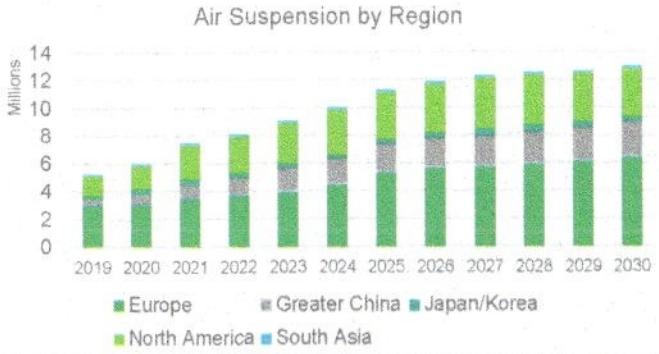

Increasing penetration of SUV = increase Air Susp penetration

 12

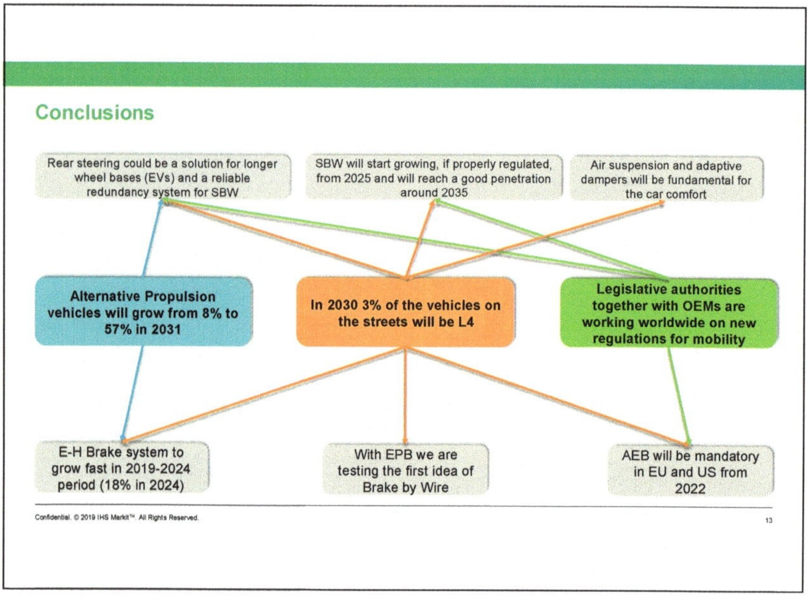
Conclusions
Rear steering could be a solution for longer wheel bases (EVs) and a reliable redundancy system for SBW
SBW will start growing, if properly regulated, from 2025 and will reach a good penetration around 2035
Air suspension and adaptive dampers will be fundamental for the car comfort
Alternative Propulsion vehicles will grow from 8% to 57% in 2031
In 2030 3% of the vehicles on the streets will be L4
Legislative authorities together with OEMs are working worldwide on new regulations for mobility
E-H Brake system to grow fast in 2019-2024 period (18% in 2024)
With EPB we are testing the first idea of Brake by Wire
AEB will be mandatory in EU and US from 2022
Confidential. © 2019 IHS Markit™. All Rights Reserved.
13

Thank you!!!
Patricio Barbale
Senior Analyst, Planning Solutions
Corso Venezia, 5 | Milan
P: +39 02 3601 7320
patricio.barbale@ihsmarkit.com
IHS Markit Customer Care
CustomerCare@ihsmarkit.com
Americas: +1 800 IHS CARE (+1 800 447 2273)
Europe, Middle East, and Africa: +44 (0) 1344 328 300
Asia and the Pacific Rim: +604 291 3600
Disclaimer
IHS Markit

Will driving still be fun in the future? Vehicle dynamics systems through the ages

Dr.-Ing. Frank Schmidt, CTO, Bosch Engineering GmbH

Dr. Lars König, Chief Expert Control Theory and Vehicle Motion, Bosch Engineering GmbH

In public discourse, the future of individual mobility and the active enjoyment of driving are frequently viewed as contradictions. This overlooks the reality that there will still be drivers in the future who attach great importance to vehicles offering supreme driving dynamics. On the one hand, drivers will benefit from a high degree of personalization thanks to new vehicle-dynamics functions that result from ever-advancing electronic connectivity, sensor and actuator development, in addition to the electrification of the powertrain. On the other hand, and especially in the supercar segment, these functions will open up entirely new dimensions in vehicle dynamics by assisting the driver right up to the physical limits of driving, thereby increasing driving stability and safety. The growing number of challenges encountered in the course of functional development calls for an integrated development process.

Introduction

Current and future vehicle development alike will be shaped by two megatrends. Firstly, there is the electrification of the powertrain, which is intended to further reduce and shift traffic-related CO_2 emissions and immissions. Concepts in this field range from purely electric powertrains right through to complex hybrid topologies. Secondly, vehicle manufacturers and their suppliers are working to automate driving functions further with a long-term objective of autonomous driving, where the vehicle assumes lateral and longitudinal control tasks entirely independently and without the driver's supervision. The primary drivers in this case are driving safety and comfort, in addition to a reduction in consumption and emissions.

On a technical level, the connectivity of electronic systems and components within the vehicle itself, with other vehicles, and with the traffic infrastructure is paving the way for both of these development trends. For example, in just a few years, we will already witness vehicle components using the so-called "internet of things" to independently establish and terminate data connections again as required; both within the vehicle and with the outside world. This super-fast, real-time data exchange between connected vehicles will assist the on-board sensor system with surroundings perception. By drawing on information of this kind, the vehicles will then be able to predictively adapt their systems to the respective traffic situation and road surface.

However, when all is said and done, there is still an emotional element to driving that must not be disregarded. In this context, the role of personalized driving functions is becoming more and more important. After all, each and every driver has their own subjective preference for how the vehicle ought to "feel". Incidentally, this also applies to lateral and longitudinal control with a view to the highly automated driving functions of the future: What one driver perceives as being superior acceleration or a comfortable cornering speed in autopilot mode might simply be too fast or sporty for someone else.

What is more, those who purchase high-performance sports cars in particular expect concepts that make the vehicle easier to master – even for inexperienced drivers – and let them brave the limits of vehicle dynamics without sacrificing safety. The challenge of applying such technology to vehicles essentially lies in guaranteeing comprehensible, reproducible, and authentic driving characteristics in any situation – right up to the very physical limits. In an ideal scenario, the driver should not even notice the intervention of the closed-loop control systems involved. This helps drivers to build trust in the dynamic capabilities of the vehicle and in their own abilities, with the result that they can even tackle unexpected situations with confidence.

All considerations mentioned above leave vehicle manufacturers and suppliers faced with the task of fulfilling a variety of occasionally opposing development requirements concerning efficiency, low pollutant emissions, comfort and safety, not to mention injecting driving with an added fun factor.

From ABS to ACC

For more than 40 years now, the development of new comfort and safety systems in the vehicle chassis has been inseparably linked with the integration of electronic control components. The launch of the Bosch ABS system in 1978 laid down the first marker. The operating principle is now a familiar one: The electronic control system detects the tendency for wheels to lock at a very early stage and precisely reduces the brake pressure. This means that the vehicle can still be steered, even with the brakes fully applied, enabling the driver to avoid obstacles in an emergency and bring the vehicle to stop safely. In a later development, the traction control system (TCS) entered volume production in 1986. This system prevents the wheels from spinning when starting off and accelerating. In 1995, the next generation of active safety systems followed in the form of the electronic stability program (ESP). ESP reduces the risk of skidding by making precise brake and engine interventions to keep the vehicle safely on track.

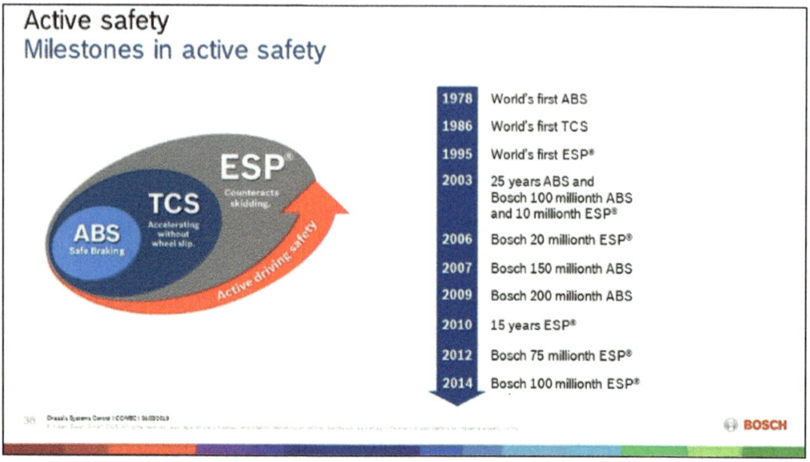

Fig. 1: Milestones in active safety systems from Bosch (© Robert Bosch GmbH)

The ABS, TCS and ESP safety systems make a significant contribution to preventing accidents, or at least minimizing their impact. For this precise reason, the systems are counted as mandatory safety equipment for new vehicles in many regions worldwide.

While first-generation ABS and ESP systems were still more or less "stand-alone" solutions that only required minimal integration with other electronic systems, such as the engine management, they have since become an integral component of vehicles' electronics architecture. New functions are nowadays implemented by numerous, varied system components that operate in an interconnected manner while complementing one another. This opens up new options for control systems, both in terms of active safety and also with regard to comfort, efficiency, and vehicle dynamics. The evolution of ESP into the adaptive cruise control (ACC) system is just one example. The vehicle's surround sensors detect the vehicle driving in front, while the ACC system uses the sensor data to calculate its speed and maintain the desired distance through brake and engine interventions.

Fig. 2: Adaptive cruise control (ACC), (© Robert Bosch GmbH)

As the functional spectrum of vehicle-dynamics control systems expands, there is also a shift in the criteria used to assess their functional quality: For example, while the NVH characteristics of the ESP might be irrelevant in safety-critical situations, drivers prefer an acoustically unobtrusive system for comfort, efficiency and vehicle-dynamics functions where the interventions made by the control system are neither palpable nor audible.

Efficiency systems for electric vehicles

Connectivity enables regeneration functions to be incorporated into hybrid and electric vehicles. When the vehicle brakes, these functions serve to reconvert kinetic energy (which would otherwise be converted into heat by the brakes) back into electrical energy before storing it. The challenge when developing these brake systems is really in the overlapping of braking and regeneration performance: This must be designed to achieve high recuperation rates while simultaneously remaining as imperceptible as possible for the driver in its regulation. During the design process, the brake force distribution, brake pedal "feel", and the control of the ABS function, among other things, are all subjected to particular scrutiny.

The hardware components of regenerative braking systems must be matched to the specific requirements of electric vehicles. This applies to the brake booster, for instance. Due to the lack of vacuum otherwise generated by a combustion engine for a conventional brake booster, braking assistance must be provided by electromechanical means. Here, the driver's braking commands are picked up by a sensor on the brake pedal and transmitted to the system's ECU. An integrated electric motor then converts its torque into the required power assistance by means of a transmission unit. The force applied by the booster and the driver is converted into hydraulic pressure in a conventional master cylinder.

Fig. 3: Vacuum-independent brake booster, (© Robert Bosch GmbH)

Electrohydraulic systems are the next step toward "brake-by-wire", where the brake system exclusively comprises E/E components. In these concepts, the transmission of the brake signal is purely electronic, with hydraulic systems used to transmit the braking force to the brake cylinders at the wheels. When the brake pedal is depressed, the ECU employs integrated sensors to calculate the driver's braking intention, from which it derives the actuation signals for the electric motor. At the same time, the system utilizes a pedal-feel simulator to generate a tangible resistance in the brake pedal for the driver's benefit. The electric motor moves the hydraulic brake piston; the pressure generated is then transmitted to the wheel brakes via the brake lines. Advantages of this system include ultra-short braking distances thanks to highly dynamic pressure accumulation, a longer range for hybrid and electric vehicles, and elevated levels of freedom in energy recuperation.

Fig. 4: Electrohydraulic brake system, (© Robert Bosch GmbH)

Alongside the brakes, other chassis components – such as the steering system – have also been progressively developed and electrified over the past few years. In an initial step, a system for hydraulic servo assistance was developed that featured an electric pump instead of a mechanical one. The second step took the shape of an electrical servo-assistance system in which the electric motor drives the steering directly via a transmission unit. As with the brake system, a purely electrical "by-wire" system without any mechanical connections – in this case, the steering column – is also the long-term development goal in this case. With a view to the vehicle dynamics, the connectivity of the electrified drive, brake, and steering systems opens up new opportunities for adapting the vehicle to changing conditions with even greater speed and precision. On the one hand, this benefits the development of personalized driving functions; on the other hand, "by-wire" systems are an essential element of advanced, fully automated driving concepts that will eventually lead to autonomous "robo-cabs".

Driving fun: sporty and safe thanks to electronic functions

Alongside vehicle safety and efficiency – and parallel to the development of automated driving functions – optimizing the vehicle dynamics is a crucial design objective for future vehicle-dynamics control concepts. In this context, the increased, cross-domain connectivity of the electronic systems will pave the way for new driving functions designed to make driving more fun by concurrently increasing the potential of the vehicle dynamics and expanding the options for personalization. In the context of a targeted development of vehicle-dynamics functions, the entire development process needs to be aligned with the requirements of system connectivity. As early as the conceptual phase, for example, a definition needs to be set for the hardware components that will need to be integrated into the overall control system as the actuators of the vehicle-dynamics functions, what is expected of the vehicle in terms of its dynamics, and which forms of personalization must be implemented. During development, the individual steps and milestones at a component and system level must then be continuously matched to the development of the overall vehicle. To this end, Bosch Engineering has drawn up an integrated development approach in which all specialists involved in the project work together on site. This shortens communication paths and leverages potential for synergy due to the entire design of the vehicle dynamics originating from a single source.

Torque vectoring, which involves the lateral dynamics being influenced by gentle brake interventions, is just one example of connectivity in software where integration is based on ESP hardware that is already in place anyway. Depending on the driving situation, speed, and the predefined system settings, the control system fine-tunes the balance (understeering and oversteering) of the vehicle while improving its stability at the limits of its dynamic range. In sports and race cars, this can be felt in the higher speeds during zig-zag driving and lane-changing maneuvers, greater driving stability when taking evasive action, improved balance when braking on bends, optimized yaw damping in the event of extreme maneuvers at high speeds, and ultimately shorter lap times on the race track.

Fig. 5: Torque vectoring through brake interventions, (© Bosch Engineering GmbH)

For electric vehicles equipped with an electric motor on each individual wheel, the torque-vectoring function can be realized by adapting the distribution of the motor torques. High data speeds and transmission rates within the electronic network are required for this type of closed-loop control to work at a sufficiently high level. Nonetheless, the capabilities of current network architectures are increasingly reaching their limits with these highly dynamic control requirements. One solution involves an electronics architecture where parts of the software are relocated from the chassis control unit to the electric-motor management system. The results of this are shorter transmission paths, faster switching times, and therefore a highly sensitive and precise torque-vectoring function.

With rear-wheel steering as a supplement to the torque-vectoring software, the lateral dynamic potential of the vehicle can be extended even further because the compromises of traditional axle developments no longer pose any restrictions. The results are a simultaneous optimization of safety and agility through improved directional stability at high speeds, further elevated vehicle dynamics in the low and medium speed range, optimized yaw damping, improved driving safety during highly dynamic maneuvers, and a smaller turning circle. The high flexibility in design also frees up space for additional optimization approaches that have not previously received much attention. As such, it is possible to use a "chauffeur mode" to better align the vehicle's driving characteristics to the needs of the passengers sitting in the rear seats, to cite one example.

Fig. 6: Bosch Engineering develops control algorithms for optimum realization of the rear-wheel steering function, (© Bosch Engineering GmbH)

While torque vectoring and rear-wheel steering serve to optimize the lateral dynamics, the function of the dynamic drive assist (DDA) system from Bosch Engineering targets the vehicle's longitudinal dynamics. In simpler terms, DDA transfers the advantages of ACC to an additional active function by regulating the distance independently of the cruise control. On open stretches of road, the driver simply sets the desired speed using the accelerator and can accelerate at any time without limitation, as usual. If the DDA detects an obstacle such as a slower-moving road user in front of its own vehicle, the system calculates the speed-dependent distance using the accelerator position and a minimum distance stored as a custom setting. If the system's own vehicle continues to approach the vehicle in front, it will be gently braked through reductions in the engine torque and interventions in the brakes. In spite of this, the driver can still continue to accelerate the vehicle using the pedal, for example, in order to overtake the slower vehicle. Acceleration only ceases to be possible in the final escalation stage, once a predefined minimum distance to the obstacle is no longer maintained. The system then initiates the deceleration of its own vehicle and no longer allows it to get any closer to the vehicle in front. In doing so, it provides assistance and relief while still maintaining dynamic handling and the fun of driving through its driver-focused approach.

As the degree of connectivity in the vehicle increases, the possibilities that open up for vehicle-dynamics control concepts in terms of safety, comfort, and (most importantly) driving performance will become more and more promising. Future developments will progress through integrated approaches for controlling a variety of actuators using one central vehicle-dynamics controller. In such a connectivity concept, the individual vehicle-dynamics systems are interpreted as one domain and linked with each other. The decisive factor for the actuation of such systems is given by the desired driving-dynamics condition of the vehicle in question. In practical application, the driving-control function derived on this basis is executed by the system presenting itself as the most suitable actuator in the respective driving situation.

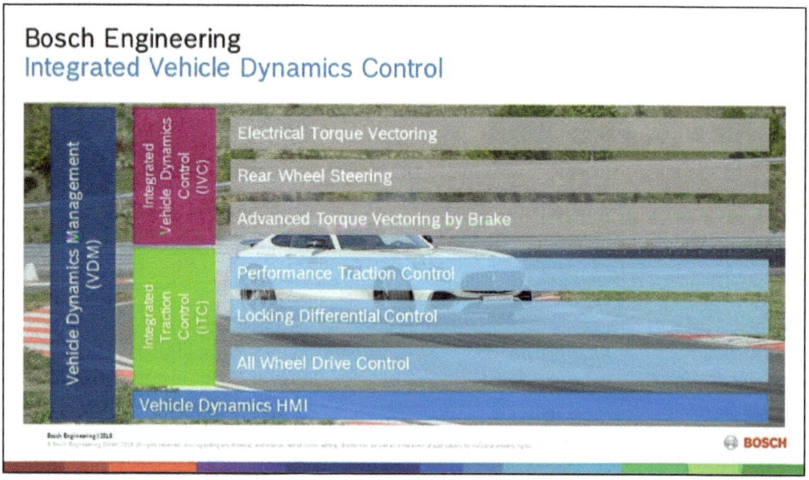

Fig. 7: Evolution of integrated driving-dynamics control systems, (© Bosch Engineering GmbH)

Bosch Engineering took the first step on this path back in 2013 with its integrated vehicle-dynamics control (IVC). The integrated lateral-dynamics control system uses the hardware systems already present in the vehicle as actuators, namely the brakes, rear-wheel steering, integral steering, active stabilizers, and differentials. Each of these components is then assigned the respective vehicle-dynamic characteristics that they are specifically able to influence. In this context, for example, the rear-wheel steering has a significant effect on a vehicle's responsiveness and required steering angle. By contrast, the balance can be most effectively altered by the brake system operating in combination with the ESP, among other things. The central vehicle-dynamics control system activates the individual actuators in order to set the desired vehicle characteristics and prevents the systems responding in parallel and interfering with one another.

Einfluss fahrdynamischer Stellglieder auf die Fahreigenschaften eines Fahrzeugs
Actuators sphere of action to influence vehicle dynamics behavior

	Aktiver Stabilisator *Active stabilizer*	Rad-individueller Elektromotor *Wheel-individual electric motor*	Aktives Differential *Active differential*	Bremse *Brake*	Hinterachs-lenkung *Rear-wheel steering*	Aktive Vorderachs-lenkung *Active front steering*
Ansprechverhalten *Steering response*	Kein Einfluss	Großer Einfluss	Kleiner Einfluss	Kein Einfluss	Großer Einfluss	Kleiner Einfluss
Gierdämpfung *Yaw damping*	Kleiner Einfluss	Großer Einfluss	Kleiner Einfluss	Kleiner Einfluss	Kleiner Einfluss	Kein Einfluss
Balance *Balance*	Kleiner Einfluss	Großer Einfluss	Kleiner Einfluss	Kleiner Einfluss	Kein Einfluss	Kein Einfluss
Lenkwinkelbedarf *Steering effort*	Kein Einfluss	Kleiner Einfluss	Kein Einfluss	Kein Einfluss	Großer Einfluss	Großer Einfluss
Anlehnen *Smooth leaning*	Kleiner Einfluss	Großer Einfluss	Kleiner Einfluss	Kleiner Einfluss	Kein Einfluss	Großer Einfluss

■ Kein Einfluss *No influence*

■ Kleiner Einfluss *Low influence*

■ Großer Einfluss *High influence*

Fig. 8: Influence of the individual actuators of the IVC on vehicle-dynamics control, (© Bosch Engineering GmbH)

Similar concepts are also being developed for traction control. For example, in a vehicle with all-wheel-drive, this might involve the interplay between the engine management, brake interventions, and inter-axle and inter-wheel differential lock. In this case, too, an innovative, model-based algorithm controls the driving function by activating the existing actuators in a coordinated manner for optimum vehicle dynamics. This enables the maximum adhesive potential of all four wheels at that moment to be utilized to best effect. Due to the fact that such a concept not only integrates the traction-relevant actuators but also their ECUs, the control system is very responsive, meaning that there is no discernible torque delay when the controller intervenes. In doing so, the system facilitates maximum speeds when exiting bends and provides the sensation of a natural driving experience with a high fun factor thanks to quicker and more precise torque control.

The next integration step might take the shape of a central controller that combines the functions of both approaches and offers an all-in-one platform for coordinating all vehicle-dynamics actuators in the vehicle. The parameters of the integrated vehicle-dynamics control system can be adapted by modifying the software. Firstly, this provides scope for efficient application in the vehicle. Secondly, a range of pre-programmed driving modes can be created with ease: At the touch of a button, the driver can then choose from a broad spectrum of vehicle settings: whether maximum comfort and safety, high dynamics and agility, or active drifting (for racing applications). Moreover, the driver can freely configure the driving characteristics in an additional mode using the fundamental parameters "stability", "balance", and "agility". This custom configuration can be saved and reactivated at any time.

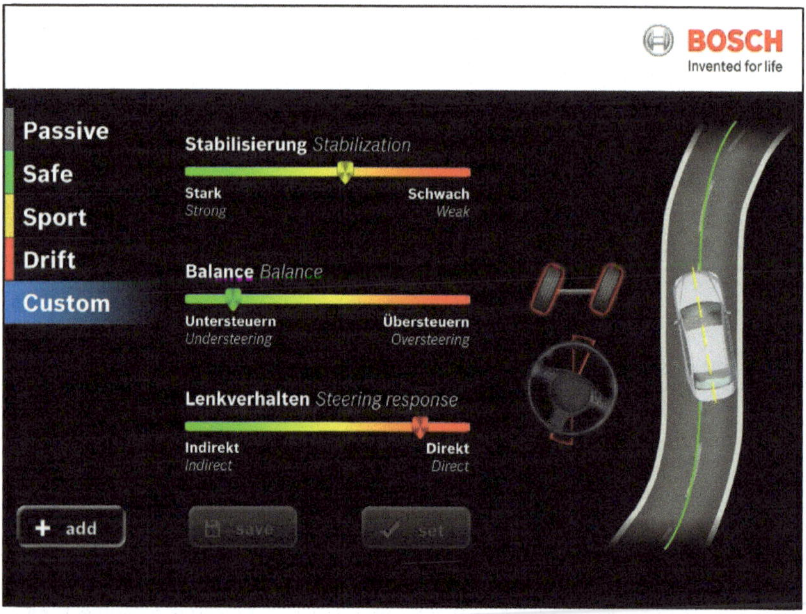

Fig. 9: Option of custom-tuning systems using the steering system as an example, (© Bosch Engineering GmbH)

Custom-tailored assistance for vehicle dynamics

In several years, automated driving functions with a SAE level of 3 and above, using which the vehicle might at least be able to continue to move for a certain time without any intervention from the driver, might even be an everyday feature. To pave the way for their introduction, new technologies for generating and processing information will be incorporated into vehicles that will enable the vehicle's current driving status in relation to its surroundings to be analyzed in great detail, thus making it possible to predict the exact conditions outside the vehicle. For instance, these might be sensors that scan the vehicle's surroundings with even higher accuracy while also supplying information on the condition of the road, such as the type of surface, asphalt temperature, and level of adhesion. The use of cloud information for vehicle-to-vehicle and vehicle-to-infrastructure communication will provide data on other road users, obstacles, accidents or speed limits on the current stretch of road, bend sequences and inclines, right through to weather conditions and the road's current friction values. New localization procedures will enable a vehicle to be tracked accurately to within just a few centimeters. At the same time, computing capacity in the vehicles' installed ECUs will rapidly increase in order to process the multitude of sensor data and other information before deriving complex control strategies from those data. Instead of rigid programming, the control units will instead operate with association patterns that are programmed in neural networks using deep-learning methods. This approach originates from the field of artificial intelligence (AI). In the future, new vehicle functions based on the connectivity between all systems and all available information inside and outside the vehicle could make accidents on the road more or less impossible. For example, current concepts make provision for ESP to be transformed from a mere skid-protection solution into a predictive accident-prevention system. Here, the appropriate control system will be constantly active in the background and monitoring the vehicle dynamics. Unlike ESP, which merely stabilizes the vehicle again through targeted brake interventions once it starts to skid, the new system is intended to work proactively and engage before a critical driving situation arises. For example, the system would precisely brake the vehicle before it entered a bend if the driving speed was too high for the selected curve radius and the adhesive conditions on the road at the time. An intervention of this kind does not take control away from drivers, rather it enables them to enjoy driving as much as possible because they can be certain that they will always be driving safely within physical limits.

One example of just how far the potential of personalized driving functions might yet extend is the concept of a race-track assist system designed for supervised training on the racing circuit and instilling drivers with a greater sense of enjoyment when driving at increased levels of safety. In technical terms, it is entirely possible to project information such as the trajectory of the racing line, braking points, speeds, or lap and split times straight into the driver's field of vision by means of a mixed-reality display. By superimposing the actual view of the circuit onto the screen displays, a race-track assist system of this kind would blur the boundary between reality and the virtual world, not to mention bridging the gap to purely computer-based racing simulators. Using an analysis tool, the driver would be able to compare various laps driven and further hone their own driving style and skills. If the concept of an assistance system like this had sufficient flexibility built into it, scalable integration of the assistance systems present in the vehicle could also be realized, meaning that the vehicle-dynamics assist functions could be fine-tuned to the level actually desired. During racing lessons or Track Day events, instructors would have the ability to actively intervene in the vehicle's steering and brakes – and adapt dynamic parameters to the circuit or the needs of the driver in question.

Driving will still be fun!

The examples described above – ranging from concepts through to existing products – are proof that taking a bold step into the world of automated driving functions does not have to mean dispensing with the fun of driving, nor active vehicle dynamics. On the contrary: Automated driving can act as a driving force for the enhanced system connectivity which is so crucial for it to work. In doing so, it can pave the way for driving functions which are designed with sheer driving fun and superior vehicle dynamics in mind. Innovative vehicle-dynamics solutions are not intended to patronize drivers, rather to ensure that they get the most enjoyment possible from driving. To this end, the systems provide drivers with the exact amount of assistance that is perceived to be helpful in the situation in question. They offer a high level of safety as they also assist the driver in highly dynamic scenarios. High-performance sports cars, whose technology is gradually applied to other segments, are frequently the jumping-off point for such developments.

At the same time, new system developments in the future will provide drivers with greater scope for adapting driving functions to their own personal preferences and mood at any given moment. This will make the vehicles of the future true all-rounders that can either be sporty, comfortable, or extremely efficient at the push of a button. Furthermore, drivers in the future will be able to choose freely whether they want to be driven by their vehicles in a fully automatic driving mode, whether to activate the vehicle-dynamics control systems in the background to act as a guardian angel while enjoying unimpeded driving freedom, or whether to simply switch the functions off and rely only on their own abilities.

As automated driving functions become more commonplace, the desire for customization of the assistance systems will grow. After all, the way in which passengers perceive speed, acceleration, or the distance to the vehicle in front varies widely from one person to another. Due to the pronounced reciprocal effects that the individual system parameters have on each other during automated driving, allowances for customization already need to be made during the conceptual phase and considered throughout the entire development process. Here, too, an integrated system-development process is absolutely imperative for function development that saves both time and costs.

KEYNOTE LECTURES III

Challenge to front-loading of the vehicle development using virtual prototyping and reformation of the process

Atsushi Tsubouchi,
Honda R&D Co., Ltd. Automobile R&D Center, Japan

This manuscript is not available according to publishing restriction. Thank you for your understanding.

An introduction to Porsche's Le Mans winning 919 Hybrid and its chassis development

N. N., Dr. Ing. h.c. F. Porsche AG

This manuscript is not available according to publishing restriction. Thank you for your understanding.

PARALLEL STRAND I

INNOVATIVE CHASSIS SOLUTIONS

A solution for future urban mobility: the Schaeffler mover – the fusion of chassis and drive

Dr. Manfred Kraus

Dr. Christian Harkort

Benjamin Wuebbolt-Gorbatenko

The process of urbanization is ongoing with emphasis in Asia [1]. The number of megacities with more than ten million inhabitants has been growing rapidly for years, particularly in Asia. The result of this process is a specific need for mobility, making special vehicle concepts necessary. One solution approach is a highly automated, i.e. driverless, passenger transportation vehicle that Schaeffler is currently developing as a feasibility study. As part of a networked mobility concept, the aim of the electrically driven Schaeffler Mover is to bring up to four passengers to the desired destination comfortably. Due to an extensive data connection with the surroundings, the passengers will be able to simply request the vehicle via a smartphone app, specify their destination, and book the trip.

One of the most important prerequisites for introducing driverless transport systems is the development of driver assistance technologies that allow for automated driving [2]. The Schaeffler study corresponds to level 5. According to studies [3], such systems will be widespread in road traffic in less than 20 years' time. Depending on the underlying scenario – between 3 and 15% of all vehicles produced will already be capable of highly automated driving in 2035.

Vehicle Concept

The technical implementation of the autonomous transport vehicle requires a completely new development process: one that is harmonized with the requirements of highly automated driving and takes advantage of the resulting technical freedoms as much as possible. For example, operating elements such as the steering wheel and cockpit can be eliminated in favor of providing the passengers with more space. In particular, maneuverability, safety, and driving dynamics stability are important development parameters for a highly automated vehicle for urban use.

On the basis of these requirements, a vehicle concept in which all four wheels can be driven and steered independently of each other already took shape during the first potential assessment. This kind of steering system would allow for a lot of agility with a high level of driving comfort, enabling all movements of the vehicle to be implemented with practically no noticeable transverse forces for the passengers.

Schaeffler has developed an innovative wheel module ("intelligent corner module") for this task, condensing the drive into a compact unit via a wheel-hub motor and wheel suspension, along with the vehicle suspension system and the actuator of the electromechanical steering: the Schaeffler Intelligent Corner Module – see **Figure 1.** On the basis of a cost-saving modular principle, identical units are installed in front and in back. With regard to control, the wheel modules open up the option of an integrated driving dynamic approach that combines the function of the ESP driving dynamics control system, force distribution between the driven wheels (torque vectoring), and all-wheel steering, thereby offering high safety reserves. In the case of a technical defect, steering and traction still remain intact if one or two wheel modules are defective, since the wheel-selective drive and the wheel-selective steering provide a large degree of redundancy.

Image 1: Wheel module of the Schaeffler Mover

Complete Vehicle

The complete vehicle concept is based on a stable frame structure that holds the wheel modules. The traction battery is housed in the underbody of the vehicle. This arrangement offers the advantage of a high degree of crash safety with good utilization of the installation space. In addition, a lower vehicle center of gravity is achieved as well. The platform concept – see **Figure 2** – opens up maximum flexibility with regard to the setup. In the simplest expansion stage, the system scope includes the chassis element, consisting of the frame, the four wheel modules, including the drive, suspension, and steering systems, the software for the driving dynamics control system, the vehicle electrical and electronic systems, and the traction battery. Additional modules supplement this substructure with outer cladding, the interior, a lighting system, and the sensor package necessary for highly automated driving, including control software for a rollable vehicle that already offers all driving functions of the transport system. Attachments and platforms, such as a cabin with windows, an air conditioning system, plus operating, display, and connectivity functions round out the operational vehicle. As an alternative, other special platforms are conceivable as well, such as for goods transport.

Image 2: Platform concept of the Schaeffler Mover

Since the vehicle is exclusively used for short distances in the urban area, a relatively small and therefore light and economical traction battery is fully sufficient for fulfilling the transport tasks according to Schaeffler's calculations. This is all the more true, since it will be possible in the future to use standstill and wait times – such as at red traffic lights – for intermediate charging via inductive charging systems.

Its turning radius of less than 5 m with four-wheel steering makes the vehicle very maneuverable in city traffic, Moreover, the wheels can be swiveled out up to 90° when the vehicle is at a standstill, even making driving perpendicular to the vehicle's longitudinal axis possible and allowing for maneuvering, parking, or turning around in very compact spaces.

Wheel Module Setup

The steering system of the wheel module is designed as an electromechanical steer-by-wire system. An actuator integrated in the wheel module coaxially to the steering axis – see **Figure 1** – turns the complete unit during a steering movement, thereby implementing the direction specified by the control unit. As described above, the module's steering angle is limited to 45° max. in normal driving mode, while it can be adjusted to 90° during maneuvering, **see Figure 3**. The actuator used for the steering function is from the Schaeffler's existing large series modular system and has proven to be extremely robust in other applications.

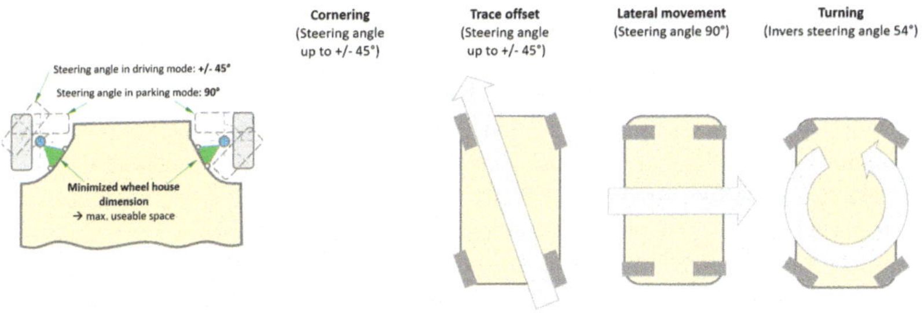

Image 3: Basic steering function and driving modes

The individual parts of the chassis are shown in **Figure 4**. It consists of a trailing arm, which is suspended on the frame via a forked carrier. On the wheel side, the trailing arm is connected with a strut, which is supported from above via a strut bearing on the fork carrier. As an option, the wheel module can be equipped with electromechanical ride height control, which raises and lowers the vehicle, such as for helping passengers to enter and exit.

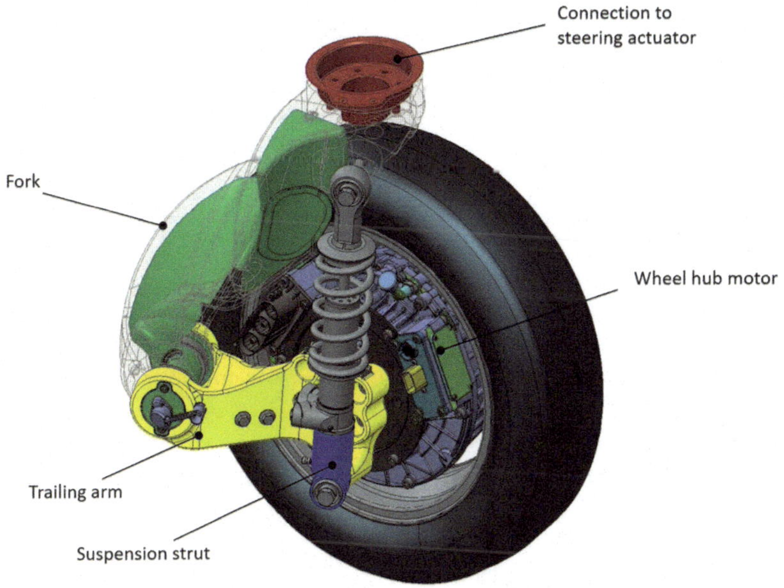

Image 4: Chassis components

Wheel-Hub Motor

The traction motor of the wheel module – **Image 5** – has been designed as a brushless, permanently-excited synchronous machine. In the current design, each of the four electric motors – with 300 V of operating voltage – supplies a continuous output of 13 kW and a temporary peak output of 25 kW. The nominal torque of 250 Nm per motor can be increased for a brief period of 60 sec. to 500 Nm max. The top speed of the electric motor is 1400/min. The drive torque of the motor is transferred to the wheel hub via planetary gear. The liquid cooling system ensures that the traction motor is thermally robust. .

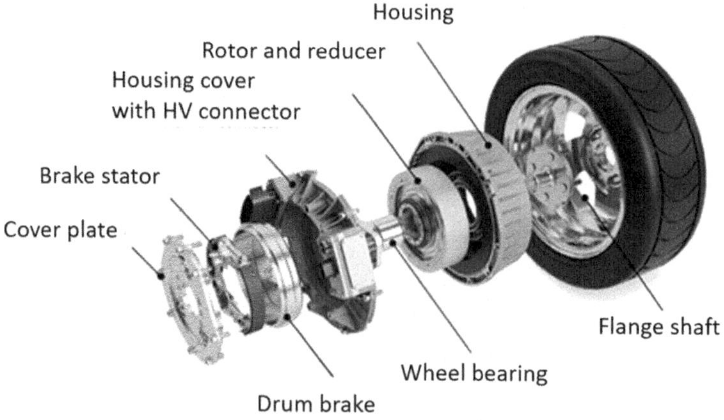

Image 5: The wheel-hub drive components

When the vehicle is braked, the traction motors work as a generator and feed electrical energy back to the vehicle battery. The system is supplemented with mechanically operating drum brakes on all four wheels. The maximum velocity is around 100 km/h, the acceleration of 50 km/h in less than 3.3 sec. It is possible to drive away on a hill with a gradient of up to 31%.

Control Algorithm for the Steer-by-Wire System and the Drive

The complex control algorithm for the steer-by-wire system and the drive constitutes a central component of the software development. On the one hand, the large functional scope resulting from the individual wheel steering needs to be reflected in the control software; on the other, it is necessary to implement numerous safety circuits and redundancies in the program. The development of the steer-by-wire control software was implemented completely within Schaeffler. **Figure 6** shows the signal flows that are relevant for activating and controlling the wheel modules. The control unit for the highly automated driving functions – marked in red in Figure 6 above – is directly connected with the controller for the complete vehicle (blue) via a CAN bus. Among other things, this controls the braking and recuperation function, along with comfort elements such as the air conditioner. During the development and test phase, a driver will be on board to monitor the driving function and intervene using a central joystick if necessary. Later on, in regular operation, this module – integrated via CAN bus in Figure 6 – will no longer be necessary. The vehicle control system has reciprocal data contact with the driving dynamics controller. The traction and steering specifications of the driving dynamics controller are implemented in the wheel modules in corresponding torques and steering angles.

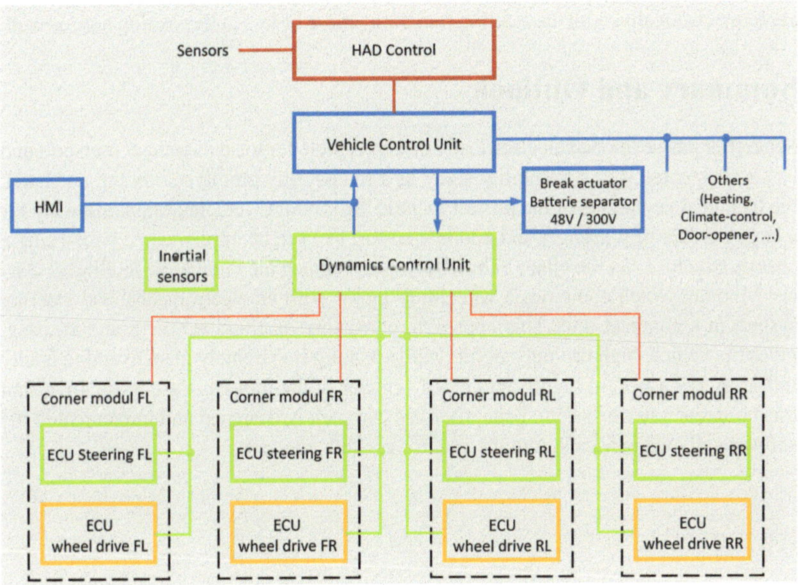

Image 6: Signal flows for controlling the wheel modules

Safety Concept

It has been possible to achieve maximum safety though a multilevel failure strategy, which – depending on the severity of the fault – involves various measures that build upon each other. If the vehicle electronics discovers a fault in a wheel module that impairs the steering, the drive, or the data communication, then the defective module will be shut off. The other three wheel modules can take over the traction and steering tasks of the defective wheel module without any problem, with the result that the function of the complete vehicle will not be impaired at all. The next safety level involves faults in the driving dynamics controller or the failure of the 48-V or 300-V vehicle electrical system. In this case, emergency braking is initiated via the hydraulically actuated drum brakes. If the failure involves the steering system, then it will be deactivated and will remain more or less in the same position, with the vehicle coming to a standstill while following the progression of the curve. The third escalation stage is activated when there is total system failure. The hydraulically activated emergency braking is then triggered mechanically, with the steering remaining more or less in its position here as well.

Summary and Outlook

Schaeffler has developed an electrically driven vehicle for local passenger transport and has implemented it in a feasibility study as a roadworthy prototype. On the one hand, the technical key points of the project include the modular vehicle design, allowing for a large number of passenger and goods transport tasks to be implemented with a single chassis structure. On the other, Schaeffler has developed the innovative Intelligent Corner Module, which combines a wheel-hub motor with an electromechanical steering system in a compact unit. This approach, consisting of all-wheel drive and steering, gives the vehicle high driving stability while making it extremely maneuverable in city traffic. A key aspect of highly automated vehicles of the future is the control algorithm for the steer-by-wire system. Schaeffler has completely designed and implemented the software all by itself, including the extensive safety concept.

[1] Sanwal, M.: Paris Agreement on Climate Change: The First Global Sustainable Development Agreement. https://sustainabledevelopment.un.org/content/documents/1017666_Sanwal_Paris%20Agreement%20on%20Climate%20Change-The%20First%20Global%20Sustainable%20Development%20Agreement.pdf. Downloaded on December 17, 2017

[2] http://www.bast.de/DE/BASt/Forschung/Forschungsfoerderung/Downloads/cedr_call_2014_2.pdf?__blob=publicationFile&v=2. Downloaded on December 17, 2017

[3] Study by Roland Berger: Autonomous Driving – Penetration rate of highly automated cars, 2018

New concept for the front and rear suspension taking advantage of the opportunity offered by the electrification process

Luca Dusini, Gaetano Battaglia

[+] Maserati S.p.A, 41100 Modena, Italy

Introduction

The electrification process which has been started in the automotive market opens a lot of opportunities to the engineers to think again to the best solution they can design in order to have the best result in terms of functionality, performance and costs. As the constraints in an battery electric vehicle change, also the answers to the questions every engineer has during the development can be different in comparison to the standard solutions which are now widespread in the market. This revolution doesn't affect only the powertrain development, the controls and the human-machine interface, but could affect all the fields present in a car development. If the engineers will have enough creativity and courage, among these, there will be also the chassis.

Maserati has developed a brand new suspension module which has as target to take advantage to the new possibilities which a battery electric vehicle architecture can give in order to improve the following areas:

- Increasing comfort
- Improving aerodynamic performance
- Managing the high torque of an electric motor
- Permitting an innovative style
- Increasing the range of the car using an high-performance regenerative braking

The solution proposed is the one shown in Figure 1.

In this paper the ground basis of this concept will be described: it will be evident that every opportunity described in this article opens the way to new solutions which in turn allow further improvements. This process starts a virtuous circle that will significantly change the architecture of the suspension, as happens in a domino.

Figure 1: innovative suspension for battery electric vehicle

How to use the space left by the combustion engine

In the modern internal combustion engine vehicles, the space in the front of the car is optimized and is used to accommodate the engine, the cooling system, the battery and the other additional accessories (Figure 2). All available space is used.

Figure 2. Layout of a front axle with internal combustion engine

With the arrival of the electric engines, the space distribution has to be revisited. The engine is smaller in comparison with a traditional one and instead in the car the space for the batteries must be found. The most simple solution is to use the free space on the front to give to the customer an additional trunk. Looking at the figure 3 it is possible to notice that the remaining parts which take up space in the vertical direction are the components of the suspension, like the dampers and the springs. The starting idea of this project is to put these component in an horizontal position, upper the electric engine, connecting them with a push-rod solution (figure 4): in this way as first result the solution will be more compact. This can give to the designers two benefits: the first one is that the frontal area can be reduced and so the drag force is significantly reduced as well the second one is the possibility to design a shape of the car completely different in comparison to an internal combustion engine car, giving to these new type of cars a different feeling to the customer at the first glance. The reduction in the height of the car bonnet is estimated, for a sedan car, in 150mm (figure 5).

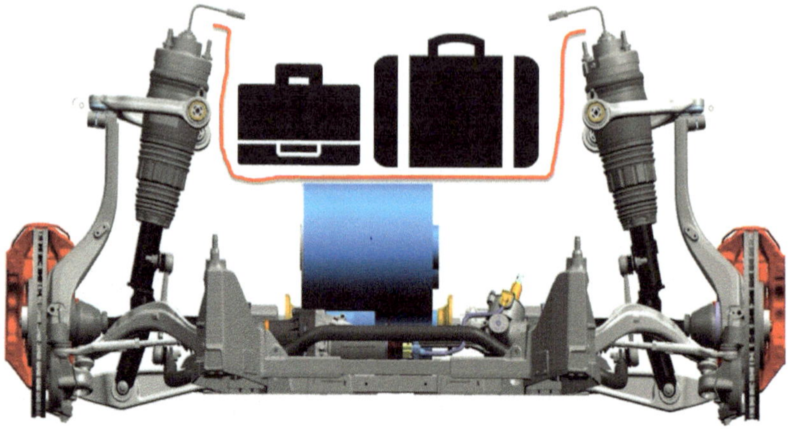

Figure 3. Usage of the space left by the internal combustion engine: additional trunk

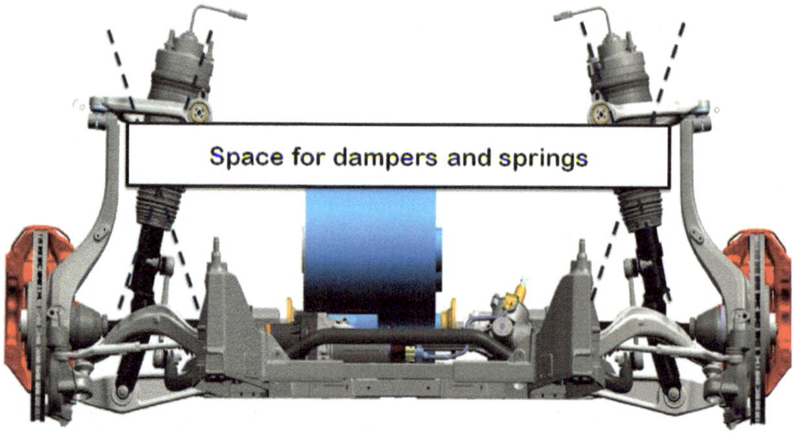

Figure 4. Solution to reduce the vertical impact of the suspension

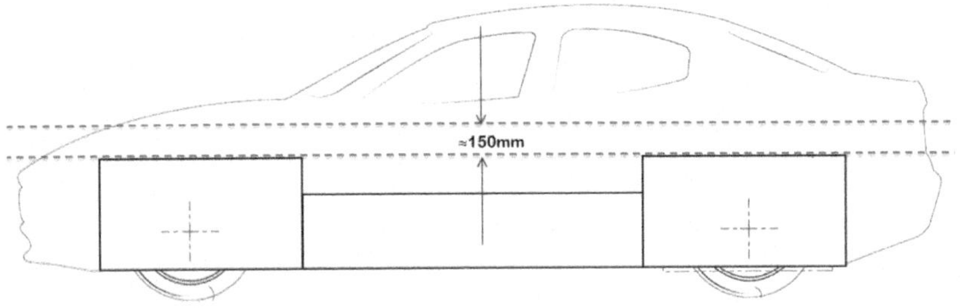

Figure 5. Possible height reduction of the bonned using the new suspension concept

In addition to what is described above, there is a secondary advantage: in the future especially for the autonomous driving, there will be the need to increase the comfort using complex active damping to reduce as much as possible the movement of the body during driving. These systems require a big space in comparison to the standard dampers and is difficult to accommodate between the links of the suspension system. Using this new concept the positioning will be less complex because the dampers are far away from the links and are positioned in an area where there aren't components with big movements. In the solution proposed in this paper, the coil spring is substituted with a composite cross leaf spring for weight reduction. For the reduction of drag force, a leveling system has been considered, acting directly on the leaf spring.

How to use the regenerative braking in order to simplify the suspension system

The regenerative braking is a very important feature for an electric vehicle in order to increase the range of the car, but also to improve the performance of a sporty BEV, reducing the temperature of the brake disks and giving more energy for the acceleration phases. If this feature is developed at the maximum level, this opens new opportunities to design in a different way the whole suspension system.

Potential of regenerative braking

In order to predict the effect of regenerative braking before having the first prototypes of an electric battery vehicle, Maserati developed a virtual environment which enables to evaluate the impact of this feature on all the components of the car, before the availability of the physical parts. In order to do that, a complex virtual model has been developed: it is able to predict not only the vehicle dynamics performance, but also the thermal status of the brakes, of the batteries and of the electric motors. In this way it is possible to simulate a realistic behaviour of the car taking in account the physical limit of the component. In order to do that, a v- model has been followed (figure 6). At the beginning all the departments build up a complex model which is fully compliant with the real behaviour of the component. Then this model is reduced so that it is able to run in real-time. Then all the single models are put in a common environment which permits to the different models to talk each other. At this point this global model is able to run in real-time and so it can be tested on the simulator in order to have the feedback of the driver and to develop the controls. Inside this model a detailed simulation environment for braking is present, with the thermal model of the brakes and the hydraulic actuation (figure 7). In the meanwhile, the tracks used for the development of Maserati has been digitalized and reconstructed at the dynamic simulator (figure 8).

After these activities, the driver follows all the durability program on the virtual tracks, like he does usually in the real world. In this way it is possible not only to perform all the durability tracks using the virtual model of the electric vehicle, but also to make some projections on what it will be possible to reach within two or three years, considering the improvement of the battery system and the electric motors.

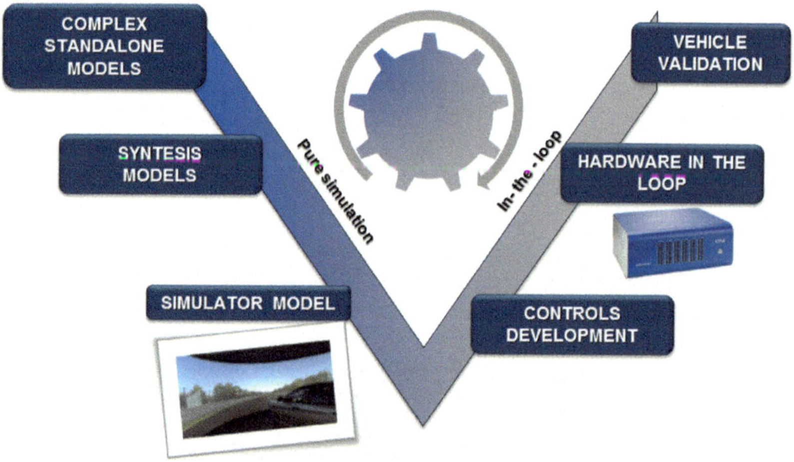

Figure 6. V-model for an integrated simulation environment

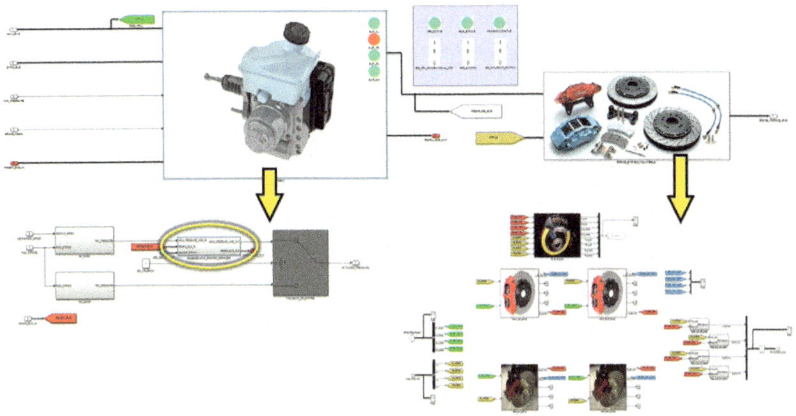

Figure 7. Model of the braking system

Using these simulation data collected during the virtual tests, the half shafts have been dimensioned using an high performance regenerative braking logic which works until ABS intervention, with a safety margin to make this simulation feasible in the real application.

Figure 8. Digital scan of the internal tracks for simulator usage

Looking at the results, it is clear that, if the target is to use as much as possible the regenerative braking, the half shafts must be designed to be capable to bear braking torque which are higher than the one they usually transmit only during the traction phase. In particular, the torque applied to the halfshaft is not only in one direction (traction), but it is positive and negative generating an alternate fatigue cycle. The overall damage on the half shaft is for these reasons increased by 60-90% compared to a standard application. (figure 9).

Another important point to be highlighted is that, using a high performance regenerative braking, the temperatures of the disks are much lower than in a standard application (figure 10). For this reason, for this kind of application, the issues regarding the cooling are not so important for the brake disks, but are transferred to the cooling of the electric motor and batteries which needs a specific cooling system, as moreover already happens in the current applications of battery electric vehicles. In this case this system has of course to be empowered.

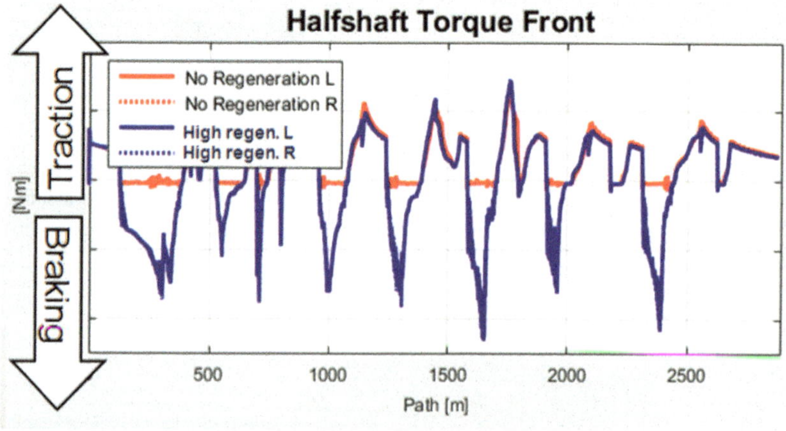

Figure 9. Increased torque on the halfshafts due to regenerative braking

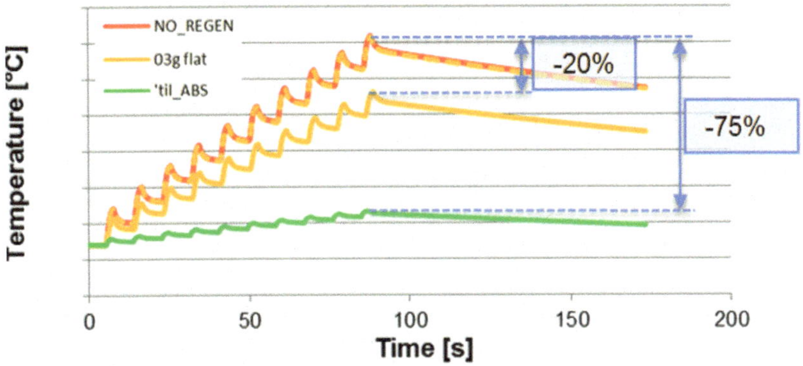

Figure 10. Simulation of an AMUS test. The temperature refers to a front disk. Using regeneration until 0.3g of deceleration (standard approach) the reduction of the temperature increasing of the front disk during the maneuver is ≈ 20%. Using the regeneration until ABS, the reduction is ≈ 75%.

Differences between the innovative suspension and the standard solution considering the high performance regenerative braking

Considering what is described in the previous chapter, if a high performance regenerative braking is used, there are the following consequences:

- The temperature of the disks is on average lower than the standard solution an the brake disks need less cooling
- The cooling must be concentrated to the electric engine and the batteries
- The half shafts must be dimensioned for alternating torque

This new condition leads as consequence that the brake disks can be moved more easily from the unsprung masses (figure 11) inboard the car in order to reach some advantages:

- Reduction of unsprung masses for an improvement of ride comfort
- Torque steer management due to the high output torque electric motors
- Simplification of the suspension
- Improvement of aerodynamic performance of the car

In this paragraph these advantages will be analyzed in detail

Figure 11. innovative suspension: the brakes are inboard like shown in the figure

Unsprung mass reduction

Moving the disk and the caliper inboard, the unsprung mass reduction can be estimated in the range of 15-20kg for every corner: this reduction shall take in account also the simplification of the knuckle which don't need to consider the fixing of the caliper. This reduction has a direct improvement on the ride comfort : the reduction can be estimated in a 15% in the range of the unsuspended masses.

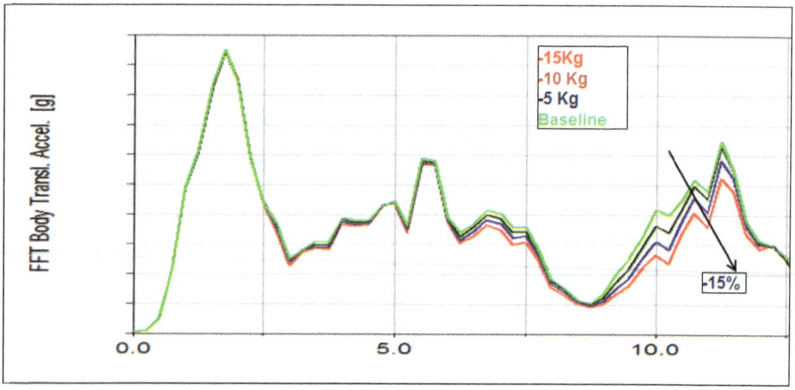

Figure 12. Simulation of ride comfort performance on an uneven road. In the range of the unsuspended masses, the reduction is about 15% having a reduction of 15Kg.

Torque steer management

If there is an high performance electric motor on the front, considering that the peak torque of the engine is very high and that the torque increases with a higher gradient in comparison to an internal combustion engine, the torque steer is a risk to be mitigated as much as possible. For this reason the kingpin offset of the suspension should be reduced consequently. In a conventional solution, the suspension must adopt a scheme with a full virtual steering axis, increasing the costs of the components, because this leads to a suspension with more links and ball joints.

Using the innovative suspension described in this article, since the braking system is no longer installed on the knuckle, there is the opportunity to occupy the portion of space left free by the calipers and the disc with the joints which determine the steering axis. The suspension thus can be simplified from a full virtual steering axis suspension (for example a full virtual double wishbone) to a more simple double wishbone with a physical steering axis passing through the wheel center reaching potentially a very small kingpin offset, even lower than the conventional solutions.

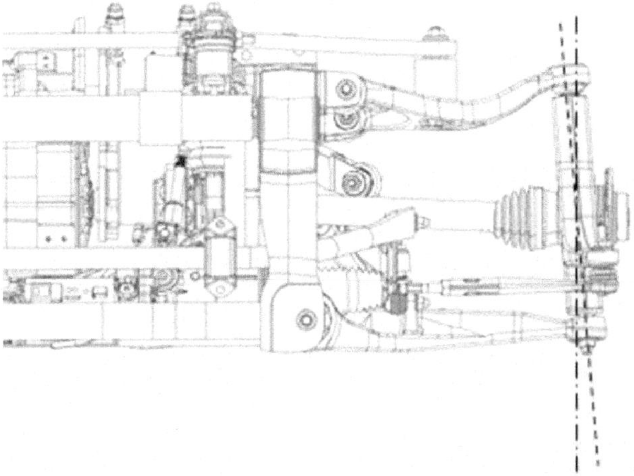

Figure 13. - - - - steering axis, —·— wheel center axis. With the innovative suspension solution the kingpin offset is ≈0mm with benefits to the torque steer

Suspension simplification

The suspension concept described in this paper allows to simplify very much the suspension scheme. As described before the knuckle can be designed without taking in account the fixing of the caliper. The absence of damper and spring between the links takes as consequence the possibility to design more simple connection between the wheel and the body. There is in addition to that another benefit coming from the inboard braking: the possibility to use a double wishbone without the need to have a big distance between the upper control arm and the lower control arm, without losing performance in terms of ride comfort.

In a conventional suspension, with outboard brakes, during braking the force generated on the brake disc causes a rotation of the knuckle and for this reason there is the need to have a distance between the upper and the lower arm as bigger as possible or to stiff the bushing with bad consequences on the ride comfort and road noise. For this reason the in the luxury sedan cars, the upper arm is usually positioned over the wheel, using a double wishbone architecture. With the inboard brakes, the braking force acts on the wheel center and doesn't cause any rotation of the knuckle. For this reason it is possible to adopt a double wishbone architecture with the upper arm positioned inside the wheel with a smaller distance between upper control arm and lower control arm, without the need to stiffer the bushings. In this way it is possible to reduce the unsprung masses and make the system lighter.

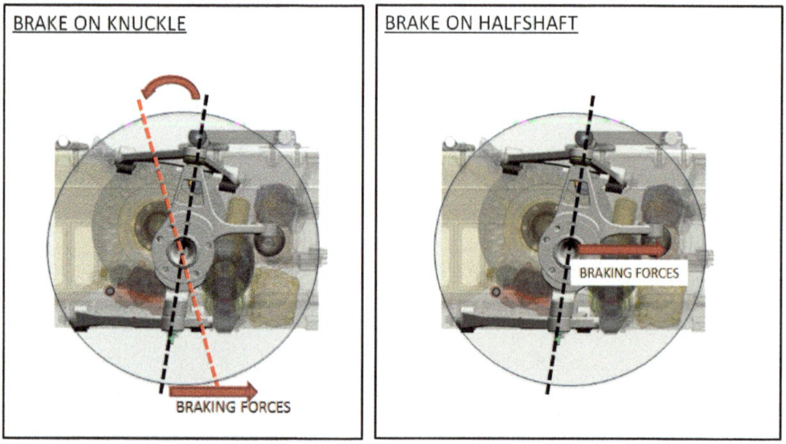

Figure 14. looking at the left side of the figure, with the brake on the knuckle, there is a caster angle loss due to the interaction between the brake disk and the knuckle during the braking actuation. In order to avoid this bad behaviour, a stiffening of the bushings is needed. On the right side, with the braking system inboard, there isn't a caster angle loss and so the bushings can be designed to improve the filtering of the road inputs without the need to stiffen them to avoid the rotation

Reduction of the aerodynamic drag force

When air flows around the wheel wells it gets disturbed by the rims of the vehicles and forms an area of turbulence around the wheel. In order for the air to flow more smoothly around the wheel well, smooth wheel covers are preferred. As the braking are inboard, there isn't the need to cool the brakes using the openings in the rim which can be completely covered increasing the aerodynamic drag of the car and extending the range.

Modularity of the innovative suspension system

The complete suspension module described in this paper is imagined as a single element containing the suspension, brake system and electric motor. It is then mounted on the central part of the car's body with screws. By varying the relative position between the suspension-motor module and the central part in vertical direction, the relative height of the wheel centre and the rest of the body can be easily modified as well as changing the distance between the driver and the wheels in the longitudinal direction. This allows to use the same solution for different type of cars (sedans, SUV and so on) without modifying nothing of the suspension module. Using this concept, it will be more simple to use the same suspension architecture also for different platforms, increasing the commonality of the parts and so reducing the costs.

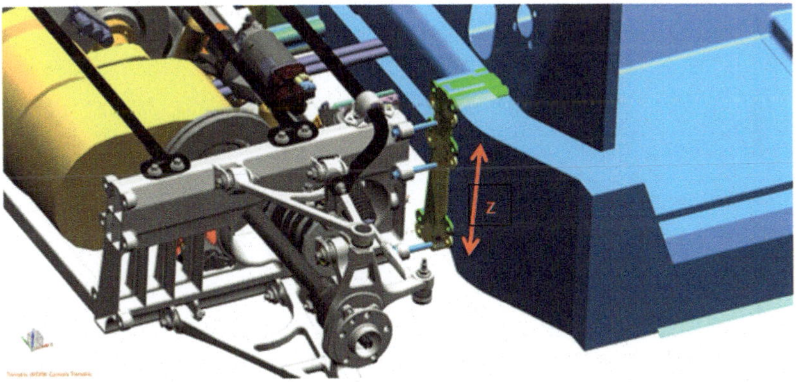

Figure 15: the module is connected to the rest of the body using screws. This allows easily to move the complete suspension+motor system in the vertical direction modifying the ride height of the car. In the same way also a longitudinal movement is possible

Conclusion

In this paper, the authors show how the electrification process which is going on in the automotive market can affect also the suspension design. Maserati developed a brand new solution which uses the opportunity given by these new degrees of freedom in order to improve performance, costs and modularity of the suspension systems of the future.

In the first chapter there is a description of the vision of the authors about how to use in a smart way the space left by the internal combustion engine. The proposal is to use it to accommodate the dampers and springs, considering also a lifting system and eliminating the shock towers. In order to reduce the space needed and for weight reduction, a composite leaf spring is considered.

In the second chapter the opportunities coming from a high regenerative braking are taken in account. In the new suspension module the braking system is inboard and this leads to many beneficial effects on the suspension system: the kingpin offset can be drastically reduced, the knuckle and the links of the suspension are simplified , the rims can be covered in order to reduce the drag force of the and the unsrpung masses are reduced for a better ride comfort performance.

The third chapter describes briefly the big opportunity given by this suspension module: communalize the suspension system for very different type of cars, increase the flexibility of the platforms in terms of wheel base variation and ride height.

Detection of wet road conditions in the new generation of the 911

Yves Billet

Abstract

Wet roads dominate the traffic accident statistics, well above all other weather-related factors. Apart from speed-induced hydroplaning situations, drivers tend to underestimate the effect of a wet pavement on braking distance or overall vehicle stability.

With the Porsche Wet Mode, the new generation of the 911 introduces a driver assistance system, designed to flag the road condition up to the driver, along with offering a stability setup for wet pavements.

The detection of the road condition relies on acoustic sensors attached to the wheel arch liners of the front axle. When driving over a water film, the sound of the road noise and the impact of the spray dispersed by the tires triggers the acoustic sensors and urges preconditioning of the chassis control systems. When the situation persists or the spray intensity increases, the instrument cluster suggests to the driver to activate a special drive mode, which facilitates driving on wet pavements.

1 Introduction

1.1 European accident statistics

The European annual accident report shows the distribution of road fatalities by weather condition.

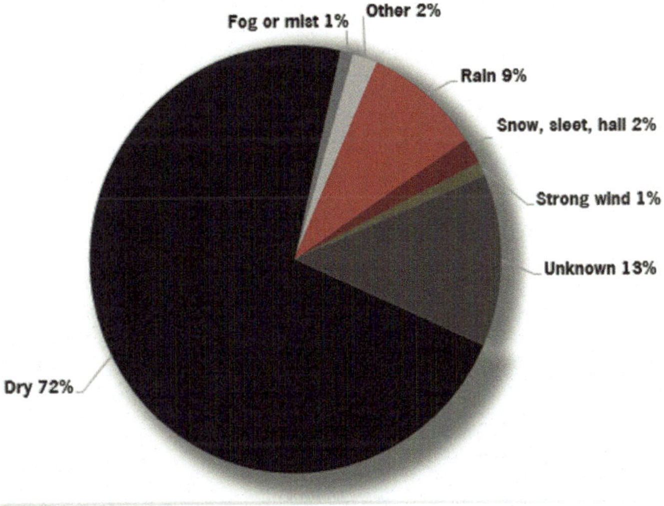

Fig. 1: Distribution of fatalities by weather condition, EU, 2014 [1]

Apart from geographical differences between northern and southern Europe, on an overall basis, rain proves to be the single most important weather influence factor throughout the year, accounting for up to 10% of all fatal accidents.

Drivers tend to underestimate the effect of reduced wet grip on the stability of the vehicle, especially when braking or driving through curves. Drivers tend to overestimate their driving skills in wet conditions, as opposed to cautious driving on snow and ice. Apart from experiencing reduced wet grip, hitting a hydroplaning situation can be a drastic surprise.

At night or in conditions of reduced visibility, the implication of a wet road surface becomes even more difficult to judge.

1.2 Requirements for wet stability in the new Porsche 911

The new Porsche 911 comes with a detection system for wet road conditions.

The detection links into a recommendation in the instrument cluster (refer to Fig. 2). The recommendation's intention is to raise the driver's awareness for the wet road surface, irrespective of the weather or the daylight situation. The recommendation balances between robustness and sensitivity, in order to avoid both over- and underwarning the driver.

The Porsche 911 is aimed to reach maximum driving stability on wet roads within the physical limits of the vehicle and the vehicle's systems. The high level of assistance and comfort on wet pavements follows a multi-stage approach, only imposing part of the stability countermeasures automatically, and making activation of the full stability setup an explicit driver's choice. The vehicle still allows for complete driving experience and agility on the racetrack, whilst offering additional stability and comfort when explicitly chosen by the driver. The wet mode as an additional driving mode is purposely not aiming at increasing race performance, but at increasing stability.

Fig.2: Driver recommendation and status icon in the instrument cluster

2 Water film thickness, spray intensity and contact properties

2.1 Choice of sensor position behind the front wheels

The detection principle in the Porsche 911 relies on acoustics and does not depend on an optical measurement, not to compromise the performance at night, in environments with increased dirt, or in conditions of reduced visibility. By placing the sensors in the wheel arch liner behind the wheel, their surface stays clean automatically whenever there is water on the road: the spray immediately washes out eventual dirt on the sensor's surface.

A wet road condition is extremely difficult to catch. Figure 3 shows a freeway in Belgium briefly after a heavy rain shower:

- The weather does not correlate directly with the road condition. Although it is not raining any more, the road is still substantially wet, and remains so for a certain time to come, depending on the initial amount of water, the amount of traffic, the outside air temperature, and so on.
- The amount of water on the road varies depending on the traffic density. The right hand lane in figure 3, which is typically the lane with the heavier truck traffic, is substantially less wet than the left hand lane with the passenger car traffic.
- The water film thickness depends on the type of pavement. The pavement of the emergency lane in figure 3 has less macroscopic texture, favoring a water film to build on its surface.
- The water film thickness may continually change within a given driving lane, depending on the type of pavement, the inclination of the road surface and the presence of ruts.

Fig. 3: Freeway driving in Belgium briefly after a heavy rain shower

The contact properties between tire and road surface essentially define the vehicle's stability on the wet road surface. To allow for a correlation with the road condition underneath the tire, the wet detection sensors in the Porsche 911 attach to the wheel arch liners behind the front wheels. In this position the sensors catch variations in the film thickness in real-time as seen by the front tires (refer to figure 4).

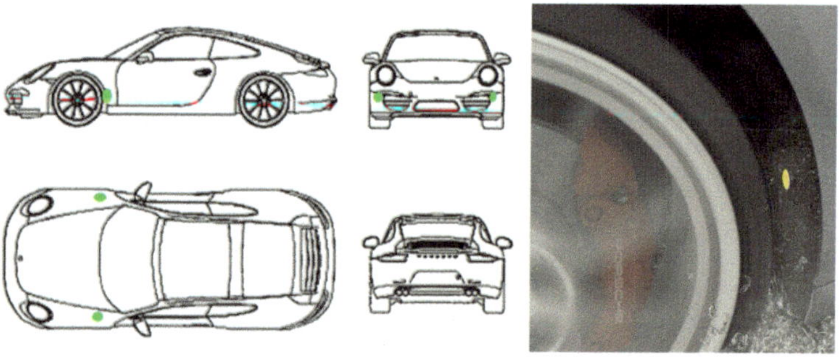

Fig. 4: Position of the acoustic wet detection sensors behind the wheel arch liners of the front axle

The sensors are located behind the front wheels as opposed to the rear wheels to yield the best possible reaction time under changing conditions.

2.2 Spray intensity as an indicator for film thickness

Fig. 5: Water film thickness as influenced by the macroscopic road roughness

The water film thickness in a given weather situation is mainly influenced by the macroscopic road roughness (refer to figure 5). With decreasing macroscopic road roughness, the water film thickness and the spray intensity of the water picked up by the tire increases, leading to an increase in the amplitude of the acoustic measurement.

Microscopic road roughness in relation to spray intensity unfortunately acts differently: pavement types like cobblestone provide good water drain. When it comes to rainy weather, they do not yield a high spray amplitude.

For a given tire type and tire condition, the wet grip depends on the type of pavement. Spray intensity is a good indicator for a reduced grip potential, as long as the grip properties of the tire-road contact depend on macroscopic roughness. Contrarily, missing spray intensity does not automatically point to a high wet grip potential, as microscopic road roughness may play a substantial role as well.

A stationary water film on the road surface encourages hydroplaning, depending on the film thickness and the vehicle speed [2]. Increasing spray intensity is an indirect indicator for increased hydroplaning probability.

2.3 Limitations of spray intensity as a road condition indicator

The average film thickness typically varies slowly when driving in or out of a rain zone. In such case, the increase or decrease in spray intensity provides a certain forecast. The acoustic sensor cannot predict sudden changes in spray intensity. Consequently, the system cannot reliably warn for hydroplaning in all situations.

The spray intensity is a qualitative indicator for water film thickness only, as tire type and tire profile are contributors as well. Tire influences on the spray intensity, as seen by the acoustic sensors, prove to be rather minor as compared to the influence of the pavement.

3 Acoustic detection of water spray

3.1 Detection principle

The wet detection sensor is a piezo-electric sensor. The sensitive element consists of a piezo-foil glued to the backside of the sensor's housing. Upon microscopic bending of the housing surface, the piezo-foil generates a voltage signal. The analogue voltage signal is pre-processed electronically first to provide the best possible signal to noise ratio, before being converted into the digital domain, where wetness as a feature is being extracted from the signal contents. The acoustic detection principle follows early investigations by Porsche in the 1990s [3].

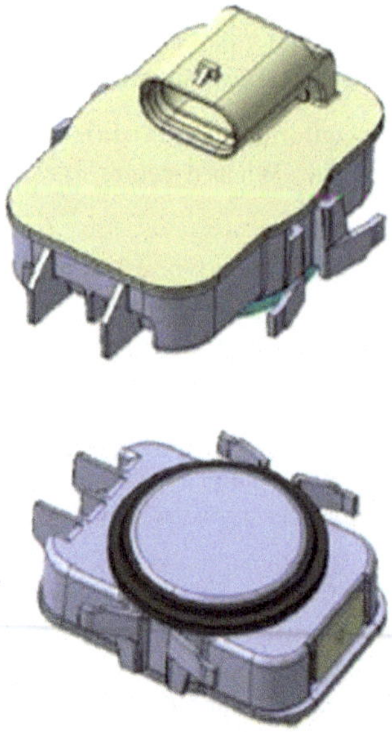

Fig.6: Wet detection sensor with its circular sensitive area

3.2 Signal contents and disturbances

The sensitive surface of the sensor housing experiences both water impact as an effective signal, and disturbances that are to be compensated for (refer to figure 7). The sensor outputs a wetness degree as a continuous time-based measurement, along with its temperature and diagnostic information.

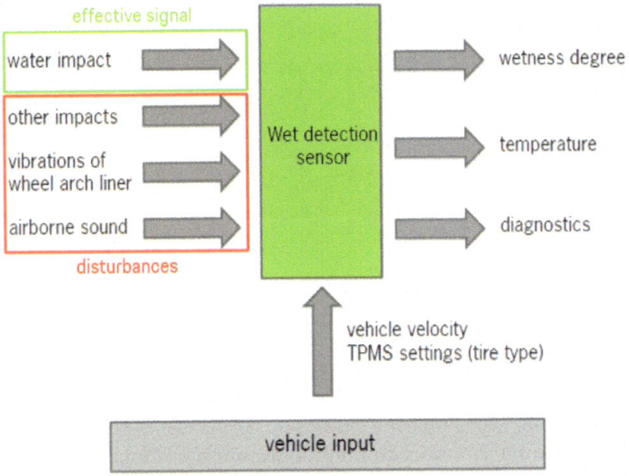

Fig.7: Signal contents and disturbances

The disturbances relate to three main categories:

- impacts other than water droplets hitting the sensitive area (sand particles, loose gravel)
- vibrations of the wheel arch liner transferred through the bracket
- airborne sound (wind noise, road noise)

Disturbances are dealt with differently depending on their nature.

3.2.1 Handling of impacts other than water droplets

Particles other than water droplets are only of interest if the tire effectively transports these particles from the road to the sensor's sensitive area. A vehicle with winter tires, for instance, tends to pick up far less particles than one with summer tires, by this inherently being less susceptible for foreign impacts.

The distribution of water droplets in the wheel arch is rather uniform and continuous, when driving on a water film, whereas gravel particles only hit the sensitive area punctually. This leads to different signal characteristics in the time domain, allowing for separation of both effects (refer to figure 8).

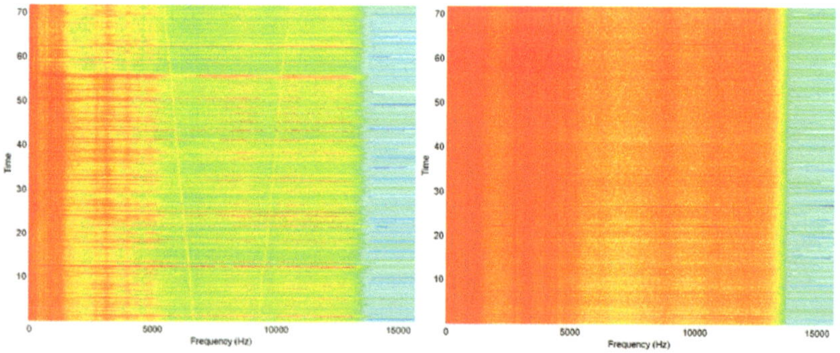

Fig. 8: Signal characteristics of gravel impacts (left) as compared to wetness impact (right)

3.2.2 Handling of vibrations of the wheel arch liner

A specific bracket design along with a rubber element provide decoupling of the vibrations of the wheel arch liner from the sensor housing. A measurement setup with accelerometers on both wheel arch liner and sensor housing validates the effectiveness of the decoupling (figure 9).

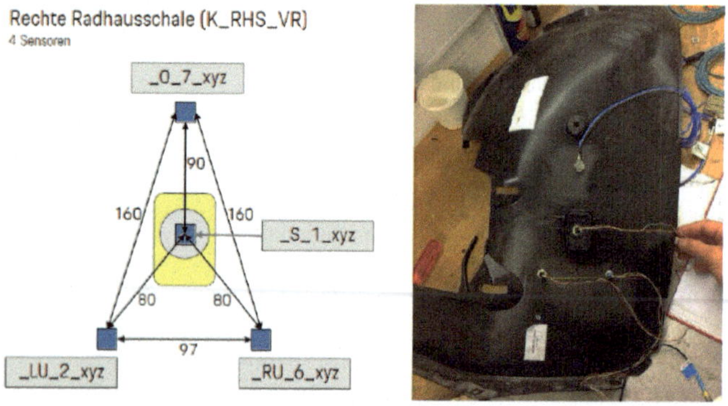

Fig. 9: Measurement setup with 3 accelerometers on the wheel arch liner in a triangular arrangement, and one accelerometer connected to the sensor housing

Results show very good decoupling above 1.5 kHz (refer to figure 10).

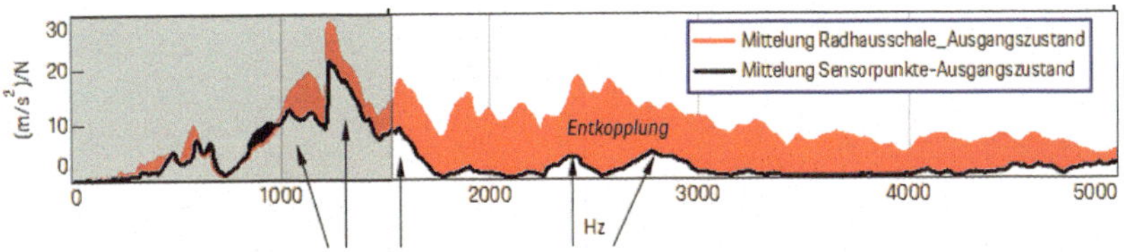

Fig. 10: Decoupling of the sensor membrane (black) from the wheel arch liner (red) in the frequency range above 1.5 kHz. The arrows indicate resonances of the sensor's housing.

3.2.3 Handling of airborne sound

Airborne sound originates from wind noise (aerodynamics) and road noise. It strongly increases with the vehicle speed. Characterization of wind noise in the windtunnel and road noise on the rollbench (refer to figure 11), in a setup with accelerometers similar to the one of figure 9, show that both disturbances have comparable frequency content and amplitude.

Fig. 11: Characterization of road noise on the rollbench (left) and wind noise in the windtunnel (right)

The frequency range of the disturbance broadens with increasing vehicle speed (refer to figure 12). The dominant signal energy content lies well below 1 kHz. The wheel arch liner does not have speed dependent resonances. To compensate for disturbances originating from road noise and wind noise, the sensor includes a high pass filter. The high pass filter in addition helps to reduce effects below 1.5 kHz that were not covered by the decoupling (refer to the grey area in figure 10).

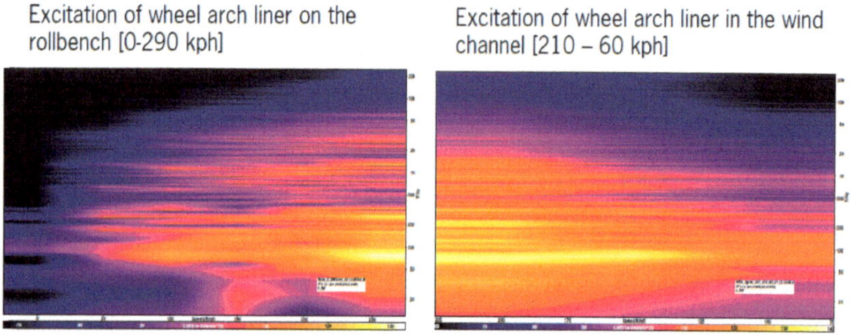

Fig. 12: Frequency contents of road noise (left) and wind noise (right), with the vehicle velocity on the x-axis and the frequency on the y-axis

A software algorithm deals with the remaining effects not fully covered by the high pass filter, using tire type and vehicle velocity as the main entry signals.

4 Processing of the sensor signals

4.1 Electronic vehicle architecture

Both wet detection sensors connect to the central chassis ECU (named "Elektronik Fahrwerks Plattform" or EFP, refer to figure 13). The ECU unites the information of both sensors and prepares a road condition state for vehicle control and display functionality.

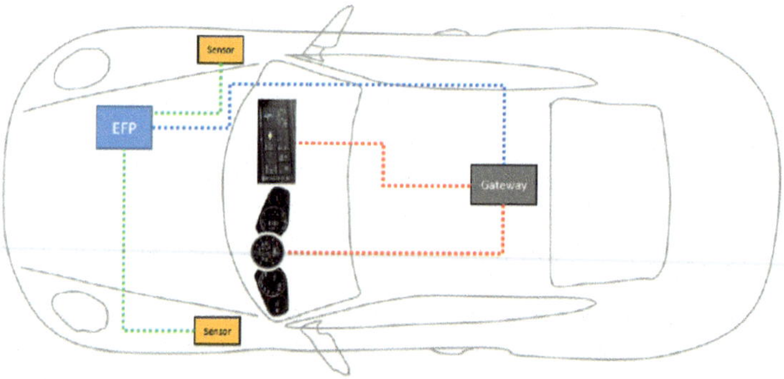

Fig. 13: Electronic vehicle architecture with wet detection sensors, EFP ECU, the vehicle gateway, the instrument cluster and the central display unit.

4.2 Driving state compensation

The EFP-ECU compensates for the driving state when processing the wetness information. Steering wheel angle, velocity, direction of travel and other driving-state related properties all have an influence on the amount of water spray hitting the wheel arch liner.

The processing of the wetness information in the ECU for the purpose of vehicle control differs from the processing for the purpose of display in the instrument cluster.

4.3 Reaction in vehicle control

Vehicle control upon detected wetness follows a two-stage approach. Upon detection of road wetness, the braking system and the all-wheel drive are preconditioned for wet pavements. When the situation persists or spray intensity in the wheel arch increases, the driver receives a recommendation to activate a drive programme preconfigured for wet roads. When doing so, additional stability countermeasures in drivetrain, chassis control and aerodynamics guarantee comfortable driving on wet roads.

5 Conclusion

The Porsche Wet-Mode introduces a first to market road condition detection mechanism based on acoustics. It acts independent of other technologies used for observing the vehicle's surroundings, offering a path for redundancy or enrichment in a sensor fusion approach for road condition monitoring. The acoustic detection method also closes a substantial gap when it comes to detecting conditions of intermediate friction under constant driving.

References

[1] CARE Database, European annual accident report, https://ec.europa.eu/transport/road_safety/specialist/statistics_en

[2] Ong, G. P. and Fwa, T.F. (2007), Study of factors influencing vehicle hydroplaning speed, Journal of the Eastern Asia Society for Transportation Studies, Vol. 7, 2007

[3] Görich, H.-J. and Jacobi, S. (1996), Verfahren zur Erfassung der Benetzung einer Fahrbahnoberfläche, Deutsches Patentamt DE 42 13 221 A1

RISK ANALYSES
AND VIRTUAL METHODS

Haftungsrechtliche Einordnungen von Simulation und Test (Liability risks in the design and execution of tests and simulations)

Dr. Philipp Ehring, branmatt || legal

I Einleitung

Simulationen und Tests liefern oftmals wichtige Erkenntnisse über Eigenschaften eines Produkts. Welche Eigenschaften überprüft werden müssen, hängt davon ab, welche Eigenschaften das Produkt überhaupt aufweisen muss. Aus rechtlicher Sicht ist der anzulegende Maßstab hier zunächst der vertraglichen Vereinbarung mit dem Käufer des Produktes zu entnehmen. Dies reicht aber nicht aus, da es oftmals weitere Erwartungen an ein Produkt bestehen: Erwartungen, die üblicherweise an das Produkt gestellt werden bzw. Erwartungen, die die Sicherheit des Produktes betreffen. Gut durchgeführte Simulationen und Tests helfen hier Haftung zu vermeiden, da sie grade im Bereich der Konstruktion ermöglichen, Fehler zu identifizieren, so dass diese nicht in die spätere Produktion bzw. Serie gelangen. Tests und Simulationen kommt daher unmittelbar eine haftungsbegrenzende Funktion zu.

Nicht jeder Konstruktionsfehler lässt sich durch Tests und Simulationen und durch eine sorgfältig durchgeführte Planungs- und Erprobungsphase verhindern. Gegen das verbleibende Risiko kann sich der Hersteller durch den Abschluss einer Produkthaftpflichtversicherung schützen. Allerdings enthalten diese regelmäßig sog. Erprobungsklauseln, nach der der Versicherer die Deckung unter Hinweis auf eine nicht ausreichende Erprobung verweigern kann. In diesem Zusammenhang hat die Durchführung und Dokumentation von Tests und Simulationen haftungserhaltende Funktion, da bei ausreichender Erprobung in Form von – dokumentierten – Simulationen und Tests der Versicherer nicht aus seiner Einstandspflicht entlassen wird.

Im folgenden Beitrag soll untersucht werden, welche Anforderungen aus haftungsrechtlicher Sicht an Tests und Simulationen zu stellen sind. Dazu sind zunächst Test- und Simulationen in die Prozesse eines Unternehmens einzuordnen (hierzu unter II.1). Sodann schließt sich eine Darstellung der rechtlichen Rahmenbedingungen an, innerhalb derer Test- und Simulationen einzuordnen sind (hierzu unter II.2). Nach der Darstellung der Anforderungen an Test und Simulationen aus haftungsrechtlicher Sicht (hierzu ausführlich unter II.3.) wird die haftungserhaltende Funktion von Test- und Simulationen innerhalb der sog. Erprobungsklauseln der Produkthaftpflichtversicherung erläutert werden (hierzu unter III.).

II Haftungsrechtliche Bedeutung von Tests und Simulationen – haftungsausschließende Funktion

1 Aufgabe von Test und Simulationen nach QM-Vorschriften

Für die juristische, insbesondere haftungsrechtliche Einordnung von Tests und Simulationen, ist es zunächst erforderlich, diese in den lebenden Organismus eines Unternehmens einzuordnen. Hierzu lohnt sich ein Blick in die Qualitätsmanagementnormen der ISO 9000:2015-11 und die ISO 9001:2015-11.

Nach Ziffer 3.11.8 der ISO 9000:2015-11 werden im Rahmen von Tests Bestimmungen vorgenommen, also eine oder mehrere kennzeichnende Eigenschaften (sog. Merkmale)[1] und ihrer Werte (sog. Merkmalswerte) ermittelt.[2] Sodann werden diese bestimmten Merkmale und Merkmalswerte mit den Anforderungen[3] an einen spezifischen beabsichtigten Gebrauch oder eine spezifisch beabsichtigte Verwendung abgeglichen. Zeigt das Ergebnis des Tests, dass die Anforderungen erfüllt sind (sog. Konformität[4]), kann es zum Zweck der Validierung verwendet werden. Nach der ISO 9000:2015-11 ist der Test also Teil des Validierungsprozesses.

Bei der Validierung handelt es sich um die Bestätigung durch Bereitstellung eines objektiven Nachweises[5], dass die Anforderungen für einen spezifisch beabsichtigten Gebrauch oder eine spezifisch beabsichtigte Anwendung erfüllt sind. Gemäß der Anmerkung 1 zu Ziff. 3.8.13 ISO 9000:2015-11 entstammt der für die Validierung erforderliche objektive Nachweis dem Ergebnis eines Tests oder einer anderen Form der Bestimmung (3.11.1), z.B. Durchführung alternativer Berechnungen oder Überprüfen von Dokumenten[6]. Auch wenn die Qualitätsnormen dies nicht ausdrücklich benennen, lässt sich unter eine andere Form der Bestimmung die Simulation subsumieren. Bei dieser handelt es sich um ein Experiment an einem Modell, um Erkenntnisse über das reale System zu gewinnen.

Von der Validierung zu unterscheiden ist die Verifizierung. Bei dieser handelt es sich gem. Ziff 3.8.12 ISO 9000:2015-11 um die Bestätigung durch Bereitstellung eines

1 vgl. Ziff. 3.10.1

2 Ziff. 3.11.1 ISO 9000:2015-11

3 Ziffer 3.6.4 ISO 9000: Erfordernisse oder Erwartungen, das oder die festlegt, üblicherweise vorausgesetzt oder verpflichtend ist.

4 Ziffer 3.6.11 ISO 9000: Erfüllung einer Anforderung.

5 Ziffer 3.8.3 ISO 9000:2015-11: Daten (3.8.1), also Fakten über ein Objekt, welche die Existenz oder Wahrheit von etwas bestätigen.

6 Ziffer 3.8.5 ISO 9000:2015-11: Informationen, einschließlich des Trägermediums, wobei Informationen Daten, also Fakten über ein Objekt (3.8.1), mit Bedeutung sind (vgl. 3.8.2).

objektiven Nachweises (3.8.3), dass festgelegte Anforderungen (3.6.4) erfüllt worden sind. Der für die Verifizierung erforderliche Nachweis kann das Ergebnis einer Prüfung (3.11.7) oder anderen Form der Bestimmung (3.11.1) sein, z.B. Durchführen alternativer Berechnungen oder Überprüfen von Dokumenten (3.8.5). Während die Validierungsgrundlage über einen Test gewonnen wird, geschieht dies bei der Verifizierung über die sog. Prüfung. Bei dieser handelt es sich gem. Ziff. 3.11.7 ISO 9000:2015-11 um die Bestimmung der Konformität mit festgelegten Anforderungen. Auch hier lässt sich unter die andere Form der Bestimmung die Simulation subsumieren.

Verifizierung und Validierung unterscheiden sich dadurch, dass bei ersterer Produkteigenschaften anhand bestimmter festgelegter Anforderungen überprüft werden, bei der Validierung hingegen wird das Produkt auf eine bestimmte Verwendung hin getestet:

Mittel	**Prüfung**	**Test**
Verfahren	Verifizierung	Validierung
Ziel	Erfüllung von definierten Anforderungen	Erfüllung einer definierten Verwendung bzw. eines beabsichtigten Gebrauchs

Beiden Verfahren kommt ihre Hauptrolle in der Entwicklungsphase eines Produktes zu. Dies ergibt sich deutlich aus Ziff. 8.3.2 lit. c) ISO 9001:2015-11, welche fordert, dass bei der Bestimmung der Phasen und Steuerungsmaßnahmen die erforderlichen Tätigkeiten zur Entwicklungsverifizierung und Entwicklungsvalidierung berücksichtigt werden müssen. Aufgabe der Steuerungsmaßnahmen ist es gem. Ziffer 8.3.4 lit. c) ISO 9001:2015-11, dass

- Verifizierungstätigkeiten durchgeführt werden, die sicherstellen, dass die Entwicklungsergebnisse die in den Entwicklungseingaben enthaltenen Anforderungen erfüllen;
- Validierungstätigkeiten durchgeführt werden, die sicherstellen, dass die resultierenden Produkte und Dienstleistungen die Anforderungen erfüllen, die sich aus der vorgesehenen Anwendung oder dem beabsichtigten Gebrauch ergeben.

Da die Simulation eine Form des Tests bzw. der Prüfung ist und diese wiederum Mittel der Validierung und Verifizierung, sollen im Folgenden nur noch die Begrifflichkeiten Test und Prüfung bzw. Validierung und Verifizierung verwendet werden.

2 Validierung/Verifizierung im haftungsrechtlichen Umfeld

Für die haftungsrechtliche Einordung von Tests und Prüfungen bzw. Validierungen und Verifizierungen ist zunächst eine kurze Skizzierung des Haftungsumfeldes erforderlich. Hierbei ist zwischen Haftungsansprüchen aus Vertrag und solchen, die nicht auf ein vertragliches Verhältnis beruhen, zu unterscheiden. Bei ersteren handelt es sich um gewährleistungsrechtliche Mangelansprüche, hauptsächlich aus den kaufrechtlichen Bestimmungen der §§ 434 ff BGB, bei letzteren um die gesetzlichen Ansprüche des Produkthaftungsgesetzes bzw. der aus §§ 823 ff. BGB resultierenden Produzentenhaftung.

a) Mangelhaftung

Anknüpfungspunkt für die Haftung im Kaufrecht ist das mangelhafte Produkt. Wann ein entsprechender Mangel vorliegt, ergibt sich aus §§ 434, 435 BGB. Unterschieden wird dabei zwischen Sach- und Rechtsmängeln. Ist das Produkt mangelhaft, stehen dem Käufer bestimmte, in § 437 BGB aufgeführte Rechte zu.

aa) Mangelbegriff

In Bezug auf den Sachmangel unterscheidet § 434 Absatz 1 BGB drei Arten: In erster Linie kommt es für die Mangelhaftigkeit bzw. -freiheit auf die zwischen den Vertragsparteien getroffene Vereinbarung an (§ 434 Abs. 1 S. 1 BGB). Entsprechend ist das Produkt frei von Sachmängeln, wenn es die vereinbarte Beschaffenheit aufweist. Mangelhaft ist es, wenn es diese Beschaffenheit nicht hat. Art und Ausmaß der Abweichung sind dabei unerheblich, so dass jede Abweichung des geschuldeten Zustandes von dem Vereinbarten einen Sachmangel begründet.[7] Lediglich Rücktritt und Schadensersatz statt der ganzen Leistung also diejenigen Rechtsbehelfe, bei denen der Käufer die mangelhafte Sache zurückgibt – sind bei unwesentlichen Pflichtverletzungen ausgeschlossen.[8] Die Beschaffenheitsvereinbarung kann dabei sowohl konkrete Eigenschaften des Produkts beinhalten, also Material, Maße, Form etc., als auch die Eignung für eine bestimmte Verwendung, etwa den Einsatz eines Bauteils in verschiedenen Klimazonen.[9]

Liegt keine vertragliche Beschaffenheitsvereinbarung vor, kommt es nachrangig auf die nach dem Vertrag vorausgesetzte Verwendung an (§ 434 Abs. 1 S. 2 Nr. 1). Eignet sich das Produkt für die nach dem Vertrag vorausgesetzte Verwendung, ist es mangelfrei. Eignet es sich nicht, ist es mangelhaft. Die Vorschrift hat Fälle im Sinn, in denen sich die Vorstellungen der Parteien nicht auf einzelne Merkmale der Beschaffenheit richten,

7 *Faust*, in: Bamberger/Roth/Hau/Poseck, BeckOK BGB,46. Edition Stand: 01.05.2018, § 434 Rdn. 38.

8 *Faust*, in: Bamberger/Roth/Hau/Poseck, BeckOK BGB,46. Edition Stand: 01.05.2018, § 434 Rdn. 44.

9 *Westermann*, in: Münchener Kommentar zum BGB, 7. Auflage 2016, § 434 Rdn. 10.

sondern darauf, dass das Produkt für eine bestimmte Verwendung tauglich sein soll. Die fehlende Verwendungseignung muss daher nicht auf fehlenden Beschaffenheitsmerkmalen des Produkts beruhen, sondern auf anderen mit dem Produkt zusammenhängenden Umständen.[10] Vertraglich vorausgesetzt im Sinne von § 434 Abs. 1 Satz 2 Nr. 1 BGB ist die nicht vereinbarte, aber von beiden Parteien übereinstimmend unterstellte Verwendung des Produkts, die von der gewöhnlichen Verwendung abweichen kann.[11] Hierbei handelt es sich also um solche Verwendungen, die grade nicht im Lastenheft aufgeführt sind, sondern sich allenfalls aus diesem ergeben, weil sie bei der Auswahl der Beschaffenheitskriterien von beiden Parteien unterstellt wurden.

Wenn nach dem Vertrag keine Verwendung vorausgesetzt wurde („sonst"), sind nach § 434 Abs. 1 S. 2 Nr. 2, S. 3 BGB die Eignung für die gewöhnliche Verwendung und die übliche und vom Käufer zu erwartende Beschaffenheit maßgeblich, die auch durch öffentliche Äußerungen des Verkäufers und des Herstellers geprägt werden kann. Was die gewöhnliche Verwendung ist, entscheidet die Verkehrsanschauung.[12] Maßgeblich ist der Erwartungshorizont eines vernünftigen Durchschnittskäufers.[13] Dieser Erwartungshorizont wird primär durch einen Vergleich mit anderen Stücken der gleichen Gattung geprägt, so dass die gewöhnliche Verwendung eines Produktes dasjenige ist, was man mit anderen Produkten der gleichen Gattung machen kann.[14] Der Erwartungshorizont wird dabei aber nicht nur durch das konkrete, gekaufte Produkt, sondern auch durch damit im Wettbewerb stehende Produkte mit gleichem Verwendungszweck und in derselben Preisklasse (Konkurrenzprodukten) geprägt.[15]

Für die Mängel im Sinn des § 434 Abs. 1 S. 2 Nr. 1 und 2, S. 3 BGB gilt ebenso, dass es nicht darauf ankommt, ob sie erheblich oder unerheblich sind.[16] Lediglich die Rechtsfolgen Rücktritt und Schadensersatz statt der ganzen Leistung sind bei unerheblichen Mängeln ausgeschlossen.

Ein Rechtsmangel im Sinne des § 435 BGB liegt vor, wenn Dritte keine oder nur die innerhalb des Kaufs übernommenen Rechte in Bezug auf das Produkt gegen den Käufer geltend machen können. Im Rahmen der Entwicklung von Produkten kommen hier

10 *Westermann*, in: Münchener Kommentar zum BGB, 7. Auflage 2016, § 434 Rdn. 46.

11 BGH, Urt. v. 16. März 2012, V ZR 18/11, Rdn. 16.

12 *Westermann*, in: Münchener Kommentar zum BGB, 7. Auflage 2016, § 434 Rdn. 24.

13 *Faust*, in: Bamberger/Roth/Hau/Poseck, BeckOK BGB,46. Edition Stand: 01.05.2018, § 434 Rdn. 57.

14 *Faust*, in: Bamberger/Roth/Hau/Poseck, BeckOK BGB,46. Edition Stand: 01.05.2018, § 434 Rdn. 58.

15 *Westermann*, in: Münchener Kommentar zum BGB, 7. Auflage 2016, § 434 Rdn. 25; *Faust*, in: Bamberger/Roth/Hau/Poseck, BeckOK BGB,46. Edition Stand: 01.05.2018, § 434 Rdn. 5.

16 *Faust*, in: Bamberger/Roth/Hau/Poseck, BeckOK BGB,46. Edition Stand: 01.05.2018, § 434 Rdn. 3.

allenfalls gewerbliche Schutzrechte in Betracht,[17] die aber in der Regel nicht Gegenstand von Validierungs- und Verifizierungsverfahren sind.

bb) Rechtsfolgen

Ist ein Produkt mangelhaft, stehen dem Käufer verschiedene Rechte zu. Allen voran hat er ein verschuldensunabhängiges und damit garantieähnliches Recht auf Nacherfüllung in Form der Nachbesserung (Reparatur) oder Nachlieferung (Ersatzlieferung). Erst wenn die Nacherfüllung gescheitert, endgültig verweigert oder nicht zumutbar ist, kann der Käufer die Folgerechte Rücktritt und Minderung sowie den verschuldensabhängigen Schadensersatz geltend machen.

Aus den Rechtsfolgen der Lieferung eines mangelhaften Produkts ergeben sich damit auch die unmittelbaren Haftungsrisiken, deren Vermeidung die Durchführung von Validierungs- und Verifizierungsverfahren dient: Nur wer sein Produkt kennt, kann im bilateralen Vertragsverhältnis die Produktbeschreibung so wählen bzw. entscheiden, ob er ein nachgefragtes Produkt überhaupt liefern kann, und damit Haftungsrisiken vermeiden.

b) Produkthaftung

Im Unterschied zur Mangelhaftung hat die Produkthaftung einen anderen Ansatzpunkt. Dabei ist zunächst zwischen der verschuldensunabhängigen Haftung aus dem Produkthaftungsgesetz und der verschuldensabhängigen Produzentenhaftung der §§ 823 ff. BGB zu unterscheiden. Beiden ist gemein, dass sie die Verletzung bestimmter Rechtsgüter (Leben, Gesundheit, Körper, Eigentum, Unversehrtheit einer Sache) schützen. Der wesentliche Unterschied zwischen beiden Haftungstatbeständen liegt darin, dass die Produzentenhaftung an ein schuldhaftes Verhalten anknüpft (vgl. § 823 Abs. 1 BGB, die Haftung aus dem Produkthaftungsgesetz hingegen an das Inverkehrbringen eines fehlerhaften Produktes (vgl. § 1 Abs. 1 iVm. Abs. 2 Nr. 1 ProdHaftG).[18]

aa) Deliktische Produzentenhaftung

Anknüpfungspunkt der deliktischen Produzentenhaftung ist die Verletzung einer Verkehrssicherungspflicht durch die Inverkehrgabe eines nicht sicheren Produktes, so dass immer dort, wo durch fehlerhafte Herstellung eines Produktes eine Gefahrenquelle geschaffen wird, der Hersteller dafür zu sorgen hat, dass diese – soweit möglich – nicht zur Entfaltung kommt.[19] Im Bereich der Warenherstellung ist für die Bereiche der

17 Str. für eine Behandlung von gewerblichen Schutzrechten als Rechtsmangel: *Westermann*, in: Münchener Kommentar zum BGB, 7. Auflage 2016, § 435 Rdn. 4; a.A. *Faust*, in: Bamberger/Roth/Hau/Poseck, BeckOK BGB,46. Edition Stand: 01.05.2018, § 435 Rdn. 11.

18 *Lenz*, in: Lenz, Produkthaftung, 1. Auflage 2014, S. 103 Rdn. 158.

19 *Lenz*, in: Lenz, Produkthaftung, 1. Auflage 2014, S. 116 Rdn. 187.

Konstruktion, Fabrikation, Instruktion und Produktbeobachtung diese allgemeine Verkehrssicherungspflicht durch die Rspr. weiter konkretisiert worden.[20] Da Validierungs- und Verifizierungsverfahren nahezu ausschließlich in der Entwicklung eines Produkts eine Rolle spielen, soll an dieser Stelle lediglich die Konstruktionspflicht näher dargestellt werden.

Aus der allgemeinen Verkehrssicherungspflicht folgt für den Konstruktionsbereich, dass nur solche Produkte für den Verrieb hergestellt werden dürfen, die von ihrer Konstruktion bzw. Zusammensetzung her gewährleisten, dass sie für den ihnen zugedachten Verwendungszweck geeignet und bei den tatsächlichen Einsatzbedingungen betriebssicher sind.[21] Entsprechend liegt ein Konstruktionsfehler vor, wenn das Produkt schon seiner Konzeption nach unter dem gebotenen Sicherheitsstandard bleibt.[22] Die Sicherheitsanforderungen an eine Konstruktion werden dabei durch das jeweils gefährdete Rechtsgut und die Größe der Gefahr bestimmt. Können etwa Körper und Gesundheit von (vielen) Menschen gefährdet werden, sind die Anforderungen besonders hoch.[23] In der sog. Airbag-Entscheidung führt der BGH hierzu aus:[24]

> *„Zur Gewährleistung der erforderlichen Produktsicherheit hat der Hersteller bereits im Rahmen der Konzeption und Planung des Produkts diejenigen Maßnahmen zu treffen, die zur Vermeidung einer Gefahr objektiv erforderlich und nach objektiven Maßstäben zumutbar sind."*

Geschuldet ist also nicht jede „denkmögliche" Sicherheitsmaßnahme im Sinne einer „optimalen" Qualität, sondern nur erforderliche Sicherheitsmaßnahmen. Sicherheitsrisiken müssen in den Grenzen des technisch Möglichen und wirtschaftlich zumutbaren konstruktiv ausgeschaltet werden.[25] Die Freiheit von sog. Entwicklungsfehler, also Gefahrenmomenten und Nebenwirkungen, die nicht ausgeschaltet werden können, weil sie nicht voraussehbar waren, ist nicht geschuldet.

Die **Zumutbarkeit** einer Maßnahme hängt dabei insbesondere von der Größe der vom Produkt ausgehenden Gefahr ab. Je größer die Gefahren sind, desto höher sind die

20 *Kullmann,* in Kullman/Pfister/Stöhr/Spindler, Produzentenhaftung, Kza 1520, S. 7; *Lenz,* in: Lenz, Produkthaftung, 1. Auflage 2014, S. 120 Rdn. 192.

21 *Kullmann,* in Kullman/Pfister/Stöhr/Spindler, Produzentenhaftung, Kza 1520, S. 8, m.w.N.Fn. 56.

22 *Förster,* in: Bamberger/Roth/Hau/Poseck, BeckOK BGB, 46. Edition, § 823 Rdn. 691; BGH, Urt. v. 5. Februar 2013, VI ZR 1/12 – Heißwasser Untertischgerät, Rdn. 13; BGH, Urt. v. 16. Juni 2009, VI ZR 107/08 – Airbag, Rdn. 15.

23 OLG Hamm NJW-RR 2011, 893 – Grillpaste

24 BGH, Urt. v. 16. Juni 2009, VI ZR 107/08 – Airbag, Rdn. 15 ff.

25 *Klindt/Handorn,* NJW 2010, 1105.

Anforderungen an die Zumutbarkeit einer Maßnahme.[26] Bei erheblichen Gefahren für Leben und Gesundheit von Menschen sind dem Hersteller deshalb weitergehende Maßnahmen zumutbar als in Fällen, in denen nur Eigentums- oder Besitzstörungen oder aber nur kleinere körperliche Beeinträchtigungen zu befürchten sind.[27]

Welche Maßnahmen **erforderlich** sind, beschreibt der BGH in der Airbag-Entscheidung:

> *Erforderlich sind die Sicherungsmaßnahmen, die nach dem im Zeitpunkt des Inverkehrbringens des Produkts vorhandenen neuesten Stand der Wissenschaft und Technik konstruktiv möglich sind und als geeignet und genügend erscheinen, um Schäden zu verhindern.*

Welche **Position** der neueste Stand von Wissenschaft und Technik in der Hierarchie der technischen Regeln, bestehend aus den allgemein anerkannten Regeln der Technik, dem Stand der Technik und – als höchstem Niveau – dem Stand von Wissenschaft und Technik einnimmt, ist nicht ganz klar.[28] Festgelegt hat sich der BGH in der Airbag-Entscheidung aber dahingehend, dass

> *(...) der insoweit maßgebliche Stand der Wissenschaft und Technik nicht mit Branchenüblichkeit gleichgesetzt werden (darf) (...) .*

Begründet hat er dies damit, dass die in der jeweiligen Branche tatsächlich praktizierten Sicherheitsvorkehrungen durchaus hinter der technischen Entwicklung und damit hinter den rechtlich gebotenen Maßnahmen zurückbleiben können.[29] DIN-Normen, VDE-Vorschriften etc. Stellen insoweit also nur einen Mindeststandard dar, der von dem neusten Stand der Wissenschaft und Technik streng zu unterscheiden ist und diesem, angesichts der Dauer eines Normierungsprozesses, regelmäßig „hinterherhinken“ dürfte.[30]

Präzisiert hat der BGH den neusten Stand von Wissenschaft und Technik in der Airbag-Entscheidung wie folgt:

> *„Die Möglichkeit der Gefahrvermeidung ist gegeben, wenn nach gesichertem Fachwissen der einschlägigen Fachkreise praktisch einsatzfähige Lösungen zur Verfügung stehen. Hiervon kann grundsätzlich erst dann ausgegangen werden, wenn eine sicherheitstechnisch überlegene Alternativkonstruktion zum Serieneinsatz reif ist. Der Hersteller ist dagegen nicht dazu verpflichtet, solche Sicherheitskonzepte*

---

26 *Förster*, in: Bamberger/Roth/Hau/Poseck, BeckOK BGB, 46. Edition, § 823 Rdn. 692.

27 BGH, Urt. v. 5. Februar 2013, VI ZR 1/12 – Heißwasser Untertischgerät, Rdn. 13.

28 *Klindt/Handorn*, NJW 2010, 1105 (1106).

29 BGH, Urt. v. 16. Juni 2009, VI ZR 107/08 – Airbag, Rdn. 16.

30 *Lenz*, in: Lenz, Produkthaftung, 1. Auflage 2014, S. 232 Rdn. 382.

umzusetzen, die bisher nur "auf dem Reißbrett erarbeitet" oder noch in der Erprobung befindlich sind."

Maßgeblich für den **neusten Stand von Wissenschaft und Technik** ist das gesicherte Fachwissen der einschlägigen Fachkreise. Dieses muss praktisch einsatzfähige Lösungen bereitstellen, also solche, die zum Serieneinsatz reif sind. Das Kriterium der Serienreife wird ergänzt dadurch, dass keine Techniken eingesetzt werden müssen, die nur in der Theorie („auf dem Reißbrett") existieren und die sich noch in der Erprobung befinden. Hierdurch wird der neuste Stand von Wissenschaft und Technik vom Stand von Wissenschaft und Technik abgegrenzt.[31] Letzterer bezeichnet die Sachkunde, die im wissenschaftlichen und technischen Bereich vorhanden ist, also die Summe an Wissen und Technik, die allgemein anerkannt ist und allgemein zur Verfügung steht.[32] Umfasst hiervon werden auch die öffentlich zugänglichen wissenschaftlich-technischen Forschungsergebnisse von Universitäten, technischen Hochschulen und Fachhochschulen, also die theoretischen Erkenntnisse der Wissenschaft.[33] Der neuste Stand von Wissenschaft und Technik erfasst damit nicht Erkenntnisse, die irgendein Ingenieur, Techniker oder Wissenschaftler auf der Welt gewonnen hat, die aber nicht öffentlich zugänglich und auch nicht abgesichert und gefestigt sind.[34]

In Abgrenzung zum bisher eingesetzten Stand von Wissenschaft und Technik sind erforderliche Sicherheitskonzepte nur solche, die als „**Alternativdesign**" gegenüber der vom Hersteller bislang verwendeten Vorgehensweise sicherheitstechnisch überlegen sind.[35] Die Neukonstruktion muss also aus Sicherheitsgesichtspunkten gegenüber der bislang eingesetzten Technik ein Mehr an Sicherheit bieten, nur dann ist sie als Alternativdesign erforderlich.

bb) Produkthaftung nach dem ProdHaftG

Die Grundsätze der Konstruktionsverantwortung gelten auch im Bereich der Produkthaftung nach dem ProdHaftG. Gem. § 3 Abs. 1 ProdHaftG hat ein Produkt einen Fehler, wenn es nicht die Sicherheit bietet, die unter Berücksichtigung aller Umstände, insbesondere seiner Darbietung, des Gebrauchs, mit dem billigerweise gerechnet werden kann, sowie des Zeitpunkts, in dem es in den Verkehr gebracht wurde, berechtigterweise erwartet werden kann.[36] Abzustellen ist dabei nicht auf die subjektive Sicherheitserwartung des jeweiligen Benutzers, sondern objektiv darauf, ob das Produkt

31 Vgl. auch *Kullmann*, in Kullman/Pfister/Stöhr/Spindler, Produzentenhaftung, Kza 1520, S. 16 f.

32 BT-Drs. 11/2447, S. 15.

33 *Lenz*, in: Lenz, Produkthaftung, 1. Auflage 2014, S. 238 Rdn. 396.

34 *Kullmann*, in Kullman/Pfister/Stöhr/Spindler, Produzentenhaftung, Kza 1520, S. 16 f.

35 *Förster*, in: Bamberger/Roth/Hau/Poseck, BeckOK BGB, 46. Edition, § 823 Rdn. 693.

36 BGH, Urt. v. 16. Juni 2009, VI ZR 107/08 – Airbag, Rdn. 12.

diejenige Sicherheit bietet, die die in dem entsprechenden Bereich herrschende Verkehrsauffassung für erforderlich hält[37] Die nach § 3 Abs. 1 ProdHaftG maßgeblichen Sicherheitserwartungen beurteilen sich grundsätzlich nach denselben objektiven Maßstäben wie die Verkehrspflichten des Herstellers im Rahmen der deliktischen Haftung gemäß § 823 Abs. 1 BGB.[38] Damit umfasst die Sicherheitserwartung im Rahmen der Produkthaftung nach dem ProdHaftG, dass das Produkt nach den gleichen Maßstäben konstruiert ist, die auch im Rahmen der deliktischen Produzentenhaftung gelten. Die Ausführungen zur Produzentenhaftung zu Inhalt und Umfang der Konstruktionspflicht gelten also auch unverändert innerhalb der Produkthaftung.

3 Anforderungen an Validierung und Verifizierung aus haftungsrechtlicher Sicht

Legt man diesen gesetzlichen Haftungsrahmen über die zur Verfügung stehenden Mittel Test, Prüfung, Validierungs- und Verifizierungsverfahren ergibt sich folgendes Bild:

a) Mangelhaftung

Innerhalb der Mangelhaftung dient die Durchführung eines Verifizierungsverfahren und damit die Prüfung dazu, konkrete Eigenschaften des Produktes zu eruieren, um so festzustellen, ob das Produkt die im Lastenheft vorgegebenen Anforderungen, die eine Beschaffenheitsvereinbarung iSd. § 434 Abs. 1 S. 1 BGB darstellen, erfüllt. Die Validierung und damit dir Durchführung von Tests setzt hingegen da an, wo die Parteien eine bestimmte Verwendungseignung zum Gegenstand ihrer Beschaffenheitsvereinbarung gemacht haben. Dabei ist es Aufgabe des Herstellers, auch die nicht vereinbarte, sondern lediglich vorausgesetzte Verwendungseignung iSv. § 434 Abs. 1 S. 2 Nr. 1 BGB zu identifizieren und den Validierungsprozess entsprechend auszurichten. Gleiches gilt für die gewöhnliche Verwendung bzw. die übliche vom Verkäufer zu erwartende Beschaffenheit im Sinne von § 434 Abs. 1 S. 2 Nr. 2 BGB.

Innerhalb der Mängelhaftung ergeben sich die Anforderungen an Tests und Prüfungen bzw. Validierungs- und Verifizierungsverfahren aus der vertraglich getroffenen Beschaffenheitsvereinbarung. Zu schauen ist, was innerhalb der verschiedenen Beschaffenheitskategorien des § 434 BGB vertraglich vereinbart wurde und ob das konstruierte Produkt diese Anforderungen erfüllt.

37 Vgl. BT-Drucks. 11/2447, S. 18; BGH, Urteil v. 17. März 2009 – VI ZR 176/08 – „Kirschtaler", Rdn. 6.

38 BGH, Urteil v. 17. März 2009 – VI ZR 176/08 – „Kirschtaler", Rdn. 6.

b) Produkthaftung

Im Rahmen der Produkthaftung ist die Auswahl und Gestaltung von Validierungs- und Verifizierungsverfahren anders vorzunehmen, da es hier keine klare vertragliche Vereinbarung mit einem Lastenheft gibt. Test- und Prüfverfahren innerhalb der Validierung und Verifizierung sind so auszurichten, dass das Produkt entsprechend seiner Eigenart und späteren Verwendung im Hinblick auf eine sichere Konstruktion untersucht werden kann. Maßstab sind also nicht die in einem Vertrag niedergeschriebenen Eigenschaften des Produkts, sondern die Sicherheitserwartung der betroffenen Verkehrskreise. Diese muss der Hersteller identifizieren und danach seine Validierungs- und Verifizierungsprozesse ausrichten. Erschwerend kommt hinzu, dass der Hersteller nicht nur den vorschriftsmäßigen Gebrauch des Produkts, sondern auch den vorhersehbaren Fehlgebrauch in seine Überlegungen miteinbeziehen muss. Die berechtigte Sicherheitserwartung geht nämlich grundsätzlich auch dahin, dass das Produkt so konzipiert ist, dass es unter Beachtung der Installations- und Gebrauchsanleitung bei bestimmungsgemäßem Gebrauch oder vorhersehbarem Fehlgebrauch gefahrlos benutzt werden kann.[39] Ist beispielsweise die Überbeanspruchung von Produkten für den Hersteller vorhersehbar, dann muss auch dies für ihn Anlass sein, die Überbeanspruchung bei der Konstruktion einzukalkulieren und ggf. Verstärkungen vorzusehen.[40]

Inhaltlich müssen Test- und Prüfverfahren so ausgestaltet sein, dass sie sowohl in zeitlichem als auch sachlichem Umfang sicher Aussagen über das Langzeitverhalten des jeweiligen Produktes ergeben und soweit irgend möglich den tatsächlichen Einsatzbedingungen des jeweiligen Produkts entsprechen.[41] Auch hier unterscheidet sich das Produkthaftungsrecht von der Mängelhaftung, da bei letzterer – soweit zwischen den Parteien nicht ausdrücklich eine abweichende Vereinbarung getroffen wird – grundsätzlich nur die Mangelfreiheit im Zeitpunkt der Übergabe des Produktes vom Verkäufer an den Käufer geschuldet wird (vgl. § 434 Abs. 1 iVm. § 446 BGB). Die Verjährungsregelungen im Produkthaftungsrecht sind aber weitaus länger, so dass das Sicherheitsniveau auch weitaus länger aufrechterhalten werden muss.[42]

39 BGH, Urt. v. 5. Februar 2013, VI ZR 1/12 – Heißwasser Untertischgerät, Rdn. 14.

40 *Kullmann,* in Kullman/Pfister/Stöhr/Spindler, Produzentenhaftung, Kza 1520, S. 13.

41 *Kullmann,* in Kullman/Pfister/Stöhr/Spindler, Produzentenhaftung, Kza 1520, S. 19.

42 Im Produkthaftungsrecht beträgt die Höchstdauer der Haftung 10 Jahre ab Inverkehrgabe des Produktes (vgl. § 13 Abs. 1 ProdHG), innerhalb der deliktischen Produzentenhaftung beträgt die absolute Verjährung je nach Verschuldensgrad und verletzen Rechtsgut zwischen 10 und 30 Jahren (vgl. § 199 BGB).

Die eingesetzten Prüfgeräte und -verfahren müssen dem neusten Stand der Technik entsprechen und die Prüfung selber durch fachkundiges Personal erfolgen.[43] Falls erforderlich müssen notwendige Prüfmittel neu geschaffen werden.[44]

Ob die Tests und Prüfung als tatsächliche Versuche unter realen Bedingungen mit Testbauteilen oder im Rahmen von virtuellen Tests (Simulationen) bzw. des sog. Frontloadings durchgeführt werden, hängt von der Zuverlässigkeit des jeweiligen Verfahrens ab. Ermöglicht das gewählte Verfahren sichere Aussagen über das Sicherheitsverhalten des Produktes selber – kurz wie langfristig – und über etwaige Nebenwirkungen bzw. gefährliche Materialeigenschaften, steht seinem Einsatz nichts entgegen. Bestehen jedoch Zweifel an der Aussagekraft des gewählten Verfahrens, ist dieses aus sicherheitstechnischer Sicht nicht geeignet, im Rahmen des Validierungs- und Verifizierungsprozesses die nötige Konstruktionsverantwortung zu erfüllen.

4 Fazit

Der Auswahl des geeigneten Test- bzw. Prüfverfahrens und damit einhergehend der Wahl zwischen Validierung oder Verifizierung kommt sowohl im Bereich der Mangelhaftung als auch im Bereich der Produkthaftung entscheidende Bedeutung zu, wenn Haftungsrisiken reduziert bzw. vermieden werden sollen. Vor allem im Bereich der Produkthaftung wird dabei regelmäßig dem Validierungsverfahren der Vorzug zu geben sein, weil nur über dieses die Wechselwirkung des Produktes mit Erwartungen der Produktnutzer abgeglichen werden kann:

Gewährleistungsrecht	Produkthaftungsrecht	Testverfahren
Vereinbarte Beschaffenheit		Verifizierung
Vereinbarte Verwendung		Validierung
Typische Verwendung		Validierung
	Sicherheitserwartungen	Validierung

Im Bereich der Mängelhaftung ergeben sich diese Erwartungen weitestgehend aus der vertraglichen Vereinbarung zwischen den Parteien. Nur in begrenztem Rahmen spielt die Beschaffenheit des Produktes für die gewöhnliche Verwendung, die üblicherweise erwartet werden darf, eine Rolle. Anders verhält sich dies im Bereich der Produkthaftung. Hier ist Maßstab die Sicherheitserwartung der Produktnutzer. Da in diesen Sicherheitserwartungen auch ein vorhersehbarer Fehlgebrauch des Produktes enthalten

43 Bamberger/Roth/Spindler, BGB, 2. Auflage, § 823 Rdn. 494.
44 *Kullmann*, in Kullman/Pfister/Stöhr/Spindler, Produzentenhaftung, Kza 1520, S. 19.

ist, kommt der Auswahl und inhaltlichen Gestaltung von Validierungsprozessen eine ganz andere und wesentlich komplexere Bedeutung zu.

III Erprobungsklauseln in der Produkthaftpflichtversicherung – haftungserhaltende Funktion

Zur Absicherung von Produkthaftungsrisiken schließen Hersteller sog. Produkthaftpflichtversicherungen ab. In diesen finden sich verschiedene Haftungsausschlüsse für Fälle, in denen der Versicherer eine Einstandspflicht ablehnt. Hierunter fällt die sogenannte Erprobungsklausel, die in Ziffer 6.2.5 der Musterbedingungen des Gesamtverbandes der Deutschen Versicherungswirtschaft e.V. (GDV) wie folgt formuliert ist:

„Nicht versichert sind Ansprüche aus Sachschäden und alle sich daraus ergebenden Vermögensschäden sowie Ansprüche aus Schäden gemäß Ziffer 4 durch Erzeugnisse, deren Verwendung oder Wirkung im Hinblick auf den konkreten Verwendungszweck nicht nach dem Stand der Technik oder in sonstiger Weise ausreichend erprobt waren.

Dies gilt nicht für Schäden an Sachen, die mit den hergestellten oder gelieferten Erzeugnissen weder in einem Funktionszusammenhang stehen noch deren bestimmungsgemäßer Einwirkung unterliegen."

Zweck der Erprobungsklausel ist es zu verhindern, dass solche Risiken vom Versicherungsschutz der Produkt-Haftpflichtversicherung erfasst werden, die bei ausreichender Erprobung des Produkts durch den Produzenten verhindert werden können, weil es nicht Aufgabe der Versicherungswirtschaft ist, den Unternehmen im Ergebnis ihre Entwicklungsrisiken nachzufinanzieren.[45] Dabei erfassen Erprobungsklauseln nur Schäden an Sachen und nicht an Personen, da die Erprobungsklausel den Versicherer vor allem vor Austauschkosten bedingt durch unerprobte Erzeugnisse schützen sollen.

Erprobungsklauseln sind deswegen schwierig in der Anwendung, da sie verschuldensunabhängig formuliert sind und auf den Stand der Technik Bezug nehmen. Stand der Technik ist der Entwicklungsstand fortschrittlicher Verfahren, Einrichtungen und Betriebsweisen zur Erprobung, der die praktische Eignung für den konkreten Verwendungszweck gesichert erscheinen lässt.[46] Dabei ist der Stand der Technik höher anzusiedeln, als die allgemein anerkannten Regeln der Technik, steht aber nicht mit dem Stand von Wissenschaft und Technik bzw. dem neusten Stand von Wissenschaft und

45 OLG Bremen, NVersZ 1999, 535 (536).

46 *Voit*, in: Prölss/Martin, Versicherungsrecht, 30. Auflage, 265 ProdHM Ziff.6, Rdn. 10.

Technik auf einer Stufe.[47] Aus der Unklarheit dieser Begrifflichkeit wird deutlich, dass – soweit möglich – Vorkehrungen getroffen werden müssen, um eine spätere Inanspruchnahme der Produkthaftpflichtversicherung nicht zu gefährden.

Vor einer Anwendbarkeit der Erprobungsklausel zu Lasten des Herstellers schützt in erster Linie die ausreichende aber auch ausreichend dokumentierte Erprobung, also die Durchführung von Tests, Prüfungen, Validierungs- und Verifizierungsverfahren. Darüber hinaus kann eine Anpassung der Klausel im Verhandlungswege angestrebt werden, die Art und Umfang der Erprobung näher bestimmt oder gar ein Verschuldenserfordernis in die Formulierung mit aufnimmt.

IV Fazit

Die Bedeutung von Tests und Simulationen, als Teil von Validierungs- und Verifizierungsverfahren, hat innerhalb der haftungsrechtlichen Verantwortung des Herstellers eine erhebliche Bedeutung. Tests und Prüfungen stehen am Anfang der Herstellungskette, so dass ihre sorgfältige Auswahl und Durchführung erhebliches Haftungsreduzierungspotential hat. Durch die fehlerhafte Konstruktion eines Produkts wird der Grundstein für eine spätere Produkthaftung bzw. einen Serienschaden gelegt. Ausreichende und verlässliche Tests und sind daher von grundlegender Bedeutung, um innerhalb der Konstruktionsphase, vor allem auch in Zeiten steigenden Wettbewerb- und Kostendrucks die Vertragsmäßigkeit und Sicherheit von Produkten zu gewährleisten und so Haftung zu vermeiden.

47 *Voit*, a.a.O.

Simplified modeling of self-excited rear axle wheel shimmy vibrations by using the Proper Orthogonal Decomposition (POD)

Sebastian Wagner, M.Sc., BMW Group

Dr.-Ing. Johannes Mayet, BMW Group

Prof. Dr.-Ing. Dr. h.c. Dieter Schramm, University of Duisburg-Essen

Abstract

Vehicles are subject to a variety of requirements, of which the technical objectives are only a subfield. In addition, the different sub-areas increasingly interact with each other, whereby the complexity of the development process of a vehicle continuously rises. In particular, components with a high degree of interaction must therefore be considered overall with regard to all influences.

The elastic rear subframe mount must not only meet acoustic objectives, but also influences the suspension stiffness and the self-excited vibration phenomenon wheel shimmy [Wagner et al. 2018] on the vehicle rear axle, which can be safety-critical. This can quickly lead to conflicts regarding targets in the design of the bushing stiffness, as the acoustic requirements among other things lead to lower bushing stiffness in contrast to the vehicle dynamic targets.

If a target conflict is uncovered at a late stage in the testing phase, changes to the vehicle are associated with high effort and high costs due to the largely fixed parameters.

In order to uncover such target conflicts early in the development process, reliable system behavior predictions must be made in the first part of the vehicle development process, the design phase [Zimmermann et al. 2017]. As there exist no final vehicle data at this early stage, the different characteristics are defined by corridors, which are increasingly narrowed. Due to this small database, no complex multibody-models can be built up at this time. Thus, simplified models are used to estimate the target values [Schramm et al. 2018].

One possibility for the automated development of simplified models is the use of mathematical reduction methods [Kerschen et al. 2005]. In this approach, a complex multibody-model is reduced to the minimum required degrees of freedom (DOFs) for a certain vibration phenomenon. Performing parameter variations in the aforementioned corridors on the detailed model and following reduction allows revealing the influences on the simplified model. However, the parameterization of the simplified models must be based on data that is already available at an early stage of development.

This paper deals with an analyzing method that is able to adequately describe the self-excited vibration phenomenon wheel shimmy already in early stages of the design phase. The resulting simplified simulation model was developed by applying Proper Orthogonal Decomposition. After setting up this simplified model once, using a similar vehicle, there is no need for a detailed multibody-model. The novelty value of this paper is the application of the mathematical reduction method to a self-excited vibration phenomenon for complex vehicle models with many DOFs.

1 Introduction

Vehicle development is characterized by a large number of requirements, of which technical aspects are only a subdivision. In particular, conflicting requirements create a field of tension in the design and can therefore lead to conflicting goals.

In the classic vehicle development process according to the V-model, see Figure 1, requirements for the system are defined and evaluated at the overall vehicle level at the early phase of development. At this point, all the different parameters are still existing as property corridors, which are increasingly being narrowed. Due to the large amount of freedom available for development, changes to the properties can be implemented in this phase without any problems or great financial expense.

At a later stage while detailing these requirements, they are broken down to subsystems and finally to the individual components. If conflicting objectives are revealed at component level, the higher-level subsystem properties are usually already fixed. Therefore, these conflicts can only be solved with significant effort and high costs.

In order to detect and address such late conflicts of objectives at the very beginning of the vehicle development process, an overall and sufficiently detailed consideration of all properties must be carried out early in the development process. This is possible with the so-called cooperative design.

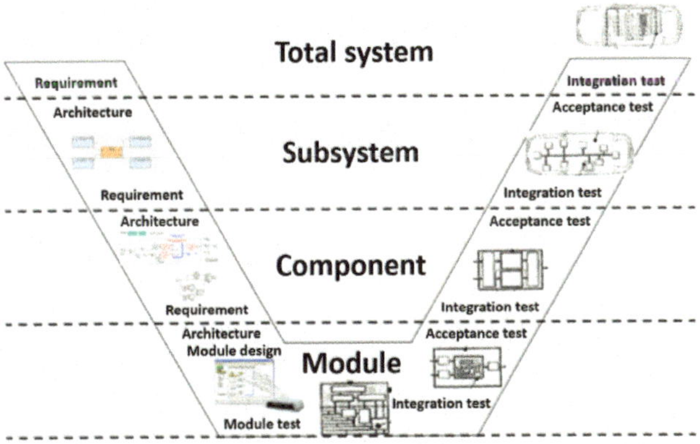

Figure 1: Design according V-model [Zimmermann et al. 2017].

In order to ensure a complete consideration of all parameters, all relevant input variables are defined at component level as so-called design variables. The system properties are evaluated at the overall system level using so-called design objectives. The mathematical linkage between design variables and design objectives is accomplished using simplified simulation models. On the one hand, this is due to the fact that in this early development phase only simplified simulation models can be built due to the few details. On the other hand, the model should be able to handle a large number of variants in a very short time such that sensitivity analysis are possible with a large number of design variables and design objectives. In order to limit the number of variants and to optimize the quality of the results, a statistical design of experiments (DoE) will be used.

2 State of the Art

2.1 POD

There are different procedures for creating simplified models. One possibility is Proper Orthogonal Decomposition (POD). POD is a mathematical procedure, which is often used in statistics and helps to present large amounts of data in a compact way. With the help of POD, the order of a system can be reduced on the one hand, while on the other hand system properties, such as the Eigenvalues of the system, remain largely unaffected in the considered frequency range. For this purpose, the output system is projected onto the reduced subspace by means of an orthogonal transformation [Kerschen et al. 2005].

A great advantage of this method is the minimization of the average square distance between the original and the reduced system. POD can also be applied to non-linear systems. However, it must be pointed out, that due to the reduction, a linearization takes place in the operating point and statements about the new system are therefore limited to a range around the operating point [Kerschen et al. 2005].

Quaranta et al. have already used POD to evaluate the local stability of periodic motions in multibody-systems (MBS) [Quaranta et al. 2004].

Strongly related to this work are also previous investigations by Kreuzer and Kust, who used POD for the "control of self-excited vibrations of long torsional strings" [Kreuzer et al. 1996].

Another example is the work of Steindl and Troger, where reduction of DOFs is considered with the help of POD [Steindl et al. 2001].

In particular, it must be noted, that the system created by POD is represented in a subspace. The individual POD modes are orthogonal to each other, see (1).

$$\hat{q}_k^T \hat{q}_n = \begin{cases} 0 & \text{for} \quad n \neq k \\ \neq 0 & \phantom{\text{for}} \quad n = k \end{cases} \tag{1}$$

In contrast, in a classical eigenvalue analysis, the eigenmodes of the system are orthogonal to each other with respect to the mass or stiffness matrix, but not necessarily to each other directly [Markert 2010], see (2).

$$\hat{q}_k^T M \hat{q}_n = \begin{cases} 0 & \text{for} \quad \omega_n \neq \omega_k \\ \hat{m}_n & \qquad\quad n = k \end{cases} \tag{2}$$

Therefore, POD modes and eigenmodes of a classical eigenvalue analysis are not directly comparable. This means that the modes of a POD-modified but non-reduced system do not match the modes of the initial system – whereas the eigenvalues of both systems do. The modes of both systems only coincide in the special case, if the mass matrix is proportional to the unit matrix [Kerschen et al. 2005].

2.2 Wheel shimmy phenomenon

Self-excited wheel shimmy vibrations occur on an elastically mounted rear axle of vehicles during braking with high deceleration [Zeller 2012]. When considering the vibration mode in the MBS, two major DOFs can be detected. On the one hand, the complete rear axle including rods and wheels, rotates as a rigid body around the vertical z-axis, see Figure 2 left side. On the other hand, the two wheels deflect in opposite directions relative to the body, see Figure 2 right side. Superimposing these two modes, the wheels describe an elliptical trajectory in the longitudinal-vertical plane [Wagner et al. 2018].

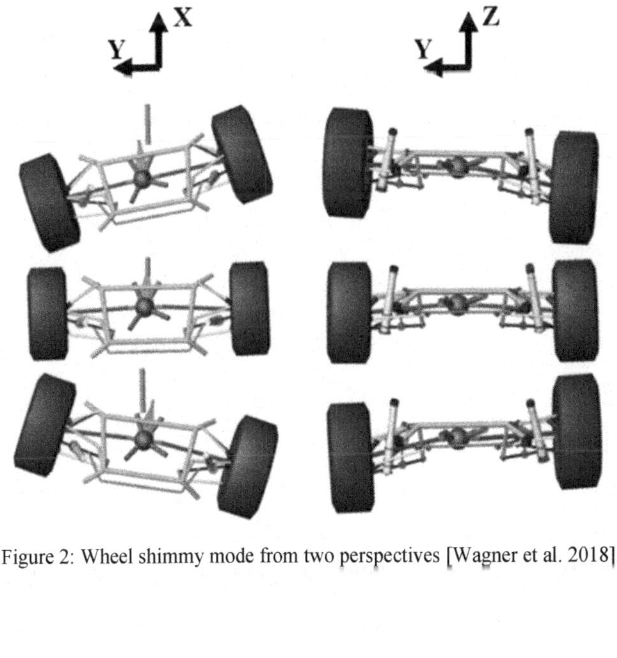

Figure 2: Wheel shimmy mode from two perspectives [Wagner et al. 2018].

The eigenfrequency of this oscillation is typically between 15 and 25 Hz. Due to its self-excited character, a single excitation, such as a pothole, is sufficient for the system to start oscillating. However, this can result in damage to the rear axle subframe bushings, which are exposed to the largest relative movement. Furthermore, it is also possible that the braking distance extends, if the wheels lose ground contact.

The description of the MBS and simplified system is possible both in the time domain and via system matrices. The self-excited character has the advantage, that the evaluation of the stability is possible directly with the help of the modal damping or the real part of the mode. In the time domain, the evaluation of the stability can therefore be done by analyzing the vibration amplitudes in the time signal. In the system matrix consideration, only the real parts of the eigenvalues need to be evaluated. The algebraic sign of the real part gives information about the system stability [Markert 2010]. In both cases, there is a clear separation with objective criteria between mathematically stable and unstable systems.

The purpose of this work is to create a simplified model from the complex MBS model with more than 200 DOFs using POD, which represents the system properties of the motion shown in Figure 2.

3 Data basis

In order to build up and parameterize a simplified model by POD, a database is required. This database is composed of the time-domain signals of the DOFs from an already existing MBS simulation. Among others, displacement, velocity or acceleration data can be used. However, especially displacement and velocity signals are superimposed by low-frequency and constant components, like rigid-body movements, which should not to be represented. The separation can be done by filters or acceleration signals can be used directly.

In the above-mentioned MBS simulation, the unstable state of the vehicle is generated by a braking maneuver. Figure 3 displays the vertical displacement of the left rear wheel. From the maneuver beginning until $t = 2$s the rear axle reaches transient oscillation. In this state, the system is in constant quasi-static deceleration and still free of vibrations, although the linearized system already has unstable eigenvalues. After a singular excitation is applied to the system at $t = 2$s, the rear axle starts oscillating, so that the unstable behavior can be recognized in the time domain signals. After $t = 4$s the system reaches a stable limit cycle due to the progressive buffers in the subframe mounts.

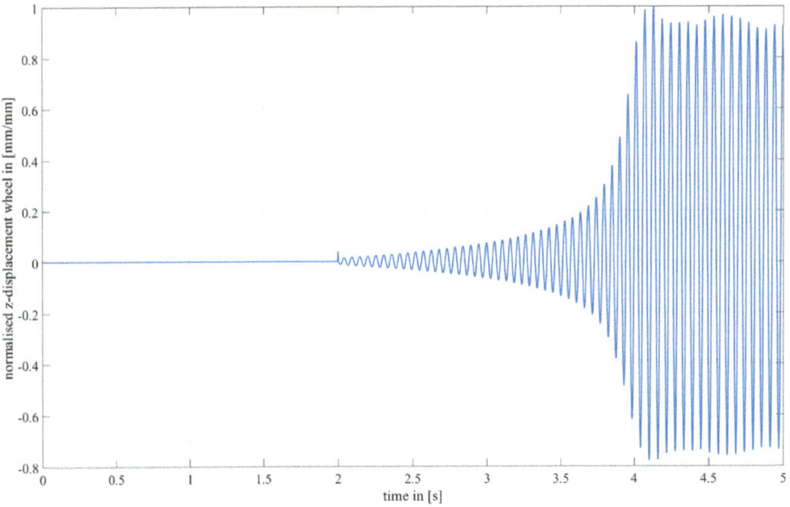

Figure 3: Instable behavior of a vertical DOF of the wheel.

3.1 MBS simulation with eigenvalue analysis

Previous investigations have shown, that the eigenvalues depend strongly on the linearization point, which changes in time. This implies, that the system is non-linear. For that reason, the results and reproducibility of the different stability evaluation methods need to be compared.

A standard stability criterion is the evaluation method of the State Matrix Eigenvalues at a given operation point. The real part of the eigenvalue defines the stability of the eigenmode. A linear system has the same eigenvalues irrespectively of the operating point.

Until $t = 2$s, as long as the system is still at quasi-static rest, the modal damping of the wheel shimmy mode is -2.36%. As soon as the system starts to oscillate, the eigenvalues vary. The variation gets major the larger the system amplitudes are. Within the first oscillation, the modal damping fluctuates between -0.92% and -2.53%, the average is at -1.88%. Half a second later, the range is already between +0.59% and -2.82%. Thus, no statement of stability is possible based on the data obtained above.

The differences in the eigenfrequency of the mode are much smaller. In the beginning, at $t = 2$s, the eigenfrequency of the unstable eigenmode is 18.89 Hz. In the first oscillation, the frequency varies between 18.88 Hz and 19.09 Hz. Half a second later, it is between 18.77 Hz and 19.40 Hz. Since the evaluation of the stability of the oscillation is achieved by modal damping, the focus of the analysis below does not lie on frequency. It is only used as a further evaluation parameter.

Both the modal damping and the eigenfrequency for this evaluation method are also displayed in Figure 4.

Since the eigenvalues of the analyzed system vary strongly depending on the operating point, a non-linear system behavior has to be considered. The first problem resulting from this is the low quantifiability of the results. The eigenvalue analysis only returns a range instead of a specific value. The second problem is the lack of comparability with other simulations when sensitivity analysis is performed, especially since the values of the sensitivities are located in similar ranges. The last problem arises from the uncertain quality regarding absolute statements. This all results in the requirement that the operating point at which the linearization is performed must be reproducible and not subject to fluctuations during parameter studies. If these criteria are not satisfied, no stability prediction can be made with an eigenvalue analysis.

3.2 MBS simulation with time-domain analysis

In order to check the validity of the information from Section 3.1 and to be able to produce comparable and objective evaluations, additionally a time domain analysis is carried out. Here, the subsequent maxima or minima are examined and the modal damping is derived from the logarithmic decrement [Markert 2010]. The DOFs, which are considered as input for the time-domain analysis, have to be characteristic for the wheel shimmy oscillation. Therefore, these DOFs must accurately reflect the oscillation behavior of the mode. The 16 characteristic DOFs of the rear axle, e.g. the vertical and longitudinal accelerations of the rods, were used for the evaluation.

Depending on the DOF, the results of modal damping vary between -1.38% and -1.67%. The average of all DOFs is -1.53%. The comparison between time domain analysis and eigenvalue analysis shows an absolute difference of 0.3%, whereas eigenvalue analysis provides the more conservative statements, see Figure 4. Due to the large differences between the two analysis methods, no absolute statement can be made at this point if only the MBS results are analyzed.

3.3 POD application

One way out of the problem is a POD. The strength of this method is to extract important information for certain boundary conditions from large amounts of data. Thus, systems with many DOFs can be broken down to simplified modelling. Initially, the basic boundary conditions will be explained in the following section.

The same time signals are used as input variables for the POD that are also evaluated in the MBS time domain analysis in Section 3.2. On the one hand, POD provides all eigenvalues of the reduced system as a result. On the other hand, the reduction matrix is also created. This allows to calculate the time domain answer of the simplified system from the curves of the unreduced original system. To create a further comparison criterion for the stability statement, a time domain analysis can be performed using this data.

Figure 4 shows the system properties of the wheel shimmy mode for the different reduction orders of the POD. The x-axis represents the number of DOFs of the reduced system. In the upper subplot the y-axis represents the modal damping and in the lower subplot the eigenfrequency. For comparison purposes, the results of the MBS model are plotted as well.

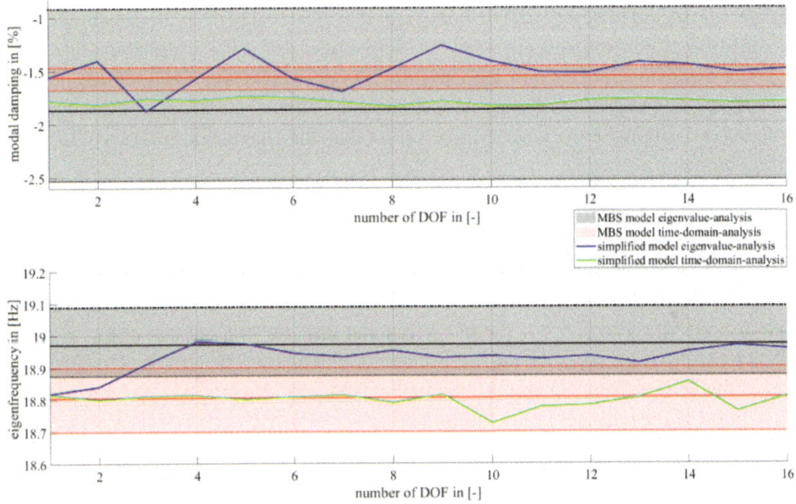

Figure 4: Comparison of models and evaluation methods for wheel shimmy vibrations. Dotted lines are lower limits, dash-dotted lines are upper limits and solid lines are time-averaged values.

The results of the POD can be viewed as a mathematical series. The higher the number of DOFs, the better is the accuracy of the reduced result corresponding to the input data. This is also shown in the results of the modal damping, see blue curve in the upper diagram of Figure 4. Since the input data for the POD are the time-domain curves of the MBS simulation, the red curve must be used here as a comparison criterion. The curve fluctuates stronger with smaller number of DOF.

For the MBS eigenvalue (gray) and time domain (red) analyses, result areas instead of concrete curves are displayed. This is due to the uncertainties already described in Sections 3.1 and 3.2. The green graph represents the time-domain analysis of the reduced model.

The comparison of the eigenfrequencies in the lower subplot shows, that all evaluation methods and models produce almost the same results. The deviations between the different methods are around 0.1 Hz and also in the simplified modelling with eigenvalue analysis, the values decrease only at small number of DOFs. The evaluation method and the model therefore have minor influence on the results of the eigenfrequency.

Larger differences can be seen in the upper subplot, where the modal damping is illustrated. First, it is noticeable, that the eigenvalue analysis of the MBS model shows a larger range in the values compared to the other three evaluation methods. The average of these shows the lowest modal damping and thus the most conservative estimate. The average of the time domain analysis of the MBS model shows 0.3% higher modal damping. The eigenvalue analysis of the simplified model varies depending on the number of DOFs in the range of MBS time domain. The time domain analysis of the simplified model shows slightly more conservative values than the eigenvalue analysis of the simplified model.

If the fluctuations from Section 3.1 are taken into account, the results of the MBS eigenvalue analysis are up to 1.6%, depending on the operating point. The time domain analysis of the MBS simulation even reduces this range to 0.15%. With the help of the data obtained from the POD, the result of the MBS simulation can even be represented with a one-DOF model. Since the POD uses the time domain signals of the upswing behavior as input signal, it is more robust against fluctuations than the eigenvalue analysis of the MBS system. Therefore, the next chapter derives a one-DOF model for the description of the wheel shimmy vibrations.

4 Simplified model

A simplified model can be developed using either a physical or a mathematical approach. Physical models are based on physical cause-effect-chain understanding of the system. The modeling is usually done by using replacement masses, stiffnesses and dampings, which are connected to each other according to their directions of motion via idealized joints and bushings. Examples can be found in [Königs et al. 2016].

In contrast, mathematical models do not reflect physical correlations, but are built or trained on the basis of statistical data. As a result, mathematical substitute models often calculate faster than physical ones. Since no more physical relations are given, both the cause-effect-chain and the physical understanding of the system can get lost. In [Mastinu et al. 2006] examples for mathematical modelling are given.

4.1 Linear regression model

In order to keep the complexity and thus also the processing time of the simplified model low, a mathematical model approach for the evaluation of the wheel shimmy vibrations is used in this study. The main criterion to evaluate the stability of wheel shimmy vibrations is the modal damping of the critical mode. For this reason, it is sufficient to consider the eigenvalues.

A simple application for a mathematical substitute model is a linear regression model, see [Rooch 2014]. It describes linear relationships between one or more independent input variables, see Table 1, and a dependent output variable, which in this case is the modal damping of the wheel shimmy mode. The mathematical formulation of the model is described by the following equation:

$$y = \beta_0 + \sum_{i=1}^{k} \beta_i (x_i - x_{0,i}) \tag{3}$$

Here, y is labeled as regressand and represents the output-value. In this case, this is the modal damping. x_i is the input variable and is called regressor. All values that can have an influence on the output variable are included here. $x_{0,i}$ is the base input value of the corresponding variable in which the regression model was linearized. β_i is the gradient to the given input variable and β_0 is the output value in the linearization point of the regression model.

4.2 Sensitivity analysis

In order to determine the influence of various parameters on the stability of the wheel shimmy vibration, a sensitivity analysis will be performed in the following. Here, the parameters, which have the greatest influence on the basis of previous experience, are varied. The parameters and the amount of variation are listed in Table 1.

Table 1: Parameter overview of the sensitivity analysis of the rear axle.

Parameter	Description	Range of variation
F_z	Wheel load	$\pm 5\%$
d_{main}	Damping of main damper	$\pm 5\%$
k_{main}	Stiffness of main spring	$\pm 10\%$
$k_{z,tire}$	Vertical tire stiffness	$\pm 5\%$
$k_{x/y/z,sf}$	Stiffness of subframe mounts	$\pm 20\%$
$d_{x/y/z,sf}$	Damping of subframe mounts	$\pm 20\%$
k_{arb}	Stiffness of Anti-Roll-Bar	$\pm 10\%$
m_{wheel}	Wheel mass	$\pm 10\%$

4.3 Parametrization

The model is parameterized based on the results gained from the sensitivity analysis. The level of detail of the input parameters can be selected as desired. In the parameter study carried out in Section 4.2, only the main influencing variables were varied. The regression model can therefore only represent the results that have been generated there.

If a finer level of detail of the simplified model is required, it can be extended by adding any parameters. The calculation time of the regression model increases only insignificantly. However, it should be noted, that the level of detail was deliberately chosen roughly, as the model is intended to contribute results to the early development phase. This means, that at this time, the vehicle properties are not available in detail and therefore only the most important vehicle properties can be parameterized.

The exact parameterization process will not be discussed in the following, as it can be assumed to be known and easy to understand. Further information can be found, for example, in [Rooch 2014].

5 Analysis and Validation

Since the focus of this paper is on the analysis of the evaluation method for the stability assessment and the corresponding results have already been sufficiently discussed in Section 3.3, a comparison between the results of the regression model and the MBS simulation will be carried out in the following. Figure 5 shows the influences of the rear axle subframe mount stiffness and damping on the modal damping, subdivided into the three directions in space. The x-shaped markings represent the MBS simulations and thus the sampling points for the parameterization of the regression model. The solid curves are generated with the help of the regression model. If the six influences are superposed, the total influence of the functional properties of the elastic rear axle subframe mounts become visible.

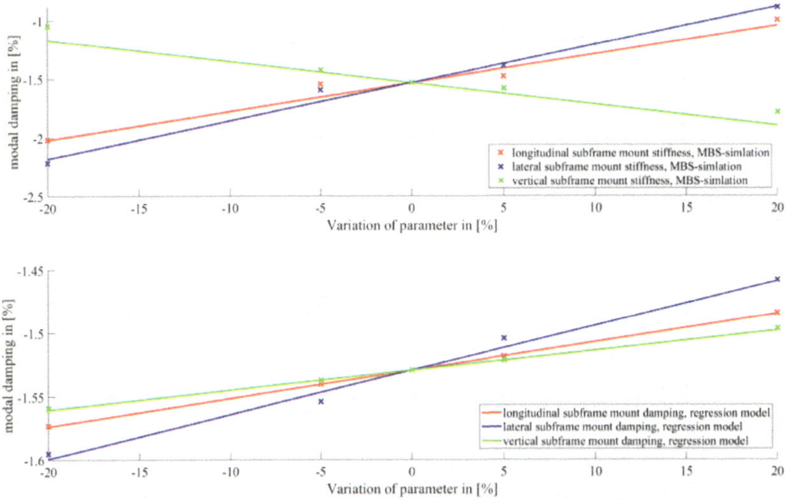

Figure 5: Comparison of MBS-simulation and regression model with different subframe mount characteristics with respect to modal damping of the wheel shimmy vibration mode.

On the one hand, it can be seen that the MBS simulation does not provide perfect linear behavior, since the crosses are not located directly on the curves. On the other hand, the regression model can reproduce the MBS simulation very well, since the regression coefficients for all individual parameters are above 98%. For the superposition of all six characteristics, the error is smaller than 5%.

6 Conclusion and Outlook

In summary, with the help of POD, concrete predictions can be made concerning the stability of self-excited and non-linear wheel shimmy vibrations. The uncertainties associated with a MBS eigenvalue analysis have been strongly reduced and the quantifiability could be increased significantly. In addition, the simplified regression model already allows statements to be made in the early concept phase.

The great advantage of the regression model is, that changes in component properties or adaptations of entire chassis concepts can be evaluated very quickly with regard to wheel shimmy vibrations by interpolation of the model. It is no longer necessary to parameterize and simulate a separate MBS model for each new chassis configuration. In addition, the simplified model opens up the possibility of high-dimensional DoE analyses. With an MBS model, these would be linked to high computing times. The enabler for this is mainly the comparability between the individual simulation results that has now been achieved. Since the simplified model only reflects linear relationships, it is essential that a new parameterization stage is subsequently backed up by a detailed MBS simulation.

As an outlook, the reduction process can also be applied directly to system matrices. Thus, the properties and degrees of freedom, that are relevant for the vibration phenomenon, could be directly derived from the reduction matrix or from the system matrices of the reduced system.

References

[1] Kerschen, G.; Golinval, J.-C.; Vakakis, A. F.; Bergman, L. A.: *The Method of Proper Orthogonal Decomposition for Dynamical Characterization and Order Reduction of Mechanical Systems: An Overview.* Nonlinear Dynamics (2005) 41: 147–169.

[2] Königs, S.; Zimmermann, M.: *Resolving Conflicts of Goals in Complex Design Processes – Application to the Design of Engine Mount Systems.* 7th International Munich Chassis Symposium, Springer Verlag, 2016.

[3] Kreuzer, E.; Kust, O.: *Analysis of long torsional strings by proper orthogonal decomposition.* Archive of Applied Mechanics, 68-80, 1996.

[4] Markert, R.: *Strukturdynamik – Skript zur Vorlesung.* Darmstadt, 2010.

[5] Mastinu, G.; Gobbi, M.; Miano, C.: *Optimal Design of Complex Mechanical Systems.* Springer Verlag, 2006.

[6] Pruscha, H.: *Angewandte Methoden der Mathematischen Statistik – Lineare, loglineare, logistische Modelle Finite und asymptotische Methoden.* Vieweg und Teubner Verlag, 1996.

[7] Quaranta, G.; Mantegazza, P.; Masarati, P.: *Assessing the local stability of periodic motions for large multibody nonlinear systems using proper orthogonal decomposition.* Journal of Sound and Vibration, Volume 271. Issues 3-5, 2004, Pages 2015 – 1038.

[8] Rooch, A.: *Statistik für Ingenieure.* Springer Verlag, 2014.

[9] Schramm, D.; Hiller, M.; Bardini, R.: *Modellbildung und Simulation der Dynamik von Kraftfahrzeugen.* Springer-Verlag, Wiesbaden, 3. Auflage, 2018.

[10] Steindl, A.; Troger, H.: *Methods for dimension reduction and their application in nonlinear dynamics.* International Journal of Solids and Structures, 2131-2147. 2001.

[11] Wagner, S.; Vena, G.; Schramm, D.: *Simplified model for self-excited rear-axle-vibrations.* Proceedings zum 18. Internationalen Stuttgarter Symposium, Springer Fachmedien Wiesbaden, 2018.

[12] Zeller, P.: *Handbuch Fahrzeugakustik.* Vieweg + Teubner Verlag, Wiesbaden, 2. Auflage, 2012.

[13] Zimmermann, M. et al.: *On the design of large systems subject to uncertainty.* Journal of Engineering Design, 1466 – 1837, Taylor and Francis, 2017.

Quasi-static and dynamic suspension measurements vs. multi-body and real-time simulation results

Sebastian Lefèvre Dr.-Ing. h. c. F. Porsche AG,

Frédéric Etienne Kracht; Dieter Schramm
Lehrstuhl für Mechatronik, Abteilung Maschinenbau
und Verfahrenstechnik, Universität Duisburg-Essen

Abstract

The aim of real-time models is, on the one hand, to improve driving dynamics and ride performance and, on the other hand, to be able to be used in driving simulators. In order to achieve these goals, not only a short computation time is necessary, but also the model accuracy is very important. For this reason, this paper investigates the simulation results of a developed real-time model and compares them with results from physical tests and a multibody simulation model. The influence of the modeling level on the simulation results is discussed in the context of the investigations carried out. For quasi-static analysis the Kinematic & Compliance (K&C) test rig and for dynamic tests a dynamic suspension test rig is used.

1. Introduction

Real-time models are becoming increasingly important in order not to lose the influence of dedicated and system-immanent elasticity effects in simulations, e.g. for driving dynamic experiments in driving simulators. Thus, the parameters of the chassis control systems for precise driving and handling performance can be determined much more precisely. Real-time models require a good accuracy and at the same time a short computing time. In order to achieve a short computation time, the level of detail of the real-time model is decisive. For dynamic analysis in particular, the model has to be as precise as possible. In addition, the requirements for modeling elastomeric bushings and tires to achieve a good correlation between the simulation and the real measurements are much higher for dynamic analyses than for quasi-static analyses. Therefore, a real-time objected-oriented modeling method (RTOOM) was developed recently [Kracht et al., 2017]. This method is physically configurable and includes elastic deformations. The modeling environment is MATLAB/SIMULINK with C-Code based S-functions.

In order to evaluate the accuracy of respective real-time models, the simulation results of the real-time model are compared with the results of a multibody simulation (ADAMS CAR) and suspension measurements on a Kinematic & Compliance (K&C) test bench. The use of the K&C test bench is an established and efficient method to accurately and reproducibly evaluate the quasi-static behavior of the suspension and therefore the handling performance. Furthermore, the simulation results of the real-time model are compared with the results of a validated multibody simulation (MBS) model in ADAMS CAR and with physical measurements on a dynamic suspension test bench (see Figure 1). The validation should be performed with appropriate dynamic tests. In the first step, the results of multibody simulation model are compared with the measurement data. Thus the reference model is valid for each test maneuver after the optimization of the parameters (curve fitting). The test bench enables dynamic excitation of the suspension up to 30 Hz in vertical direction [Will, 2018] and enables the analysis of the impact harshness of the suspension [Haberzettl et al., 2015]. In this paper, the maximum frequency considered is 15 Hz, which is a common frequency range for driving simulators. This frequency limit is justified by the eigenfrequency of the driving simulator itself [Brems et al., 2015]. This is also a typical application for real-time models. In the second step, the output of the models, the accelerations of the real-time model simulation $\ddot{\boldsymbol{r}}_{\mathrm{RT}}$ and forces of the bushings $\boldsymbol{F}_{\mathrm{RT}}$ of the different subparts in the suspension system are compared with the results of the validated multibody simulation model. The comparison is performed with the rear suspension of a Porsche 911.

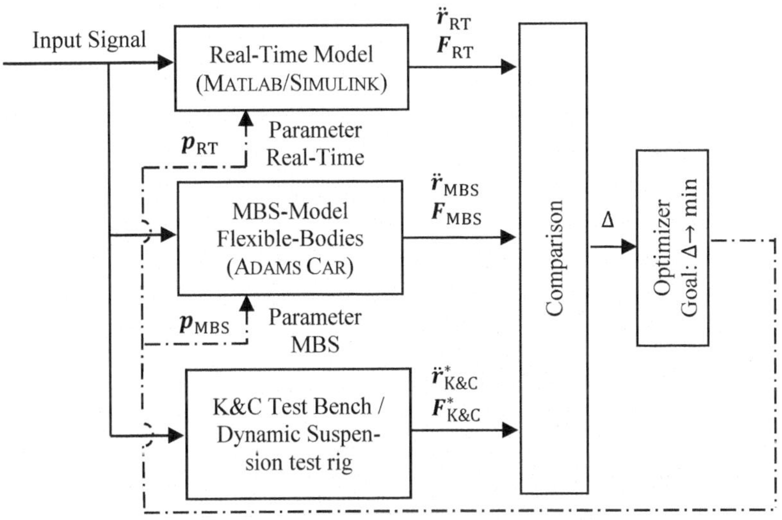

Figure 1: Validation Process

2. Kinematic & Compliance (K&C) Test Rig

The K&C test rig is a common test rig for measuring the quasi-static behavior of a suspension and thus the driving performance. On this test rig, the car is fixed to the ground. Instead of the tire and a rim, an ideal stiff wheel adapter is mounted at each corner. Each wheel adapter is connected to a cylinder (see Figure 2). This cylinder is capable of moving and loading the suspension with forces in all directions. Typically a displacement is tested in the vertical direction and a force or torque is applied in the horizontal directions and the axis of rotation. The movement is quasi-static. This means that the movement and load increase very slowly. There are no viscous damper effects which affect the results.

Characteristic results of such a test are, for example, the toe or camber reaction of a suspension under the action of a lateral force or the toe reaction under the action of an acceleration [Caputo et al., 2003] or a brake force [Holdmann et al., 1998]. The lateral force can act simultaneously in the same direction (in phase) or against each other (out of phase). This allows analyzing the stiffness of the structural components of the suspension, e.g. the subframe. The caster trail in both cases is typically 25 mm. In addition, the toe or camber reaction can be measured during a parallel vertical movement of the wheel adapters [Bell et al., 1987]. Within this paper, these tests were performed with a real prototype and with both simulation models (MBS and RTOOM). The results were later compared with each other.

Figure 2: K&C Test Rig [Will, 2018]

3. Dynamic Suspension Test Rig

A dynamic suspension test rig is used for dynamic measurements. This test rig is mounted on air bearings and has a mass of approx. 26 t. The advantage of these air bearings is that the eigenfrequency of the test rig can be adjusted from 1.05 Hz to 1.6 Hz in parallel vertical direction and from 1.75 Hz to 2.3 Hz in opposite vertical direction. The suspension is placed in a rigid rack. The rigid rack is fixed to the ground. On each side of the suspension is a hydraulic cylinder which can pulse against the mounted tire in vertical direction (see Figure 3) [Schütz, 2014].

In this paper, a swept sine with a displacement amplitude of 2 mm is used on each side. The swept sine starts near 0 Hz and raises the frequency up to 15 Hz. During this test, the wheel carrier accelerations in vertical and horizontal direction and the lateral link forces are measured. The measurement results from physical test are compared with the results of the RTOOM and MBS model.

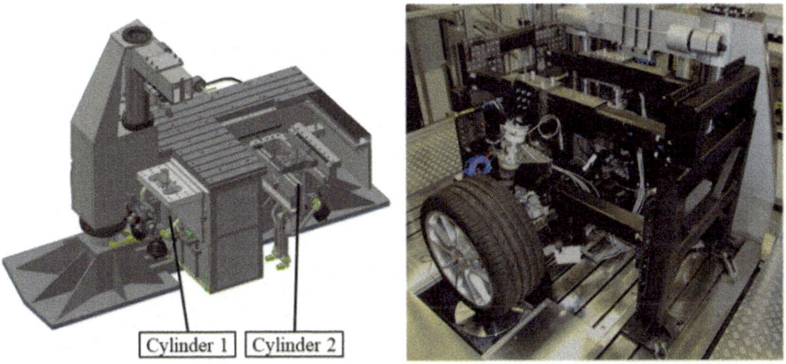

Figure 3: Dynamic Suspension Test Rig [Heine & Haberzettl, 2014]

4. Multibody Simulation Model

The multibody simulation model is developed in ADAMS CAR. The coordinates of the hard points correspond to the rear suspension of the Porsche 911 (see Figure 4). This multilink suspension has 5 rods on each side. Each link is connected to a bushing to the wheel carrier on one side and to the subframe on the other side. The subframe is attached to the chassis without bushings.

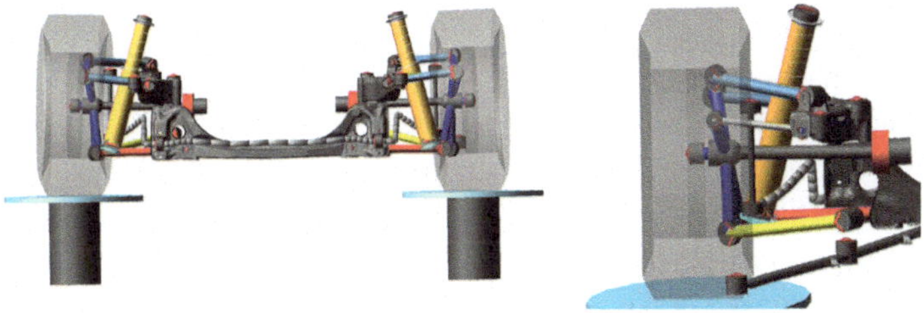

Figure 4: Porsche 911 Rear Suspension on the Virtual (MBS) Dynamic Suspension Test Rig

Each bushing stiffness was previously measured on an elastomeric component test rig and the characteristics were implemented in the model. The standard bushing model (a Kelvin-Voigt model) was used for the quasi-static K&C-tests and a frequency- and an amplitude-dependent model [Scheiblegger et al., 2011] was used for the dynamic suspension test rig tests. The tire is modelled as a Pacejka tire model [Pacejka, 2009]. The links between the bushings are assumed to be ideally stiff. The subframe is flexible and implemented as a modal neutral file (MNF-file). The spring is a conventional steel spring and the semi-active damper is fixed during simulation t with a constant damping level.

5. Real-time Objected-Oriented Modeling (RTOOM)

The real-time simulation model is developed in MATLAB/SIMULINK using an object-oriented method [Kracht et al., 2018]. The entire system is broken down into its individual physical components. For the separated parts, in turn, the equations are explicitly converted according to the desired outputs. These parameterizable blocks are then stored in a model library and combined to form an overall system [Drogies, 2006]. It also includes the spring-damper, bushings and tire models.

In contrast to other modeling techniques used in ADAMS CAR, DYMOLA or other software, algebraic equations are consistently avoided here. Because these, which in combination with differential equations would lead to differential algebraic equations (DAE). To solve these equations, special DAE solvers were needed, which work iteratively and thus cannot guarantee hard real time, since the step sizes can deviate. This can result in very long computing times. The RTOOM system consists only of ordinary differential equations, which can be solved with standard explicit solvers, like the Runge-Kutta 3 method with a fixed step size. The step size, however, is depended on the stability area of the numerical solver. With numerically very stiff systems, which can result from high bushing stiffness, sometimes only very small step sizes are permissible, which can lead to an extended computing time. To counteract this, additional stiffnesses and filter techniques are used, so that larger step sizes are numerically stable for frequency-dependent applications. A shortening of the computing time usually leads to a deterioration of the accuracy. Therefore, this method achieves an application-dependent compromise between computing time and accuracy.

In this subchapter the physical principles are first adapted to the Porsche 911 rear suspension based on [Kracht et al., 2018]. Furthermore, the bushings and other force elements are modeled. The individual components are then assembled to form the Porsche 911 rear suspension. In the last step, additional stiffnesses and filters are used to allow larger step sizes. In order to guarantee a high accuracy of the model, optimization methods are used.

5.1 Modeling of the Rigid Parts

The dynamics of the rigid-body system is described using the Newton-Euler equations [Kracht et al., 2017] along with kinematic differential equations. These equations are needed for all bodies that have a mass m and an inertia tensor $\boldsymbol{I}_{CM}$, e.g. the suspension links or the upright. The degrees of freedom of the rigid body are calculated using the transformation matrices and cardan angles. Since each individual component is connected with bushings, no reaction forces arise. Inputs of the respective component are only the impressed forces and torques of the bushings or the spring-damper element. Therefore, in contrast to [Kracht et al., 2018], the equations are resolved in such a way that the impressed forces and torques are inputs and the resulting translational and rotational accelerations are outputs (see Figure 5).

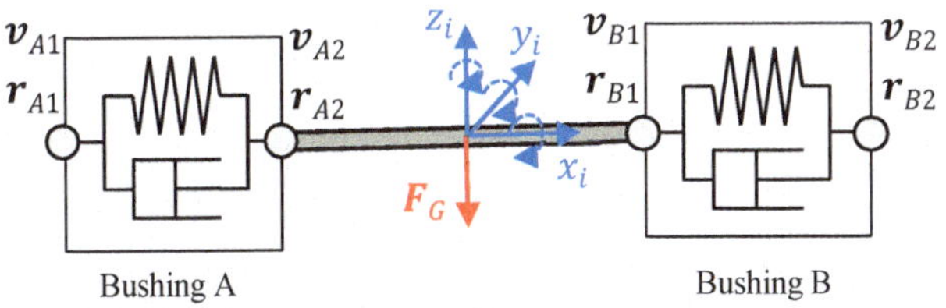

Figure 5: Rigid Body Object – 2 Forces Input

5.2 Modeling of Bushings, Spring-Damper System and the Tire

The model uses a 3-D non-linear bushing, a spring-damper system and a simplified tire model (see Figure 6). The stabilizer was modelled according to [Schramm et al., 2018].

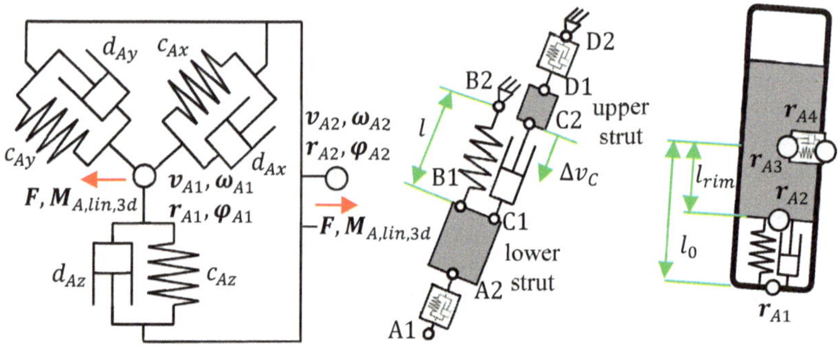

a) 3-D bushing model b) Spring-damper system c) Tire model

Figure 6: Different Force-/Torque-Elements

The bushings used within the suspension system have a translational and rotational stiffness and damping and are modeled as spring-damper elements with a three-dimensional orientation (see Figure 6a). In [Kracht & Schramm, 2019] it is proven that this type of modeling guarantees the highest numerical stability at low computing times.

A further component of the wheel suspension is the spring-damper system (see Figure 6b). The spring-damper element consists of six different components. The lower bushing is connected to the upright and lower strut. The lower strut has one rotational and one translational degree of freedom. At the lower strut the linear spring and the upper strut are connected. The damper is connected to the upper strut, which in turn is attached with a bushing.

Another component is the tire (see Figure 6c). This is simplified here with a spring-damper element that leads into the spindle and is connected in the center of the wheel via a bushing (wheel bearing) to the upright. The spring-damper force of the tire $\boldsymbol{F}_{tire}$ is

$$\boldsymbol{F}_{tire} = -\left(c_{tire}\left(l_{tire}(t) - (l_0 - l_{rim})\right) + d_{tire}(\boldsymbol{v}_{A2} - \boldsymbol{v}_{A1}) \cdot \frac{\boldsymbol{r}_{A2} - \boldsymbol{r}_{A1}}{\|\boldsymbol{r}_{A2} - \boldsymbol{r}_{A1}\|_2}\right) \frac{\boldsymbol{r}_{A2} - \boldsymbol{r}_{A1}}{\|\boldsymbol{r}_{A2} - \boldsymbol{r}_{A1}\|_2} \tag{1}$$

with the vertical stiffness c_{tire} and the vertical damping d_{tire}. l_{rim} is the radius of the rim and l_0 is the unloaded radius.

5.3 Modeling of Complete Rear Suspension System

To generate the individual elements, the equations are formulated in MAPLE as C-Code. This C-Code is used in MATLAB/SIMULINK S-functions. S-functions have the advantage of being more efficient and versatile than MATLAB-embedded functions, for example.

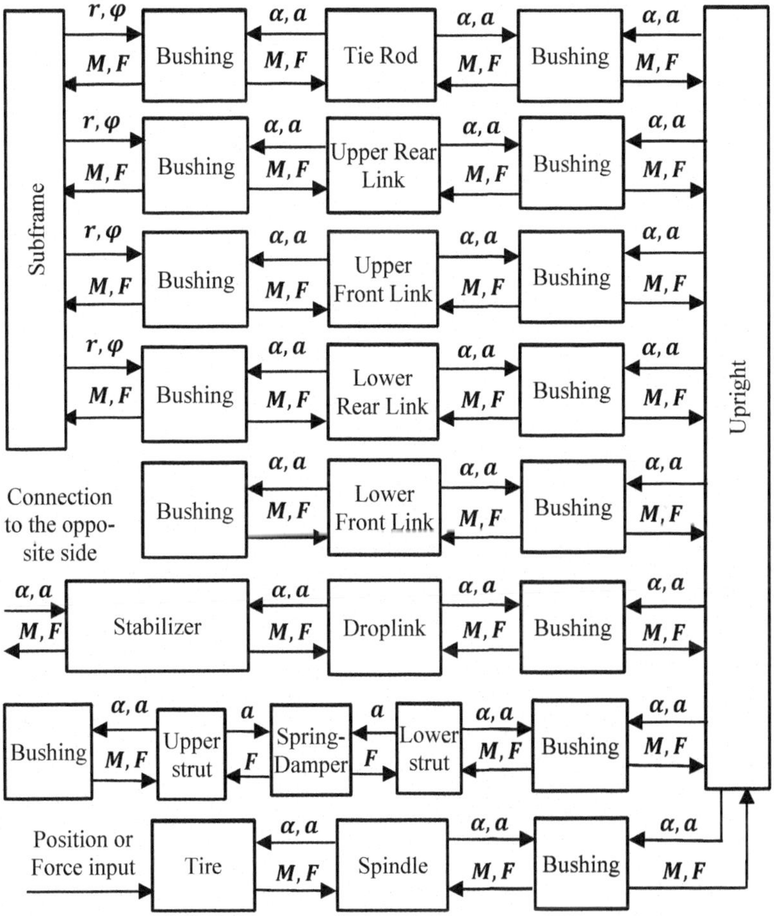

Figure 7: Complete Suspension System in RTOOM

All individual parts are parameterized according to the CAD specifications that were also used in the ADAMS CAR model. This includes the bushing, spring-damper and tire parameters on the one side and the parameters of the rigid components on the other side. The geometry, the inertia tensor and the mass of the components are used. The individual components are then linked to form the overall system, whereby the connection states like the forces and accelerations are transformed into global coordinate systems.

One side of the rear multilink suspension of the Porsche 911 is shown in Figure 7. The input of the system is either the position of the wheel-road contact point or the forces applied at the contact point. All states of the individual components can be an output and thus monitored for post-processing.

5.4 Runtime Optimization

The high stiffness of the bushings used ($c_{Bush,max} \approx 42.000$ N/mm) makes the associated model system very stiff. These stiff differential equations lead to instabilities if the step size of the numerical solver is too large. If the numerical integration method is retained, only the step size can be reduced to stabilize the system. However, the smaller step size increases the computation time. In order to reduce the step sizes, the stiffness of the system must be reduced without compromising the accuracy of the system too much or, even worse, to make the solution completely impossible.

In order to achieve this, the scope of application must first be taken into account. The model should have an application range of 0-15 Hz. Moreover, a simple explicit integration method with a constant step size h should be used. The Bogacki-Shampine (ode3) integration method is a robust and fast method that is also be used here. According to the Nyquist-Shannon sampling theorem, the maximum step size of the numerical solver is determined to simulate the frequencies of the signal. For $f_{max} = 15$ Hz the largest possible step size is $h_{max} = 1/(2f_{max}) = 1/30$ s $= 0.033$ s. To increase the numerical stable step size of the solver, a PT_1 behavior of the build-up of impressed forces is used and corresponds to a 1st order low-pass filter. The physical equivalent is the dynamic friction in the bushing. The time constant τ is determined by the desired cut-off frequency f_C, so it must be $\tau < 1/(2\pi f_c) = 0.01061$ s. The phase shift is $-\pi/4$ and the gain is $1/\sqrt{2} \approx -3$ dB.

As already mentioned, the overall system is very stiff due to the large number of very stiff bushings. Therefore, the parameters of the bushings must be selected in such a way that the accuracy is sufficient and the system is nevertheless hard real-time capable. For this purpose, an optimization process is the obvious choice. In order to enable larger step widths, the bushings must be equipped with additional rigidity in the next step. Since the forces of the bushings should remain constant, the additional springs and dampers are installed in series. This reduces the stiffness c_{sub} and the damping constant d_{sub}.

Due to the non-linearity of the system, the additional stiffness and damping are determined using optimization methods. The Levenberg-Marquardt method is used for all 14 bushings (see Figure 7). This results in 168 stiffness and damping parameters, which must be optimized so that the measurement results are achieved. The objective function is the deviation between the acceleration of a measuring point on the upright simulated with the RTOOM model and the measurements of the dynamic suspension test rig.

6. Comparison of the Results from Simulation and Experiments on K&C Test-Rig

First, the quasi-static measurement results from the K&C test rig are compared with the results of the MBS and the RTOOM model. In Figure 8 the upper bar shows the result of the MBS model (ADAMS CAR) and the lower bar indicate the result of the RTOOM model divided by the result from the K&C test rig, represented with the equation

$$z_i = 1 - \frac{x_{sim}}{x_{test\ rig}}. \tag{2}$$

The deviation between the results of the MBS model and the measurements on the K&C test rig is 6 % for the toe and 3 % for the camber change on parallel wheel travel (kinematics). The deviation of the camber change between the results of the MBS model and real measurements on the K&C test rig is due to the effect of a lateral force in the phase below 1 %. The toe change in this test has a deviation of around 20 %, but the toe change is under 1 minute per kN and the absolute deviation is 0.2 minute per kN. This is within the measurement tolerance of the K&C test rig. The percentage deviation is therefore very high. The toe and camber change is within a tolerance of 10 % under the influence of a phase-shifted lateral force, an acceleration or a braking force.

The comparison of the results from the RTOOM to the measurements from the K&C test rig show significantly better results. The kinematic toe and camber change lies within a tolerance range of approx. 6 % compared to measurements on the K&C test rig. The toe and camber change in the RTOOM model differs from the results in MBS. Furthermore, the results of the toe and camber change from the RTOOM model represent also better the measurements results from the K&C test rig as the results of the MBS while acting a lateral force in phase. The deviation of the toe change is about 3 % and the camber change about -2 %. During acting a lateral force out of phase, an acceleration or a brake force, the toe and camber changes are within 10 % compared to measurements from the K&C test rig.

However, it was expected that the results from the MBS and the RTOOM model are nearly the same. One reason is that some subparts in the RTOOM model have been modelled in a simplified or another way in order to have a shortened computing time. These include the stabilizer and the subframe. In the RTOOM model, the stabilizer is modelled simplified as torsion spring as mentioned in [Schramm et al., 2018] without a bushing between the droplink and stabilizer. In MBS the stabilizer is discretized as a spline with different reference points which represent the real part. The subframe is modelled in RTOOM as a rigid component, while the MBS model contains an elastic subframe (MNF-file). This was necessary to guarantee real-time capability. It indicates randomly a better correlation between the results of the RTOOM model and the measurement results on the K&C test rig.

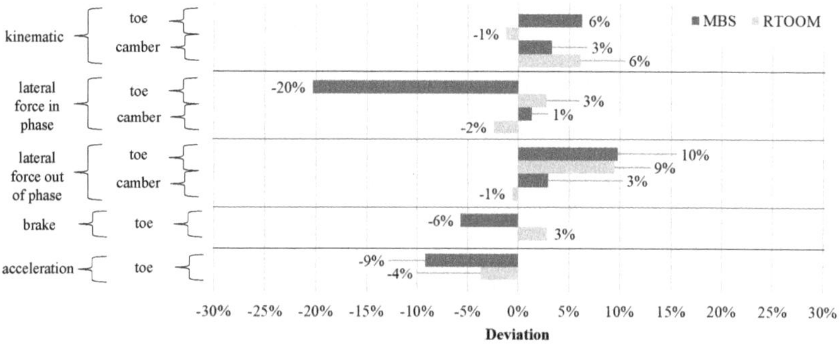

Figure 8: Comparison of Measurements on the K&C Test Rig and Simulation Results (MBS and RTOOM Model)

7. Comparison of the Results from Simulation and Experiments on Dynamic Suspension Test-Rig

The next step, the results of both simulation models (MBS and RTOOM) are compared with the measurement results of the dynamic suspension test rig. Therefore, a Fast Fourier Transformation (FFT) is used to transform the time-dependent simulated and measured signals into frequency-dependent signals. Figure 9 compares the vertical (left) and longitudinal (right) wheel carrier acceleration. All values are normalized to the maximum vertical wheel carrier acceleration at 15 Hz. The position of the acceleration sensor was the same in both simulation models and was tested by measuring the coordinate positions to a reference point using a CAD tool.

The comparison shows that the maximum amplitude in vertical and longitudinal direction of the simulation results from the MBS model are lower than the measured ones. The deviation increases with higher frequencies and has a maximum of around 20 % at 15 Hz. The virtual damping of the used tire model in the MBS model is too high in higher frequency range. The results of the RTOOM simulation model show almost the same results in vertical direction as the measured ones. The deviation in longitudinal direction is also increasing with higher frequencies and amounts to a maximum of about 7 %.

The main reason for the differences between the results of the RTOOM and the MBS model are the different bushing stiffnesses and damping coefficients. The solver used has the advantage that it is real-time capable, but has the disadvantage that the simulated hard point displacement shift differs from the reality or to the results of the MBS model. This leads to different stiffness and damping values of the bushings from the frequency- and amplitude-dependent model used. In order to achieve almost the same results as with the measurements or the MBS model, the stiffnesses must be decreased. Furthermore, the tire model differs between both models and the RTOOM does not use a flexible subframe (MNF-File).

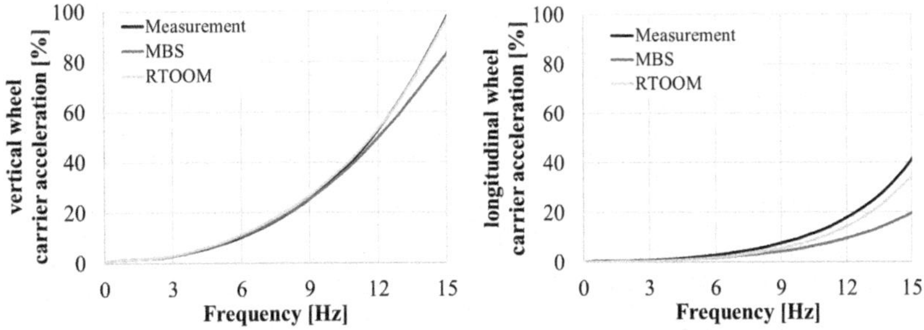

Figure 9: Comparison of Measurements on the Dynamic Suspension Test Rig and Simulation Results (MBS and RTOOM Model) of Vertical and Longitudinal Wheel Carrier Acceleration

The next step is to compare the measured and simulated forces of the lower lateral link (see Figure 10). The simulated forces of the MBS model are less than the measured ones between 12 Hz and 15 Hz (deviation around 20 % at 15 Hz). Below 12 Hz the maximum deviation is 5 %. The simulated results of the RTOOM fit better to the measured ones and differ around 10 % over a wide frequency band. The reason for the differences between the measurements and the results of the MBS model is the same as described above.

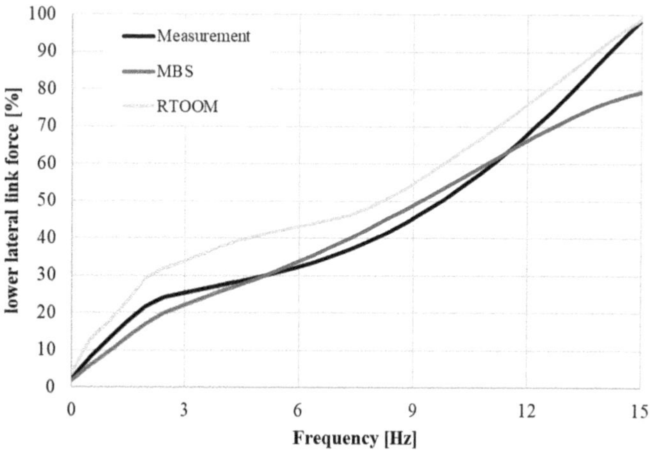

Figure 10: Comparison of Measurements on the Dynamic Suspension Test Rig and Simulation Results (MBS and RTOOM Model) of Lower Lateral Link Forces

8. Conclusion

Real-time models are becoming increasingly important. These models are not only suitable for use within driving simulators. With such models, it will also be possible to increase driving dynamics and ride performance. Within this paper, the results of a developed real-time objected oriented model (RTOOM) in MATLAB/SIMULINK are compared with results from a multibody simulation (MBS) in ADAMS CAR and measurements on a K&C test rig to enhance driving dynamics and a dynamic suspension test rig to consider ride performance. The results of the developed real-time model correlate well with the results of physical tests on a K&C test rig. The toe and camber change for quasi-static tests is within a tolerance of 10 % for the investigated rear suspension of the Porsche 911. The results of the multibody simulation model show slightly higher deviations. These, however, are within the measurement accuracy of the K&C test rig. The difference between the results of the real-time model and the multibody simulation model can be explained by not using the subframe as a flexible part in the real-time model to optimize calculation time and by the optimization of the stabilizer stiffness.

The comparison of the simulation results from the RTOOM, the MBS model and measurements on the dynamic suspension test rig up to 15 Hz shows that the simulated vertical and longitudinal wheel carrier acceleration and the lower lateral link forces of the MBS model differ slightly from the measured ones. The higher the frequency, the greater the deviation (maximum 20 % at 15 Hz). The reason for this is that the damping of the tire model used is higher than the real one. The simulated results of the RTOOM fit better to the measured ones than the simulated results of the MBS model. However, by optimizing the RTOOM model, which is necessary of real-time capability, it is possible to obtain almost the same results as the measurements. Thus, the RTOOM model as a good possibility to be used in driving simulators.

In the future the real-time model will be extended to simulate the overall car. The advantage is to adapt the mechatronic systems on the actual driving situation or to use the model e.g. in driving simulators.

9. References

Bell, S., Ellis, J., Garrott, W. & Liao, Y., 1987. Suspension testing using the suspension parameter measurement device. *SAE 870577, doi: 10.4271/870577.*

Brems, W., Van Doornik, J., De Vries, E. & Wiedemann, J., 2015. *Frequency response and latency analysis of a driving simulator for chassis development and vehicle dynamics evaluation,* S. 109-116: DSC Europe - Driving Simulation Conference Exhibition.

Caputo, A., Spina, M. & Guglielmino, E., 2003. *Sensitivity of Suspension System Performance to Bushing Stiffness Variation - An Evaluation Methology.* s.l.:SAE Technical Paper 2003-01-0237, doi: 10.4271/2003-01-0237.

Drogies, I. S., 2006. *Objektorientierte Modellbildung des fahrdynamischen Verhaltens mit MODELICA in Fahrdynamik-Regelung.* s.l.:Springer, pp. 71-91.

Haberzettl, S., Zschocke, A. & Gauterin, F., 2015. *A new method for studying the longitudinal dynamic behaviour of a suspension on a test rig.* s.l.:Proceedings of the Institution of Mechanical Engineers, Part D: Journal of Automobile Engineering, doi:10.1177/0954407015601265.

Heine, J. & Haberzettl, S., 2014. *Achsentwicklung am Fahrwerk-Identifikations-Prüfstand.* Baden-Baden: 17. Kongress Simulation und Erprobung im Fahrzeugbau, Berechnung, Prüfstands- und Straßenversuch, 18.-19. Nov, 2014, pp. 871-884.

Holdmann, P., Köhn, P., Möller, B. & Willems, R., 1998. *Suspension kinematics and compliance - measuring and simulation.* s.l.:SAE Technical Paper 980897, doi: 10.4271/980897.

Kracht, F. E., Dandekar, R., Bruckmann, T. & Schramm, D., 2017. *Echtzeitfähige Objektorientierte Modellbildung am Beispiel einer Fahrzeug-Radaufhängung, "Real-time object-oriented modeling with the example of a vehicle wheel suspension".* presented at the Dritte IFToMM D-A-CH Konferenz TU Chemnitz: s.n.

Kracht, F. E., Saba, M. & Schramm, D., 2018. *Real-time Calculation of Reaction Forces and Elasticities in Vehicle Wheel Suspensions.* Beijing, China: presented at the 14th Interna-tional Symposium on Advanced Vehicle Control (AVEC'18), 16-20 Juli 2018.

Kracht, F. & Schramm, D., 2019. *Real-Time Capable Calculation of Reaction Forces of Multibody Systems Using Optimized Bushings on the Example of a Vehicle Wheel Suspension (in press).* Duisburg: ECCOMAS Multibody Dynamics Conference.

Pacejka, H., 2009. *Tyre and Vehicle Dynamics.* Oxford: Butterworth-Heinemann Elsevier Ltd. ISBN: 978-0-7506-6918-4.

Scheiblegger, C., Pfeffer, P., Karrer, H. & Geiger, N., 2011. *Modellierung von Elastomerlagern und Hydrolagern zur Simulation von Fahrkomfort und Fahrdynamik.* S. 247-267: VDI: Reifen-Fahrwerk-Fahrbahn.

Schramm, D., Hiller, M. & Bardini, R., 2018. *Vehicle Dynamics: Modeling and Simulation.* Berlin: Springer-Verlag GmbH Deutschland ISBN: 978-3-662-54482-2.

Schütz, H., 2014. *Validierung eines mehraxialen Fahrwerkprüfstands.* Karlsruhe: Diplomarbeit.

Will, V., 2018. *Entwicklung einer fachbereichsübergreifenden Standardauswerteroutine für dynamische Prüfstandsuntersuchungen von Fahrwerken.* Stuttgart: Masterarbeit.

Model-based parameter identification of a virtual vehicle model – time-effective step-up for steering performance development

Alessandro Contini, Timo Schöning

Vehicle Test & Development Department
Hyundai Motor Europe Technical Center GmbH,
Rüsselsheim, Germany

Abstract

In the current paper a model-based approach for vehicle parameter identification will be proposed. In order to increase the quality of the simulated vehicle behaviour (responses like lateral acceleration, yaw rate and especially the steering rack force) the best fitting combination of the vehicle mass inertias, the metacentric height as well as some selected tire parameters (like cornering stiffness, pneumatic trail, etc. [4, 9, 10]) needs to be identified.

All these relevant parameters, usually difficult to be measured in the car, are varied systematically and simultaneously to generate an empirical model describing the correlation between the influencing vehicle model parameters and the simulated vehicle responses [2], [3].

For example, the simulated steering rack force, necessary requirement to generate trustable steering wheel torques, correlates with the pneumatic trail. Hence, this tire parameter can be used to adjust the simulated steering rack force to the measurements on the real-road.

The model-based approach is already implemented in an automated manner for the steering feel tuning with a hardware-in-the-loop (HiL) test bench [1] and has now been extended to the vehicle parameter identification process. This new process reduces the time necessary to parametrise and to validate a vehicle model by increasing the simulation quality at once. This validated vehicle model is later used to get a trustable result for the model-based steering feel tuning at the HiL test bench.

Based on the new time-effective step-up, the subjective tuning process on the real-road can be supported earlier.

Keywords: Model-based Parameter Identification · Design of Experiments (DoE) · Vehicle dynamics simulation · Steering-in-the-Loop · Steering feel tuning · Steering system test bench.

1 Introduction and overview of steering tuning process

The introduction explains the steering tuning process at the Hyundai Motor Europe Technical Center and describes why it is a challenge to determine certain vehicle parameters.

1.1 Overview of steering tuning process at Hyundai Motor Europe Technical Center

In [1] it has already been described how the steering tuning is supported by a hardware-in-the-loop test bench (Steering-HiL) at the Hyundai Motor Europe Technical Center in Rüsselsheim (Germany). The introduced methodology can be divided into seven separate steps, as shown in Figure 1.

Step 1: Target Setting for Steering Behavior

At the beginning of each project, the objective for the final steering performance is set. In a working group of colleagues from product planning, marketing, engineering design and application team, subjective goals that describe the steering feel are worked out together. In the further course of development, these subjective criteria result in objective criteria that are important for steering performance tuning at Steering-HiL.

Step 2: Build-up of virtual vehicle model

For the application at the Steering-HiL it is necessary to create virtual surroundings. Vehicle, environment and driving maneuvers are simulated. The steering system including the ECU is available as real hardware on the test bench and works as in the real vehicle. An input on the steering wheel results in an adjustment of the rack, which in turn causes a steering angle on the wheels. The resulting lateral force on the wheels causes rack force based on the axial elasto-kinematics characteristics. The rack force is transmitted to the steering wheel by the steering train, resulting in a steering wheel torque. This publication now supplements [1] with the description of this step: Build-up of a virtual vehicle model.

Step 3: Validation of Virtual Vehicle Model and real steering system at the Steering-HiL

At the Steering-HiL it must be ensured that the behaviour of the steering system is the same as in the real vehicle. For this purpose, the mechanical and electrical/electronic design of the steering system is validated on the test bench. The corresponding driving manoeuvres will be explained in more detail later. Only after successful validation may be continued with the next step:

Step 4: Start of automated steering tuning

The steering performance tuning on the Steering-HiL happens analogous to the tuning in the real driving test, with the difference that objective parameters are used on the Steering-HiL to describe the steering behavior. The test bench computer is connected to the steering ECU via a fieldbus interface (CAN, FlexRay, …). Through this interface, ECU variables can be applied in such a way that the desired steering behavior is achieved by running a driving maneuver catalog. For the application of the ECU variables, a design-of-experiment (DoE) is created which is run automatically. After this run a behavior model is generated to get the best fitting ECU variables to reach the wished steering behavior (s. a. Figure 5).

The DoE methodology is also used for the parameter identification described here and will be explained in more detail later in section “3 Introduction into the model-based approach”.

Step 5: Subjective validation of HiL-tuning result

In step 4 objective target parameters are reached. These must be confirmed subjectively in the driving test. If necessary, a further tuning loop can be carried out on the Steering-HIL in which the feedback from the driving test is considered.

Step 6: Fine tuning in the real car

After a HiL-tuning has been subjectively confirmed in a real driving test, the fine-tuning of the steering behavior can be continued in the vehicle. The HiL methodology does not replace the driving test, but complements it in order to guarantee an increase in steering performance with a reduction in time expenditure - just think of the variety of vehicle variants that is to be mastered in the modern vehicle development.

Step 7: Confirmation of vehicle dynamics/steering performance

After the successful development of a vehicle project, the dynamic behavior of the final product is measured on a proving ground under defined conditions. The objective target parameters defined at the beginning of the project (step 1) are to be confirmed. These measurements are also used to verify and further develop the virtual development methodology.

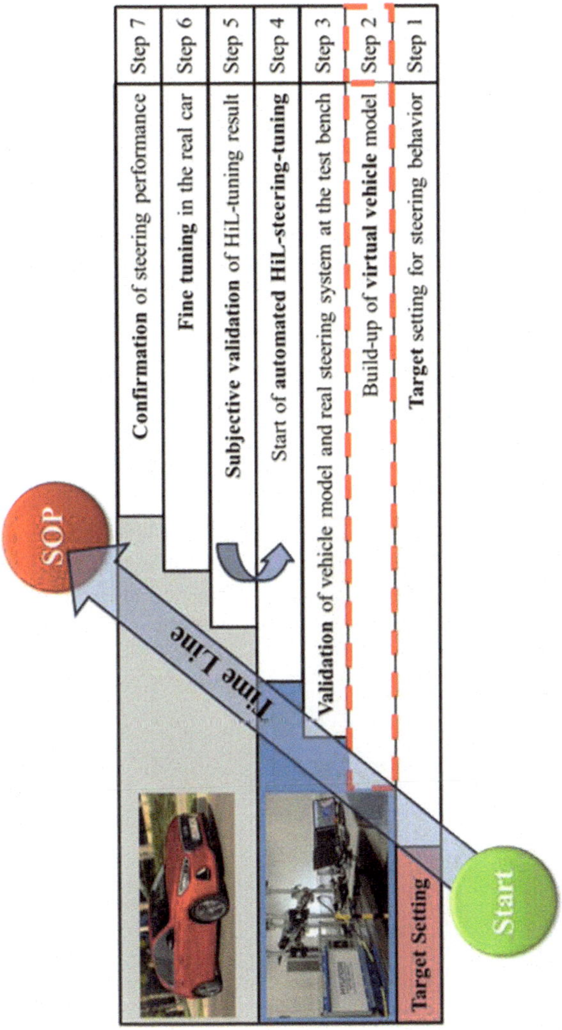

Figure 1: Steering performance tuning process at HMETC

1.2 Challenges in parameter identification for a vehicle model

In the description of step 2, the setup of a virtual vehicle model was described. The lateral vehicle dynamics behavior is of central importance for its parameterization.

The kinematics and compliance measurement machines measures the quasi-static suspension properties. This measurement data are used to parametrize the vehicle suspension. This approach leads to a very accurate simulation of the vehicle ride and handling. Further vehicle data (masses, anti-roll bar stiffnesses, spring stiffnesses, damping values,...) are available internally for the parameterization of the virtual vehicle model. But some data are also not available or difficult to be measured. Typically, the following data are not always easy to obtain in sufficient quality:

- Height of the center of gravity (e.g. for roll behavior)
- Mass inertia around the three axes (e.g. for frequency response)
- Magic-formula tire parameters (like cornering stiffness, pneumatic trail, relaxation length, etc., influencing the entire vehicle dynamics behavior)

All these relevant parameters need to be identified. The new thing about this work is that the identification process is performed in an automated manner. The parameters are varied systematically and simultaneously to generate an empirical model describing the correlation between the influencing vehicle model parameters and the simulated vehicle responses.

2 Validation process of a Virtual Vehicle Model for lateral vehicle dynamics

In [5] the validation of vehicle models is comprehensively explained. With the help of objective, characteristic driving dynamics variables (objective parameters), the comparison of simulation data to real reference driving data is given. In this example, three driving maneuvers are used for validation:

- Slowly Increasing Steer (ISO 13674-2, [7], s. a. Figure 2)
- Weave Test (ISO 13674-1, [6], s.a. Figure 3)
- Frequency Response Test (ISO 7401, [8], s.a. Figure 4)

Figure 2 shows important plots for the characterization of the Slowly Increasing Steer. In this figure, the real reference driving data (Target) shall be approximated by a systematic variation of the vehicle and tire parameters. Setting 1 and 2 represent exemplary driving dynamics behavior, which is caused by the parameter variations. To validate the Slowly Increasing Steer the following characteristic values are considered:

- Roll Behaviour
- Slip Angle Behaviour
- Self-Steering Behaviour
- Steering Rack Force

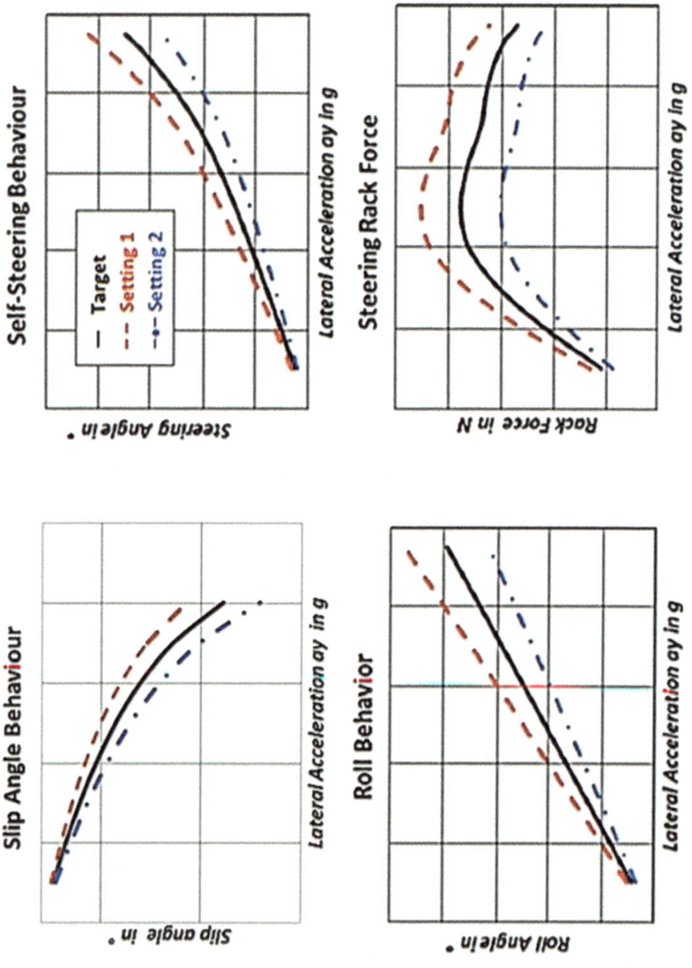

Figure 2: Vehicle and Tire Parameters are varied until the target vehicle behaviour is reached (Roll-, Self-Steering-, Slip Angle-, Steering Rack Force Behaviour for a Transient Steering Manoeuvre)

For the validation of the weave test three gradients are used in the present work, which characterize the vehicle dynamics. These are:

- Yawrate Gain
- Lateral Acceleration Gain
- Tierod Force Gain

Figure 3 shows an example of the gradient that is calculated from the Lateral Acceleration Gain.

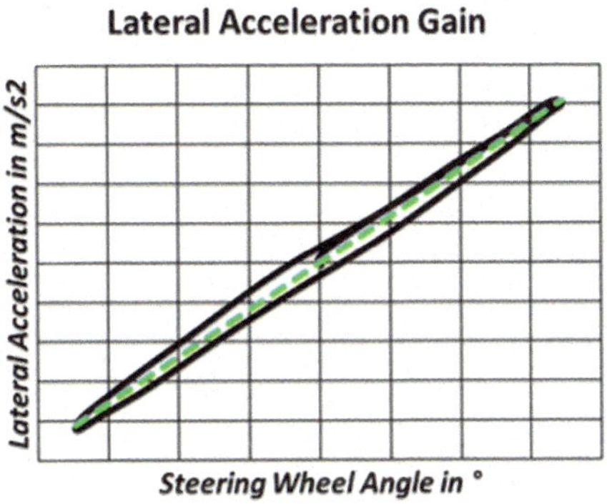

Figure 3: Validation of Weave Test (ISO 13674-1) by calculating the Gradient (– – –)

The Frequency Response Test is used to determine the lateral dynamic response of a vehicle in the linear range. By evaluating the transient behavior of the vehicle in the frequency domain, the frequency-dependent amplitude and phase response are obtained (s. a. Figure 4).

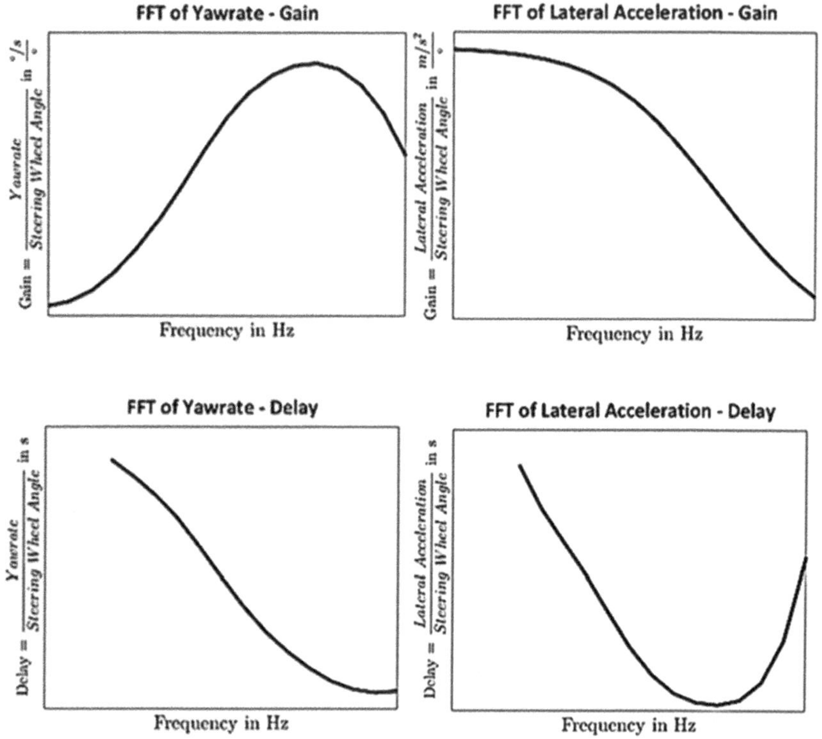

Figure 4: Validation of Frequency Response (ISO 7401) by calculating the Gradients/Delays at different frequencies

Table 1 shows an overview of the used objective parameters in this work describing vehicle dynamics behavior.

Table 1: Objective parameter for lateral vehicle dynamics

No	Objective Parameter	Unit	Manoeuvre
1	Roll Behaviour	°	Slowly Increasing Steer, Step Series
2	Slip Angle Behaviour	°	
3	Self-Steering Behaviour	°	
4	Steering Rack Force	N	
5	Yawrate Gain	(°/s) / °	Weave Test, Single Sine Test
6	Lateral Acceleration Gain	(m/s²) / °	
7	Tierod Force	N / °	
8	Yawrate Gain	(°/s) / °	Frequency Response
9	Yawrate Delay	s	
10	Lateral Acceleration Gain	(m/s²) / °	
11	Lateral Acceleration Delay	s	

3 Introduction into the model-based approach

Figure 5 shows the automated parameter identification by using a model-based approach. First, a DoE is created in which the vehicle parameters to be identified (s. a. Table 2) are varied ("controllable input factors"). The objective parameters used to validate the vehicle dynamics serve as the response (s. a. Table 1). The vehicle parameters are systematically varied and for each vehicle parameter combination the same driving maneuvers (Slowly Increasing Steer, Weave Test, Frequency Response Test) are simulated and evaluated so that the vehicle dynamics parameters can be read in directly. This procedure results in an empirical behavior model that can predict the corresponding objective parameters for a defined vehicle parameter combination, even if this combination has not been tested.

Basically, the more vehicle parameters are varied, the more tests have to be performed and the greater the computing power. For this reason, it is of particular interest to keep the sum of the vehicle parameters to be identified as small as possible.

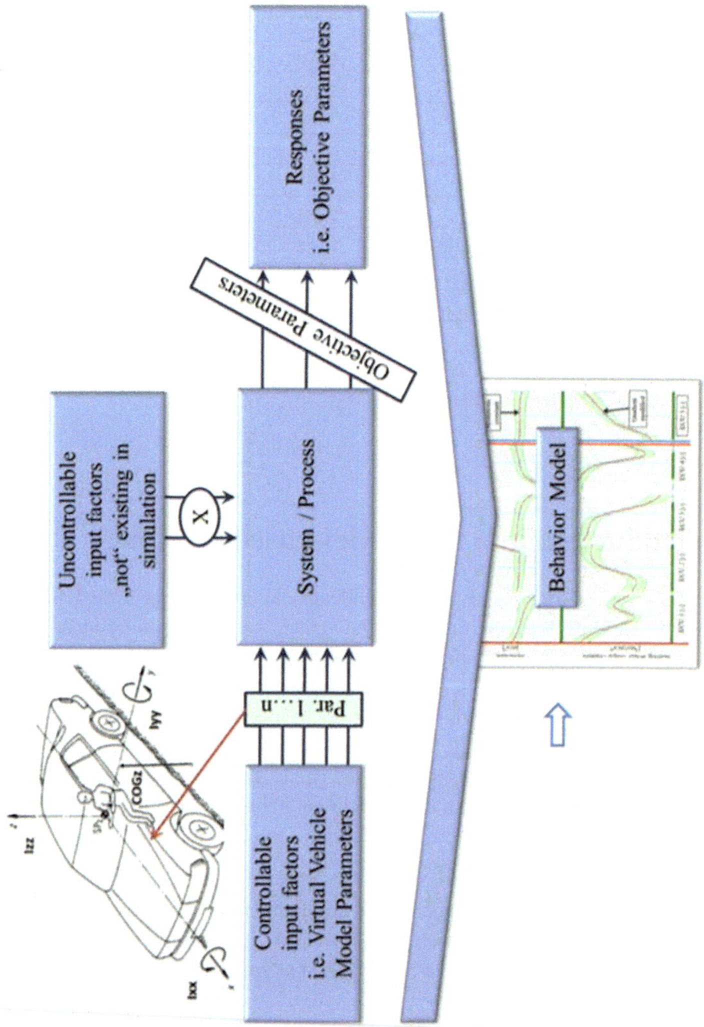

Figure 5: Automated parameter identification by using a model-based approach

Table 2: Vehicle Model Parameter

No	Param	Description	Remark
1	COG_z	Vehicle overall centre of gravity z	Vehicle
2	I_xx	Body inertia in x	
3	I_yy	Body inertia in y	
4	I_zz	Body inertia in z	
5	QDZ1	Peak trail Dpt" = Dpt*(Fz/Fznom*R0)	Aligning Coefficient
6	PKY1	Maximum value of stiffness Kfy/Fznom	Lateral Coefficient
7	PTY1	Peak value of relaxation length SigAlp0/R0	
8	LCY	Scale factor of Fy shape factor	Scaling Coefficient
9	LEY	Scale factor of Fy curvature factor	
10	LGAY	Scale factor of camber for Fy	
11	LGAZ	Scale factor of camber for Mz	
12	LHY	Scale factor of Fy horizontal shift	
13	LKY	Scale factor of Fy cornering stiffness	
14	LMUY	Scale factor of Fy peak friction coeffic.	
15	LRES	Scale factor for offset of residual torque	
16	LTR	Scale factor of Peak of pneumatic trail	
17	LVY	Scale factor of Fy vertical shift	
18	LYKA	Scale factor of alpha influence on Fx	

4 Sensitivity analysis – influence of vehicle model parameters on the vehicle behavior

It has already been described that it is of interest to keep the number of vehicle model parameters to be identified as low as possible. For this reason, a sensitivity analysis has been included in this paper in order to investigate the influence of each vehicle model parameter. Depending on the influence strength of each parameter, only those vehicle model parameters are used in the second test run which have the greatest influence on the driving dynamics (objective parameters, see also Table 1).

Table 3: Sensitivity Analysis for Frequency Response Test

	Vehicle Model Parameter																	
Objective Parameter	**LKY**	**PKY1**	**I zz**	**COG z**	**LTR**	**QDZ1**	I xx	LGAY	LMUY	I yy	LYKA	LVY	LCY	LHY	LRES	PTY1	LEY	LGAZ
Lat. Acc. Delay @ 0.5Hz	0.4	0.3	0.1	0.0	0.0	0.0	0.1	0.1	0.0	0.0	0.0	0.0	0.0	0.0	0.0	0.0	0.0	0.0
Lat. Acc. Delay @ 1.0Hz	0.4	0.3	0.0	0.1	0.0	0.0	0.0	0.1	0.0	0.0	0.0	0.1	0.0	0.0	0.0	0.0	0.0	0.0
Lat. Acc. Delay @ 1.5Hz	0.0	0.0	0.2	0.2	0.0	0.0	0.1	0.1	0.1	0.0	0.2	0.0	0.0	0.0	0.0	0.0	0.0	0.0
Lat. Acc. Gain @ 0.5Hz	0.3	0.3	0.0	0.0	0.1	0.0	0.0	0.0	0.1	0.1	0.0	0.0	0.0	0.0	0.0	0.0	0.0	0.0
Lat. Acc. Gain @ 1.0Hz	0.4	0.3	0.1	0.0	0.0	0.0	0.0	0.0	0.0	0.0	0.0	0.0	0.0	0.0	0.0	0.0	0.1	0.0
Lat. Acc. Gain @ 1.5Hz	0.5	0.3	0.1	0.0	0.0	0.0	0.0	0.0	0.0	0.0	0.0	0.0	0.0	0.0	0.0	0.0	0.0	0.0
Yawrate Delay @ 0.5Hz	0.3	0.2	0.2	0.1	0.1	0.0	0.0	0.0	0.0	0.0	0.0	0.0	0.0	0.0	0.0	0.1	0.0	0.0
Yawrate Delay @ 1.0Hz	0.3	0.3	0.2	0.0	0.1	0.0	0.0	0.0	0.0	0.1	0.0	0.0	0.0	0.0	0.0	0.0	0.0	0.0
Yawrate Delay @ 1.5Hz	0.3	0.3	0.2	0.1	0.1	0.0	0.0	0.0	0.0	0.0	0.0	0.0	0.0	0.0	0.0	0.0	0.0	0.0
Yawrate Gain @ 0.5Hz	0.0	0.1	0.0	0.2	0.3	0.2	0.0	0.0	0.0	0.1	0.0	0.0	0.0	0.0	0.1	0.0	0.0	0.0
Yawrate Gain @ 1.0Hz	0.0	0.2	0.0	0.1	0.2	0.1	0.0	0.2	0.0	0.0	0.0	0.0	0.2	0.0	0.0	0.0	0.0	0.0
Yawrate Gain @ 1.5Hz	0.3	0.2	0.3	0.0	0.1	0.1	0.0	0.0	0.0	0.0	0.0	0.0	0.0	0.0	0.0	0.0	0.0	0.0

Table 3 shows the sensitivity analysis for the Frequency Response Test. All vehicle model parameters that have a significantly smaller influence on the objective parameters than 20% were excluded from the second test run. For the frequency response test only the first 6 vehicle model parameters were used (LKY - QDZ1). The same procedure was also used for the Weave Test and for the Slowly Increasing Steer. The result is that the vehicle model parameters to be identified could be reduced from 18 to 11. The 11 resulting vehicle model parameters finally used can be found in Table 4.

5 Parameter identification of a Virtual Vehicle Model

Through the sensitivity analysis, important insights have been gained which in future will only require a variation of the most influential vehicle model parameters. This can significantly reduce the testing effort, whether automated or manual.

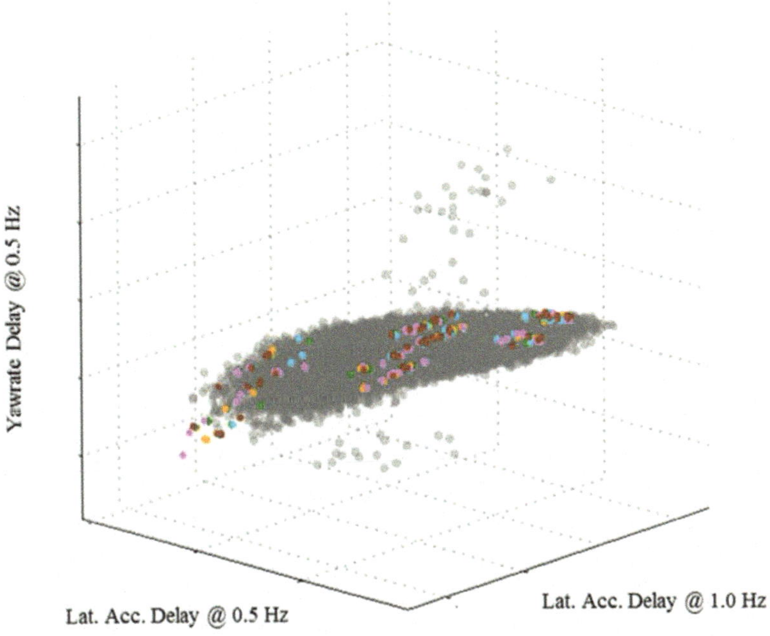

Figure 6: Pareto optimization used for parameter identification to get the ideal Objective Parameter (= vehicle responses)

In the second, "reduced" test run, the same driving maneuvers are tested again. By creating an empirical behavior model after the test run, the objective parameters from the simulation are to be optimized in such a way that they correspond to those from the real driving test. After the objective parameters from the real driving test have been set as the target, the vehicle model parameters are also optimized so that the objective parameters from the virtual driving test correspond to those from the real driving test. This optimization also runs automatically.

Figure 6 shows a multi-criteria optimization. Solving an optimization problem with multiple goals can lead to conflicts. In this example, it had to be decided whether the objective parameters of the frequency response test determined at 0.5 Hz could be optimized by compromising the quality of the objective parameters determined at 1.0 Hz. The results of Model-based Parameter Identification of a Virtual Vehicle Model can be seen in Figure 7.

For the Slowly Increasing Steer, the Weave Test and the Frequency Response Test at 0.5 Hz, it is generally worth mentioning a significant improvement in the objective parameters. In the simulation with the initial vehicle model parameters, the Yawrate Delay for the Frequency Response Test at 0.5 Hz had a relative difference of about 45% to the real driving test. Through parameter identification, the difference could now be significantly improved to a level of about 11%.

At the same time, the Figure 7 also shows that compromises were also made for the frequency response test at 1.0 Hz. The objective parameters "Yawrate Gain" and "Ay Delay" show a better behavior with the initial values. – After parameter identification these still reach a relative difference of 8% to the real driving test and are therefore still reasonable.

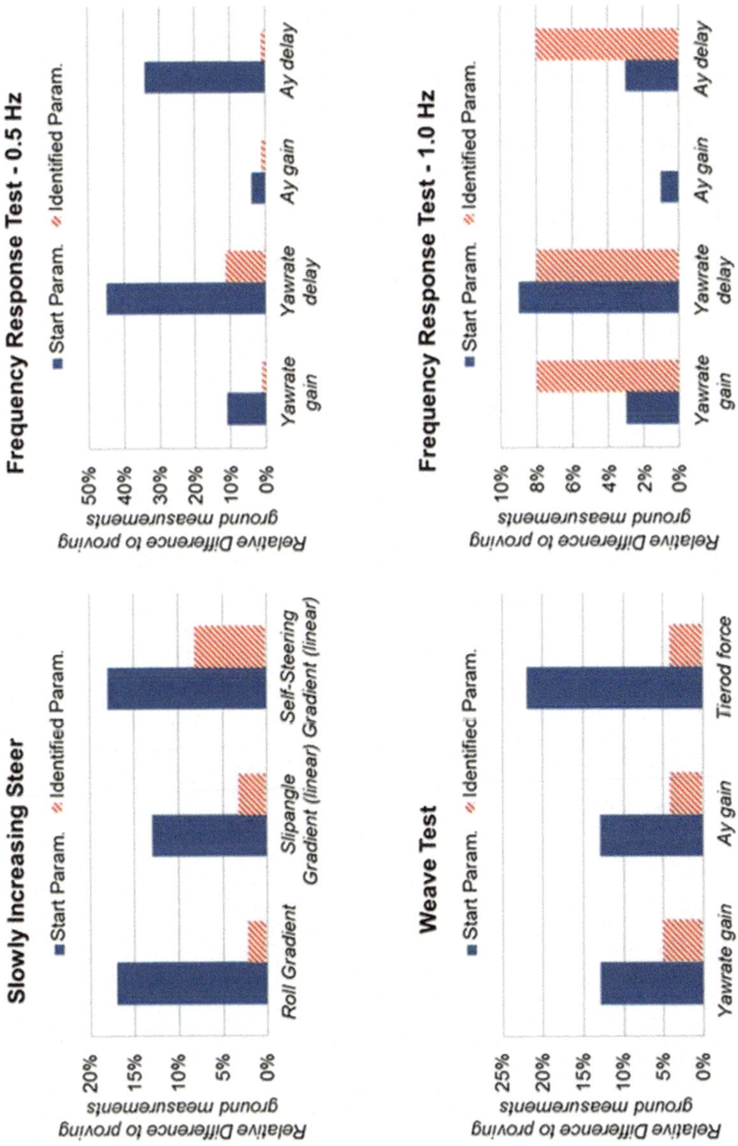

Figure 7: Results of Model-based Parameter Identification of a Virtual Vehicle Model

Finally, it is of interest to make a statement on the degree to which the initial values for the vehicle model parameters had to be changed compared to the newly identified vehicle model parameters (s. a. Figure 7). The identified vehicle model parameters vary significantly from the initial values (1% - 30%), from which it can be concluded that the methodology presented here is a helpful tool for parameter identification of a virtual vehicle model.

Table 4: Relative differences to the initial values of the vehicle model parameters

Remark	**Parameter**	**Relative Difference of Identified vs. Start Values**
Vehicle	COG_z	11%
	I_xx	4%
	I_yy	14%
	I_zz	1%
Tire (Magic formula)	QDZ1	26%
	PKY1	30%
	LCY	30%
	LHY	29%
	LKY	14%
	LMUY	29%
	LTR	25%

6 Outlook

With the help of the methodology presented, the vehicle model parameters can now be determined in such a way that the optimum target behaviour can be achieved in the simulation.

References

[1]: A.Contini, T.Schöning, The new Hyundai i30's steering feel- integrated performance development by HiL tests and DoE, Chassis Tech 2017, Munich

[2]:B.Schick, Michael Paulweber, Model-based Development Methods- what can Chassis and Powertrain Development learn from Each Other? , Chassis Tech 2015, Munich

[3]: Daniel Vilela, Roberto Spinola Barbosa, Analytical Models Correlation for Vehicle Dynamic Handling Properties, J. of the Braz. Soc. of Mech. Sci. & Eng October 2011

[4]: Pacejka H. B., "Tyre and Vehicle Dynamics", Delft University of Technology, Delft, The Netherlands

[5]: Unterreiner M., "Modellbildung und Simulation von Fahrzeugmodellen unterschiedlicher Komplexität", Universität Duisburg-Essen, Germany

[6]: ISO 13674-1 Road vehicles – Test method for the quantification of on-centre handling Part 1: Weave test.

[7]:ISO 13674-2 Road vehicles – Test method for the quantification of on-centre handling – Part 2: Transition test.

[8]: ISO 7401: Road vehicles – Lateral transient response test methods.

[9]: Lundquist C., Schön T. B., "Recursive Identification of Cornering Stiffness Parameters for an Enhanced Single Track Model", Division of Automatic Control, Department of Electrical Engineering, Linköpings universitet, Linköping, Sweden, April 2009

[10]: Gerdes J. C., Y-H Judy Hsu, "A Feel for the Road: A Method to Estimate Tire Parameters Using Steering Torque", Dept. of Mechanical Engineering, Stanford University, USA

PARALLEL STRAND II

AI IN THE CHASSIS

Vehicle side-slip angle estimation with deep neural network and sensor data fusion

M. Sc. Yuran Liang, BMW Group

Prof. Dr.-Ing. Steffen Müller, Technical University of Berlin

Daniel Rolle, BMW Group

Dieter Ganesch, BMW Group

Immanuel Schaffer, BMW Group

1 Introduction

Modern chassis control systems, advanced driver assistance systems (ADAS) and automated driving systems that demand a precise vehicle localization or a reasonable trajectory planning desire a highly accurate and reliable vehicle state estimation. However, the traditional methods such as Kalman and RLS filter, which based on the vehicle dynamic model, mainly rely on the differential equations that approximate the vehicle behaviour in reality [1, 27, 31]. The vehicle dynamics is such a nonlinear and multidimensional system with numerous parameters, which makes it very difficult to adapt the parameters in different situations and figure out appropriate model equations. The vehicle state estimation with a kinematic model instead of vehicle dynamic model can reduce the uncertainty of vehicle parameters, e.g. mass, slip stiffness and wheelbase [2, 5]. Due to the systematic errors such as offset and drift of inertial sensors, a pure kinematic method cannot achieve high accuracy and reliability in the situation where sensor errors are not adequately modelled. Recently there are some approaches using environmental sensors to handle with these problems of traditional approaches [17, 22, 24]. However, they are prone to lack of consideration of the vehicle dynamics.

This paper adopts a novel artificial intelligent method to estimate side-slip angle in linear and nonlinear situations through the fusion of the data from classical vehicle dynamic sensors and environmental sensors such as radar, lidar and camera. An innovative Network architecture based on the combination of Convolutional Neural Network (CNN), Long Short-Term Memory cell (LSTM) is proposed to fuse the raw sensor data. The neural network is trained with data from the measurement of a vehicle equipped with environmental sensors and standard vehicle dynamic sensors. A highly precise Automotive Dynamic Motion Analyzer (ADMA) has also been implemented to label the ground truth. A deep CNN has been implemented to distinguish stationary and dynamic objects with two consecutive frames. In order to accelerate the convergence and reconstruct the state space, the network loss function is modified by some model-based state variables. We discuss the side slip angle estimation with six sections. In this section, we introduce the background of our research and introduce the layout of this paper. In the next section, we analyse the problem statement of the side slip angle estimation and its importance as well as the challenges of the estimation. In the third section, we summarize the state of the art of the state estimation and the sensor data fusion for the state estimation. Our multi-view-fusion approach will be discussed in the fourth section which consists of the network architecture, the idea to identify stationary objects, the LSTM framework to overcome the time synchronization error, the training model and the methodology to improve the generalization capability. In the fifth section, we present the experimental results of our proposed system in comparison with different sensor setups. The model quality has been evaluated with respect to the accuracy and the generalization error. In the last section, we conclude the benefits along with the promising results as well as the deficits of our research for further works. To summary, we make the following contributions to the side slip angle estimation:

- Estimation without knowing the wheel-speed-based velocity and vehicle parameters.
- An end to end trainable neural network architecture for the low-level-fusion with inertial sensors, radar, lidar and camera to handle with the state estimation problem.
- A method to modify the loss function considering the vehicle dynamics to seek for a reasonable model quality.

2 Side slip angle estimation problem

On the state estimation problem, the movement of a vehicle can be described as a model with six degrees of freedom as shown in Figure 1.

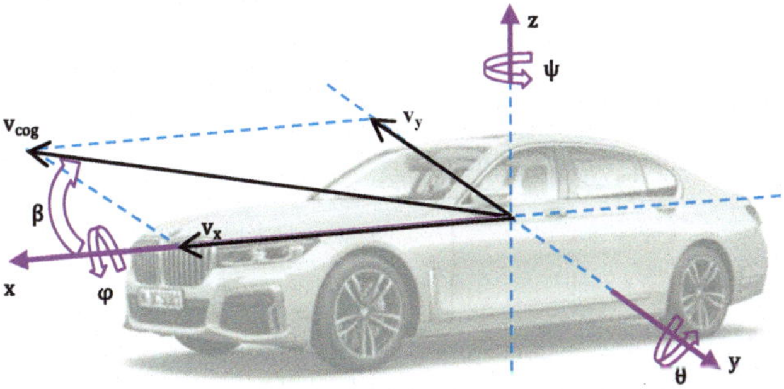

Figure 1. Six degrees of freedom vehicle kinematic model.

In the body-fixed coordinate system, the vehicle has translations along the longitudinal, lateral and vertical axis x, y, z and the rotations yaw ψ, pitch θ, roll φ around each axis. The vehicle velocity at the center of gravity v_{cog} can be decomposed into a longitudinal and lateral velocities v_x and v_y. The side slip angle β is the angle between the vehicle longitudinal axis and the direction of the v_{cog} [32]. In the kinematic model, β can be formulated as

$$\beta = \tan^{-1}\left(\frac{v_y}{v_x}\right). \tag{1}$$

Another aspect of analysing vehicle dynamics is to simplify it into a double track or single track model that considers the vehicle moving on a two-dimensional plane with three degrees of freedom x, y, and yaw ψ. The state sapce $[\dot{\psi} \quad \beta]^T$ is decribed by force and moment equilibrium that contains the vehicle parameters [32].

The side slip angle β along with the yaw rate $\dot{\psi}$ are important physical quantities to identify the vehicle stability and handling characteristics such as understeering, oversteering, neutral steering and unstable state [30]. Therefore, an accurate, reliable and robust estimation of the state varibles is the precondition to enable driving dynamic control. The yaw rate can now be directly meansured by the inertial sensors while the the side slip angle can only be estimated by the observation using avalible sensor data. The transient behaviour of a vehicle is so distinct nonlinear that its modelling becomes very challenging. There are a bunch of parameter-based and non-parameter approachs that contribute to the approximation of this nonlinearity. In the next section, we will concretely discuss some previous works in this field.

3 Related works

In this section, we summarize the recent researches about the side slip angle estimation and the potential of using environmental sensors for the vehicle state estimation. As environmental sensor data provide more redundancy and complementarity of the information about the egomotion, they can improve the estimation of vehicle state and characteristics.

3.1 Vehicle state estimation

The convention of the state estimation for chassis control systems is based on the processing of measurable variables from vehicle dynamic sensors, e.g. wheel speed sensors, inertial sensors, and steering angle sensor [9]. The side slip angle can be estimated based on a vehicle dynamic model or a kinematic model.

Göbel [23] estimates the side slip angle with an extended Kalman filter (EKF) by a double track lateral dynamic model and considers the effect of rear wheel steering. The non-linear chassis and tire model can approximate the complexity of vehicle well, but the model accuracy mainly relies on the estimation of vehicle parameters such as mass, wheel-base, axle track and wheel slip stiffness, which are different on different vehicle variances and changing while driving. Because of the linearization of transfer function with Taylor series, the EKF is also prone to implausible when the truncation errors are unneglectable.

Bechtoff et al. [3] and Selmanaj et al. [33] adopt a kinematic model instead of parameter driven vehicle dynamic model to overcome the problem of model uncertainty. These approaches achieve higher robustness and comparable accuracy. Nonetheless, the heuristic model can implicitly contain the vehicle parameters since an empirical tire model or a heuristic scheduling of state-force relations has been implemented.

Graber et al. [13] investigate a hybrid method with Recurrent Neural Network (RNN) and a kinematic model to estimate the side slip angle. RNN can work similar to Kalman filter which also considers the estimation of previous time step in a recursive way. Moreover, the combination with the physical model can help the neural network achieve a better generalization performance since the network is trained to depict the reality of the kinematic model instead of treating it as a black box. However, the RNN can also be overfitted, as it can learn the vehicle parameters implicitly since the steering angle and tire model based longitudinal velocity are also inputs of the network. Moreover, the uncertainty of neural network is not taken into account that may cause a significant impact on the functional safety if unreliable outputs are predicted.

Recently, there are also some researches which focus on the combination of artificial neural network and Kalman filter to deal with the trade-off between reliability, accuracy and robustness. Boada et al. [4] propose an observer which predicted the side slip angle through the state space of bicycle model and treat the pseudo-side slip angle from the one-layer fully connected neural network (FC) as a sensor signal, and finally correct the Kalman filter in the update phase. Novi et al. [28] adopt a similar algorithm but use a kinematic model rather than a vehicle dynamic model and correct the output of the neural network with the output state of the observer. Both approaches can handle with the uncertainty of statistical model well, but the proposed neural network itself does not take the recursive effect of the real system into account and is also too simple to generalize the complexity of vehicle dynamics in the reality due to the network architecture.

3.2 Using environmental sensors for vehicle state estimation

Conventional state estimation using vehicle dynamic sensors especially inertial sensors, however, needs to overcome the systematic error of the sensors themselves. As the development progress of ADAS and automated driving, more and more vehicles are equipped with environmental sensors such as camera, ultrasonic sensor, radar and lidar.

There are different strategies using environmental sensors to improve the performance of state estimation. Kellner et al. [18] develop a method to estimate 3-DOF Ego-Motion using multiple Doppler radars and achieve a higher performance than the pure wheel-base vehicle odometry concept. This method is, however, prone to indeterministic cycle time because of the online learning algorithm and deficient accuracy and reliability if the detected points mainly consist of dynamic objects.

Zindler et al. [36] estimate the state variables through the fusion of vehicle dynamic sensors and multi-lidar. Velocity profiles are extracted from a detected object list associated with landmarks which can be obtained from HD-digital maps. The landmark-based algorithm is active only if some landmarks have been detected and are available in the digital map which results in limited availability.

Wang et al. [35] investigate stereo camera based large-scale direct sparse visual odometry estimation through simultaneous localization and mapping (SLAM) without loop closure optimization. Their research achieves not only real-time tracking and mapping capability but also promising accuracy compared to typical visual SLAM such as ORB-SLAM2 [26] and Stereo LSD-SLAM [8]. However, the algorithm may reach its limitation under bad weather and light condition.

Instead of a pure camera or lidar SLAM, Graeter et al. [14] combined the benefits of lidar's high-accurate range detection and the powerful tracking capabilities of the cameras to obtain more robust and accurate odometry. Both kinds of research, however, do not take the correction by vehicle dynamics into account, which plays the main role in vehicle state estimation and delivers a reliable way to check and validate the algorithm.

By the convention of visual SLAM, feature engineering has a great impact on the performance of algorithms, hence we have to design tricky manipulations in order to get useful filtered feature maps. Clark et al. [6] propose a deep learning solution to reduce the expenditure of feature engineering and fuse the velocity profiles from camera's optical flows and the raw data from a 6D-inertial measurement unit (IMU). The network can be trained end to end without any labelling of optical flows, which are extracted by a pre-trained convolutional neural network (CNN) named FlowNet. The error of mono camera due to the lack of depth information can be well compensated with the inertial sensors but the camera-based approach cannot overcome the deficit of camera that the performance depends significantly on the weather and light condition.

4 Multi-view-fusion approach

As mentioned above environmental sensors provide the possibility that we can estimate the state variables independently of vehicle parameters and redundantly to the inertial sensors. In this section, we focus on our novel multi-view-fusion approach fusing the raw data of inertial sensor, camera, radar and lidar for the side slip angle estimation as an example.

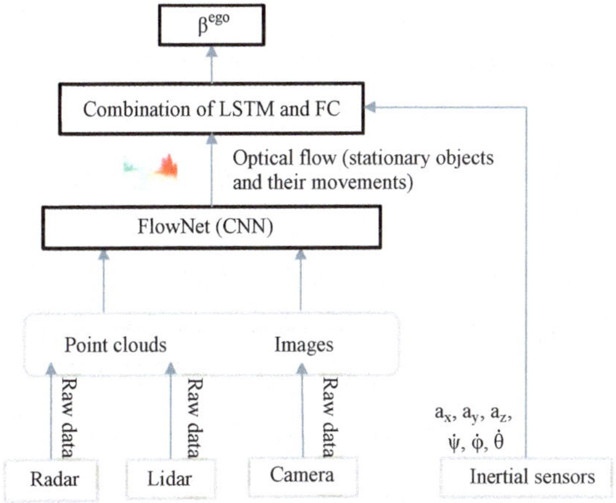

Figure 2. System overview of the multi-view fusion concept.

In the fusion concept, the raw data of the camera, lidar and radar are pre-processed and fed into a so-called FlowNet [11], a pre-trained CNN, to estimate optical flows between two frames. These are pixels of stationary objects and their relative movements to the vehicle. In the end, the flows are fused with the featured data from inertial sensors with the combination of LSTM-cell and a regression head using a fully connected layer to estimate the side slip angle.

4.1 The Neural Network architecture

As shown in Figure 3, the total network architecture can be trained end to end without any labelling for intermediate results such as optical flows of camera images and point clouds.

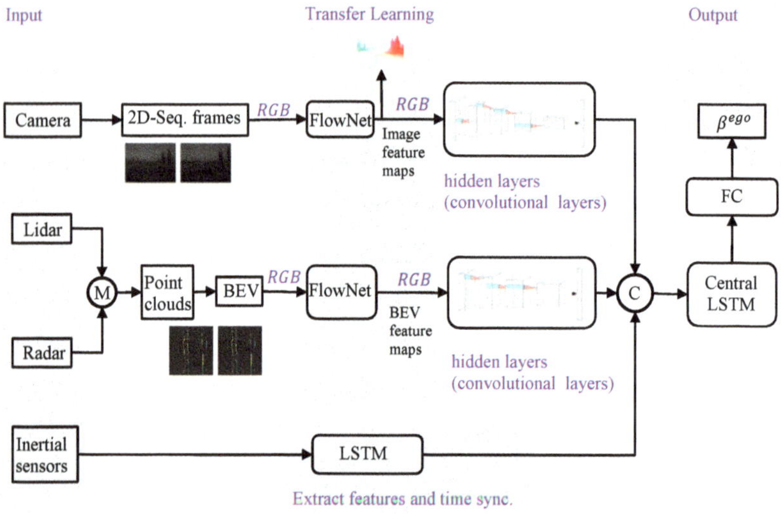

Figure 3. End to end trainable Neural Network architecture. Estimation of optical flows through transfer learning of a pre-trained CNN.

There are three kinds of inputs of the proposed network: the image pairs of frontal view from the camera, the image pairs of bird's-eye-view (BEV) [21] generated by the point clouds from the lidar as well as the radar and the inertial sensors signals from the vehicle bus. The lidar and radar receive the reflected electromagnetic waves (laser or radio), which are emitted by themselves, from the vehicle environment and generate the point clouds of the surrounding objects. These point clouds are then transformed into a common Euler-coordinate system to generate the BEV. The output of the FlowNet is the feature maps with two channels: the first channel is the direction of the velocity and the second channel is magnitude of the velocity.

Several down sampling CNN layers [20] follow the FlowNet to refine the feature maps and reduce their dimension. At the meantime, an LSTM layer processes the inertial sensor data in order to get the features of the vehicle motion. The LSTM layer is also designed to synchronize the initial sensor data between two frames of images, as the synchronization is necessary due to the dynamic sampling rates of the environmental sensor systems. The output of the LSTM network is then concatenated with the output of the CNN. All these features are then fed into the central LSTM layer which provides an intermediate output as a vector. Finally, a standard fully connected Neural Network (FC) reduces the vector dimension from the feature size of the previous LSTM layer to the output dimension and outputs the final estimation of the longitudinal and lateral velocity and the side slip angle. The longitudinal and lateral velocities are predicted additionally in order to refine the estimation of the side slip angle and modify the loss function using the kinematic model in order to accelerate the convergence and achieve higher generalization performance.

4.2 Computing velocity profiles through optical flow estimation: transfer learning of the FlowNet

Optical flow is the pattern of apparent motion of pixels in images due to the relative movement between the environment and a sensor which provides the images. Each flow is calculated from two consecutive images in a time sequence. The intensity of the same objects in two frames are assumed to be constant, and only the positions of these objects in the pixel frame will change if they are moving. The optical flow concept works under this assumption and therefore, can estimate the motion of objects. At the continuous-time system, consider I(x, y, t) and I(x+dx, y+dy, t+dt) as the intensities of the same pixel in two frames, the optical flow assumption can be described as follows

$$I(x+dx, y+dy, t+dt)= I(x, y, t). \tag{2}$$

The optical flow model can be approximated and simplified as a first-order Taylor series expansion

$$\frac{\partial I(x, y, t)}{\partial x} v_x + \frac{\partial I(x, y, t)}{\partial y} v_y + \frac{\partial I(x, y, t)}{\partial t} = 0 \tag{3}$$

where v_x and v_y are the velocity of the object and $\frac{\partial I(x, y, t)}{\partial x}$, $\frac{\partial I(x, y, t)}{\partial y}$ and $\frac{\partial I(x, y, t)}{\partial t}$ are the Jacobian matrices of the intensity I along the horizontal axis, the vertical axis, and time. [29]

Calculating optical flow with only one pixel is therefore under-determined. There is a variety of methods such as the Lucas-Kanade method [7] and dense optical flow [10] designing constraints to make the problem to be over-determined. In our network, we implement the transfer learning of a state of the art algorithm for optical flow estimation named FlowNet which was proposed to estimate the motion of dynamic objects in a video [11, 16]. In our application, we are not interested in the moving objects but the stationary objects with which the ego velocities can be derived because the relative velocities of motionless objects are the reverse velocities of the vehicle. Since the sensors are moving while driving, objects estimated as dynamic by the FlowNet can be, on the contrary, stationary objects in the real world such as trees, lane markings and traffic signs. However, they can also be dynamic objects moving relatively to the Ego-vehicle. Furthermore, the values in the pixels coordinate frame are projected from the world coordinate system with the affine transformation. Due to the lack of depth data, the backward calculation of the affine transformation from pixel to world coordinates with a mono camera image is under-determined [34]. Therefore, the estimation of vehicle velocities only with a mono camera without inertial sensors is not an elegant idea.

As shown in Figure 4, the total FlowNet consists of convolutional and deconvolutional levels. In the convolutional phase, the image pairs are processed by a ten-layer pyramid-shape CNN. In the first three layers, two images share the same weights, and then the correlation between their feature maps are computed with a similarity function. The upsampled pyramid is the refinement of the predicted flows with four levels [11]. Considering that the computational effort and the storage space for millions of weights of the FlowNet are challenging for the implementation of our algorithm in the test vehicle, we only use the first level of the refinement as the output of the FlowNet. Besides, the residual layers are also taken away in order to lose the weight of the whole network. As mostly happened in the fine-tuning of a neural network, a tradeoff has to be made between the accuracy and the model size.

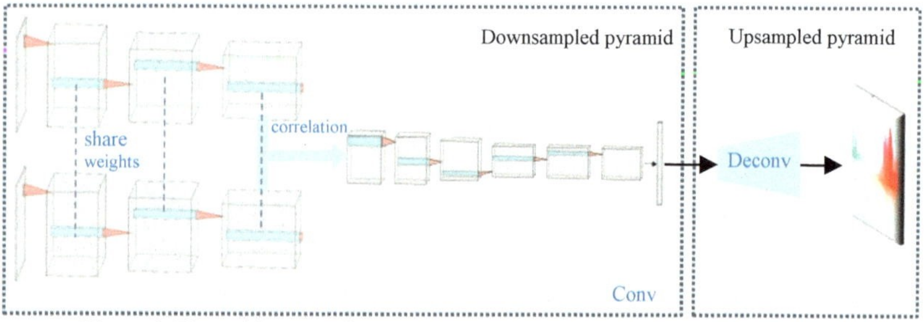

Figure 4. The architecture of the FlowNet: downsampled and upsampled pyramids.

4.3 Feature extraction and time synchronization of inertial sensors using LSTM layer

The bottom level of a multi-sensor system, as shown in Figure 2, is comprised of different sensor systems which are working on different rates. The time reference chosen for the network is the camera timestamps for the reason of high signal availability. The sampling rate of each sensor system is also dynamic and changing due to the error of the time synchronizer and the delay of the data logger. The signals of the inertial sensors are recorded over FlexRay with a quasi-constant sampling rate. E.g. in Figure 5 the sample time interval of the camera $[t_i, t_j]$ differs from $[t_j, t_k]$ resulting in a different number of samples from inertial sensors. We implement an LSTM layer to process various sequence lengths of the input and extract the motion features. RNN takes the recursive transfer of the previous time step into account, which can reconstruct the reality of vehicle motion and accommodates to this dynamic input dimension while the feed-forward network can only deal with the problem with certain sequence length [15].

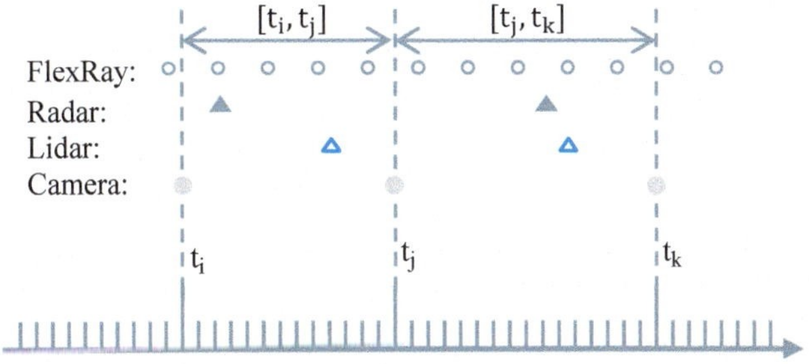

Figure 5. Multi-sampling rates of sensor systems. Camera timestamps as the baseline for the network.

4.4 Loss function for optimization

For a promising generalization capability and to avoid overfitting of the hybrid network, we modify the loss function with vehicle dynamic context. The hybrid loss function for the training is composed of two aspects: the basic and the vehicle state related loss function. Since the side slip angle estimation is a regression problem the basic loss function is calculated with the mean square error (MSE)

$$L_{base}=\frac{1}{n}\sum_{i=1}^{n}\left(\hat{y}_i-y_i^{label}\right)^2 \tag{4}$$

where $\hat{\mathbf{y}}_i$ denotes the prediction of the network as a matrix and $\mathbf{y}_i^{\mathbf{label}}$ denotes as the ground truth at the ith sample. As already mentioned, the network also predicts the longitudinal and lateral velocity intermediately to refine the final output. Hence, $\hat{\mathbf{y}}_i$ and $\mathbf{y}_i^{\mathbf{label}}$ are vectors consisting of longitudinal velocity, lateral velocity and side slip angle.

In the kinematic model, β can be calculated with the equation (1), so in addition to the direct prediction, we estimate a calculated kinematic side slip angle

$$\hat{\beta}_{kin} = \arctan\left(\frac{\hat{v}_y}{\hat{v}_x + v_{x,\,start}}\right). \tag{5}$$

We can get the $\hat{v}_x$ and $\hat{v}_y$ from the prediction of the network. $v_{x,\,start}$ is a hyperparameter which avoids zero division error and reduces the error caused by $\hat{v}_x$ since the relative error of longitudinal velocity raises at low speed. The cost of the network caused by the kinematic model error is also defined by the MSE

$$L_{kin} = \frac{1}{n}\sum_{i=1}^{n}\left(\hat{\beta}_{kin,\,i} - \beta_i^{label}\right)^2. \tag{6}$$

Under the condition that the longitudinal acceleration a_x is close to zero and the lateral acceleration a_y is commonly under $4 m/s^2$, we can say that the vehicle dynamics can be modelled with linear differential equations and the slip stiffness can be treated as a constant value. The model uncertainty decreases under this driving condition, which makes the approximation simple and easy. Consequently, we can estimate a linear side slip angle β_{lin} through the single-track model and set the tolerance $tol_{\beta_{lin}}$ to 30%. The loss function increases when the prediction is out of the tolerance.

$$\begin{aligned}
&\beta_{lin_{lb}} = \beta_{lin} - tol_{\beta_{lin}} * |\beta_{lin}| \\
&\beta_{lin_{ub}} = \beta_{lin} + tol_{\beta_{lin}} * |\beta_{lin}| \\
&\text{if } \hat{\beta} > \beta_{lin_{lb}} \text{ and } \hat{\beta} < \beta_{lin_{ub}} \\
&\qquad L_{\beta_{lin}} = 0 \\
&\text{else if } |a_x| < 1.5 m/s^2 \text{ and } |a_y| < 4\ m/s^2 \\
&\qquad L_{\beta_{lin}} = MSE(\hat{\beta}, \beta_{lin}).
\end{aligned} \tag{7}$$

The linear part of the loss function is only active under the condition described in equation (7), where $\hat{\beta}$ is an output of the network. In the non-linear situation, the linearity assumption is no longer satisfied. We modified the loss function instead using a non-linear algorithm of a double track mode-based EKF and set $tol_{\beta_{nlin}}$ to 10%. Equation (8) shows a similar algorithm of the loss function for the nonlinear vehicle state

$$\beta_{nlin_{lb}}=\beta_{nlin}-tol_{\beta_{nlin}}*|\beta_{nlin}| \quad (8)$$

$$\beta_{nlin_{ub}}=\beta_{nlin}+tol_{\beta_{nlin}}*|\beta_{nlin}|$$

$$\text{if } \hat{\beta}>\beta_{nlin_{lb}} \text{ and } \hat{\beta}<\beta_{nlin_{ub}}$$

$$L_{\beta_{nlin}}=0$$

$$\text{else if } |a_y|>4 \text{ m/s}^2 \text{ and } v_{x_{bus}}>30\text{km/h}$$

$$L_{\beta_{nlin}}=MSE(\hat{\beta},\beta_{nlin}).$$

Due to the inaccuracy of the tire model at low speed we constraint the activation additionally with the wheel speed-based longitudinal velocity $v_{x,\,bus}$ that can be obtained in the vehicle bus.

All previously described loss functions are superimposed to form an overall loss function, to summary defined as follows

$$L= L_{base}+f_p*(L_{kin}+L_{\beta_{lin}}+L_{\beta_{nlin}}). \quad (9)$$

As the ground truth is labelled with a highly precise measurement platform, the largest influence of the optimization is the basic loss function while the influence of the further loss functions can be controlled by a factor f_p (e.g., f_p can be increased if precise sensor data or vehicle parameters are known). The superimposition has the advantage that the modified term can not only compensate for the error of the labelled ground truth but also increase the generalization capability since the network is trained to depict the vehicle dynamics.

4.5 End to End Optimization of the Proposed Network

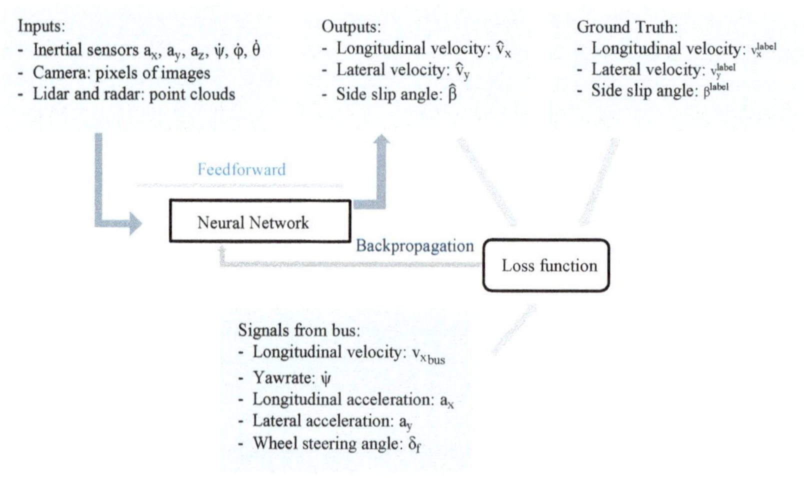

Figure 6. The training process of the network: feedforward of predictions and backpropagation of errors.

The proposed network is designed to be trainable end to end since the labelling of optical flows is expensive and requires extra sensors such as multi-layer lidar with high resolution. Figure 6 shows the schematic representation of the training process. The inputs, the signals from vehicle bus and the data of the ground truth are the measurement from our test vehicle. The networks are trained with the well-known feedforward and backpropagation algorithm [12]. The parameters of the network at the first epoch are randomly initialized to calculate the prediction (feedforward). The error is then computed with the loss function defined in the previous paragraph. The training of the network can be transferred to the following optimization problem

$$\underset{\theta,\ \lambda}{\text{minimize}}\ J = L(\theta,\ \lambda, X, Y) \tag{10}$$

J is the objective function to optimize, θ denotes the weights and bias of the network, λ is the hyperparameters, which are iteratively fine-tuned with validation dataset. And X, Y are the inputs and labels.

The prediction errors are propagated backwards and their gradients along θ can be derived with the chain rule enabling the analytical optimization with gradient descent (backpropagation). The weights and bias can, therefore, be updated with every feedforward and backpropagation until converging to the local optimum.

5 Experimental Results

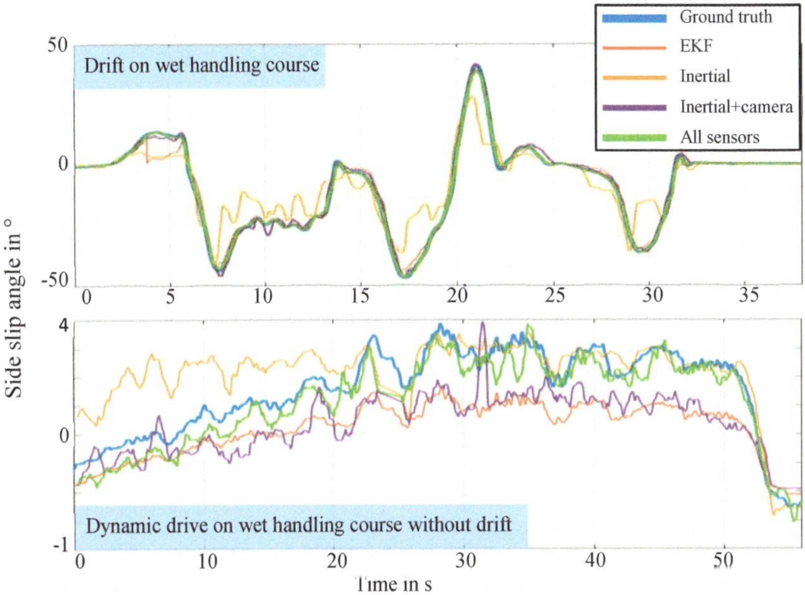

Figure 7. Comparison of ground truth, model-based EKF and different fusion setups in dynamic and highly dynamic driving conditions.

This section presents the results of the proposed network with various sensor setups compared with a traditional model-based EKF lateral dynamics estimation. The networks are trained with the measurements of a test vehicle under diverse open-loop and closed-loop handling maneuvers. The dataset is randomly split into 85% training set, 5% development set for the fine-tuning of hyperparameters and 10% test set. The networks are trained with an advanced gradient descent algorithm [19]. The training algorithm also takes the momentum of the gradients into account to alleviate the oscillation of the weights and bias over epochs.

Moreover, the learning rate α is decaying while training. These two algorithms help the network to converge faster and make the training not so sensitive to the learning rate, which means that we need less time to fine-tune the α. Figure 7 compares the results of the side slip angle estimation between inertial sensors, inertial sensors plus camera, all sensor setups, EKF, which is adapted and calibrated with the vehicle parameters, and the ground truth. The fusion of inertial sensors with all the environmental sensors yields to a promising performance, since it delivers comparable performance as the calibrated EKF in the highly nonlinear situation and an even more accurate estimation than the EKF in the transition region between linear and nonlinear situation. The accuracy decreases if only the camera and inertial sensors are fused. The estimation only with the inertial sensors is prone to the worst performance due to the initialization of LSTM without any information of velocities. A potential improvement of this estimation is to increase the sequence length of the data for the training and neglect the outputs at the early time steps on the prediction.

The generalization capability is also an important criterion to evaluate the model quality. Figure 8 shows the learning curves of the three concepts with the training set and test set. High loss on the training set means high bias while on the test set means high variance [25]. The optimization of a neural network is, therefore, to seek for low bias and low variance. The network using only the inertial sensors data results in a high bias and high variance while both fusion concepts yield to low training error. The network with inertial sensors and camera but without point clouds, however, has a high variance comparable with the inertial concept, which means that it has a limited generalization performance.

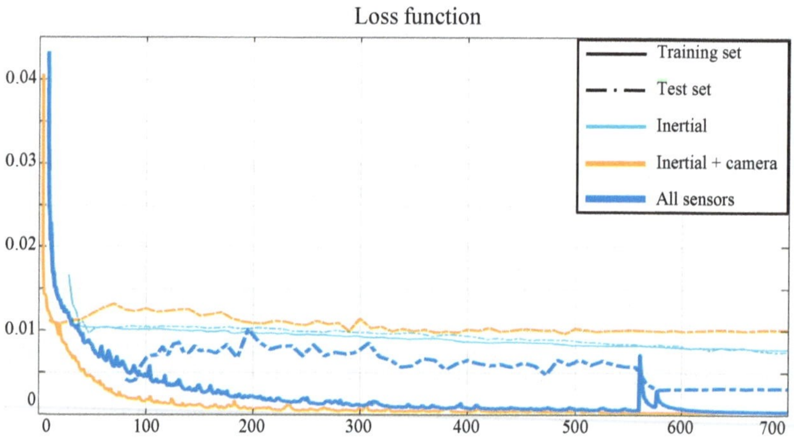

Figure 8. Comparison of the loss function to evaluate network performance.

6 Conclusion and Future Work

In this paper, we showcase a multi-view fusion concept to research on the accurate and robust side slip angle estimation problem. The promising results in the previous section indicate that using environmental sensors improves the estimation significantly. Furthermore, our approach enables an estimation without knowing any vehicle parameters or the longitudinal velocity using the wheel speed sensors. Therefore, the velocity signal integrity can be increased. Compared to traditional methods, our approach has the advantage of being adaptive to overcome the time synchronization error. The end to end trainable network architecture can also be transferred to learn other vehicle state estimation problem. Due to the limitation of computing power, our network is not trained with all the available data. We intend to investigate further studies on training the network with more data to improve and validate the generalization capability further. For the implementation on customer vehicles, we propose to optimize the network architecture to make it compact and fast enough for the computational platform. Besides, it is also worthwhile to combine the neural network with traditional methodologies such as Kalman filter to reduce the model uncertainty of neural network and therefore improve the reliability.

References

[1] Anton Obermüller, Modellbasierte Fahrzustandsschätzung zur Ansteuerung einer aktiven Hinterachskinematik, Dissertation, TU München, 2012.

[2] Baumgarten Goetz, Verfahren zur Erzeugung eines Schwimmwinkelsignals, 14.06.2012.

[3] J. Bechtoff, L. Koenig, and R. Isermann, "Cornering Stiffness and Sideslip Angle Estimation for Integrated Vehicle Dynamics Control," IFAC-PapersOnLine, vol. 49, no. 11, pp. 297–304, 2016.

[4] B. L. Boada, M.J.L. Boada, and V. Diaz, "Vehicle sideslip angle measurement based on sensor data fusion using an integrated ANFIS and an Unscented Kalman Filter algorithm," Mechanical Systems and Signal Processing, 72-73, pp. 832–845, 2016.

[5] A. Brunker, T. Wohlgemuth, M. Frey et al., "Odometry 2.0: A Slip-Adaptive EIF-Based Four-Wheel-Odometry Model for Parking," IEEE Transactions on Intelligent Vehicles, p. 1, 2019.

[6] R. Clark, S. Wang, H. Wen et al., "VINet: Visual-Inertial Odometry as a Sequence-to-Sequence Learning Problem," 1/29/2017.

[7] B. D. Lucas and T. Kanade, "An Iterative Image Registration Technique with an Application to Stereo Vision (IJCAI)," vol. 81, 1981.

[8] J. Engel, J. Stuckler, and D. Cremers, "Large-scale direct SLAM with stereo cameras," in 2015 IEEE/RSJ International Conference on Intelligent Robots and Systems (IROS), pp. 1935–1942, IEEE, 28.09.2015 - 02.10.2015.

[9] A. Eskandarian, Handbook of Intelligent Vehicles, Springer London, London, 2012.

[10] G. Farnebäck, "Two-Frame Motion Estimation Based on Polynomial Expansion," in Image Analysis, J. Bigun and T. Gustavsson, Eds., pp. 363–370, Springer Berlin Heidelberg, Berlin, Heidelberg, 2003.

[11] P. Fischer, A. Dosovitskiy, E. Ilg et al., "FlowNet: Learning Optical Flow with Convolutional Networks," 4/26/2015.

[12] I. Goodfellow, Y. Bengio, and A. Courville, Deep learning, The MIT Press, Cambridge, Massachusetts, London, England, 2016.

[13] T. Graber, S. Lupberger, M. Unterreiner et al., "A Hybrid Approach to Side-Slip Angle Estimation with Recurrent Neural Networks and Kinematic Vehicle Models," IEEE Transactions on Intelligent Vehicles, p. 1, 2018.

[14] J. Graeter, A. Wilczynski, and M. Lauer, "LIMO: Lidar-Monocular Visual Odometry," 7/19/2018.

[15] S. Hochreiter and J. Schmidhuber, "Long Short-Term Memory," Neural Computation, vol. 9, no. 8, pp. 1735–1780, 1997.

[16] E. Ilg, N. Mayer, T. Saikia et al., "FlowNet 2.0: Evolution of Optical Flow Estimation with Deep Networks," 12/6/2016.

[17] Jonathan Horgan, Patrick McDaid, and Petros Kapsalas, Verfahren zum Bestimmen eines Bewegungsparameters eines Kraftfahrzeugs durch Auffinden

von invarianten Bildregionen in Bildern einer Kamera des Kraftfahrzeugs, Kamerasystem und Kraftfahrzeug, 28.12.2012.
[18] D. Kellner, M. Barjenbruch, J. Klappstein et al., "Instantaneous ego-motion estimation using multiple Doppler radars," in 2014 IEEE International Conference on Robotics and Automation (ICRA), pp. 1592–1597, IEEE, 5/31/2014 - 6/7/2014.
[19] D. P. Kingma and J. Ba, "Adam: A Method for Stochastic Optimization," 12/22/2014.
[20] A. Krizhevsky, I. Sutskever, and G. E. Hinton, "ImageNet classification with deep convolutional neural networks," Communications of the ACM, vol. 60, no. 6, pp. 84–90, 2017.
[21] J. Ku, M. Mozifian, J. Lee et al., "Joint 3D Proposal Generation and Object Detection from View Aggregation," 2018.
[22] M. Maaref, J. Khalife, and Z. M. Kassas, "Lane-Level Localization and Mapping in GNSS-Challenged Environments by Fusing Lidar Data and Cellular Pseudoranges," IEEE Transactions on Intelligent Vehicles, p. 1, 2018.
[23] Mirek Göbel, Verfahren zum Bestimmen des Fahrzustands eines zweispurigen Fahrzeugs durch SchwimmwinkelSchätzung, 02.03.2006.
[24] T. Miyasaka, Y. Ohama, and Y. Ninomiya, "Ego-motion estimation and moving object tracking using multi-layer LIDAR," in 2009 IEEE Intelligent Vehicles Symposium, pp. 151–156, IEEE, 6/3/2009 - 6/5/2009.
[25] P. Munro, H. Toivonen, G. I. Webb et al., "Bias Variance Decomposition," in Encyclopedia of Machine Learning, C. Sammut and G. I. Webb, Eds., pp. 100–101, Springer US, Boston, MA, 2010.
[26] R. Mur-Artal and J. D. Tardos, "ORB-SLAM2: an Open-Source SLAM System for Monocular, Stereo and RGB-D Cameras," IEEE Transactions on Robotics, vol. 33, no. 5, pp. 1255–1262, 2017.
[27] K. Nam, S. Oh, H. Fujimoto et al., "Estimation of Sideslip and Roll Angles of Electric Vehicles Using Lateral Tire Force Sensors Through RLS and Kalman Filter Approaches," IEEE Transactions on Industrial Electronics, vol. 60, no. 3, pp. 988–1000, 2013.
[28] T. Novi, R. Capitani, and C. Annicchiarico, "An integrated artificial neural network–unscented Kalman filter vehicle sideslip angle estimation based on inertial measurement unit measurements," Proceedings of the Institution of Mechanical Engineers, Part D: Journal of Automobile Engineering, vol. 14, 095440701879064, 2018.
[29] N. Paragios, Y. Chen, and O. Faugeras, Handbook of Mathematical Models in Computer Vision, Springer-Verlag, s.l., 2006.
[30] R. Rajamani, Vehicle dynamics and control, Springer, New York, NY, 2012.
[31] Rudolf Ertlmeier, Modellbasierte und fahrdynamikunterstützte Überschlagserkennung, Dissertation, Otto-von-Guericke-Universität Magdeburg, 2013.
[32] D. Schramm, M. Hiller, and R. Bardini, Vehicle dynamics: Modeling and simulation, Springer, Berlin, 2018.

[33] D. Selmanaj, M. Corno, G. Panzani et al., "Robust Vehicle Sideslip Estimation Based on Kinematic Considerations," IFAC-PapersOnLine, vol. 50, no. 1, pp. 14855–14860, 2017.
[34] S. Thrun, W. Burgard, and D. Fox, Probabilistic robotics, MIT Press, Cambridge, Mass., 2006.
[35] R. Wang, M. Schwörer, and D. Cremers, "Stereo DSO: Large-Scale Direct Sparse Visual Odometry with Stereo Cameras," 8/25/2017.
[36] K. Zindler, N. Geiss, K. Doll et al., "Real-time ego-motion estimation using Lidar and a vehicle model based Extended Kalman Filter," in 17th International IEEE Conference on Intelligent Transportation Systems (ITSC), pp. 431–438, IEEE, 10/8/2014 - 10/11/2014.

Potentials of modern active suspension control strategies – from model predictive control to deep learning approaches

Guru Bhargava Khandavalli, M.Sc.
Institut für Kraftfahrzeuge, RWTH Aachen University (ika)

Dr.-Ing. Marcus Kalabis
Ford-Werke-GmbH

Dipl.-Ing. Daniel Wegener
Institut für Kraftfahrzeuge, RWTH Aachen University (ika)

Univ.-Prof. Dr.-Ing. Lutz Eckstein
Institut für Kraftfahrzeuge, RWTH Aachen University (ika)

Abstract

The active suspension system has always been a topic of interest because of its ability to influence the ride quality by exerting independent forces on the suspension by the usage of separate actuators. Various strategies have been proposed over the years in order to estimate the appropriate control action. These strategies are typically feedback oriented and dependent on many factors like the control objective, the frequency of the excitation and system non-linearities that result in the formulation of a complex problem. A simulation-based study using a comprehensive quarter car model with an active suspension is beneficial to summarize the character of each of these approaches by comparing attributes like formulation, performance, robustness, tunability and requirements for implementation on real physical systems.

Since active suspensions use an on-board computer and sensor measurements to determine the control action, the enactment of a strategy is limited by the computational requirements and available measurements. This study aims to foresee such challenges when implementing modern control strategies like H_∞ control or preview based approaches like Model Predictive Control (MPC) and Deep Learning methods. The latter will focus on both Supervised Learning (SL) and Reinforcement Learning (RL) approaches. These methods are firstly developed in a virtual environment and subsequently implemented on a physical quarter car setup excited by a servo-hydraulic actuator. Finally, a comparison of the performances of the different control approaches is presented.

1 Introduction

One of the most important tasks [1] of a suspension system is to enhance ride comfort experienced by the passengers. Active suspensions play a key role in attenuating the movements of the vehicle structure when excited by the road unevenness. Their functioning is dependent on a control logic which requires information from various sensors on the vehicle. Since different driving situations require different control actions, identification of the optimal control action is very challenging. Eventually, many control strategies have been put forward to control suspension systems in motor vehicles [2]. Each strategy is based on certain assumptions and certain empirical entities which should be optimized to conform with the model parameters. Of the numerous control approaches, the potentials of three modern control systems, namely H_∞ control, Model Predictive Control using Supervised Learning and Reinforcement Learning are presented in this paper. Though these strategies are widely studied, they are often validated only on a simulation basis. In the current study, the potentials of these strategies are validated by implementing on a physical quarter car test bench.

1.1 Literature Review

The H_∞ control problem [3], also known as the disturbance attenuation problem, has been discussed as an applicable method for suspension systems in the context of stability and robustness. Application of H_∞ control to suspension systems has been proven to assure performance on a simulation basis in [4], [5] and [6]. In [7], the H_∞ control law was implemented in an experimental vehicle but the formulation of actuator constraints during the controller synthesis was not explicitly addressed. For the current study, the approach proposed in [5] is adopted since the approach considers constrained H_∞ control imposing time domain constraints on entities like suspension deflection, actuator force and dynamic wheel load.

Model Predictive Control has been discussed in the field of control engineering for a long time for its advantages of handling the model and the constraints explicitly as well as generating high performance optimization law at every step. However, the slow nature of MPC due to the complexity of the optimization problem reduced its suitablility for fast dynamic processes like suspension systems. To make it suitable for fast processes, explicit MPC methods [8] [9] [10] [11], where the control law is established based on offline calculations and polyhedral state space regions, have been put forward and have shown good promise with respect to offering performance as well as respecting the time domain constraints. However, traditional explicit approaches scale badly with increasing prediction horizon and constraints decreasing the practical applicability of these approaches. In [12], an explicit MPC approach, where ANNs are used to approximate the control law, was proposed for a semi-active suspension system. The controller offered a decent benefit over passive system when simulated using a full vehicle assembly for various control objectives focussing on ride comfort and ride safety. For the current study, this approach was extended by considering road preview in developing the control law for an active system.

Ever since the application on Atari Games [13], Reinforcement Learning [14] has gained attention in the field of control engineering and robotics. The application of model based RL algorithms to suspension systems has been discussed in [15], [16], [17], and [18] focussing on development of integrated approaches for control of suspension systems. However, the drawback of these approaches is the complexity involved in formulating a model-based selection of a particular action during the learning process. In order to investigate model-free approaches, the current study considers Q-learning algorithm where the controller is modelled as an agent that explores the action space based on ε-greedy policy [14]. The agent acquires a reward for improving ride comfort and is penalized for violating the time domain constraints. Through Reinforcement Learning, the agent learns to make use of the available range of actuator force to enhance ride comfort while being considerate to the constraints.

1.2 Research Approach

In the current study, the controllers are developed in a virtual environment using simulation models during the design phase and later employed in a real system to verify their potentials. As a result, the study is concerned with two main aspects namely controller development and implementation on the testrig as shown in Figure 1. The controller development phase begins with modelling of the suspension dynamics and the road excitation. Utilising the equations of motion, the system is represented using state space equations. This process is described in Chapter 2.

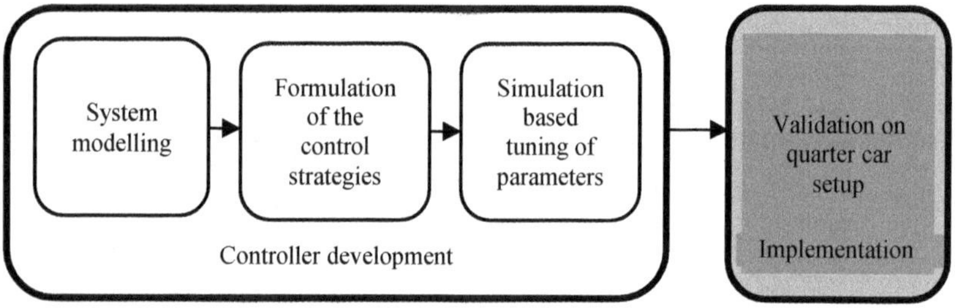

Figure 1: Research Approach

After defining the system dynamics, the formulations of the control strategies which comprise the system model as well as control-related parameters are studied to identify the tuning parameters and also to investigate the online applicability of these strategies. The identified tuning parameters are fine-tuned based on quarter car simulations so that the approaches comply with the control objectives and physical constraints. The definition of the control objectives and the procedure followed to devise the controllers for online implementation are described in Chapter 3.

Following the synthesis of controllers, the potentials of each controller are verified by evaluating their performances on a real system. To do this, a comprehensive quarter car testrig is setup where a hydraulic cylinder with a frequency bandwidth of 0-120 Hz is used for road excitation. The actuator is mounted in the damper path of a Ford independent wheel suspension. The setup comprises multifarious sensors to facilitate required measurements. A comparison of the controllers' performances with the reference passive and skyhook systems is discussed in Chapter 4.

Finally, the work is briefly summarized in Chapter 5. An insight into the possible extensions or improvements for the controllers is also provided.

2 System Modelling

System modelling forms the base of the controller development process, especially because the development process includes simulations. Since mathematical models are mostly not exact, it is important to address the system formulation and the assumptions. This chapter describes the modelling of suspension dyamics and the road data as the source of excitation, which are the key elements of the system.

2.1 Suspension Model and State Space

Since the current study aims at implementing the control strategies over a quarter car test setup, a linear quarter car vehicle model is considered, as it is proven to form a good basis for controller design. The linear quarter car model owing to its simple structure and low dynamic order offers significant advantages such as intuitive understanding of the vertical dynamics and low computational requirements. [19] While various studies [12] [20] suggest usage of models with nonlinearities, simulations with the current parametrisation showed that the required additional computational effort significantly outweighed the performance gain obtained by including nonlinearities and friction components. As a result, a linear quarter car model is used to represent the vertical dynamics in the control approaches as shown in Figure 2. The active suspension component is an active hydraulic actuator generating the force input (u).

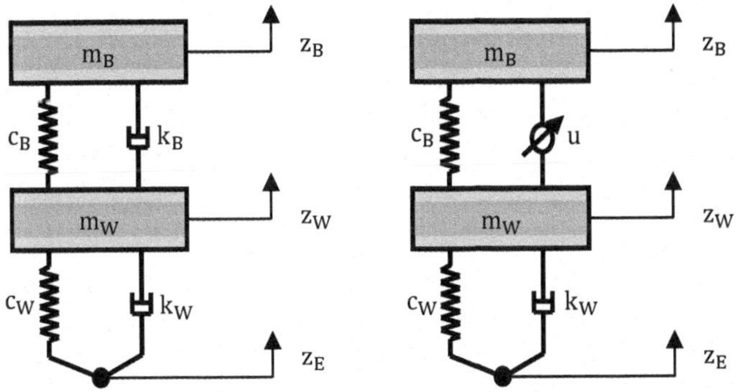

Figure 2: Passive suspension system (left), Active suspension system (right)

The equations of motions for the active suspension system are:

$$m_B \ddot{z}_B = -c_B(z_B - z_W) - u \tag{1}$$

$$m_W \ddot{z}_W = -k_W(\dot{z}_W - \dot{z}_E) - c_W(z_W - z_E) + c_B(z_B - z_W) + u \tag{2}$$

where m_B, m_W, k_B, k_W, c_B and c_W represent the sprung mass, the unsprung mass, the passive damping coefficient, the tire damping coefficient, the suspension spring stiffness and the tire stiffness respectively. If $u = k_B(\dot{z}_B - \dot{z}_W)$, then (1) and (2) represent the equations of motion for the passive suspension system.

The system can be represented by a Linear Time Invariant (LTI) state space in the form of (3), (4) and (5). In the equations, the column vectors are represented using bold characters whereas other entities are represented using normal characters.

$$\dot{\mathbf{x}}(\mathbf{t}) = \mathrm{A}\mathbf{x}(\mathbf{t}) + \mathrm{B}_1\mathrm{u}(\mathrm{t}) + \mathrm{B}_2\mathrm{z}(\mathrm{t}) \tag{3}$$

$$\mathrm{y}_1(\mathrm{t}) = \mathrm{C}_1\mathbf{x}(\mathbf{t}) + \mathrm{D}_{1\mathrm{u}}\mathrm{u}(\mathrm{t}) + \mathrm{D}_{1\mathrm{z}}\mathrm{z}(\mathrm{t}) \tag{4}$$

$$\mathrm{y}_2(\mathrm{t}) = \mathrm{C}_2\mathbf{x}(\mathbf{t}) + \mathrm{D}_{2\mathrm{u}}\mathrm{u}(\mathrm{t}) + \mathrm{D}_{2\mathrm{z}}\mathrm{z}(\mathrm{t}) \tag{5}$$

The state vector $\mathbf{x}(\mathbf{t})$ is given by $\mathbf{x}(\mathbf{t}) = [x_1 \quad x_2 \quad x_3 \quad x_4]^T$ where x_1 is the suspension deflection, x_2 is the sprung mass (body) vertical velocity, x_3 is the tire deflection and x_4 is the unsprung mass (wheel) vertical velocity. Even though, tire deflection is a difficult measurement to obtain, it is chosen to befit the H_∞ controller formulation which is described in [5]. However, ANN based approaches offer a significant advantage considering required measurements such as wheel acceleration instead of tire deflecition to enforce the control laws, which is addressed in Chapter 5.

The vector $y_1(t)$ is performance measurement output and $y_2(t)$ is the constrained output. For the current study, the measured vertical acceleration of the sprung mass is selected to quantify ride comfort performance. The suspension deflection is chosen as the constrained output because of the physical limitations on the amount of possible deflection.

The model assumes the actuator to be ideal, i.e. the actuator delay component is neglected. In consequence, including actuator dynamics in system modelling will be a potential future activity.

2.2 Road Data

Road unevenness, in general, is an excitation that comprises different amplitudes and wavelengths at irregular time periods. This stochastic unevenness can be described mathematically with the help of Power Spectral Density (PSD), which can be approximated using the degree of unevenness and the waviness of the road [1]. Exploiting this approximation, a new method considering the usage of benchmark road data as an excitation signal is introduced. In this method, two road excitation signals namely training road data and the testing road data are considered as shown in Figure 3.

The testing road data consists of measurements obtained by physically measuring the road unevenness height whereas the training road data is a virtual excitation signal obtained by using approximated values of degree of unevenness and waviness of the testing data. The intention behind using this method is to challenge the generalization capability of the controllers. Therefore, the strategies discussed in subsequent sections use the training road data as the excitation signal during the design phase. Later, after the controllers are synthesized, they are validated using the testing road data. This procedure ensures that the controllers are tested with new and realistic inputs and not just with the inputs used during the controller synthesis.

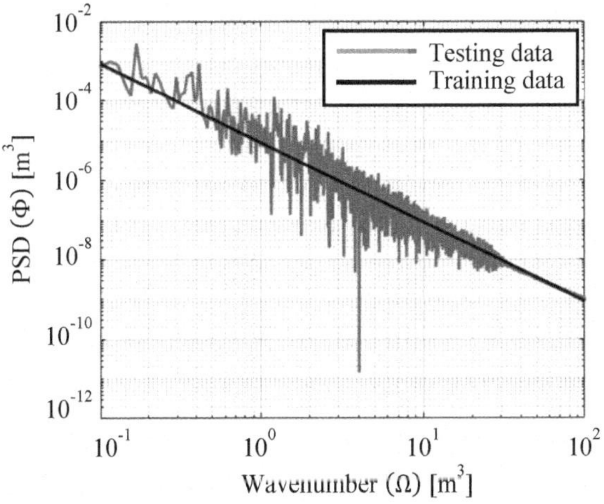

Figure 3: Road data used for training and testing purposes

Further, neural networks have a tendency to memorize the values when over-trained [21]. In such cases, they exhibit poor performance when exposed to inputs which are not encountered during the learning process. By using the above-described method, the SL and RL controllers are exposed to new inputs necessitating robustness of the controllers with regard to road excitation.

3 Conception and Synthesis of the Control Systems

The current chapter discusses the development of each controller, the simulation-based fine-tuning process and the factors to be considered for online implementation. In order to identify the key factors affecting the controllers' performance, a clear understanding of the strategies is necessary. For this purpose, the controller formulations are briefly described.

3.1 H∞ Control

The H∞ control is applicable to linear systems to minimize the H∞ norm of the closed loop transfer function between the performance output y_1 and the road disturbance z defined in the frequency domain as [22]:

$$\|G\|_\infty := \sup_\omega \sigma_{max}[G(j\omega)] \tag{6}$$

where G is the stable transfer matrix as shown in Figure 4 and σ_{max} is the maximum singular value.

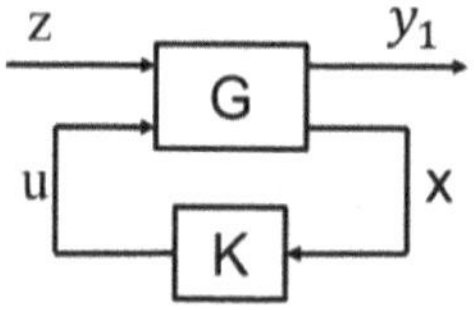

Figure 4: State feedback closed loop system

As it can be seen in Figure 4, the control action u is a constant gain on the state measurements described in section 2.1.

3.1.1 Formulation of H∞ Control

The control law, considering a state feedback system, is defined as:

$$u = Kx \tag{7}$$

For a given $\gamma > 0$, internal stability and the H∞ norm of the performance output to the input disturbance less than γ is guaranteed by existence of matrices $Q = Q^T > 0$ and Γ such that [5]:

$$\begin{bmatrix} AQ + QA^T + B\Gamma + \Gamma^T B^T & * & * \\ B_1^T & -\gamma I & * \\ C_1 Q + D_{1u}\Gamma & D_{1z} & -\gamma I \end{bmatrix} < 0 \tag{8}$$

where $*$ represents the transpose of the element across the diagonal.

The feedback gain K is defined as:

$$K = \Gamma Q^{-1} \tag{9}$$

The satisfaction of the time domain constraints can be guaranteed when, in addition to the above Linear Matrix Inequality (LMI)(8), for some symmetric matrices U and Y, the following LMIs are also valid for given weights β_1 and β_2.

$$\begin{bmatrix} \frac{1}{\beta_1}U & \Gamma \\ \Gamma^T & Q \end{bmatrix} \geq 0, \text{ with } U_{ii} \leq u_{i,max}^2 \tag{10}$$

$$\begin{bmatrix} \frac{1}{\beta_2}Y & C_2Q + D_{2u}\Gamma \\ * & Q \end{bmatrix} \geq 0, \text{ with } Y_{ii} \leq y_{2i,max}^2 \tag{11}$$

where $u_{i,max}$ and $y_{2i,max}$ are the maximum actuator force and the maximum suspension deflection respectively.

3.1.2 Solution of the H∞ Problem with Time Domain Constraints

The LMI problem is solved using Semidefinite Programming (SP). The optimization problem is formulated as:

$$\min_{\gamma,Q,\Gamma,U,Y} \gamma \qquad \text{s.t.} \tag{12}$$

LMIs (8), (10) and (11) are satisfied with

$$Q = Q^T > 0 \tag{13}$$

$$U = U^T \tag{14}$$

$$Y = Y^T \tag{15}$$

Many tools are available to solve optimization problems associated with LMI formulations. Because of the ease offered in the LMI optimization problem formulation, CVX, a package for specifying and solving convex programs [23] [24], is used. For a feasible solution to the optimization problem, it is possible to arrive at some values of Q, Γ, U and Y for given β_1 and β_2. The values of β_1 and β_2 are established by conducting a set of simulations using the quarter car model described in section 2.1. For different values of $\beta_1 \in [0,1]$ and $\beta_2 \in [0,1]$, the LMI optimization problem was solved which resulted in a set of controller gains. Employing each controller gain, the quarter car model was virtually excited by the training road data described in section 2.2. For each simulation, the maximum value of the actuator force requested by the controller as well as the maximum value of the suspension deflection were recorded as shown in Figure 5.

The shaded regions in the graphs indicate the range of values for which constraint violations occur whereas for the values of β_1 and β_2 in the unshaded region, the controller respected the time-domain constraints.

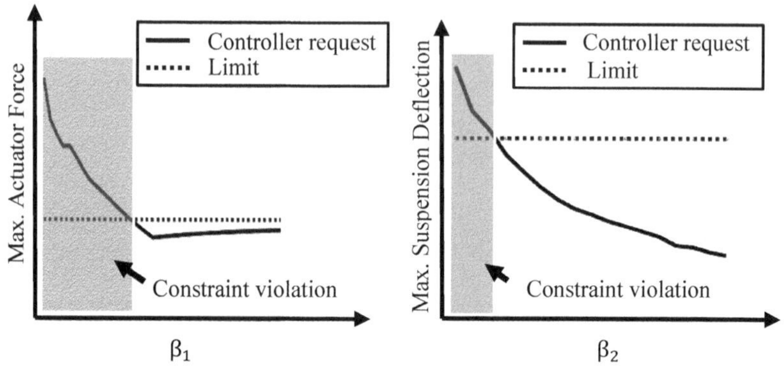

Figure 5: Simulation results showing the dependence of constraint violations on the weights β_1 and β_2

For the current study and the considered model parameters, the values of β_1 and β_2 were chosen to be 0.015 and 0.3 respectively to solve the optimization problem (12) and subsequently evaluate the feedback gain K as in (9).

3.2 Model Predictive Control

Model Predictive Control is a preview based setup where the control action sequence is evaluated by solving the finite horizon optimization problem at each time step and supplying the first value of the trajectory as the control action [8] [25].

3.2.1 Formulation of MPC

The discrete form of the state space represented by (3), (4) and (5) is expressed as:

$$\mathbf{x(k+1)} = A_d\mathbf{x(k)} + B_{1d}u(k) + B_{2d}z(k) \tag{16}$$

$$y_1(k) = C_1\mathbf{x(k)} + D_{1u}u(k) + D_{1z}z(k) \tag{17}$$

$$y_2(k) = C_2\mathbf{x(k)} + D_{2u}u(k) + D_{2z}z(k) \tag{18}$$

The formulation of the MPC optimization problem begins with the expression of the performance outputs over the prediction horizon $\hat{\mathbf{Y}}$ in terms of the current state measurement $\mathbf{x(k)}$, current control action u(k), current road disturbance z(k), the sequence of control actions $\hat{\mathbf{U}}$ and the road disturbance over the prediction horizon $\hat{\mathbf{Z}}$.

$$\hat{\mathbf{Y}} = \begin{bmatrix} y_1(k+1) \\ y_1(k+2) \\ \vdots \\ y_1(k+N_p) \end{bmatrix}, \hat{\mathbf{U}} = \begin{bmatrix} u(k+1) \\ u(k+2) \\ \vdots \\ u(k+N_c) \end{bmatrix} \text{ and } \hat{\mathbf{Z}} = \begin{bmatrix} z(k+1) \\ z(k+2) \\ \vdots \\ z(k+N_p) \end{bmatrix}$$

where N_p is the number of samples in the prediction horizon and N_c is the number of samples in the control horizon. By concatenating the state space matrices, the following form of the equation for $\hat{\mathbf{Y}}$ is obtained in the following form:

$$\hat{\mathbf{Y}} = F\mathbf{x}(\mathbf{k}) + \Theta_k u(k) + \Theta_{N_c}\hat{\mathbf{U}} + \Phi_k z(k) + \Phi_{N_p}\hat{\mathbf{Z}} \tag{19}$$

The values of the matrices F, Θ_k, Θ_{N_c}, Φ_k and $\Phi_{N_p}$are suitably obtained. The minimization problem is given by

$$\min_{\hat{\mathbf{U}}}(\hat{\mathbf{Y}}^T Q \hat{\mathbf{Y}} + \hat{\mathbf{U}}^T R \hat{\mathbf{U}}) \qquad \text{s.t.} \tag{20}$$

$$u_{min} \leq \hat{\mathbf{U}} \leq u_{max} \tag{21}$$

$$|y_2| \leq ss_{max} \tag{22}$$

where u_{min} and u_{max} are the minimum and maximum values of the possible control action and ss_{max} is the maximum possible suspension deflection. The matrices Q and R are the weight matrices for the performance output and the control action respectively.

3.2.2 Substitution of MPC with Supervised Learning

The optimum solution for the convex optimization problem, minimizing the cost function value for a given state and preview data, is found by using the interior-point method. To catalyse the optimization process, the analytical gradient of the cost function is supplied to the solver. Using the Matlab Optimization toolbox [26], for a sample time of 10 ms and a horizon of 250 ms, the online implementation of MPC converges to the optimal control sequence in 500 ms on average (with Intel(R) Core(TM) i7-6700 CPU @3.40 GHz, 32 GB RAM).

Practically, this convergence time is not suitable for online implementation keeping the fast dynamics of a suspension system in mind. Therefore a metamodel using neural networks is used to approximate the MPC optimization scheme [12] to realise online implementation. Here, the idea of Supervised Learning is applied to train a neural network. The training data is obtained by offline calculation of optimal solutions for various initial states subject to various road conditions. For the neural network, the state measurements as in (3) are supplied as inputs and the corresponding first control actions from the optimal sequences of control are set as targets as shown in Figure 6.

The neural network tuning procedure was performed using the Matlab Deep Learning Toolbox [27] where various architectures corresponding to different hyperparameters were studied. Using Bayesian regularization, the network converged to acceptable performance when trained using 10000 samples with an architecture comprising 4 hidden layers with 10 neurons each.

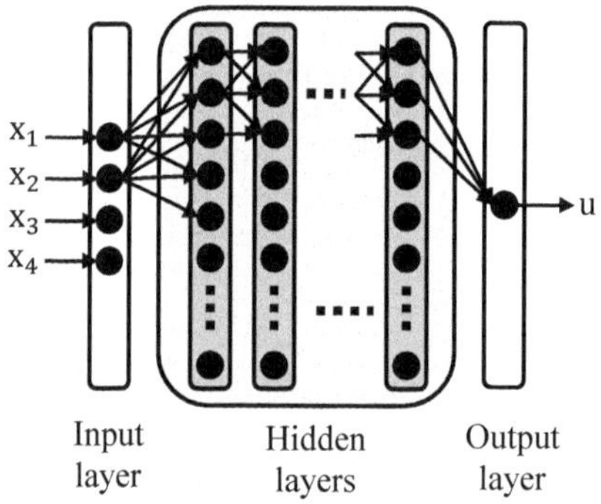

Figure 6: Neural Network architecture with the state measurements as inputs and the actuator force as output

Each neuron in the hidden layers has a purely linear transfer function whereas the output neuron is configured to have a sigmoid transfer function to comply with actuator force constraints.

3.3 Reinforcement Learning as Control System

Reinforcement Learning is a form of machine learning where there is an agent and an environment. The agent explores and exploits the state-action space to maximize a cumulative reward. Based on various trials, the agent detects the actions that return the highest rewards [14].

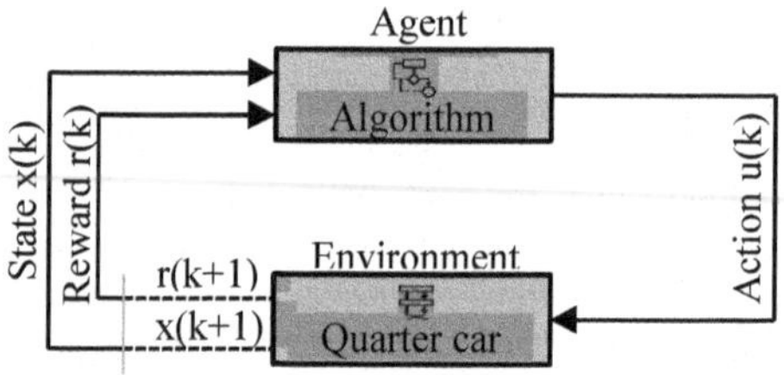

Figure 7: Framing of Reinforcement Learning

Most of the RL approaches use an environment modelled as a Markov Decision Process (MDP) i.e. the future state is assumed to be independent of the past, given the present.

3.3.1 Q-learning as RL Algorithm

Q-learning is a model free, off-policy temporal difference control algorithm appropriate for models with stochastic transitions [14]. The key idea of this algorithm is that a cumulative value called Q-factor is stored for each state-action pair in the system [28] as shown in Figure 8.

Action→ / State ↓	u_1	u_2	...	u_n
$\mathbf{x_1}$	$Q(\mathbf{x_1}, u_1)$	$Q(\mathbf{x_1}, u_2)$	...	$Q(\mathbf{x_1}, u_n)$
$\mathbf{x_2}$	$Q(\mathbf{x_2}, u_1)$	$Q(\mathbf{x_2}, u_2)$	...	$Q(\mathbf{x_2}, u_n)$
⋮	⋮	⋮	...	⋮
$\mathbf{x_m}$	$Q(\mathbf{x_m}, u_1)$	$Q(\mathbf{x_m}, u_2)$	...	$Q(\mathbf{x_n}, u_n)$

Figure 8: Q-Table

In order to fill the values Q-Table, a simulation setup is developed where the quarter car model is excited using the training road data for a certain number of episodes. As the simulation progresses through each episode, the agent explores the action space to learn the actions which result in minimized body accelerations. The exploration of the action space is modelled using the ε – greedy policy [14]. When a greedy action is taken, the agent's knowledge is exploited where the agent tends to take the action with the maximum Q-value. Otherwise, the agent tends to explore by taking a random action and updates the knowledge base. At each sample time, for a given state and action, the immediate reward is measured and the corresponding Q-value is updated according to:

$$Q(\mathbf{x_k}, u_k) \leftarrow (1-\alpha)Q(\mathbf{x_k}, u_k) + \alpha[r_{k+1} + \gamma \max_u Q(\mathbf{x_{k+1}}, u)] \qquad (23)$$

The term $r_{k+1} + \gamma \max_u Q(\mathbf{x_{k+1}}, u)$ is called the learned value where k is the sample index. $\alpha \in (0,1]$ is the learning rate which controls the precedence levels of newly learned information. If $\alpha = 0$, the agent learns nothing but only exploits the previous knowledge whereas $\alpha = 1$ makes the agent consider only the newly acquired knowledge. r_i is the immediate reward acquired by moving from $\mathbf{x_k}$ to $\mathbf{x_{k+1}}$ and $\gamma \in [0,1]$ is the discount factor which determines the importance of the future rewards.

3.3.2 Q-learning for Continuous State Space

Since the suspension state space is continuous in nature, the standard Q-learning method might not be directly applicable. Therefore, to generalize the model for continuous state action space, Q-learning with a linear approximator is implemented. The basic purpose of this approximator is to accept the state measurements as the input and return the Q-values for each possible control action as shown in Figure 9. The approximation function ($\widehat{Q}_\theta(\mathbf{x}, u)$) is the replica of a neural network which is dependent on various features (f_i) and corresponding weights (θ_i). Each feature is a function of either the state or the action or both.

$$\widehat{Q}_\theta(\mathbf{x_k}, u_k) = \sum_i \theta_i f_i(\mathbf{x_k}, u_k) \approx Q(\mathbf{x_k}, u_k) \tag{24}$$

The values of the weights (θ_i) are updated during each learning episode according to the rule yielding from the stochastic gradient descent method [29] till the convergence criteria are satisfied.

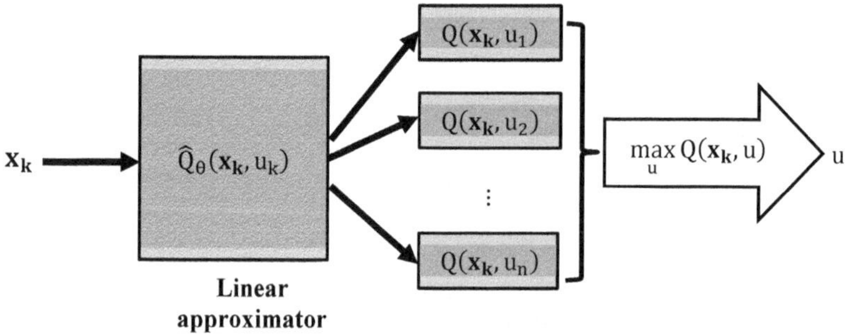

Figure 9: Implementation of Q-learning with linear approximator

After sufficient learning, for a given state, the action resulting in the maximum Q-value is supplied as the control action to the suspension system.

4 Evaluation of the Controllers

The H_∞ control, the SL control and the RL control are implemented on the quarter car testrig where the servo-hydraulic actuator is fed with the testing road data signal (as excitation signal), as described in section 2.2. The control laws established with the help of simulations are used to request the actuator force for given state measurements. To quantify the controllers' performances, certain criteria have to be defined which are described in this chapter along with the comparison of the controller performances.

4.1 Performance Measurement

To quantify the performance of the controllers, the root mean square (RMS) value of the sprung mass acceleration measurements from the sensors is considered.

$$|\ddot{z}_B|_{rms} = \sqrt{\frac{1}{T}\int_0^T \ddot{z}_B^2 \, dt} \tag{25}$$

Further to take the human sensitivity to vibrations into account, an entity called ISO score is introduced. The measurement signals are passed through the 5th order ISO-2631 filter [30] as shown by (26) and then the RMS value of the filtered signal is considered to evaluate the ISO score (27). It can be understood that a lower ISO score implies better performance.

$$W(s) = \frac{87.72s^4 + 1138s^3 + 11336s^2 + 5453s + 5509}{s^5 + 92.6854s^4 + 2549.83s^3 + 25969s^2 + 81057s + 79783} \tag{26}$$

$$\text{ISO score} = |W(s)\,\ddot{z}_B|_{rms} \tag{27}$$

The RMS values and the ISO scores of the body accelerations corresponding to each of the controllers are compared with that of the passive system and a skyhook system to assess the benefit offered by them. The skyhook controller used as reference controller consists of a constant gain on the body vertical velocity, in order to settle the body motion. It is tuned to utilize the actuator without overloading it.

4.2 Comparison of the Controller Performances

The control strategies, as described in Chapter 3, are implemented on the quarter car testrig. Based on the sprung mass acceleration measurements, a potential comparison of the strategies is established. The power spectral densities of the sprung mass acceleration for the three controllers in comparison with the passive system and the skyhook system are shown in Figure 10. When comparing the performance near the body eigenfrequency region (~ 1.4 Hz), the skyhook controller outperformed all the other controllers. In the low frequency range (between 1 Hz and 4 Hz) where the SL and RL controllers decrease the sprung mass acceleration to a substantial extent, the benefit offered by H_∞ controller appears to be trivial. This is depicted by the curve of H_∞ controller concurring that of the passive system in Figure 10 (left).

According to the ISO scores in Table 1, all the controllers show noteworthy decrease in the body vertical acceleration as shown in Figure 10 (right) where the SL controller offers the best performance (~36% benefit over passive and ~13% benefit over skyhook).

Despite being model-free, the RL controller offers similar benefit (~35% benefit over passive and ~12% benefit over skyhook) in the frequency region of highest human sensitivity.

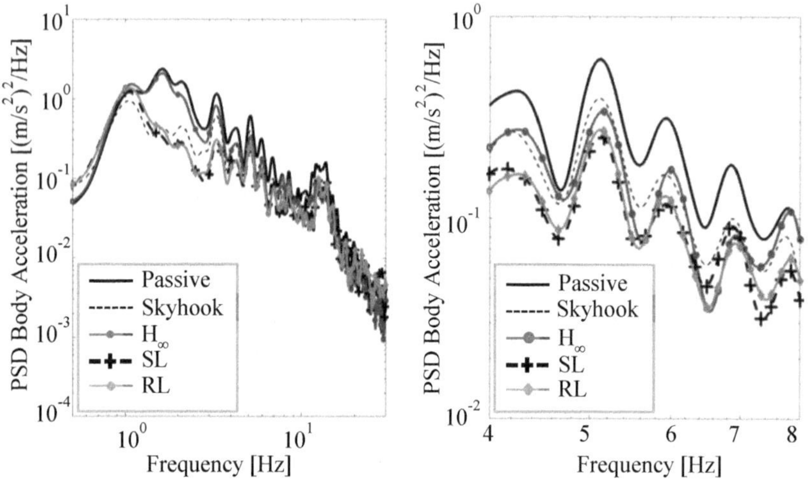

Figure 10: Power Spectral Density of sprung mass acceleration over the frequency range of 0.5 Hz to 30 Hz (left), Power spectral density of sprung mass acceleration over the frequency range of 4 Hz to 8 Hz (right).

Considering overall performance, all three controllers proved to guarantee performance improvement when compared with the passive system. Nevertheless, with the available range of actuator force, the RL controller and the SL controller showed significant advantage over the H_∞ controller when compared with that of the skyhook system.

Table 1: Overall RMS values of the sprung mass acceleration and the corresponding ISO scores of the passive and active systems

	Passive	Skyhook	H_∞	SL	RL
$\lvert\ddot{z}_B\rvert_{rms}$ in $\left[\frac{m}{s^2}\right]$	2.30	1.70	2.05	1.56	1.59
ISO score in $\left[\frac{m}{s^2}\right]$	1.69	1.24	1.39	1.08	1.09

Further, the benefits offered by the SL and RL controllers attest the potential of the neural networks as function approximators. Despite being trained on a different dataset, the controllers prove efficient in handling new input signals (or state measurements).

5 Summary, Discussion and Outlook

Three modern active suspension control strategies namely H_∞ control, Model Predictive Control using Supervised Learning and Reinforcement Learning were examined with regard to ride comfort enhancement. With the help of simulations, a special emphasis was laid on the tuning factors and the requirements for online implementation. In this process, several factors affecting the controller performance were identified as listed in Table 2.

Table 2: Factors affecting controller performance

Controller	Factors
H_∞	• State space approximation of the quarter car suspension • Weighting factors β_1,β_2
SL	• State space approximation of the quarter car suspension • MPC Objective function weights (Q, R) and horizon length • Neural network hyperparameters (architecture, algorithm, data)
RL	• Q-learning hyperparameters (α, γ) • Environment and reward modelling (r_k) • Approximation function/neural network parameters(θ_i, f_i)

After tuning the controllers, their potentials were analysed by comparing their performances with that of the passive and skyhook systems on a comprehensive quarter car testrig. The SL and the RL controllers offered significant enhancement of ride comfort which proves the high potential of Deep Learning methods in the field of active suspension control. The H_∞ control performance appeared inferior to that of the skyhook, SL and DL controllers which can be improved by introducing additional tuning channels such as shaping functions.

Looking into the methodologies, since H_∞ and SL controllers oblige state space formulation, it is a concern that the chosen state measurements might not be available in production vehicles. However, using neural networks in conjunction with the MPC model offers some freedom in the selection of required measurements i.e., when training the neural network, the inputs in the training data need not be the same as the state measurements. Instead, the state measurements can be replaced with measurements which can be measured with the help of stock sensors. On the other hand, the RL controller does not involve the state-space approach to formulate the control objectives. This offers great freedom in using complicated system models such as nonlinearities making use of look-up tables. Such models can be readily incorporated to make the approach more realistic and obtain a more intelligent agent.

The future work might deal with strengthening of these approaches by using frequency dependent shaping functions to model the human sensitivity to vibrations. The performance can be selectively enhanced by introducing these shaping functions while modelling the controller. Further, the approach can be consolidated by taking actuator dynamics into account. However, including shaping functions and actuator dynamics necessitate additional state measurements which might make the approach complex by necessitating the introduction of state estimators. Moreover, the approaches can be extended to full vehicle models where the effects of pitch and roll on the vertical dynamics can be modelled and analysed. Reinforcement Learning, which is rather new in the field of suspension controls, offers a lot of scope in validating the potential obtainable by other algorithms like SARSA, Actor-Critic methods as well as other model-based RL algorithms.

References

[1] L. Eckstein, Vertical and Lateral Dynamics of Vehicles, Aachen, 2014.

[2] S. Savaresi, C. Poussot-Vassal, C. Spelta, O. Sename and L. Dugard, Semi-Active Suspension Control Design for Vehicles, Elsevier, 2010.

[3] G. Zames, "Feedback and Optimal Sensitivity: Model Reference Transformations, Multiplicative Seminorms, and Approximate Inverses," IEEE Transactions on Automatic Control, pp. 301-320, April 1981.

[4] MathWorks Inc., "Robust Control of an Active Suspension," 2019. [Online]. Available: https://de.mathworks.com/help/robust/gs/active-suspension-control-design.html.

[5] H. Chen and K.-H. Guo, "Constrained H∞ Control of Active Suspensions: An LMI Approach," IEEE Transactions on Control Systems Technology, pp. 412-4221, 2005.

[6] N. M. Ghazaly, A.-N. Sharkawy, A. S. Ali and G. Abdel-Jaber, "H∞ Control of Active Suspension System for a Quarter Car Model," International Journal of Vehicle Structures and Systems, pp. 81-87, January 2016.

[7] M. Yamashita, K. Fujimori, K. Hayakawa and H. Kimura, "Application of H∞ control to active suspension systems," IFAC Proceedings Volumes, pp. 87-90, 1993.

[8] A. Alessio and A. Bemporad, "A Survey on Explicit Model Predictive Control," in Nonlinear Model Predictive Control, Berlin, Heidelberg, Springer, 2009, pp. 345-369.

[9] K. Kouramas, N. Faísca, C. Panos and E. Pistikopoulos, "Explicit/multi-parametric model predictive control (MPC) of linear discrete-time systems by dynamic and multi-parametric programming," Automatica 47, pp. 1638-1645, 2011.

[10] L. H. Cseko″, M. Kvasnica and B. Lantos, "Analysis of the explicit model predictive control for the semi-active suspension," Periodica Polytechnica, pp. 41-58, 2010.

[11] J. Theunissen, A. Sorniotti, P. Gruber, S. Fallah, M. Dhaens, K. Reybrouck and C. Lauwerys, “Explicit model predictive control of active suspension systems,” Proceedings of the International Conference on Advanced Vehicle Powertrains, pp. 344-362, 2017.

[12] R. Dessort and C. Chucholowski, “Explicit model predictive control of semi-active suspension systems using Artificial Neural Networks (ANN),” 8th International Munich Chassis Symposium, 2017.

[13] V. Mnih, K. Kavukcuoglu, D. Silver, A. Graves, I. Antonoglou, D. Wierstra and M. Riedmiller, “Playing Atari with Deep Reinforcement Learning,” arXiv preprint arXiv:1312.5602, 2013.

[14] R. S. Sutton and A. G. Barto, Reinforcement Learning: An Introduction, London, England: The MIT Press, 2017.

[15] G. Frost, T. J. Gordon, M. N. Howell and Q. H. Wu, “Moderated Reinforcement Learning of Active and Semi-Active Suspension Control Laws,” Proceedings of the Institution of Mechanical Engineers, Part I: Journal of Systems and Control Engineering, p. 249–257, 1 November 1992.

[16] M. Howell, G. Frost, T. Gordon and Q. Wu, “Continuous action reinforcement learning applied to vehicle suspension control,” Mechatronics, pp. 263,276, 1997.

[17] S. Tognetti, S. M. Savaresi, C. Spelta and M. Restelli, “Batch Reinforcement Learning for Semi-Active Suspension Control,” 18th IEEE International Conference on Control Applications, pp. 582-587, 8-10 July 2009.

[18] I. O. Bucak and H. Oz, “Vibration control of a nonlinear quarter-car active suspension by reinforcement learning,” International Journal of Systems Science, 2011.

[19] G. Koch, E. P. S. Spirk and B. Lohmann, “Design and Modelling of a Quarter-Vehicle Testrig for Active Suspension Control,” Technical Reports on Automatic Control, 21 July 2010.

[20] M. F. Soong, R. Ramli and A. Saifizul, “Between simplicity and accuracy: Effect of adding modeling details on quarter car vehicle model accuracy,” PLoS ONE 12(6), 2017.

[21] C. Zhang, O. Vinyals, R. Munos and S. Bengio, "A Study on Overfitting in Deep Reinforcement Learning," CoRR, 2018.

[22] J. C. Doyle, K. Glover, P. P. Khargonekar and B. A. Francis, "State-Space Solutions to Standard H2 and H∞ Control Problems," IEEE Transactions on Automatic Control, pp. 831-847, August 1989.

[23] M. Grant and S. Boyd, "CVX: Matlab Software for Disciplined Convex Programming, version 2.1," March 2014. [Online]. Available: http://cvxr.com/cvx.

[24] M. Grant and S. Boyd, "Graph implementations for nonsmooth convex programs," in Recent Advances in Learning and Control, Springer-Verlag Limited, 2008, pp. 95-110.

[25] L. Wang, Model Predictive Control System and Design Implementation using MATLAB, Springer, 2008.

[26] MathWorks, Inc., "Optimization Toolbox," 2019. [Online]. Available: https://de.mathworks.com/help/optim/index.html.

[27] MathWorks, Inc., "Deep Learning Toolbox," 2019. [Online]. Available: https://de.mathworks.com/help/deeplearning/.

[28] A. Gosavi, Simulation-Based Optimization: Parametric Optimization Techniques and Reinforcement Leanring, New York, NY: Springer, 2014.

[29] F. Gustafsson, Control of Inverted Double Pendulum using Reinforcement Learning, 2016.

[30] L. Zuo and S. Nayfeh, "Low order continuous-time filters for approximation of the ISO 2631-1 human vibration sensitivity weightings," Journal of Sound and Vibration 265, pp. 459-465, 2003.

[31] E. Frazzolli, 6.241 Dynamic Systems and Control, Lecture 25: H∞ sysnthesis, Lecture Slides, Massachusetts Institute of Technology, 2011.

Deep learning for lateral vehicle control – an end-to-end trained multi-fusion steering model

Johann Haselberger
University of Applied Sciences Kempten

Dr. Jian Chen
Porsche Engineering Services GmbH

Prof. Bernhard Schick
University of Applied Sciences Kempten

Abstract

Deep Learning based behavior reflex methods found their way into modern vehicles. To model the human driving behavior it is not sufficient to rely solely on individual, non-contiguous camera frames without taking vehicle signals or road specific features into account. In this work four temporal fusion methods are evaluated based on three different Deep Learning models. The proposed spatio-temporal Mixed Fusion model extends the present end-to-end models and consist of multiple levels of fusions. The raw image data from a single front facing camera is mixed with recorded vehicle data and a map based predicted road bank angle gradient vector. The model accesses multiple time axes: temporal features of multiple image frames are extracted through a combination of Convolution and LSTM layers while it can also make assumptions about the future road condition with the use of upcoming Ground Truth road bank angle changes. Experiments are performed on a recorded data set of real world drivings. Results show, that this approach leads to an accurate imitation of the human driver with an inference capability of more than 60 FPS.

1 Introduction

Due to a number of interlocking trends, there has been much progress in the field of Deep Learning powered driver assistance systems and highly automated pilot functions in recent years. On the basis of an exploding number of innovative deep learning methods especially in the field of computer vision and the availability of automotive high performance computing platforms, these methods found their way into modern vehicles. Those systems can be grouped into two major categories - mediated perception and behavior reflex approaches.

Traditionally established systems are based on a clear division of the driving function into several separate sub-tasks and are summarized as mediated perception approaches [1]–[3]. These partial tasks include sensor fusion, perception of the environment, localization and planning of the maneuvers or trajectories to follow. Rule-based algorithms are used to compute control commands from generated environment models with predefined parameters. While this framework ensures smooth vehicle control, it is challenging to determine the needed, optimal rules [4], [5].

Especially in the area of visual data processing, Deep Learning supports the conventional, humanly engineered computer vision approaches. A good example poses the estimation of precise lane positions using a neural network as presented by Gurghian et al. [6]. In addition in [7] a novel multi network approach is introduced, where pixel-wise lane segmentations are mapped to instance aware lane fits.

Besides finding lane markings, it is also important to detect, classify and track other road vehicles, pedestrians, traffic lights or objects on the road. Many Deep Learning based classifiers have shown to be able to match or even outperform human visual capabilities in particular tasks.

In contrast to that, so-called behavior reflex approaches [1] follow a completely different way of thinking. Instead of dividing an autonomous driving function into sub-tasks, system inputs are used directly to derive the desired control output based on a data learned policy. As these systems are often trained on the raw sensor inputs without any further pre- or postprocessing, the term end-to-end learning is commonly used. This concept is closer to human behavior, as drivers make their decisions heavily based on visual impressions. Therefore the phrase visuomotor control is applied in the field of robotics.

These data driven model-free methods can be transferred more easily to other environments or tasks in contrast to the rule-based mediated perception techniques, if sufficient training data is present [8]. In addition, uncertainties or data noise can be handled internally [9].

In this work, the focus is set to the appliance of end-to-end trained steering policies for lane-keeping on country roads and highways. The majority of existing models are based on a mapping of spatial visual features of merely a single, individual camera frame to a steering angle command. Especially changing lighting and road conditions on curvy roads however require the inclusion of temporal dependencies in form of multiple contiguous frames. Therefore a knowledge transfer from the nearby research area of video classification is performed. In addition, particular vehicle signals and predictive map based road bank angle changes are integrated into the end-to-end learned framework.

The reminder of this paper is organized as follows. In Section 2 a brief overview of topic-related preliminary work is given. The followed methodology is described in Section 3. Sections 4 and 5 cover the introduced steering neural networks as well as the generated and used data set. Afterwards, the achieved results are presented in Section 6. This work is completed by a resume in Section 7.

2 Related Work

Surprisingly, the pioneering work for automotive behavior reflex methods has already been done more than three decades ago. In 1989 Pomerleau et al. [10] built the ALVINN system, Autonomous Land Vehicle in a Neural Network, which was based on a fairly simple MLP architecture with only 29 hidden units. Steering commands were derived directly from the system inputs. This idea was taken up by NVIDIA, which presented a functional demonstrator, the DAVE-2 system [11], by using a CNN architecture, archiving approximately 98 % autonomy.

Based on this, multiple subsequent works focused on end-to-end trained steering policies. In [4] the system is extended to a multi-task application for steering and velocity inference. While the speed prediction can utilize temporal informations through a single LSTM layer, the steering prediction is done using an adapted Version of PilotNet without any temporal feature extraction. Also [12], [13] and [14] used a PilotNet derivate to offline train an end-to-end steering controller. In addition Jhung et al. [12] applied Reinforcement Learning to finetune the network feasible of handling partially observable circumstances during online usage.

While these approaches overall form their steering decisions exclusively on the basis of the current observation of the situation, in several subsequent works the temporal features are also considered. Therefore a combination of Convolution and LSTM layers is often adopted [15]–[20]. A special feature in [17] is the application of feature mimicking from auxiliary networks to improve the accuracy and robustness of the steering inference. Yuan et al. extend the Conv-LSTM architecture by mapping the network parameters similar to a state-space-controller to different driving situations [21].

All the models mentioned above have in common that they are strictly based on visual network inputs. To extend this, in [22] the vehicle wheel speeds are used together with a single front facing camera to mimic the behavior of a model predictive controller for dynamic offroad driving both for steering and accelerating. Hubschneider et al. [23] demonstrated the positive impact of including the vehicle's indicator signals into an end-to-end trained steering model for obstacle avoidance or lane changes. Instead of a temporal inclusion in form of multiple time-related frames, a spatial history was used. In [24] a 360° surround-view is archived by placing multiple cameras on top of the vehicle roof. In addition route planner information realized through a number of stacked GPS points or screenshots of the rendered planned route are added to the models input, while the image data only (cameras and map screenshots) approach tends to work better. Another, purly signal based method for coupled longitudinal and lateral control is explored in [25]. However no temporal features were extracted here within the proposed MPL and CNN architectures for the trajectory tracking task.

When it comes to extracting temporal dependencies between sequential images, video classification is the closest research area. In [26] Karpathy et al. summarizes the four fusion methods to extend the connectivity of convolutional neural networks in the time axis: Single Frame, Late Fusion, Early Fusion and Slow Fusion. A visual comparison is given in Figure 1. Pierre et al. [8] apply these four methods to perform robotic visuomotor control using the example of leader following. In contrast to this work experiments were performed within a simplified simulation and laboratory environment with mobile robots. In addition, the network input consists solely of image data.

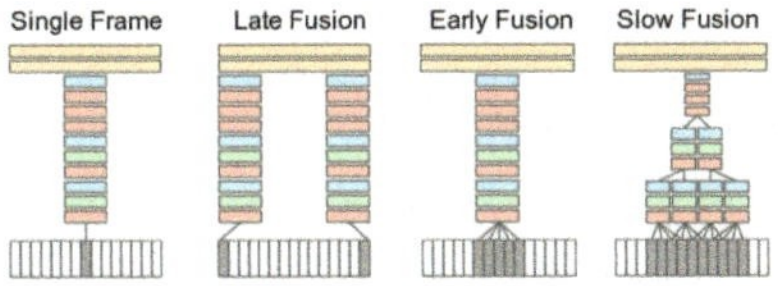

Figure 1: Overview of the temporal fusion methods presented by [26]. The different types of layers are represented with red, green and blue boxes. The fusion takes place within the yellow layers.

Due to the fact, that a lane keeping system or the modeling of human steering behavior deals with the vehicle's lateral movement, all external factors which apply lateral forces on the car should be considered. In [27] the impact of the road bank angle on the lane keeping quality is extensively derived. Results show, that the inclusion of the measured bank angle into the control loop of a given controller increases the system accuracy. However only the currently observed banking angle was used.

In this paper, however, this is extended by including a vector of future road bank angle changes based on a Ground Truth map and selected vehicle dynamic signals into the end-to-end trained steering model. This step is based on the assumption, that a human driver monitors the course of road changes and adapts the steering requests in a feed forward control manner. This form of a mixed fusion of spatio-temporal image features, vehicle signals and precise road characteristics, especially the bank angle, has not been covered in the research area of end-to-end learned steering models yet.

3 Methodology / Approach

At a high level perspective, the trained network can be interpreted as a function that allows to predict the control output Y_t at a given timestep t throughout its learned network parameters Θ based on the current observation state X_t:

$$Y_t = f(\Theta, X_t) \tag{1}$$

The parameters are learned by gradient descent backpropagation using the loss function L and the learning rate λ,

$$\Theta \rightarrow \Theta - \lambda \frac{\delta L\left(\tilde{Y}; f\left(\Theta, \tilde{X}\right)\right)}{\delta \Theta} \tag{2}$$

utilizing tuples $\left\{\tilde{X}, \tilde{Y}\right\}$ of the recorded training dataset.

While a restricted visual, single frame based mapping implies, that the network input X_t consists solely of the image i_t, in this work a more extended composition is assumed:

$$X_t = \{I_t, S_t, \dot{Q}_t\} \tag{3}$$

where I_t is defined as a vector of buffered frames $(i_t, i_{t-s}, \ldots, i_{t-s(\omega-1)})$, S_t as a collocation of the selected vehicle signals: velocity, lateral acceleration and yaw rate $(v_t, a_{y,t}, \dot{\psi}_t)$ and $\dot{Q}_t$ as the predictive road bank angle gradient $(\dot{q}_t, \dot{q}_{t+s_q}, \ldots, \dot{q}_{t+s_q(\omega_q-1)})$. As a result the network input consists of both past and future observations. The time window of this review or preview is defined by the parameters ω and ω_q. To introduce a strided sampling of the vectors S_t and $\dot{Q}_t$, s and s_q can take positive integer values. The network output Y_t on the other hand is simply set to the target control signal z_t, the steering rack position. In Figure 2 the whole system overview is illustrated. It should be emphasized that the two blocks feature extraction and rack position estimation are combined into a joined Deep Learning Model and trained in a end-to-end manner.

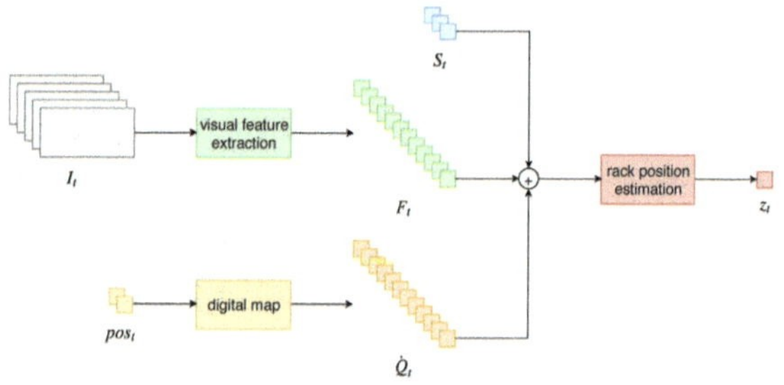

Figure 2: Global overview of the proposed end-to-end trained steering model. A defined number of camera frames are bundled within I_t, $\dot{Q}_t$ defines the vector of predicted road bank angle changes based on location related map data. The vehicle signals v_t, a_t and $\dot{\psi}_t$ are grouped within S_t. The proposed network fuses the feature extraction and rack position estimation into a single end-to-end trained model with the input $X_t = \{I_t, S_t, \dot{Q}_t\}$.

The whole steering angle prediction problem can either be defined as a classification or regression task. If the choice comes down to classification, the steering signal has to be quantized and grouped with a given granularity, while a regression network directly outputs the inferenced steering signal. Experiments in [28] have shown, that the regression approach outperforms the classification method - therefore this realization is followed.

When it comes to regression, there are basically two possible validation and loss metrics, the mean squared error (MSE) and mean absolute error (MAE):

$$MSE = \frac{1}{n}\sum_{i=1}^{n}(z_i - \tilde{z}_i)^2 \tag{4}$$

$$MAE = \frac{1}{n}\sum_{i=1}^{n}|z_i - \tilde{z}_i| \tag{5}$$

By squaring the errors, MSE assigns relatively high importance to large deviations between the recorded steering rack position $\tilde{z}_i$ and the predicted position z_i. Intuitively, it seems logical that during a real drive it is harder to recover from a few big errors than from many minor ones. For this reason, MSE is our loss function of choice for the gradient update of all proposed models and training iterations.

Besides MSE and MAE two more evaluation metrics are used which are more related to real in-car usage of the trained models. As the predicted steering rack position z has to be processed further, eg. by some sort of controller, output fluctuations of directly consecutive presented network inputs X_t and X_{t+1} should be minimized. Therefore the metric mean absolute fluctuation (MAF) is calculated to:

$$MAF = \frac{1}{n-1}\sum_{i=1}^{n-1}|z_{t+1} - z_t| \tag{6}$$

When it comes to real time performance, the needed time to map the system input X_t to the defined output Y_t has to be considered. As an evaluation criterion, the mean processing time (MPT) and its reciprocal value, the mean processing frequency (MPF) is used:

$$MPF = \frac{1}{MPT} = \frac{1}{\frac{1}{n}\sum_{i=1}^{n} t_{X_n \to Y_n}} \tag{7}$$

In this paper the following approach is chosen: first the feature extraction block is examined by evaluating the different image fusion methods Single Frame, Early Fusion, Slow Fusion and Late Fusion. After that the influences of inclusion of the road bank angle gradient $\dot{Q}_t$ and the vehicle signal vector S_t are investigated independently. Finaly a Multi Fusion model is derived which utilizes spatio-temporal image data in combination witch the vehicle signals S_t and the road data $\dot{Q}_t$.

4 Baseline Models

In this work, three different main architectures are introduced, whose global layer configurations stay constant for all single frame, late fusion, early fusion and slow fusion experiments. As a convention, all dimensions are represented in the form $T \times h \times w \times c$, where h and w define the image height and width. Channels are denoted with c, T marks the number of frames for the slow fusion approach. However, in case $T = 1$ this dimension is dropped.

The common objective of all networks is to map the current observation state X to a steering rack position z based on the learned human behavior. For the following descriptions of the three DNN variations, the general simplification is made, that the network input consists only of a single RGB camera frame I with the shape $66 \times 198 \times 3$ - thus only the single frame models are illustrated as a baseline.

Besides this, the networks share further features. Every convolution and dense layer is followed by an RELU activation function and a batch normalization layer. Exception to this is the last layer - as it forms the subsystem output neither of the two is used. Overall, there were no Dropout layers applied within any of the network architectures.

4.1 PilotNet

The first model to be described is the PilotNet architecture, introduced by Bojarski et al. for their autonomous DAVE-2 system [11]. This model was chosen not only because it was one of the first modern end-to-end trained steering models, but also because it provides a well known comparative baseline. The network consists of five convolution and three fully connected layers in total. The whole architecture and the layer output shapes are unrolled in Table 1. The top three 2D convolutions extract the image features with quadratic filter kernels K of size 5 and a stride S set to 2. For the last convolution layers the kernel size was reduced to 3 and the stride set to 1.

Similar to classic CNNs the convolution layers are followed by fully connected layers, which could be interpreted as the steering controller function. But as mentioned by Bojarski et. al. a clear cut between feature extraction and control could not be made as the system is trained end-to-end. Counting up all the network parameters leads to 253 thousand trainable weights and biases.

4.2 SimpleNet

The second architecture with about 151 thousand trainable parameters is much simpler than PilotNet, thus called SimpleNet in the following, and partly inspired by the CG23

layer	parameters	output shape
Input I		$66 \times 198 \times 3$
2D Conv	$F = 24, K = 5, S = 2$	$31 \times 97 \times 24$
2D Conv	$F = 36, K = 5, S = 2$	$14 \times 47 \times 36$
2D Conv	$F = 48, K = 5, S = 2$	$5 \times 22 \times 48$
2D Conv	$F = 64, K = 3, S = 1$	$3 \times 20 \times 64$
2D Conv	$F = 64, K = 3, S = 1$	$1 \times 18 \times 64$
Flatten		1152
Dense	$N = 100$	100
Dense	$N = 50$	50
Dense	$N = 10$	10
Dense	$N = 1$	1

Table 1: Single frame PilotNet architecture

layer	parameters	output shape
Input I		$66 \times 198 \times 3$
2D Conv	$F = 16, K = 7, S = 3$	$20 \times 64 \times 16$
2D Conv	$F = 32, K = 5, S = 3$	$6 \times 20 \times 32$
2D Conv	$F = 64, K = 3, S = 2$	$2 \times 9 \times 64$
Flatten		1152
Dense	$N = 100$	100
Dense	$N = 20$	20
Dense	$N = 1$	1

Table 2: Single frame SimpleNet architecture

Net [29] which was proposed in the Udacity Self-Driving Challenge. Notwithstanding, some changes have been made. Instead of a consitent kernel size of 3 a larger kernel of size 7×7 is implemented. As mentioned by Yang et al., lager kernels improve the model's ability to capture lanekeeping relevant environment features [4]. Per convolution layer the parameter K is decreased by 2. The full architecture is listed in Table 2.

Similar to [8], no Max Pooling layers are used intentionally. Two dimensional max pooling does not only reduce the spatial size of a given input tensor, it is also used in classification networks to introduce translation invariance. However, the lane keeping task heavily depends on horizontal shifts within the input frames. To downsample the activation maps, each convolution layer uses a stride $S = 3$. Figure 3 illustrates the transformation of the input image to a more and more vector like representation. The author of CG23 Net mentioned, that the original model suffers from overfitting, as a large number of parameters are nested within the first fully connected layer. To overcome this, SimpleNet consists of two dense layers with only 100 and 10 neurons instead of one large 1024 neurons wide layer.

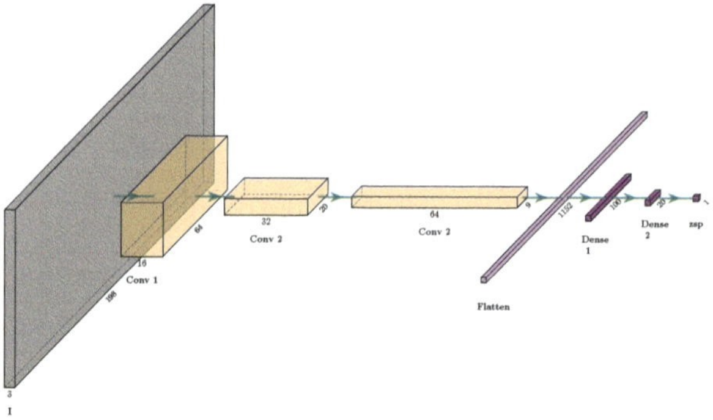

Figure 3: Layer output dimensions of the single frame **SimpleNet** architecture. Only the shapes of input, convolution, flatten, dense and output layers are shown.

4.3 MiniInceptionNet

As PilotNet and SimpleNet are both straight sequential models without any inner branches, a third architecture based on the famous InceptionV1 network [30] with parallel grouped convolutions is introduced. As InceptionNet originally was designed to solve classification problems with more than 1000 different classes, like on the ImageNet dataset, a much reduced architecture presented by Zhangb et al. [31], which was adapted for the CIFAR10 dataset, containing only 10 different classes is used within this survey. In the following, this model is referred as MiniInceptionNet.

The whole model consists of a combination of the three building blocks - a Convolution, Inception and Downsampling module. The Convolution block consits of a single 2D convolution followed by a batchnormalization and activation layer. In order to create a deeper network, that can extract more profound image features, more convolution layers have to be added to a given model. But instead of simply chaining these layers in a sequential manner, multiple convolution layers get combined within the Inception block in parallel. Letting filters operating on the same level makes the resulting model slightly wider instead of deeper, thus helping the gradient updates pass through the entire network.

While four up to seven convolution layers are used in the InceptionV1 architecture, only two of them with fixed kernel sizes $K_1 = 1$ and $K_2 = 3$ are used in the adapted MiniInceptionNet version. No stride was used within the Inception modules. The Downsample module of original MiniInceptionNet has been modified by dropping the Max Pool operation completely - following the same reasoning as mentioned in the description of SimpleNet. To end up with a closely related output shape, the number of filters are

layer	parameters	output shape
Input I		$66 \times 198 \times 3$
2D Conv	$F = 32, K = 3$	$64 \times 196 \times 32$
Inc Block	$F = 16$	$64 \times 196 \times 32$
Inc Block	$F = 16$	$64 \times 196 \times 32$
Downsize	$F = 64$	$31 \times 97 \times 64$
Inc Block	$F = 32$	$31 \times 97 \times 64$
Inc Block	$F = 32$	$31 \times 97 \times 64$
Downsize	$F = 128$	$15 \times 48 \times 128$
Inc Block	$F = 64$	$15 \times 48 \times 128$
Downsize	$F = 128$	$7 \times 23 \times 128$
Avg Pool	$P = (3 \times 5)$	$2 \times 4 \times 128$
Flatten		1024
Dense	$N = 100$	100
Dense	$N = 1$	1

Table 3: Single frame MiniInceptionNet architecture

adjusted, as seen in Table 3. The filter sizes F_1 and F_2 are chosen to be identical, resulting in $F_1 = F_2 = F$. To reduce the model complexity even further, two of four middle and one of the last inception blocks are skipped. With about 480 thousand trainable parameters, MiniInceptionNet still poses the biggest DNN-model in this work.

5 Driving Dataset

In this section, the dataset used to train the described models is presented. Although there are already open data sets available, the decision was made to record a own training data set. This choice is based on the fact that the image material and the vehicle data is highly vehicle-specific - inconsistent camera setups could lead to non-transferable results [23].

5.1 Data Acquisition

For recording the NVIDIA PX2 platform was used. In order to maintain the series condition of the vehicle, only an additional front camera with a 60° horizontal field of view was installed. It was not possible to access the raw sensor frames of the existing camera on the PX2. The camera frames were recorded with the highest available framerate of 30 fps, an image resolution of 1928×1208 pixels and using the RGB colorspace. All other signals were captured from the internal FlexRay bus.

Summing up all test drives results in roughly 7 hours of unfiltered driving data under various lighting and weather conditions. The majority consists of driving on country roads and motorways, inner city roads were not considered.

5.2 Dataset generation

Due to the fact, that the network target signal, the steering rack position z, can be recorded directly, no time consuming data labeling is required. However the raw driving data is analyzed to filter out some unusable training images, like overexposed frames after tunnel exits or unexpected avoidance maneuvers on construction sites.

As the vehicle signals are recorded with a higher rate, these signals get downsampled to the camera framerate to provide a consistent time axis. Within the next step, the driving data gets filtered. First all images marked as unusable and frames that belong to situations with activated indicators are discarded. In addition, frames recorded at a velocity below 40 km/h are dropped as well, in order to capture only relevant driving conditions.

While the network inputs I_t ans S_t can be directly derived from the measurement recordings, the vector $\dot{Q}_t$ has to be calculated based on the current vehicle position pos_t. Prerequisite for this is a digitized road map. In a previous work, grid data for selected road sections with equidistant measuring points was created [32]. As a result, for road points with a given distance between each other, the road bank angle is known. In order to transfer the path-based map data into the time-based road bank angle preview $\dot{Q}_t = (\dot{q}_t, \dot{q}_{t+s_q}, \dots, \dot{q}_{t+s_q(\omega_q-1)})$, the current vehicle speed v_t and its acceleration $a_{x,t}$ are used.

There are mainly two kinds of strides and time windows used within the data pipeline. First, the image data within each training example is aranged as a vector of contiguous camera frames $I_t = (i_t, i_{t-s}, \dots, i_{t-s(\omega-1)})$, therefore s defines the amount of time between each frame. Based on the human reaction time of 100 ms [19], this parameter is set to 0.1 s. The time window ω defines the amount of frames used within a single network input X_t and is set to a maximum value of 26, resulting in 2.5 s available for a single training example. For the predictive road bank angle gradient s_q is set to 0.25 s and ω_q to 12 elements. Therefore the prediction covers 2.75 s.

While these strides are applied for the elements within each network input X_t a global stride s_{FPS} is used to sample the datapoints itself. After recording, the datapoints have uniform spacings of 0.033 s, which corresponds to 30 FPS. It is common to reduce the global framerate to under 10 FPS, as the visual changes within two successive frames recorded with 30 FPS are too minor to use them for training [11], [19], [24]. In this work s_{FPS} is set to 5 to generate a framerate of 6.

A simple, commonly used technique to even out the amount of right and left courves is to flip the images horizontally and invert the steering signal respectively [15], [19], [33], [34]. This approach minimizes the risk of introducing a bias towards a specific steering direction within the dataset itself. However, it does only make sense to perform this data manipulation on driving situations with clear lanemarkings on both sides. Otherwise, the learned steering controller would not be able to differ between right-hand and left-hand traffic, which could result in undesired behavior.

Besides flipping, multiple image augmentation approaches to enhance the system robustness under varying lighting circumstances are adopted. Therefore artificial camera effects like brightness changes, color shifts, random noise, gaussian blur or sharpening are applied randomly with alternating, normally distributed, magnitudes. Nevertheless, their maximum strength must be carefully chosen. As proofed by Du et al. heavy image manipulations bear the risk of negatively influencing the training process [15]. Therefore, only up to two camera effects are applied simultaneously on 25 % of the training data.

Even after applying the horizontal flips, the resulting dataset is highly imbalanced as large steering angles are rare compared to driving straight ahead. To overcome this, the whole dataset is clustered into multiple bins of related steering rack position values. The maximum number of elements n_{max} within the bin containing the highest (or lowest) values of z is used as reference. From each bin n_{max} elements are sampled randomly, which led to a balanced dataset with no bias towards any steering direction.

Normalization is a important step towards a good model training fit, therefore all the network inputs are scaled to be in range $[-1.0; 1.0]$. Since some pixels at the bottom and top of the camera frame contain parts of the cars engine hood and the sky, a region of interest crop is performed, because these pixels do not contribute to the network output calculation. To reduce the computational effort even further, the camera frames are then resized to 66×198 RGB images.

Finally, the dataset gets split randomly into three sub-data sets with 85 % for training, 10 % for validation and the remaining 5 % for testing. The training and validation data is also shuffled in order to reduce variance and the chance to overfit.

6 Experiments

All experiments are based on the same training, validation and test dataset. In addition every training run uses the same hyperparameters. As optimizer ADAM [35] is used with an adjusted learning rate of 5e−5, the parameters β_1 and β_2 were set unchanged to 0.9 and 0.999. The batch size n was defined to 128 samples. Training was stopped after 240 epochs to ensure every model has the chance to convergence. After every epoch the models get tested using the validation set in order to save only the best model parameters in terms of the validation error.

Since the resulting steering commands of the regression task are not homogeneously over time by their very nature, but suffer from many small jumps, single exponential smoothing [36] is applied similar as in [4], [29], [37]:

$$\hat{z}_t = \alpha z_t + (1 - \alpha)\hat{z}_{t-1} \tag{8}$$

where $\hat{z}_t$ represents the current smoothed steering rack position based on the network output z_t, the previous smoothed position $\hat{z}_{t-1}$ and the smoothing factor α. In our experiments, α is set to a constant value of 0.25. As the smoothed rack position only depends on the current and past positions, this type of filter is also implementable for real time car usage, since no future values are needed.

However, a smoothing in general is only usable, if the underlying signal is presented in its right chronological order, which is not the case for the shuffled validation set. Based on this, single exponential smoothing is only applied for the test set, used for the graphical interpretation - thus the validation results summarized in each experiment table are not affected. All performance measurements are carried out on the NVIDIA Drive PX2 platform.

6.1 Single Frame

For the single frame experiment, the parameter ω is set to 1, thus only the camera image I of the most current observation x_0 of the input vector X is used. For this reason, the three models could be used unchanged as described in chapter 4. The results are summarized in Table 4.

model	**MSE** mm^2	**MAE** mm	**MAF** mm	**MPF** $1/s$
PilotNet	**0.233**	**0.354**	0.259	99
SimpleNet	0.269	0.383	**0.220**	**136**
MiniInceptionNet	0.263	0.391	0.278	61

Table 4: Results of the single frame experiment

For this experiment, PilotNet has outperformed all other models in terms of averaged accuracy, while SimpleNet provides the lowest MAF. The overall relatively high fluctuations underline the necessity of filtering the output values. The visual comparison also shows that the different network can imitate the overall human behaviour well.

However, as stated in [18] and [38] a 100 % model fit is impossible to achieve. One of the reasons is that the prospective intention of the human driver is unknown to the DNN-model. On the other hand, the network training process tries to find a global optimized solution, thus minimizes the MSE-loss function on the entire training data set. In contrast to that, the actual driving behaviour of the human being does not always

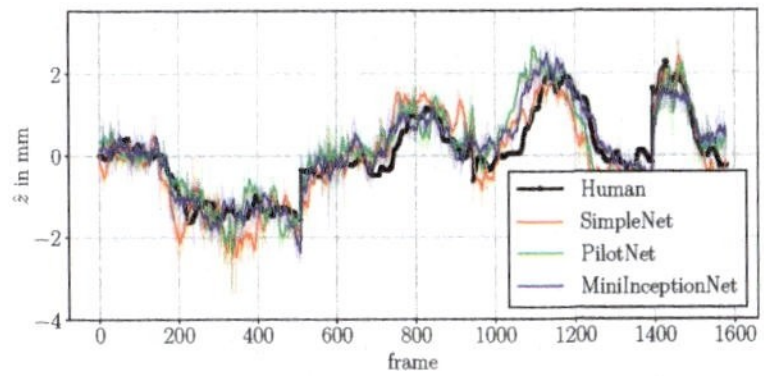

Figure 4: Comparison of the steering rack positions z of the human driver and the DNN **Single Frame** inferences of PilotNet, SimpleNet and MiniInceptionNet.

represent a global ideal solution. These contrary aspects apply not only to the Single Frame experiment, but also to all other test series and should be kept in mind.

On the basis of a limitation to 30 FPS on the camera side, all three models could easily predict the steering rack positions in realtime. SimpleNet thereby sets the benchmark with 136 frames per second, which corresponds to an average processing time of 7.35 ms. Due to the increased number of layers, MiniInceptionNet suffers from it's increased complexity. With an MPF value of 61 this architecture needs more than twice as long for a single inference step.

6.2 Early Fusion

The basic procedure behind early fusion is to extend the temporal input view of the networks by concatenating a number of subsequent input frames. The quantity is defined by the parameter ω. To give the models a suitable decision range, the resulting time window is set to half a second. Therefore ω has to be set to five concerning a data set related, predefined stride of $s = 3$. However, ω is not extended to higher values as Heylen et al. [28] summarizes in previous experiments, that a larger input window leads only to small improvements or even declining results. There is also a potential risk of a loss of the system-responsiveness.

The network input now consists of the first five elements of the data set vector X. The images of the observation states are then concatenated in the channel dimension c, resulting in a network input shape of $h \times w \times \omega c$. Stacking up five frames of $66 \times 198 \times 3$ pixels results in a input tensor of shape $66 \times 198 \times 15$.

Compared to the single frame approach, only the channel depth of the network input has changed, no new dimensionality has been added. This allows the baseline models to stay nearly unchanged, except of the first 2D convolution layer. As its filter kernels taking the whole input depth into account, these filters expand by factor ω accordingly, which leads to a small increase of the trainable network parameters for all the models.

All the introduced temporal information is fused at this layer - time related characteristics can then be propagated within the generated feature maps to the next layers. The evaluation results of the early fusion approach are listed in Table 5.

model	MSE mm^2	MAE mm	MAF mm	MPF 1/s
PilotNet	0.143	0.278	0.183	86
SimpleNet	**0.129**	**0.267**	**0.134**	**110**
MiniInceptionNet	0.223	0.357	0.232	55

Table 5: Results of the early fusion experiment

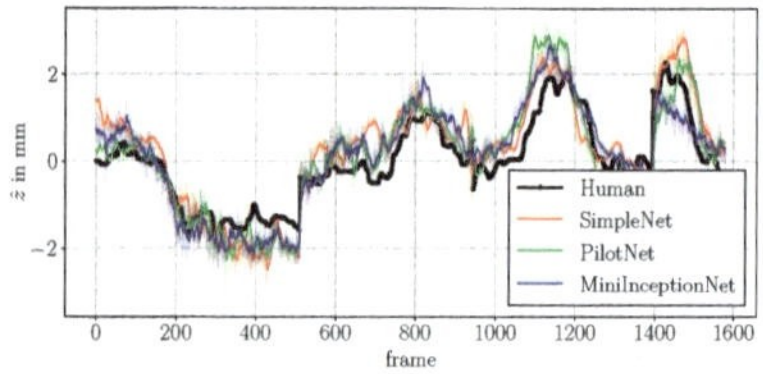

Figure 5: Comparison of the steering rack positions z of the human driver and the DNN **Early Fusion** inferences of PilotNet, SimpleNet and MiniInceptionNet.

Compared to the Single Frame results, all three model architectures clearly benefit from the additional time information in the form of several input pictures. Interestingly, this time the SingleFrame model beats the stacked PilotNet version and and can more than bisect the *MSE* value. The global fluctuations of all models have been decimated, which is a clear sign, that the models can make profound sense of the temporal features. However, this increase in performance is bought with a slight increase in computing time. At this point MiniInceptionNet falls below the 60 FPS limit.

6.3 Slow Fusion

The slow fusion approach poses a balanced middle course between late and early merging of input camera frames. Similar to the previous experiment ω was set to a length of five elements. But instead of concatinating the frames in a channel wise manner, a fourth tensor dimension is added - the expected network input shape now turns to $T \times h \times w \times c$, in this case $5 \times 66 \times 198 \times 3$.

The introduction of an additional dimension requires a drastic change of all the proposed network architectures. Every 2D convolution layer is replaced by its three dimensional counterpart. To be consistent, all the spatial dimensions as described in section 4 are kept.

With the use of 3D convolutions, the temporal information could be fused progressively throughout the network with different levels of granularity as the their filter kernels providing a additional „time"-dimensionality. Hierarchical representations of spatio-temporal data is created directly [39]. In addition Pierre et al. [8] points out, that this approach offers the model some amount of invariance of translational and scale changes along the temporal axis.

In the PilotNet model, the kernel shapes of the convolution layers were adjusted to $(2 \times 5 \times 5)$, $(2 \times 5 \times 5)$, $(2 \times 5 \times 5)$, $(2 \times 3 \times 3)$ and $(1 \times 3 \times 3)$. Together with a constant temporal stride of one, the output length of the Flatten layer stays unchanged at 1152.

The convolution layers of the SimpleNet has been changed to $(3 \times 7 \times 7)$, $(2 \times 5 \times 5)$ and $(1 \times 3 \times 3)$, also no temporal stride is applied. Similar to PilotNet, the resulting Flatten output length is 1152.

Within the MiniInceptionNet architecture no additional temporal stride is used either. Every Inception module applies a constant temporal kernel size of one. However, to downsample the time axis only the first convolution layer and every Downsampling module is adjusted. The convolution layer uses a kernel size of $(3 \times 3 \times 3)$, the Downsampling modules $(2 \times 3 \times 3)$, $(2 \times 3 \times 3)$ and $(1 \times 3 \times 3)$. For consistency, these values are chosen to result in the same Flatten output size of 1024 as for the single frame model.

The remaining Dense layers staying untouched at their baseline configurations. Table 6 contains all results for the slow fusion experiment.

model	**MSE** mm^2	**MAE** mm	**MAF** mm	**MPF** 1/s
PilotNet	**0.130**	**0.270**	0.173	71
SimpleNet	0.198	0.327	**0.171**	**96**
MiniInceptionNet	0.201	0.340	0.236	40

Table 6: Results of the slow fusion experiment

While in theory a slow fusion of the temporal features is a more balanced way to merge the multiple frames, the measurements show no significant deviation from the Early Fusion results. However the MiniInceptionNet can undercut its previous *MSE* values, while being a bit more noisy. On the test set, Slow Fusion seems to be more trustworthy than the Single Frame and Early Merging fits, when it comes to the MiniInceptionNet.

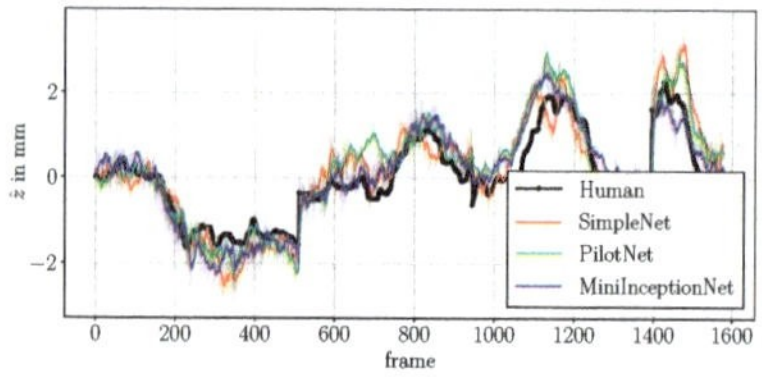

Figure 6: Comparison of the steering rack positions z of the human driver and the DNN **Slow Fusion** inferences of PilotNet, SimpleNet and MiniInceptionNet.

6.4 Late Fusion

While early and late fusion both provide the captured temporal information right at the network input space and propagate the extracted features to the output, late fusion only merges the time axis at subsequent layers. The base line models are reused to extract features of independently captured observations, shown frame by frame.

To merge these separate inferenced features, the lower part of the networks below the Flatten layers are replaced by a simple recurrent network, consisting only of one LSTM [40] with 512 neurons and two Dense layers with 16 neurons and one output neuron. Instead of simply concatenating the gathered feature vectors and performing inference with a pure MLP, a LSTM layer is introduced, as it does not ignore the temporal structure of the provided data and in theory helps to capture long-term interdependencies [39], [41].

In case of PilotNet and SimpleNet, the LSTM layer expects the input to be in shape $(\omega \times 1152)$, for the MiniInception architecture $(\omega \times 1024)$ is used instead. Similar to the experience of Heylen et al. [28], training these Conv-LSTM networks form scratch produced poor results.

Therefore the trained single frame models are used and the Flatten layer outputs are saved besides the inference results - both for the training and validation set, which leads to a new data set. During this task, no gradient based parameter update was performed to keep the models unchanged. Afterwards, the simple recurent network is trained on the new data set. The results are shown in Table 7.

The Late Fusion experiment ended up with the best test and evaluation results by far. For SimpleNet, the *MSE* dropped to only 18.22 % of the original value - which poses a significant improvement. Also the *MAF* and *MAE* values reduced noticeable. However, all three models overestimate the necessary steering amount in the situation from frame 450 upwards.

model	MSE mm^2	MAE mm	MAF mm	MPF $1/s$
PilotNet	0.056	0.166	0.058	61
SimpleNet	**0.049**	**0.155**	**0.040**	**69**
MiniInceptionNet	0.083	0.210	0.0935	41

Table 7: Results of the late fusion experiment

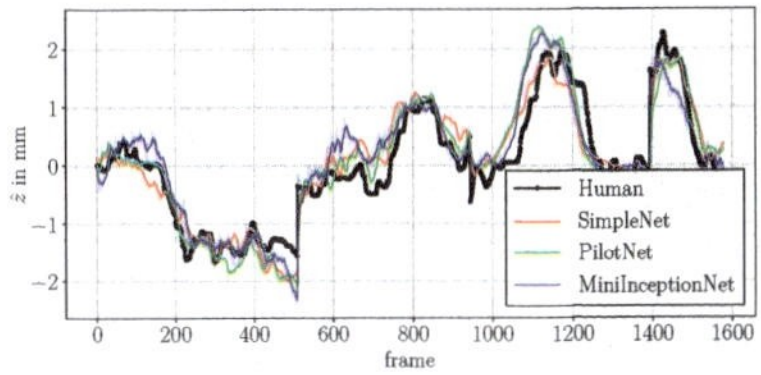

Figure 7: Comparison of the steering rack positions z of the human driver and the DNN **Late Fusion** inferences of PilotNet, SimpleNet and MiniInceptionNet.

6.5 Inclusion of road bank angle

To include a non optical vehicle signal a similar approach to late fusion is followed. Like before, the flattened extracted feature vector of the preceding convolution layers is fed into a small LSTM unit with 512 neurons. After that the resulting 512 elements wide LSTM output $F_{LSTM,t}$ and the road bank angle gradient prediction vector $\dot{Q}_t$ are con catenated together. The whole network input for this experiment is denoted by:

$$X_{Q,t} = \{I_t, \dot{Q}_t\} \tag{9}$$

The parameter ω is set to 5. The prediction horizon of $\dot{Q}$ is defined to cover three seconds in total - with s_q set to 0.25 s, ω_q is calculated to 12 elements. As baseline model, the Single Frame version of PilotNet is used, to focus only on the impact of the bank angle prediction.

The generated vector is then further processed by two fully connected layers to calculate the steering rack position z_t. The architecture can be seen in Figure 8. Due to the concatination, there is a slight increase of trainable parameters within the Dense layer. In view of calculation times, however, this impact can be neglected.

Surprisingly, after network training no difference could be detected in the validation metrics as well as in the visual comparison of the pre-defined and calculated steering steering rack positions. This effect can be traced back to vanishing weights in the subsequent neurons. To overcome this, the technique of Hubschneider et al. [23] is followed by scaling $\dot{Q}_t$ with the constant factor λ_Q.

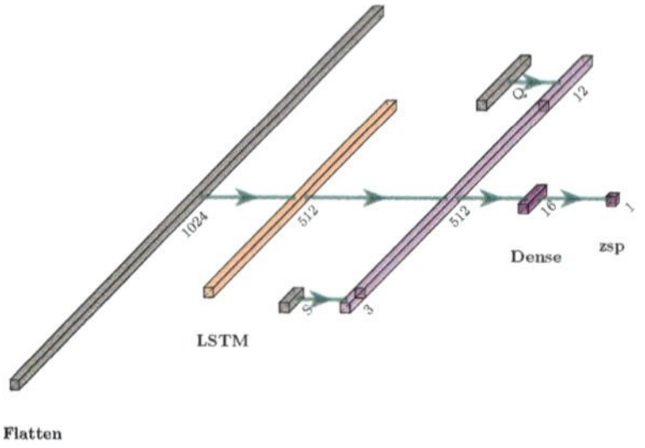

Figure 8: Layer output dimensions of the adjusted late fusion part including the road bank angle prediction horizon

To test the impact of $\dot{Q}$ in combination with multiple variations of λ_Q, a banked left curve, followed by straight section was selected to pose the test set. Figure 9 shows the optical influence of various factors λ_Q - still single exponential smoothing with $\alpha = 0.25$ is applied. For validation, a subset of the original dataset is used, as Ground Truth map data is not available for all recorded roads.

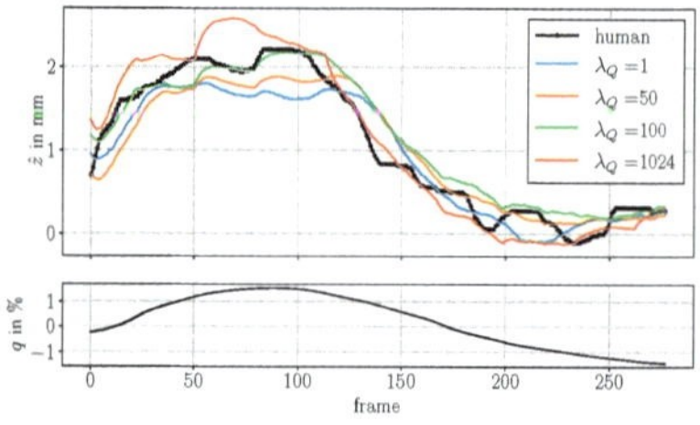

Figure 9: Comparison of the steering rack positions z of the human driver and the DNN predicted values for different factors λ_Q. The solid black line represents the recorded human behavior. The diagram below shows the corresponding road bank angles.

It is clearly visible that rising values of the scaling factor λ_Q lead to an increasing influence of $\dot{Q}$ on the steering rack position estimation. Mainly three effects can be captured. First, at the beginning of the test sequence, large deviations of the different steering rack position offsets can be observed. Interestingly, these are not in a clear order that can be assigned to the factors λ_q. For medium values of the scaling factor, the model tries to smooth out the pendulous behaviour at the curve exit, while increasing the necessary steering request during the left curve. Furthermore, it can be noted that too high scaling factors lead to an overshooting in both driving directions. With $\lambda_q = 50$

the lowest validation loss can be obtained.

The measurable relative decrease of the MSE could be measured to 8.5 %, the MAE dropped by 8 %. These comparatively small changes can be traced back to a simple reason. As seen in Figure 9 the model tries to smoothen out the harsh, oscillating steering behaviors of the human driver. As the distance between the predicted rack position and the recorded human steering effort grows, also the MSE metric increases.

6.6 Inclusion of vehicle signals

An almost identical experiment is repeated for the vehicle signals S_t - therefore the network input is adapted to:

$$X_{S,t} = \{I_t, S_t\} \tag{10}$$

The rest of the parameters remain unchanged. Again, every element of S_t is linear scaled with the constant factor λ_S. The impact on the test set can be visually examined in Figure 10.

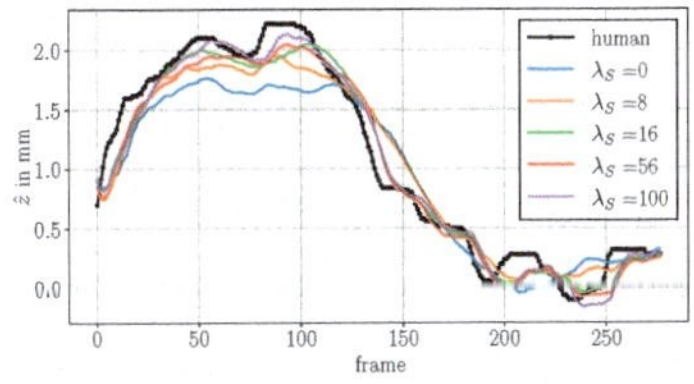

Figure 10: Comparison of the steering rack positions z of the human driver and the DNN predicted values for different factors λ_S. The solid black line represents the recorded human behavior.

Without the influence of v_t, $a_{y,t}$ and $\dot{\psi}_t$ the model underestimates the necessary steering-amount significantly. Similar to the previous experiment, increasing scaling factors λ_S reduce the discrepancy between human driving behavior and the calculated steering rack positions. In addition, the early withdrawal of the steering request (seen in the region of frame nr. 150) is improved. However, compared to the inclusion of $\dot{Q}_t$ with high values of λ_q, the oscillating human steering impulses from frame 200 upwards are not flattened out. The lowest validation loss could be determined with λ_s set to 56. As a result MSE dropped by 54 %, MAE dropped by 30 %. These high values prove, that the accuracy of a human steering modelling depends not only on the pure optical perception but also strongly on experienced accelerations, yaw rates and velocities.

6.7 The final steering model

Combining all the experience gained in the performed experiments the final steering model is constructed as followed. As the positive impacts of both the vehicle signals S_t and the predictive road bank angle vector $\dot{Q}_t$ has been showed, the full network input:

$$X_t = \{I_t, S_t, \dot{Q}_t\} \tag{11}$$

is utilized with $\lambda_q = 50$ and $\lambda_s = 56$. To extract spatio-temporal features, a Mixed Fusion approach, in form of a combination of Early Fusion followed by Late Fusion is implemented in order to join the two best approaches together. Early Fusion is chosen, as within this method, the lowest *MAF* and *MAE* values could be achieved besides Late Fusion. A visual comparison between the used Early and Late Fusion methods is presented in Figure 11.

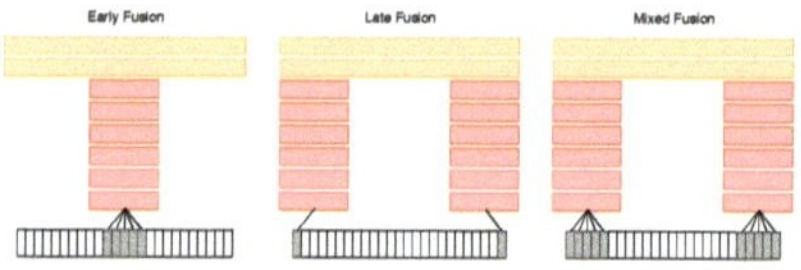

Figure 11: Evolution of the Early and Late Fusion method to a combined Mixed Fusion. The Illustration is based on [26].

At time t the sub-input I_t consists of $\omega = 14$ time-contiguous frames with an equidistant time spacing s of 100 ms. Similar to a sliding window approach, these are further processed within mini batches containing five frames. Thereby the window is shifted by one image at a time, ten times in total.

The Early Fusion variant of SimpleNet is used to extract the flattened feature vector of all the ten batches. Concatenated together, this leads to an input vector of shape 10×1152 for the following LSTM layer. The subsequent layer configurations are equal to the road bank angle experiment, except S_t is also used as system input. This approach leads to an effective time horizon of 1.4 seconds in total, whereas only 0.5 s are processed by the CNN based feature extracting part. The greater time history is formed within the LSTM layer. Larger sizes of the input window ω do not lead to better results.

With this architecture a *MSE* value of 0.0038 mm^2 could be archived on the validation-subset, which contains also road bank angle data. Compared to the baseline results without the additional inputs S_t and Q_t this represents a reduction of 66.78 %. *MAE* could be reduced to 0.0464 mm. These values extend the addition of the encapsulated effects if only one of S_t or Q_t is used. The Mixed Fusion method is able to perform inference with 69 FPS. Therefore this approach represents the most accurate driver steering modeling, while still be usable for real time car usage.

7 Conclusion

This work proposes a methodology based on a spatio-temporal Mixed Fusion to train an imitating model of the human steering behavior. Multiple raw frames of a front facing camera are mapped together with selected vehicle driving dynamics signals and Ground Truth based road bank angle gradient predictions. In order to analyze temporal dependencies within the image data efficiently, different fusion methods were evaluated. Results show, that the proposed combination of Early and Late Merging delivers a high accurate mimicking of the recorded human steering. It is also shown, that the inclusion of the road bank angle, in synergy with vehicle signals has a remarkable impact on resulting modeling errors. The final steering model is able to process the network inputs with more than 60 FPS. Therefore a real time usage is possible.

While this behavior reflex approach is based on supervised learning, future research could focus on influence of the road bank angle for online fine tuning of the Mixed Fusion model in a reinforced manner. In addition, as Deep Learning models are mostly black boxes, direct vehicle control is not purposeful. In particular, the safeguarding of such a controller is currently not given. Therefore the symbiosis of behavior reflex and mediated perception is the most promising approach.

8 Acknowledgements

The author would like to thank Porsche Engineering Services GmbH for funding the research project and providing a test vehicle. Special thanks goes also to the colleagues in the Driver Assistance Systems departments at Porsche Engineering and the University of Applied Sciences Kempten.

References

[1] C. Chen, A. Seff, A. Kornhauser, and J. Xiao, "DeepDriving: Learning Affordance for Direct Perception in Autonomous Driving", Tech. Rep., 2015. [Online]. Available: http://deepdriving.cs.princeton.edu.

[2] J. Levinson, J. Askeland, J. Becker, J. Dolson, D. Held, S. Kammel, J. Z. Kolter, D. Langer, O. Pink, V. Pratt, M. Sokolsky, G. Stanek, D. Stavens, A. Teichman, M. Werling, and S. Thrun, "Towards Fully Autonomous Driving: Systems and Algorithms", Tech. Rep. [Online]. Available: https://www.ri.cmu.edu/wp-content/uploads/2017/12/levinson-iv2011.pdf.

[3] D. Neven, B. De Brabandere, S. Georgoulis, M. Proesmans, L. Van Gool Esat-Psi, and K. U. Leuven, "Fast Scene Understanding for Autonomous Driving", Tech. Rep. arXiv: 1708.02550v1. [Online]. Available: https://youtu.be/55ElRh-g{_}7o..

[4] Z. Yang, Y. Zhang, J. Yu, J. Cai, and J. Luo, "End-to-end Multi-Modal Multi-Task Vehicle Control for Self-Driving Cars with Visual Perceptions", [Online]. Available: https://arxiv.org/pdf/1801.06734.pdf.

[5] J. E. Naranjo, C. González, R. García, and T. De Pedro, "Lane-change fuzzy control in autonomous vehicles for the overtaking maneuver", *IEEE Transactions on Intelligent Transportation Systems*, vol. 9, no. 3, pp. 438–450, 2008, ISSN: 15249050. DOI: 10.1109/TITS.2008.922880.

[6] A. Gurghian, T. Koduri, S. V. Bailur, K. J. Carey, and V. N. Murali, "DeepLanes: End-To-End Lane Position Estimation using Deep Neural Networks", Tech. Rep. [Online]. Available: https://www.cv-foundation.org/openaccess/content{_}cvpr{_}2016{_}workshops/w3/papers/Gurghian{_}DeepLanes{_}End-To-End{_}Lane{_}CVPR{_}2016{_}paper.pdf.

[7] D. Neven, B. De Brabandere, S. Georgoulis, M. P. Luc, V. Gool Esat-Psi, and K. Leuven, "Towards End-to-End Lane Detection: an Instance Segmentation Approach", [Online]. Available: https://arxiv.org/pdf/1802.05591.pdf.

[8] J. M. Pierre, "Spatio-temporal deep learning for robotic visuomotor control", *Proceedings - 2018 4th International Conference on Control, Automation and Robotics, ICCAR 2018*, pp. 94–103, 2018. DOI: 10.1109/ICCAR.2018.8384651.

[9] K. Kawaguchi Leslie Pack Kaelbling and Y. Bengio, "Generalization in Deep Learning", Tech. Rep. arXiv: 1710.05468v4. [Online]. Available: https://arxiv.org/pdf/1710.05468.pdf.

[10] D. A. Pomerleau, "Alvinn, an autonomous land vehicle in a neural network", 1989.

[11] M. Bojarski, D. D. Testa, D. Dworakowski, B. Firner, B. Flepp, P. Goyal, L. D. Jackel, M. Monfort, U. Muller, J. Zhang, X. Zhang, J. Zhao, and K. Zieba, "End to End Learning for Self-Driving Cars", 2016. arXiv: arXiv:1604.07316v1. [Online]. Available: https://arxiv.org/pdf/1604.07316.pdf.

[12] J. Jhung, I. Bae, J. Moon, T. Kim, J. Kim, and S. Kim, "End-to-End Steering Controller with CNN-based Closed-loop Feedback for Autonomous Vehicles", *IEEE Intelligent Vehicles Symposium, Proceedings*, vol. 2018-June, no. Iv, pp. 617–622, 2018. DOI: 10.1109/IVS.2018.8500440.

[13] H. Yu, S. Yang, W. Gu, and S. Zhang, "Baidu driving dataset and end-To-end reactive control model", *IEEE Intelligent Vehicles Symposium, Proceedings*, no. Iv, pp. 341–346, 2017. DOI: 10.1109/IVS.2017.7995742.

[14] C. Innocenti, H. Linden, G. Panahandeh, L. Svensson, and N. Mohammadiha, "Imitation learning for vision-based lane keeping assistance", *IEEE Conference on Intelligent Transportation Systems, Proceedings, ITSC*, vol. 2018-March, pp. 425–430, 2018. DOI: 10.1109/ITSC.2017.8317915. arXiv: 1709.03853.

[15] S. Du, H. Guo, and A. Simpson, "Self-Driving Car Steering Angle Prediction Based on Image Recognition", Tech. Rep. [Online]. Available: http://cs231n.stanford.edu/reports/2017/pdfs/626.pdf.

[16] L. Chi and Y. Mu, "Deep Steering: Learning End-to-End Driving Model from Spatial and Temporal Visual Cues", Tech. Rep. arXiv: 1708.03798v1. [Online]. Available: https://github.com/gy20073/BDD.

[17] Y. Hou, Z. Ma, C. Liu, and C. C. Loy, "Learning to Steer by Mimicking Features from Heterogeneous Auxiliary Networks", Tech. Rep. [Online]. Available: www.aaai.org.

[18] H. Xu, Y. Gao, F. Yu, and T. Darrell, "End-to-end Learning of Driving Models from Large-scale Video Datasets", [Online]. Available: https://arxiv.org/pdf/1612.01079.pdf.

[19] E. Yiqi, Hou, S. Hornauer, and K. Zipser, "Fast Recurrent Fully Convolutional Networks for Direct Perception in Autonomous Driving", [Online]. Available: https://arxiv.org/pdf/1711.06459.pdf.

[20] H. M. Eraqi, M. N. Moustafa, and J. Honer, "End-to-End Deep Learning for Steering Autonomous Vehicles Considering Temporal Dependencies", [Online]. Available: https://arxiv.org/pdf/1710.03804.pdf.

[21] W. Yuan, M. Yang, H. Li, C. Wang, and B. Wang, "End-to-end learning for high-precision lane keeping via multi-state model", *CAAI Transactions on Intelligence Technology*, 2018, ISSN: 2468-2322. DOI: 10.1049/trit.2018.1025.

[22] Y. Pan, C.-A. Cheng, K. Saigol, K. Lee, X. Yan, E. A. Theodorou, and B. Boots, "Agile Autonomous Driving using End-to-End Deep Imitation Learning", Tech. Rep. arXiv: 1709.07174v4. [Online]. Available: https://arxiv.org/pdf/1709.07174.pdf.

[23] C. Hubschneider, A. Bauer, M. Weber, and J. M. Zollner, "Adding navigation to the equation: Turning decisions for end-to-end vehicle control", *IEEE Conference on Intelligent Transportation Systems, Proceedings, ITSC*, vol. 2018-March, no. October, pp. 1–8, 2018. DOI: 10.1109/ITSC.2017.8317923.

[24] S. Hecker, D. Dai, and L. V. Gool, "Learning Driving Models with a Surround-View Camera System and a Route Planner", [Online]. Available: https://arxiv.org/pdf/1803.10158.pdf.

[25] G. Devineau, P. Polack, F. Altché, and F. Moutarde, "Coupled Longitudinal and Lateral Control of a Vehicle using Deep Learning", *2018 21st International Conference on Intelligent Transportation Systems (ITSC)*, pp. 642–649, 2018. DOI: arXiv:1810.09365v1. arXiv: 1810.09365. [Online]. Available: http://arxiv.org/abs/1810.09365.

[26] A. Karpathy, G. Toderici, S. Shetty, T. Leung, R. Sukthankar, and L. Fei-Fei, "Large-scale Video Classification with Convolutional Neural Networks", Tech. Rep. [Online]. Available: http://cs.stanford.edu/people/karpathy/deepvideo.

[27] D. Heide and A. O. Ag, "Banked road estimation and compensation algorithm for Lane Keeping Assist and its integration in GM ' s Software Architecture", pp. 38–42, 2017.

[28] J. Heylen, S. Iven, B. De Brabandere, J. Oramasm, L. Van Gool, and T. Tuytelaars, "From pixels to actions: Learning to drive a car with deep neural networks", *Proceedings - 2018 IEEE Winter Conference on Applications of Computer Vision, WACV 2018*, vol. 2018-Janua, pp. 606–615, 2018. DOI: 10.1109/WACV.2018.00072.

[29] *UdacitySDC-Challenge2 cg23*, 2016. [Online]. Available: https://github.com/udacity/self-driving-car/tree/master/steering-models/community-models/cg23 (visited on 04/24/2019).

[30] C. Szegedy, W. Liu, P. Sermanet, S. Reed, D. Anguelov, D. Erhan, V. Vanhoucke, and A. Rabinovich, "Going deeper with convolutions", Tech. Rep. arXiv: 1409.4842v1. [Online]. Available: https://arxiv.org/pdf/1409.4842.pdf.

[31] C. Zhang, S. Bengio, G. Brain, M. Hardt, B. Recht, O. Vinyals, and G. Deepmind, "UNDERSTANDING DEEP LEARNING REQUIRES RE-THINKING GENERALIZATION", Tech. Rep. arXiv: 1611.03530v2. [Online]. Available: https://arxiv.org/pdf/1611.03530.pdf.

[32] D Schneider, B Huber, H Lategahn, and P. B. Schick, "Measuring method for function and quality of automated lateral control based on high-precision digital "Ground Truth" maps", Tech. Rep.

[33] Z. Yang, Y. Zhang, J. Yu, J. Cai, and J. Luo, "End-to-end Multi-Modal Multi-Task Vehicle Control for Self-Driving Cars with Visual Perception", 2018, ISSN: 10514651. DOI: 10.1109/ICPR.2018.8546189. arXiv: 1801.06734.

[34] T.-d. Do, M.-t. Duong, Q.-v. Dang, and M.-h. Le, "Real-Time Self-Driving Car Navigation Using Deep Neural Network", pp. 7–12, 2018.

[35] D. P. Kingma and J. Lei Ba, "ADAM: A METHOD FOR STOCHASTIC OPTIMIZATION", Tech. Rep. arXiv: 1412.6980v9. [Online]. Available: https://arxiv.org/pdf/1412.6980.pdf.

[36] R. Hyndman, A. Koehler, K. Ord, and R. Snyder, *Forecasting with Exponential Smoothing*. Springer Berlin Heidelberg, 2008. DOI: 10.1007/978-3-540-71918-2. [Online]. Available: https://doi.org/10.1007/978-3-540-71918-2.

[37] J. Kim and J. Canny, "Interpretable Learning for Self-Driving Cars by Visualizing Causal Attention", Tech. Rep. arXiv: 1703.10631v1. [Online]. Available: https://arxiv.org/pdf/1703.10631.pdf.

[38] Z. Chen and X. Huang, "End-to-end learning for lane keeping of self-driving cars", in *2017 IEEE Intelligent Vehicles Symposium (IV)*, IEEE, 2017, pp. 1856–1860, ISBN: 978-1-5090-4804-5. DOI: 10.1109/IVS.2017.7995975. [Online]. Available: http://ieeexplore.ieee.org/document/7995975/.

[39] J. Carreira, A. Zisserman, Z. Com, and. Deepmind, "Quo Vadis, Action Recognition? A New Model and the Kinetics Dataset", Tech. Rep. arXiv: 1705.07750v3. [Online]. Available: https://arxiv.org/pdf/1705.07750.pdf.

[40] S. Hochreiter and J. Urgen Schmidhuber, "LONG SHORT-TERM MEMORY", *Neural Computation*, vol. 9, no. 8, pp. 1735–1780, 1997. [Online]. Available: http://www7.informatik.tu-muenchen.de/{~}hochreithttp://www.idsia.ch/{~}juergen.

[41] J. Donahue, L. A. Hendricks, M. Rohrbach, S. Venugopalan, S. Guadarrama, K. Saenko, and T. Darrell, "Long-term Recurrent Convolutional Networks for Visual Recognition and Description", [Online]. Available: https://arxiv.org/pdf/1411.4389.pdf.

LIGHTWEIGHT DESIGN AND DRIVEABILITY

Challenges and benefits for lightweight chassis components focusing on rear axle development

Dr. Thomas Kersten, Volkswagen AG

Dr. Sierk Fiebig, Volkswagen AG

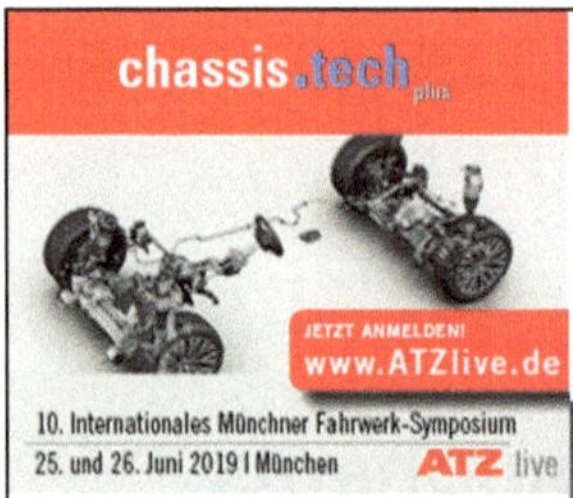

Challenges and benefits for lightweight chassis components focusing on rear axle development

Dr. Thomas Kersten,
Volkswagen AG
Dr. Sierk Fiebig,
Volkswagen AG

page 1

Agenda

- Introduction – chassis requirements
- Rear axle development @ Volkswagen
- Conceptual lightweight design in rear axle system
  - GRP-X Axle
- Examples of lightweight design in rear axle development
  - Lightweight design using simulation tools
  - Manufacturing process (twist beam axle with TRB, MLR, Tailored Tubes)
  - Effective of manufacturing processes (hybrid components)
- Summary and conclusion

page 2

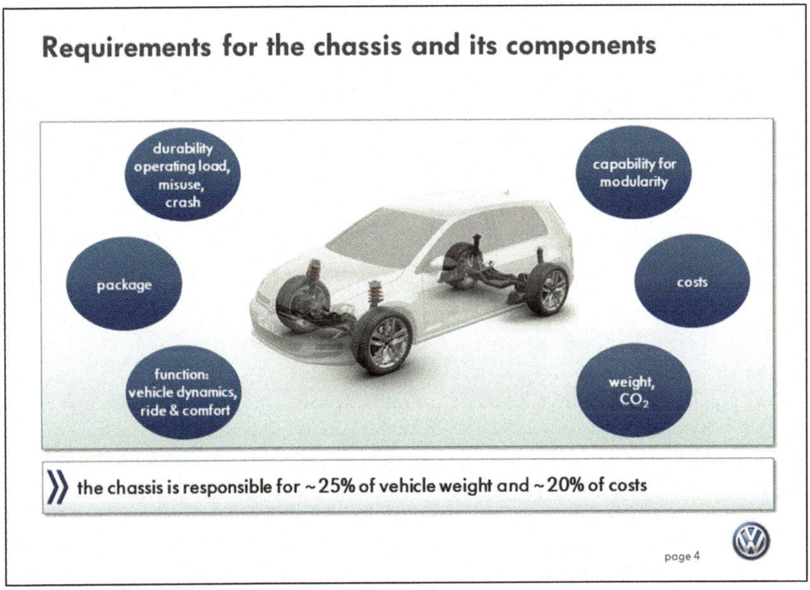
Requirements for the chassis and its components
durability
operating load,
misuse,
crash
capability for
modularity
package
costs
function:
vehicle dynamics,
ride & comfort
weight,
CO_2
the chassis is responsible for ~25% of vehicle weight and ~20% of costs
page 4

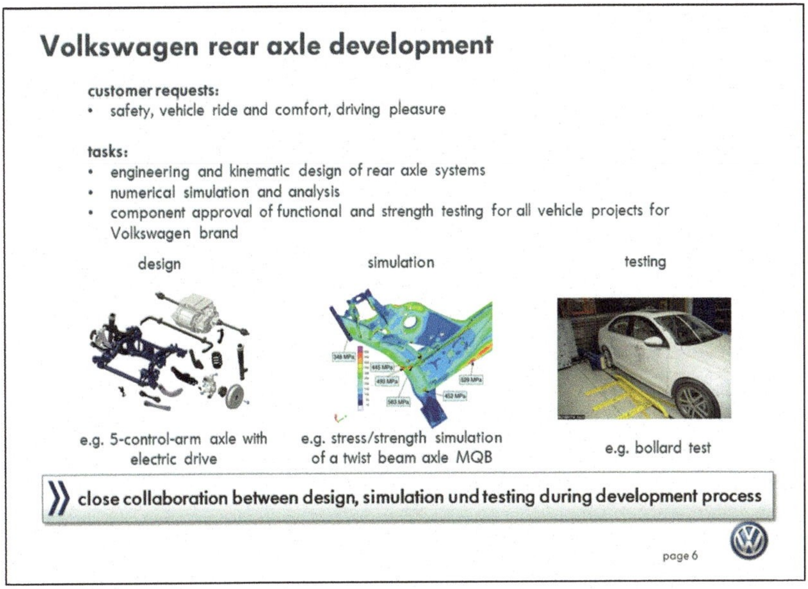
Volkswagen rear axle development
customer requests:
• safety, vehicle ride and comfort, driving pleasure
tasks:
• engineering and kinematic design of rear axle systems
• numerical simulation and analysis
• component approval of functional and strength testing for all vehicle projects for Volkswagen brand
design
simulation
testing
e.g. 5-control-arm axle with electric drive
e.g. stress/strength simulation of a twist beam axle MQB
e.g. bollard test
close collaboration between design, simulation und testing during development process
page 6

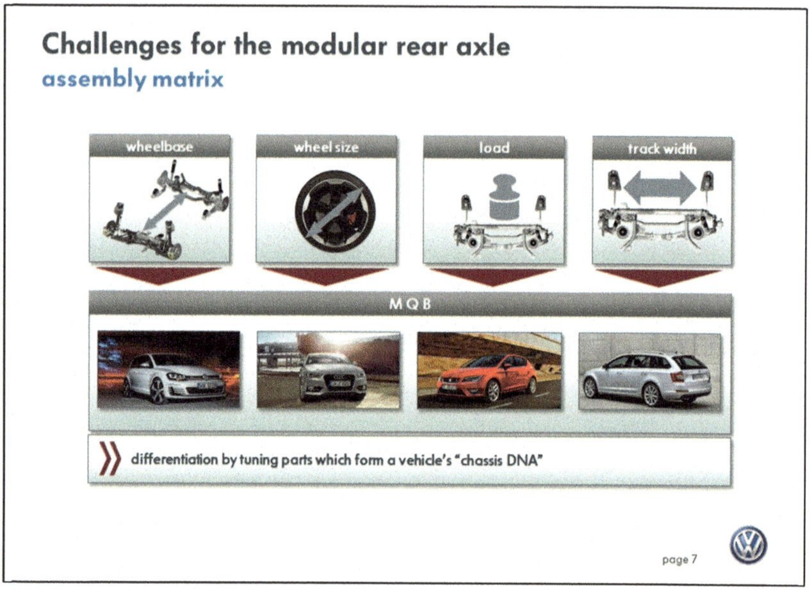
Challenges for the modular rear axle
assembly matrix
wheelbase
wheel size
load
track width
MQB
differentiation by tuning parts which form a vehicle's "chassis DNA"
page 7

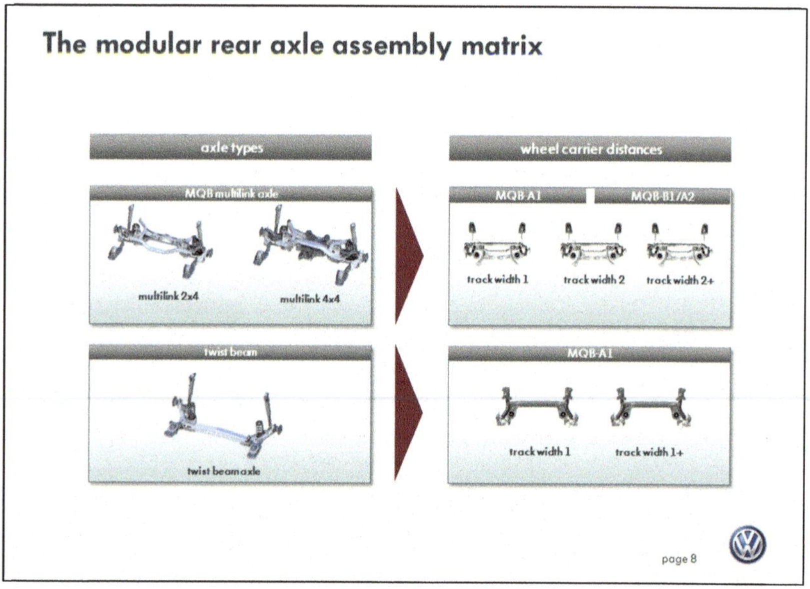
The modular rear axle assembly matrix
axle types
wheel carrier distances
MQB multilink axle
multilink 2x4
multilink 4x4
MQB A1
MQB-B1/A2
track width 1
track width 2
track width 2+
twist beam
twist beam axle
MQB-A1
track width 1
track width 1+
page 8

MQB multilink axle (2x4)

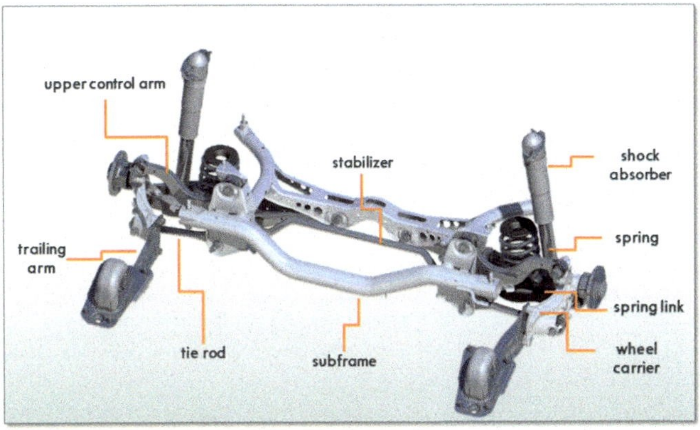

page 9

MQB Twist beam rear axle

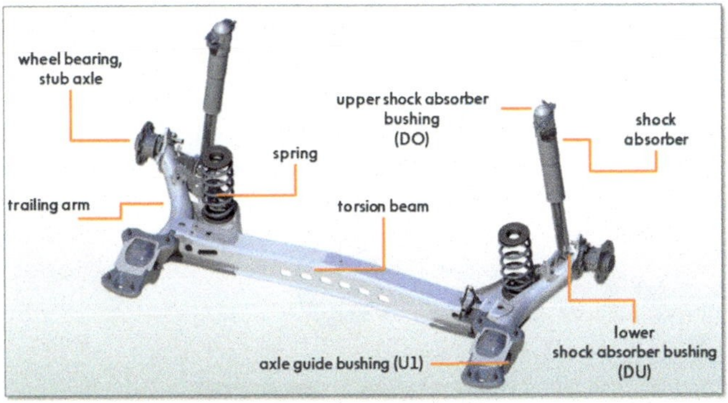

page 10

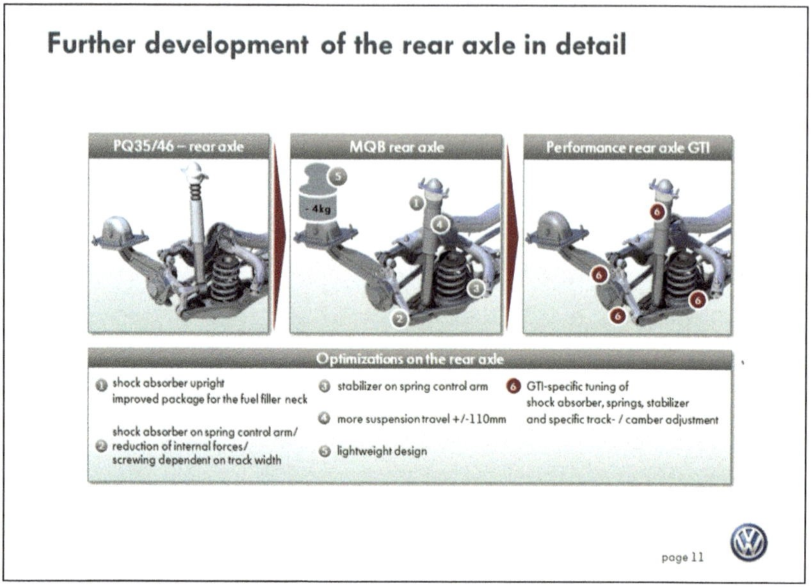
Further development of the rear axle in detail
PQ35/46 – rear axle
MQB rear axle
Performance rear axle GTI
- 4kg
Optimizations on the rear axle
1 shock absorber upright improved package for the fuel filler neck
2 shock absorber on spring control arm/ reduction of internal forces/ screwing dependent on track width
3 stabilizer on spring control arm
4 more suspension travel +/-110mm
5 lightweight design
6 GTI-specific tuning of shock absorber, springs, stabilizer and specific track- / camber adjustment
page 11

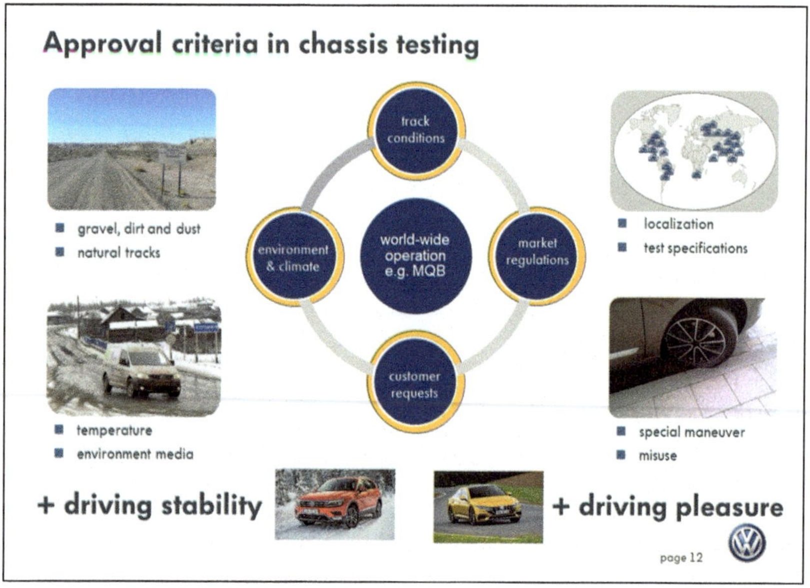
Approval criteria in chassis testing
gravel, dirt and dust
natural tracks
temperature
environment media
track conditions
environment & climate
world-wide operation e.g. MQB
market regulations
customer requests
localization
test specifications
special maneuver
misuse
+ driving stability
+ driving pleasure
page 12

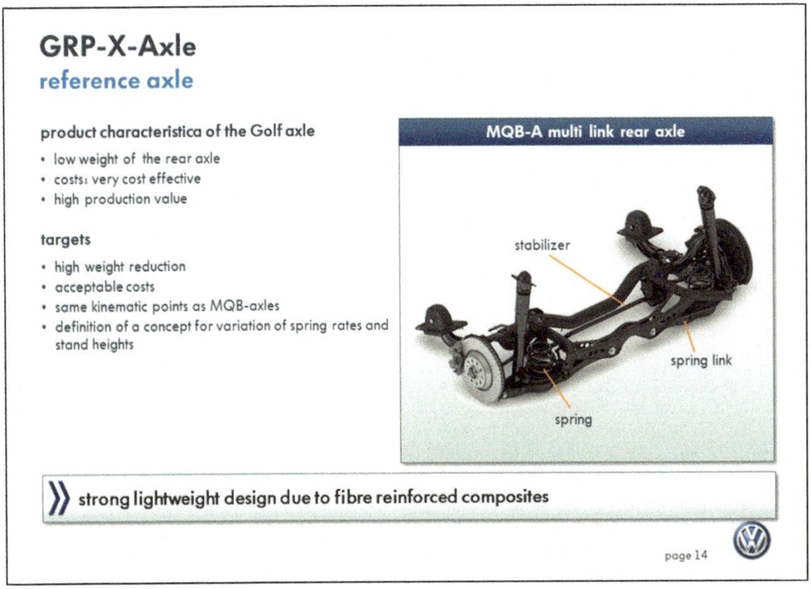
GRP-X-Axle
reference axle
product characteristica of the Golf axle
• low weight of the rear axle
• costs: very cost effective
• high production value
targets
• high weight reduction
• acceptable costs
• same kinematic points as MQB-axles
• definition of a concept for variation of spring rates and stand heights
MQB-A multi link rear axle
stabilizer
spring link
spring
strong lightweight design due to fibre reinforced composites
page 14

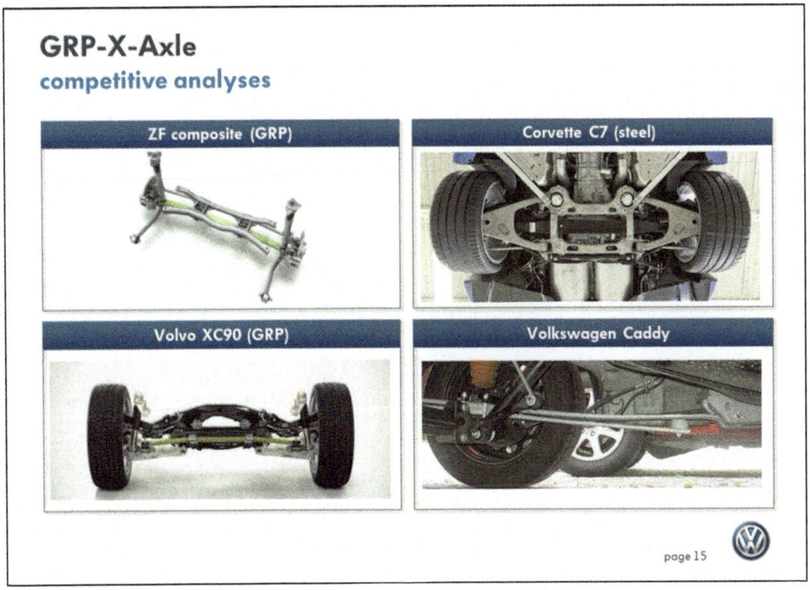
GRP-X-Axle
competitive analyses
ZF composite (GRP)
Corvette C7 (steel)
Volvo XC90 (GRP)
Volkswagen Caddy
page 15

GRP-X-Axle

fiber reinforced plastics in chassis-parts

technic

- substitution of
  - springs
  - spring links
  - anti-roll bar
- variation of springs by different geometries and fiber volume content

potential

- weight reduction up to 54%
- unique feature:
  first car with wheel-piloting GRP-springs
- acceptable costs

GRP-X-springs for Golf GTI

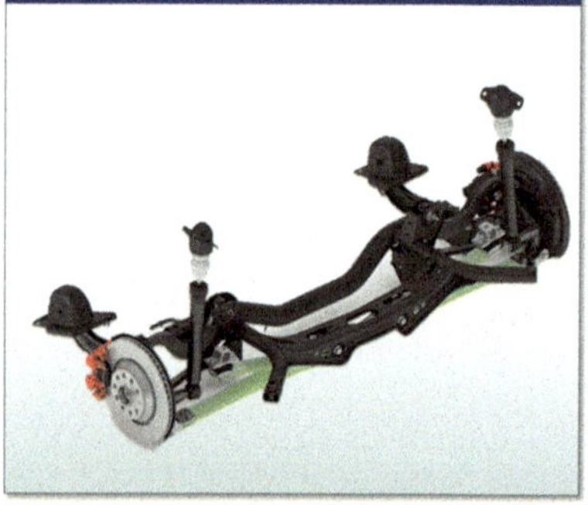

» new concept for axles with potential of dramatical weight reduction while taking equal costs

page 16

MQB GRP-X-Axle

detailed design

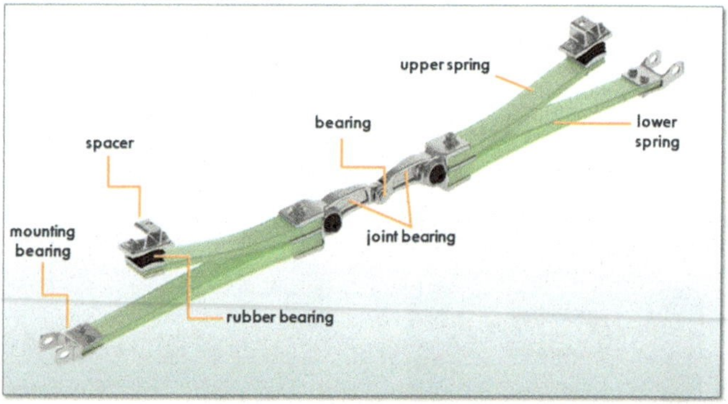

page 17

GRP-X-Axle
weight comparison

series-concept Golf	
total weight	100%
stabilizer	14%
2x springs	32%
2x spring links	38%
coupling bar + several small parts	16%

X-spring-concept Golf	
total weight	47%
GRP springs	27%
joint-bearing	8%
mounting bracket	3%
spacer and rubber bearing	9%

›› the GRP-X-Spring design is a combination of strong lightweight design and robustness

page 18

GRP-X-Axle
function analysis

load case 01 – extreme curves

- one side compression
- without rotation of joint bearing around U3-bearing-point
- most loads carried by lower spring

one side compression

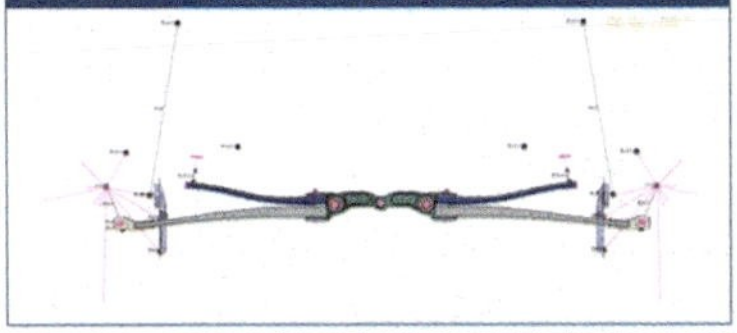

load case 02 – vertically kick

- two side compression
- joint baerings rotate around U3-bearing-point
- most loads carried by upper spring

two side compression

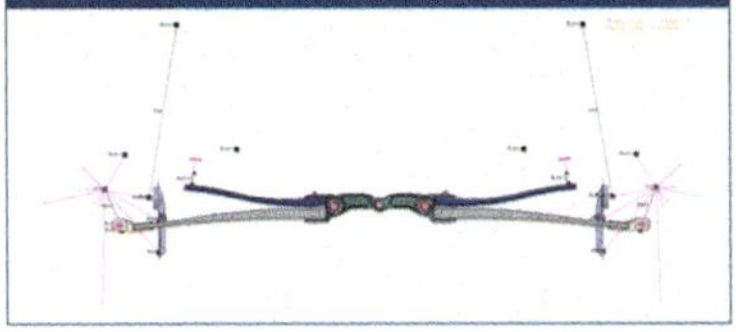

page 19

GRP-X-Axle
test status

test in test rig

- ✓ test of load cases used for design
- ✓ measurement of spring rate and rate of anti-roll bar
- ✓ successful endurance test

test in car

- ✓ build-up of prototype car
- ✓ determination of kinematics
- ✓ cold climate testing
- practical test of comfort and driving dynamics

page 20

GRP-X-Axle
conclusion

successful predevelopment of an axle with fiber-glass reinforced plastics material

- same cost level as current series
- weight reduction of ~7kg
- improvement of corrosive issues

successful testing of the axle

- result at axle-test-rig are ok
- positive feedback at first driving experiences

GRP-X-springs for Golf GTI

» transferability of concept into other vehicle classes is proved

page 21

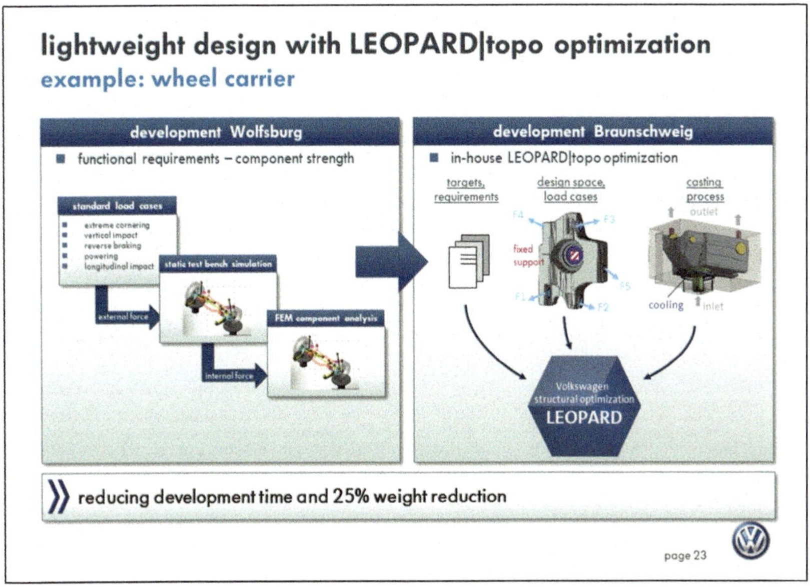
lightweight design with LEOPARD|topo optimization
example: wheel carrier
development Wolfsburg
functional requirements – component strength
standard load cases
extreme cornering
vertical impact
reverse braking
powering
longitudinal impact
static test bench simulation
external force
FEM component analysis
internal force
development Braunschweig
in-house LEOPARD|topo optimization
targets, requirements
design space, load cases
casting process
outlet
F4
F3
fixed support
F5
F1
F2
cooling
inlet
Volkswagen structural optimization
LEOPARD
reducing development time and 25% weight reduction
page 23

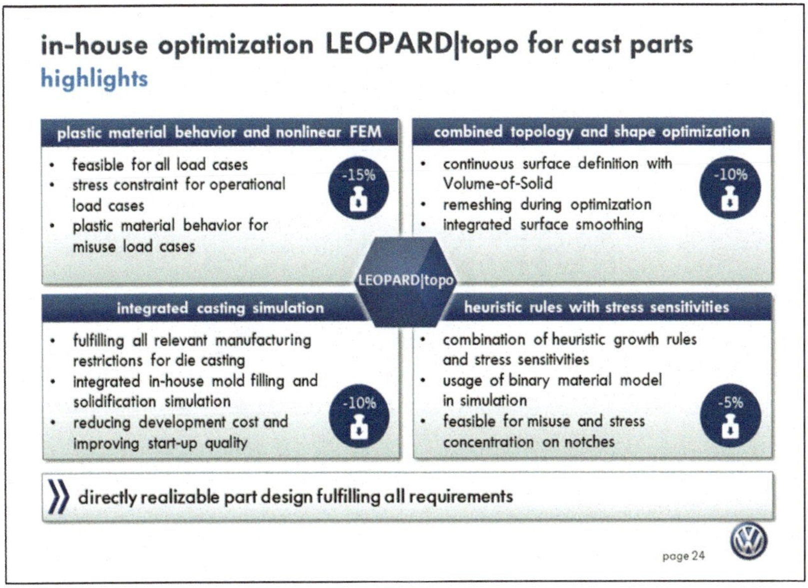
in-house optimization LEOPARD|topo for cast parts
highlights
plastic material behavior and nonlinear FEM
feasible for all load cases
stress constraint for operational load cases
plastic material behavior for misuse load cases
-15%
combined topology and shape optimization
continuous surface definition with Volume-of-Solid
remeshing during optimization
integrated surface smoothing
-10%
LEOPARD|topo
integrated casting simulation
fulfilling all relevant manufacturing restrictions for die casting
integrated in-house mold filling and solidification simulation
reducing development cost and improving start-up quality
-10%
heuristic rules with stress sensitivities
combination of heuristic growth rules and stress sensitivities
usage of binary material model in simulation
feasible for misuse and stress concentration on notches
-5%
directly realizable part design fulfilling all requirements
page 24

in-house optimization LEOPARD|topo for cast parts

wheel carrier optimization

commercial software

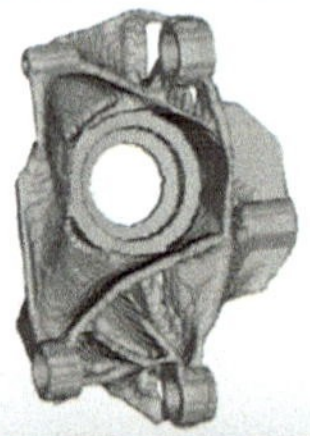

- no misuse
- not castable
- no integrated shape optimization

LEOPARD structural optimization

- ✓ all mechanical requirements fulfilled
- ✓ castable without geometrical changes
- ✓ integrated smoothing and shape optimization

》 clear advantages in the development process with in-house optimization software

page 25

MQB multilink rear axle

overview

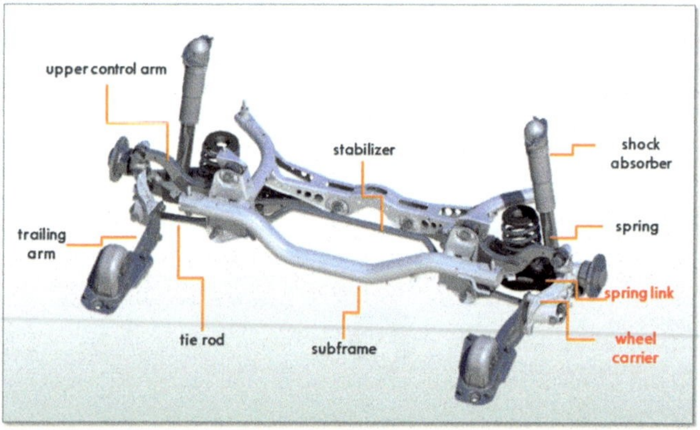

page 26

Spring link optimization

rear axle kinematic

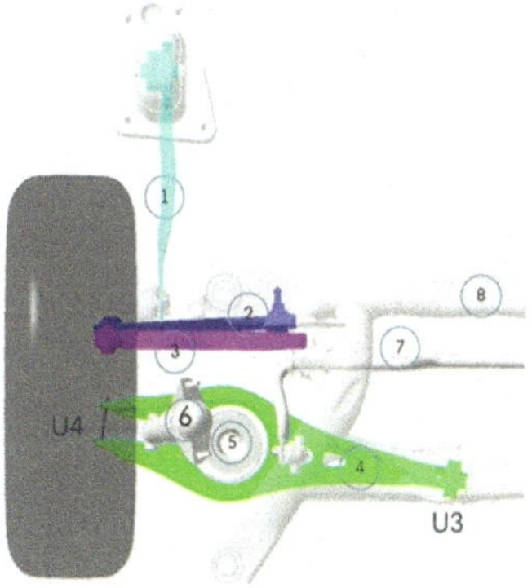

- 1 trailing arm
- 2 tie rod
- 3 upper control arm
- **4 spring link**
- 5 spring
- 6 shock absorber
- 7 stabilizer
- 8 subframe

page 27

Spring link (loadcase longitudinal impact)

Golf V vs. Golf VII

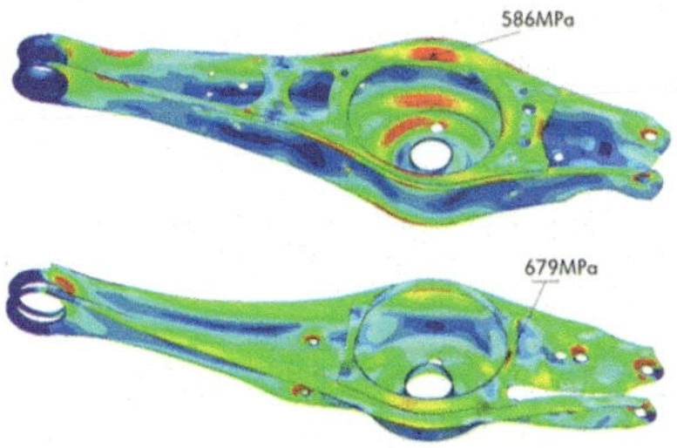

v. Mises stress comparison (FEM)

- **above:** Golf V Weight: 5.0 kg/vehicle
- **below:** Golf VII Weight: 4.0 kg/vehicle
- **lightweight design measures:** shape optimization, manufacturing concept, material selection
- **potential:** 1.0 kg/vehicle, ~20 % of component **...although shock absorber and stabilizer forces must be absorbed additionally!**

» 20% weight reduction through structural changes despite higher load!

page 28

Cooperation with suppliers

Mubea
light.efficient.global.

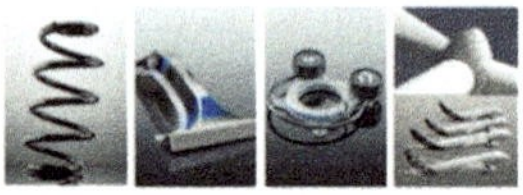

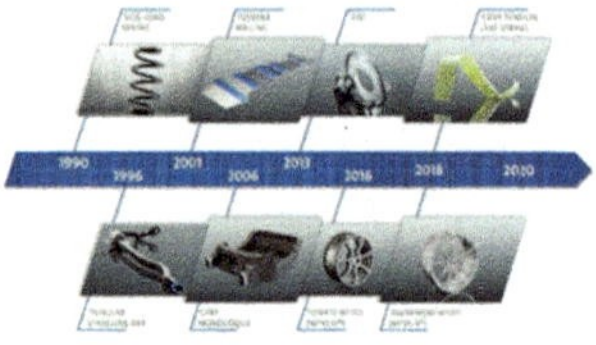

- founded as family business in 1916
- 2.4 billion €, 15,000 employees at 50 sites
- market and innovation leader for automotive lightweight parts
- maximum vertical integration from raw material to product
- global footprint with advanced technology worldwide according to customer locations

source:
Mubea

Torsion Beam Axle in Tailor Rolled Blanks TRB

Mubea
light.efficient.global.

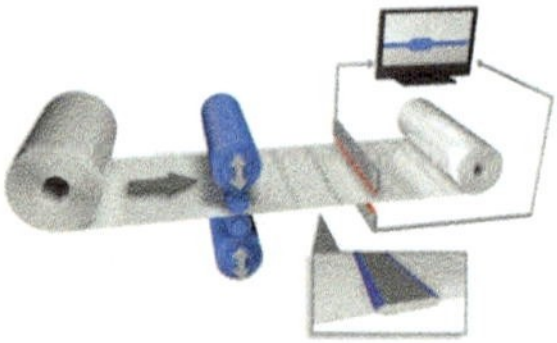

- Tailor Rolled Blanks is based on a cold rolling process with variable rolling gap
- Flexibility of the process enables the use of one part design for different vehicle derivates
- Objective: lightweight design with -20% mass reduction for twist beam axles at same or improved performance

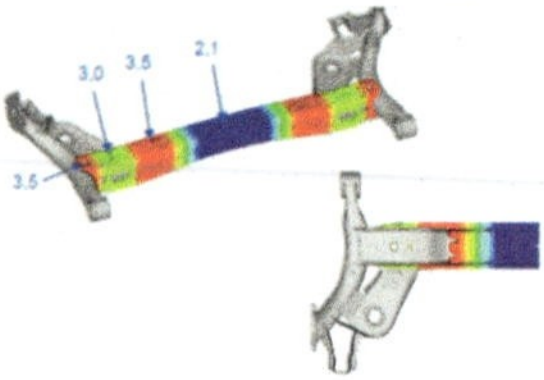

Innovation & future concepts

- Design of a holistic axle concept for two different roll rates
- Perpetuation of all constraints: package, stiffness, strength, ...

weight		re-calculation VW-series production	Mubea TRB-profile	
axle beam	[kg]	8,11	6,22	-1,89 (-23,3%)
insert plates	[kg]	1,95	1,73	-0,22 (-11,3%)
sum	[kg]	10,06	7,95	-2,11 (-21%)

source:
Mubea

Coil Springs & Coil Spring Systems

Mubea
light.efficient.global.

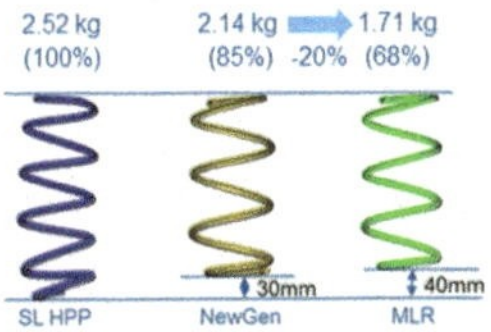

- Lightweight and functional optimization thru advanced material design and processes
- Optimal value added: wire, coil spring, isolator
- Objective: Solutions for CASE: high loads/spring rates, NVH

Innovation & future concepts

- 15% mass reduction through SLM2150 technology
- Enhanced surface protection due to adopted coating systems
- Elimination critical spring area through dura-connect (bonded isolator)
- Additional 15 % mass reduction potential by MLR package
- Enhanced cold forming capabilities tailored to high spring rates
- NVH as part of an integrated system design for coil springs

source:
Mubea © Muhr und Bender KG

Stabilizer Bar Systems

Mubea
light.efficient.global.

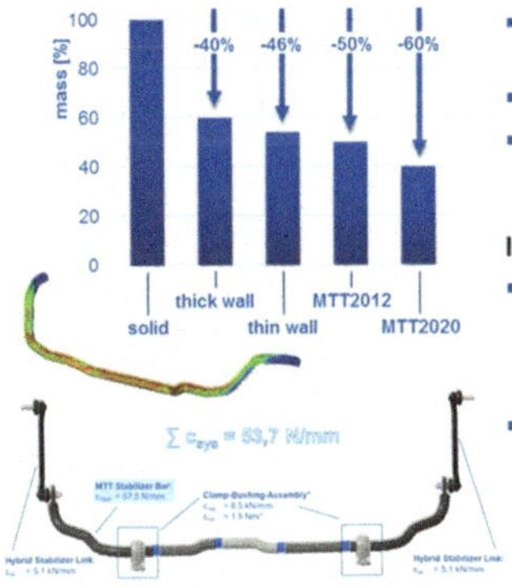

- Lightweight and functional optimization thru advanced material design and processes
- Optimal value added: tube, stab bar, clamps, bushings, links
- Objective: optimal stab bar system design

Innovation & future concepts

- Lightweight Stab Bar - MTT2020
    - 20% mass reduction over MTT2012
    - Tailor Tube - MTT featuring three wall thicknesses
    - Highest material (tube) yield strength - SBT1900
- Stab Bar Systems
    - Integrated design of stab bar systems (iMOG)
    - Integrated design of stab bar-link joint
    - System testing with static and dynamic rate, NVH-analysis and road load data under development

source:
Mubea © Muhr und Bender KG

Chassis Composites

Mubea
light.efficient.global.

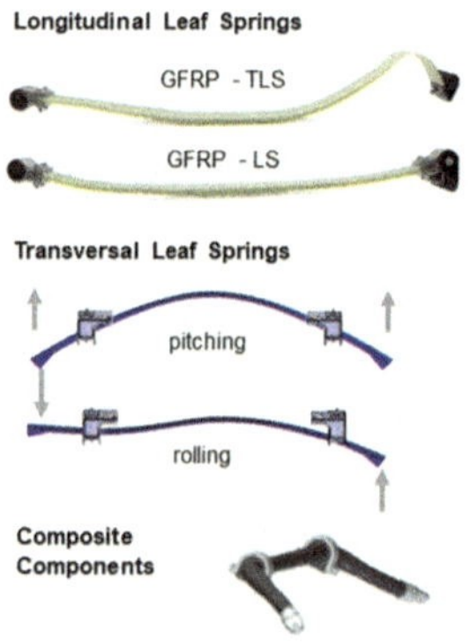

- Combined suspension knowhow and GFRP (Glas Fiber Reinforced Plastics) material expertise
- Objective: 70%+ mass reduction over steel leaf springs (up to 50kg reduction for pick-up trucks), robustness, ride performance advantages

Innovation & future concepts

- GFRP Leaf Springs (LS) with linear characteristics and Tension Leaf Springs (TLS) with progressive characteristics, safe failure and improved NVH
- Very competitive lightweight oncost
- Worldwide first development of a GFRP Transversal LS with pitching and rolling characteristics – high customer interest!
- Advanced development for multiple chassis components i.e. CFRP (Carbon Fiber) Stab Bar

source:
Mubea

Spring link concepts (e-vehicle)

concept comparison

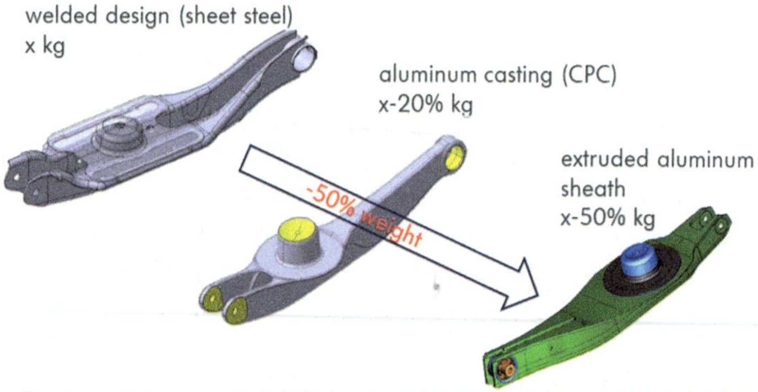

» 50% weight reduction through material and manufacturing process selection

page 36

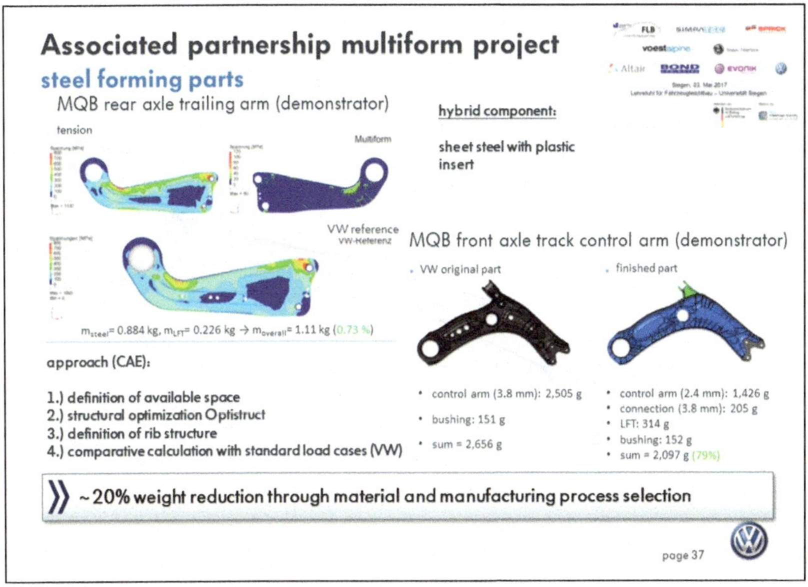
Associated partnership multiform project
steel forming parts
MQB rear axle trailing arm (demonstrator)
tension
Multiform
VW reference
VW-Referenz
m_{steel} = 0.884 kg, m_{LFT} = 0.226 kg → $m_{overall}$ = 1.11 kg (0.73 %)
approach (CAE):
1.) definition of available space
2.) structural optimization Optistruct
3.) definition of rib structure
4.) comparative calculation with standard load cases (VW)
hybrid component:
sheet steel with plastic insert
MQB front axle track control arm (demonstrator)
VW original part
finished part
control arm (3.8 mm): 2,505 g
bushing: 151 g
sum = 2,656 g
control arm (2.4 mm): 1,426 g
connection (3.8 mm): 205 g
LFT: 314 g
bushing: 152 g
sum = 2,097 g (79%)
~20% weight reduction through material and manufacturing process selection
page 37

chassis.tech plus
JETZT ANMELDEN!
www.ATZlive.de
10. Internationales Münchner Fahrwerk-Symposium
25. und 26. Juni 2019 I München
ATZ live
Challenges and benefits for lightweight chassis components focusing on rear axle development
Dr. Thomas Kersten,
Volkswagen AG
Dr. Sierk Fiebig,
Volkswagen AG
Thank you!
page 40

Innovative use of composites in suspension systems

P. Kuhn, C. Loy, M. Kuehnel, W. Pinch, C. Ebel
SGL Technologies GmbH

Abstract

The use of composite materials e.g. carbon fibre reinforced polymers in suspension systems can contribute to lighter and therefore more energy-efficient passenger cars and commercial vehicles. Besides ensuring the technological feasibility, the implementation of the corresponding large-scale production is essential to compete with classical metal structures. Within the chassis, both lateral and longitudinal leaf springs made from composite materials have been found to meet these requirements. Single- and multi-axis struts, produced in a spatial winding process, are a further promising example of load-path optimized chassis components showing high potential for automotive series applications.

In the present paper, both composite leaf spring and strut applications are presented with a focus on the holistic development approach besides their respective technical features.

1 Introduction

Fibre reinforced polymers offer outstanding weight specific strength and stiffness properties. In combination with good corrosion resistance, energy absorption capacity and material damping characteristics, they can play a key role in the development of energy-efficient designs in automotive industry. Starting from minor applications in body shell parts, polymer composites are more and more used also in highly loaded structures. This can be attributed to huge improvements in the understanding of the material behaviour and in the structural analysis capabilities in recent years. Fig. 1 shows applications of composite materials and components in a passenger car.

To compete with traditional automotive materials e.g. steel and aluminium alloys, a holistic approach is needed to exhaust the lightweight potential of fibre reinforced polymers and to reduce development and manufacturing costs and time. Herein, the most important issues are both determining a load path optimized design considering the specific composite material properties and the selection of the suitable manufacturing strategy. Major improvements in automatic manufacturing processes (e.g. automatic fibre placement, spatial winding) and high-performance materials (e.g. pre-impregnated carbon fibres, fast curing resins) enable new possibilities for serial production of composite components with consistent high quality.

The technical requirements of both transversal and longitudinal leaf springs can be well fulfilled by the orthotropic material properties of glass fibre reinforced polymers (GFRP) and corresponding series production processes (e.g. resin transfer moulding) are already proven. The production of the lateral leaf springs for the VOLVO Scalable

Product Architecture (SPA) platform with a volume of 500.000 parts per year is a good example showing the high applicability for series production for this kind of composite parts. Straining pre-impregnated carbon fibres, so called TowPreg, between guiding pins or load introduction elements is another example of load path optimized use of composites in combination with automated manufacturing process technology (fibre lay-up, consolidation, curing). Application scenarios in suspension systems for both composite leaf springs and struts are described in the following chapters.

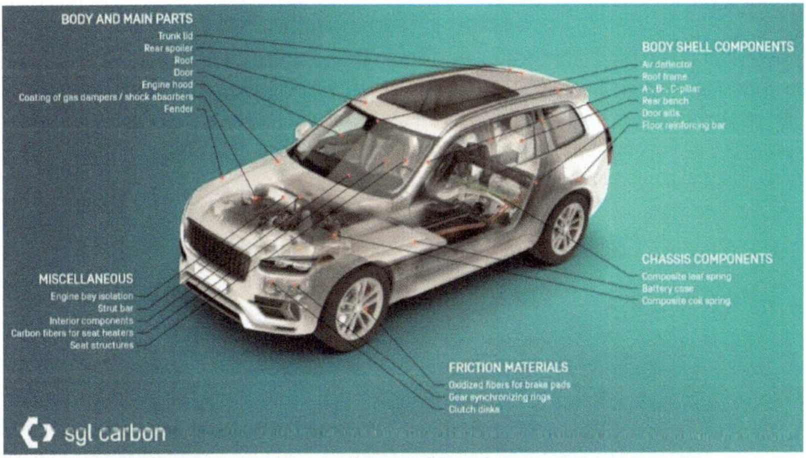

Fig. 1: Automotive applications of composite materials and components.

2 Composite leaf spring applications

Based on the general requirements for passenger and commercial vehicles (e.g. fuel economy, ride performance, pay load) the most important component targets to be fulfilled for a composite leaf springs are:

- Ride rate
- Minimum mass
- Durability performance

Further, transversal leaf springs can fully or partly undertake stabilizer bar functions, resulting in additional requirements on the roll rate and the roll/ride rate ratio. As longitudinal leaf springs are typically used in high load applications (e.g. mid-size and full-size trucks), a progressive spring rate behaviour is required commonly, considering the significant vehicle weight difference between curb and maximum payload.

2.1 Lateral leaf spring

Low profile and compact suspension types, like Short-Long Arm (SLA), Integral Link, 4- and 5-Link system, enable the realization of innovative vehicle configurations (e.g. for autonomous and electric vehicles) having high packaging efficiency. Using these suspension types can contribute to increased cargo and/or interior volume and higher load-through widths. Realizing a suspension system being compact in both vertical (Z) and fore-aft (X) direction, the replacement of classical steel coil springs by a lateral leaf spring is a promising approach. Fig. 2 shows the implementation of a lateral leaf spring in an Integral Link rear axle suspension system.

Fig. 2: Low profile and compact suspension system with lateral leaf spring (rear axle)

As apparent from Fig. 2 the use of a lateral leaf spring instead of a coil spring is beneficial not only regarding space requirements in Z-direction. Due to leaner package design, the spring mounting point on the Lower Control Arm (LCA) can be shifted more outwards in horizontal (Y), closer to knuckle joint hard-point, resulting in an improved spring linkage ratio. Further, the assembly situation of the spring itself offers different tuning options to meet the requirements of different car configurations within a small package space and by using the same leaf spring part. Desired vehicle ground can be adjusted at the connection point of the spring and the LCA. The ride rate of the spring itself is mainly defined by its thickness profile and the length ratio between outer and inner bushings. Hence, the position of the inner bushings can be used to set the spring rate without any significant change of the packaging or using additional parts, as illustrated in Fig. 3.

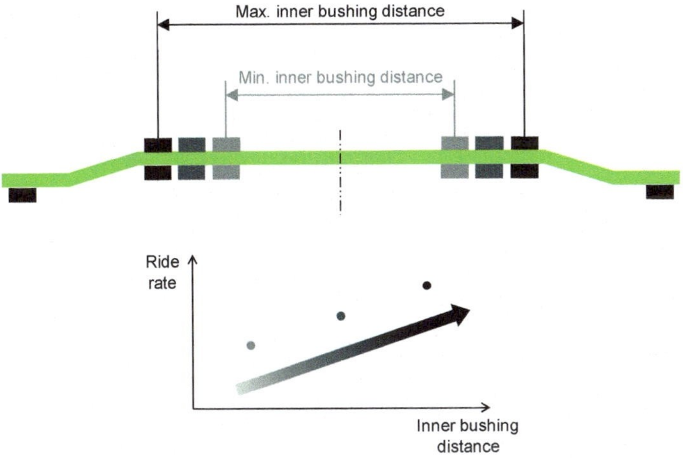

Fig. 3: Effect of inner bushing distance on ride rate.

Considering the load scenario of the lateral leaf spring shows the high suitability using a fiber reinforced polymer material. The leaf spring can be interpreted mechanically as a four-point bending beam and is mainly loaded along the spring longitudinal axis, therefore. Transverse and torsional loads can be minimized by choosing suitable hard point configuration of the LCA. So, the lightweight potential of composite materials can highly be used while putting the fibers mostly along the longitudinal direction of the spring. Polymer composites made of glass fibers fulfill the required strain to failure properties at a simultaneously sufficient stiffness level.

2.2 Longitudinal leaf spring

For high payload applications in commercial vehicles, the Hotchkiss solid axle suspension type, as illustrated in Fig. 4, is a very good choice regarding driving performance and packaging requirements. Efficient lateral packaging is enabled by springs mounted under or alongside longitudinal frame rails. Installation of the longitudinal leaf spring in over- or underslung configuration offers further design flexibility. In contrast to classical lateral leaf spring applications, the longitudinal leaf spring has a wheel-controlling function, as the front half-length acts as trailing arm.

Fig. 4: Hotchkiss solid axle suspension system with leaf spring in overslung configuration [1].

In the classical steel version, stacked leaves of different lengths are commonly used, following the bending moment curve of a centre loaded three-point bending beam. Progressive spring behavior is realized in form of an additional helper spring, coming in contact to the main leaf package at a certain load level. Fig. 5 shows the characteristic nearly bi-linear load-deflection curve with an abrupt rate transition behavior. A first step towards a more weight optimized design is the replacement of the steel helper spring by a lighter composite part, resulting in an unchanged overall spring rate curve. However, to fully exhaust the light weight potential, complete composite design can be used. In combination with an intelligent shackle concept, the main leaf package and the helper spring can be replaced by a single GFRP spring, offering a potential weight reduction of more than 50%. The progressive load-deflection curve of the mono-leaf composite spring is plotted in Fig. 5, additionally. Compared to the helper spring configuration, the mono-leaf design enables a more desired smooth rate build-up without any distinctive rate transition behaviour. Further, a tailored thickness profile of the GFRP spring can be realized, e.g. enabling variable design of the front- and rear-arm to take different wind-up and wind-down load cases into account. Using a mono-leaf spring design, interleaf friction is eliminated, resulting in reduced wear and a more efficient use of the shock absorber. The mounting hardware at the front and rear bushings can be maintained when replacing the longitudinal steel spring by a GFRP design.

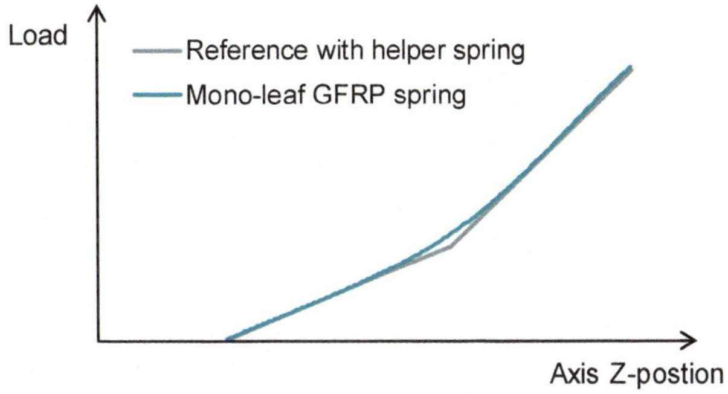

Fig. 5: Load-deflection curves of different longitudinal leaf spring configurations.

2.3 Composite leaf spring development process

The development process of both transversal and longitudinal composite leaf springs can be separated in four phases, illustrated in Fig. 6.

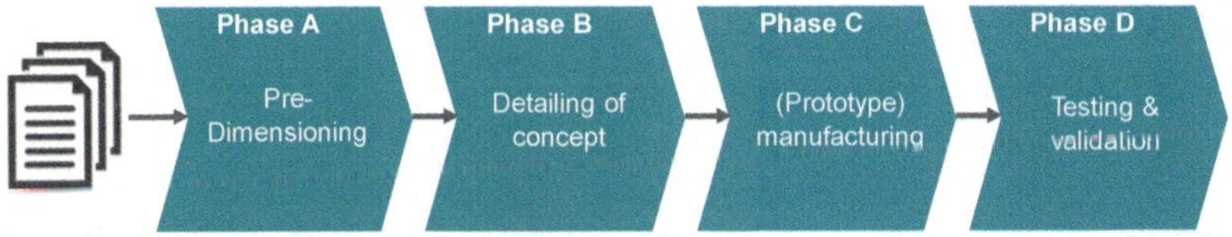

Fig. 6: Phases of the leaf spring development process.

2.3.1 Phase A: Pre-Dimensioning

Starting from the desired spring targets and boundary conditions, general feasibility is checked and a first design approach is developed. In this initial phase of the development process, using analytical tools based on the beam theory and simple Finite Element models enable an efficient study of different concepts.

2.3.2 Phase B: Detailing of concept

In phase B, detailed Finite Element models are used for an in-depth study of the spring concepts. Focus is not only on the spring performance itself but also on the interaction between the spring and the attachments (e.g. bushings) to ensure that the requirements of the subsystem are fulfilled. The mechanical feasibility of the spring design is evaluated in a lifetime analysis, e.g. applying phenomenological damage models. Therefore, a detailed understanding of the used GFRP material is necessary, which is typically got from quasi-static and dynamic tests on coupon level.

2.3.3 Phase C: Manufacturing

Choice of the appropriate manufacturing process depends on many factors (e.g. volume, design, material availability) and, therefore, must be made for each composite spring individually. In general, resin transfer molding (RTM) and prepreg pressing are well suitable manufacturing routes for high volume spring production, already proven in automotive serious projects. Fig. 7 and Fig. 8 show the composite leaf spring process flow of the RTM and prepreg pressing technology, respectively.

The main technological difference between the two processes is that in the RTM process a resin system is injected within a mold into a dry preform, while during prepreg pressing an already pre-impregnated preform is pressed and cured. Due to the high injection pressure, RTM tools are more complex than molds for prepreg pressing and a dosing unit for the resin system is needed, in addition. However, RTM offers high automation potential and is therefore well suitable for high volume composite leaf spring scenarios.

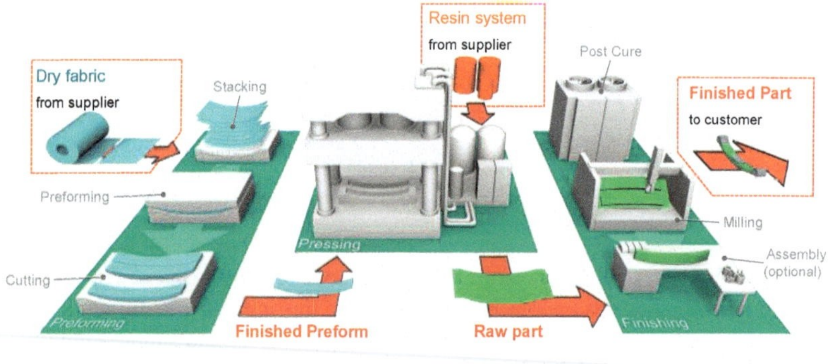

Fig. 7: RTM manufacturing route for composite leaf springs.

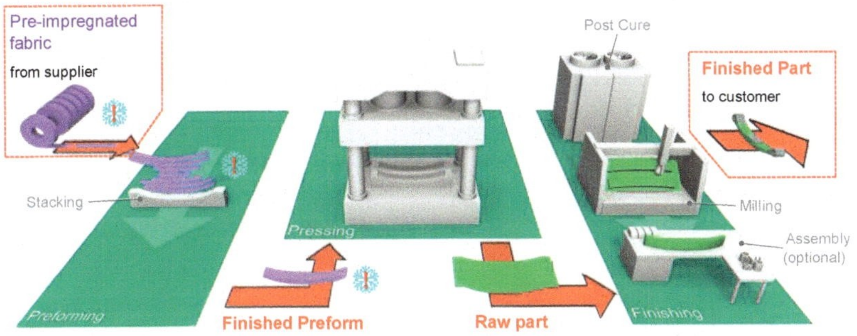

Fig. 8: Prepreg pressing manufacturing route for composite leaf springs.

2.3.4 Phase D: Testing & validation

Mechanical tests on component (GFRP spring) and subcomponent level (spring and attachments) are typically performed. Stiffness tests of the spring itself enable the evaluation of the spring design and are important for the validation of the simulation models (see subchapters 2.3.1 and 2.3.2). Further, stiffness measurements are an important part within the quality control and can be integrated relative easily to series production processes. Focus on the subcomponent tests is to check the durability requirements under representative boundary conditions (e.g. load spectrum, climate, media). Considering the attachments allows a realistic loading scenario of the spring and enables the evaluation of the subsystem properties. As polymer composites are more sensitive to temperature and humidity effects compared to metals, fatigue tests under so called hot/wet conditions are found to be the most relevant for leaf springs made of GFRP.

3 TowPreg winding struts

With the innovative and lean TowPreg spatial (3D) winding process high quality and high volume automotive components meeting their structural and economical requirements can be realized. Based on the experience of the successful development and prototyping of structural components, realized on a serial production equipment, the technology will be described in the following chapter.

The usage of single pre-impregnated fibres (TowPreg) allows load path optimized and extremely light structural components and chassis parts. The manufacturing process is characterized by straining tows between guiding pins or load introduction elements which get consolidated and cured afterwards. The combination of an automated fibre lay-up by a robot with a modified high-performance epoxy resin featuring rapid curing (3 min cycle time) makes the process highly suitable for ultra-light automotive components with consistent quality and waste-free production.

The lean manufacturing process is characterized by only three process steps, also illustrated in Fig. 9:

1. Automated Winding

   A robot arm steers the guiding pins through the end-effector of the winding eyelet. Thereby the TowPregs is unspooled and laid around the pins.

2. Isothermal Pressing

   After the winding preform is completed, it gets placed in a heated press-tool. Closing the tool consolidate the preform and cures the component (Fig. 10).

3. Post Treatment

   Due to the tool design, only a manual post treatment step is necessary to remove the burr.

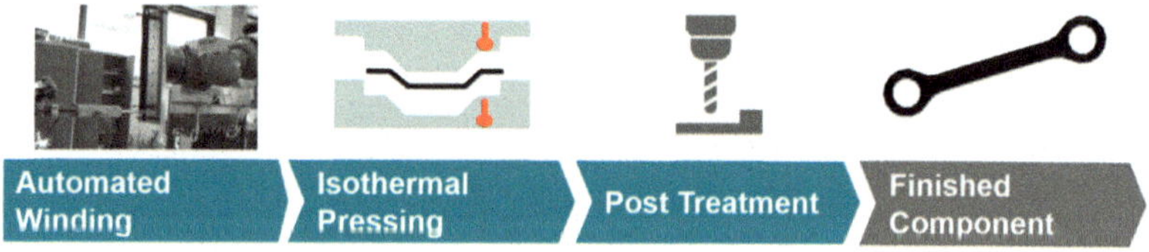

Fig. 9: Manufacturing approach for TowPreg winding components.

a) b)

Fig. 10: TowPreg preform (a) and the consolidated and cured component (b).

The usage of a single pre-impregnated fibers (TowPreg) is one of the characteristics of the manufacturing approach. With the in-house development and production of TowPregs an individual adjustment of the fiber type and the resin system based on the application requirements is possible. Another benefit is the high-quality consistency, such as resin content or tack, as well as the avoidance of a release film or paper and its clean processing.

To reach process time in the minute range TowPregs with a snap-cure resin system can be used. For example, the resin system E420 is a modified high-performance epoxy resin featuring rapid curing at elevated temperatures. The curing can take place within 3 minutes at 150°C and even shorter press cycles can be achieved if a suitable post-cure is applied. The high glass transition temperature (140°C - 150°C) and its fast build-up allows for a hot demoulding of the composite part. The combination with a tailored tack for automated processing makes TowPregs based on E420 the ideal choice for high-volume serial productions. The long out-life at room temperature (up to one month) permits a simple processing during production.

One of the most noteworthy feature of the process is the possibility to direct integrate load introduction inserts. The form-locking between the inserts and the load bearing fibres leads to a direct load introduction without any stiffness-loss connection such as an adhesive bounding or phase-delayed load transmission. Nevertheless, the design of the insert is a critical for the implementation in the manufacturing process as wells as the mechanic of the component. Other aspects such as the differences in the thermal expansions or the corrosion issues between fixing point and insert or insert and the fibres, need to be investigated. Nevertheless, there are solutions which already have been proven. To utilize the performance to the utmost, design freedoms for the inserts are required, enabled by using additive manufacturing methods.

For components in the suspension systems, chassis or body-in-white components several requirements need to be achieved. That is the reason why the design and component development is the key factor for working concepts. Only with known stress conditions and strength dimensioning resulting from the prevailing load spectrum a targeted design is possible. This light construction method is characterized by a minimum use of material, which only utilize the material needed for the dominated load case. This can be achieved with specific dimensioning methods based on resolving the component on roving level or field-based properties.

(a) **(b)**

Fig. 11: Test sample for material characterization (a) and roving-based simulation approach (b).

Complex and multi-axis loadings must be reduced to single oriented conditions, to meet the requirements of the monolithic and highly anisotropic structures. However, hybrid designs composed of skeleton structure and a multi-axis substructure can be designed. Furthermore, the manufacturing conditions and limitations must be already considered during the design and development to get an architecture which is ready for automated manufacturing. This is one of the key drivers of the technology.

Concluding, the new developed manufacturing technology with its closed development approach shows great potential for suitable application at an attractive cost level. Especially when current state of the art technology reaches their limits e.g. functional integration, cross-section variations or design flexibilities, 3D TowPreg winding might be a smart solution. With a holistic development approach that considers the application's and process specific boundary conditions enables the "first-time-right" manufacturing of components.

4 Summary

In this paper, the general suitability of using innovative composite design in suspension applications was demonstrated using the examples of GFRP leaf spring and single- and multi-axis strut applications. Herein, presented the holistic development approach, combining design and manufacturing concepts, is essential for harvesting the light-weight potential of polymer composites and to enable their effective use in automotive series production.

Bibliography

[1] http://maybach300c.blogspot.com/2012/09/rigid-and-semi-rigid-crank-axle.html

Design of an innovative concept of a front wheel car suspension with a metal-elastic triangular control arm

Christina Sax, University of Applied Sciences Munich

Dr. Ralf Stroph, BMW Group

Norbert Deixler, BMW Group

Abstract

In strut front suspensions, a sickle-shaped triangular control arm is often used as a lower suspension arm for small and mid-sized vehicles. With the growing importance of electric vehicles, the demands for driving comfort and acoustics are increasing. These requirements force a reduction of the rubber bushing stiffness, which cannot be properly realized with the current bushing arrangement of the sickle-shaped triangular control arm. For a solution to the goal conflicts it is necessary to clearly separate the lateral stiffness and longitudinal elasticity functions in the lower suspension arm. Previous implementations use the resolution of the triangular control arm in two separate suspension arms. To reduce costs, a metal-elastic triangular control arm is developed which combines the functions mentioned above in one component. The concept consists of a compression strut region, which is responsible for the absorption of longitudinal and proportional lateral forces and a transverse link area, which transmits lateral forces along its alignment, but allows a displacement of the front wheel suspension in the vehicle longitudinal direction by its metal elasticity.

In this paper, based on the state of the art, the procedure of the design of the metal-elastic triangular control arm is explained. The characteristics of the front wheel suspension with the sickle-shaped and the metal-elastic triangular control arm are shown in a functional data comparison.

1 Introduction

With the change in mobility and the growing demand for electric vehicles, new challenges arise for established suspensions. The missing background noise of a conventional combustion engine direct the attention of the vehicle occupants to the acoustic, generated from or transmitted by the chassis, which is gaining considerable importance. Not only the use of new drive concepts, but also the constant cost pressure requires cost-effective and functionally optimized suspension systems at the same time. This research work focuses on the strut front suspensions of small and mid-sized vehicles, usually having deficits in the areas of driving comfort and vehicle acoustics. A metal-elastic suspension control arm is integrated into a strut front suspension in order to improve driving comfort and acoustics of the vehicle. This makes it possible to fulfil the demands and challenges arising from the electrification of vehicles and also provides a cost advantage compared to previous solutions. With a prototype the function of the concept will be proven in a test vehicle. This research work is carried out in cooperation with the vehicle technique and technologies department of the BMW Group.

2 State of the Art

2.1 One-joint strut front suspension

The one-joint strut front suspension is an independent wheel suspension and is often used in the small and mid-sized vehicle segment due to its simple and space-saving design [1]. Figure 1 illustrates the wheel guidance components of the one-joint strut front suspension of a BMW series car.

Figure 1: One-joint strut front suspension with sickle-shaped triangular control arm

The suspension strut consists of a vertical damper which is surrounded by a coil spring and is mounted to the vehicle body. The steering knuckle underneath is fixed to the suspension strut and connected to a sickle-shaped triangular control arm with a ball joint at the lower end. This control arm is elastically mounted with rubber bushings at two points to the front subframe in the longitudinal direction of the vehicle. For the transmission of the steering movement there is a tie rod between steering knuckle and the steering gear. The steering system is positioned behind the wheel centre.

2.2 Sickle-shaped triangular control arm

The sickle-shaped triangular control arm used in the front suspension described above is one of the most common types of control arms in strut front suspensions. The implementation usually consists of a single or double-shell sheet metal forming part. The rear bushing, seen in the direction of travel, is additionally connected to the vehicle body by its mounting. Figure 2 shows a current series BMW double-shell triangular control arm.

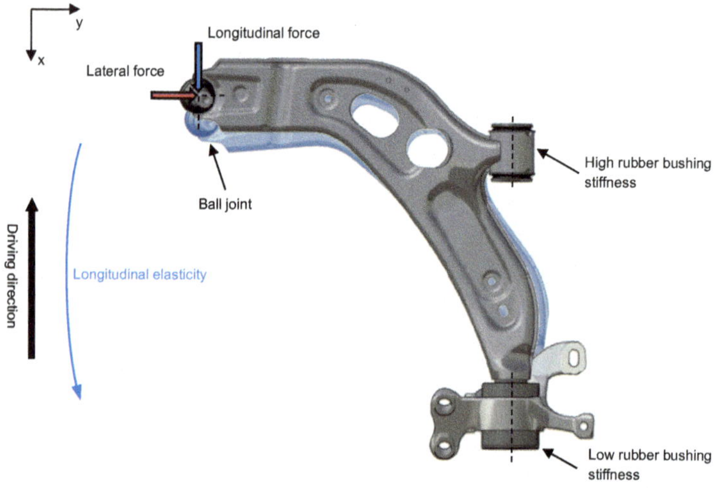

Figure 2: Sickle-shaped triangular control arm

High driving comfort is achieved by an appropriate longitudinal elasticity of the front wheel suspension under longitudinal forces. This means that slight, compliant movements caused by the wheel forces, along the longitudinal vehicle axis, are permitted (shown semi-transparent in Figure 2). These deflections can be used to compensate small potholes, edges and comparable road unevenness. In the concept of the sickle-shaped triangular control arm, the longitudinal elasticity is generated by a low stiffness of the rear rubber bushing. In order to achieve the softness of the rear rubber bushing, the front rubber bushing has to be a multiple harder to absorb loads in axial direction [2]. A higher rubber bushing stiffness results in a reduced decoupling potential and a higher noise level inside the vehicle [3]. A conflict between comfort and acoustics arises. In order to improve the acoustic properties, the stiffness of the front rubber bushing has to be reduced, which is not feasible due to the bushing arrangement.

3 Metal-elastic triangular control arm

3.1 Conception

To resolve the conflict between comfort and acoustics it is necessary to clearly separate the „lateral stiffness“ and „longitudinal elasticity“ functions in the lower suspension arm level of the strut front suspension. Previous implementations use the resolution of the sickle-shaped triangular control arm in two separate suspension arms. This leads to a second component which has to be mounted additionally and generates further costs. A common example is the two-joint strut front suspension, typically used in the next higher vehicle segment.

For this reason, a metal-elastic triangular control arm is developed that combines the functions mentioned above in one component. The concept consists of a compression strut region and a metal-elastic transverse link area. The compression strut transmits longitudinal and proportional lateral forces, the metal-elastic transverse link transfers lateral forces along its orientation and allows compliant longitudinal movements of the front suspension by its metal-elasticity. To save costs and to simplify the component, the compression strut region and the transverse link area is thought to be manufactured form a single sheet metal plate. Figure 3 shows the conception of the metal-elastic triangular control arm.

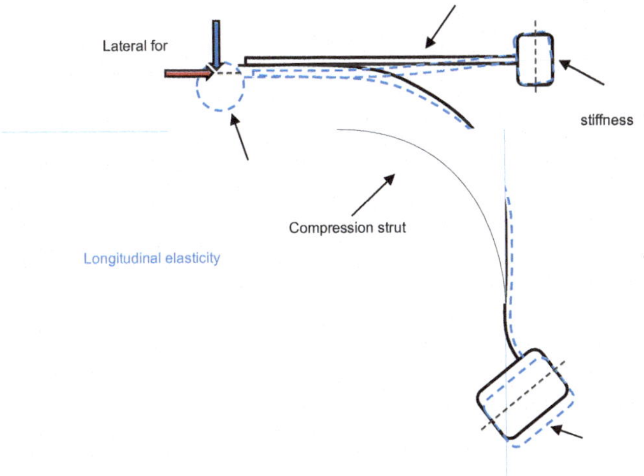

Figure 3: Conception of the metal-elastic triangular control arm

The metal-elastic transverse link, with rectangular cross-section, works according to the Euler-Bernoulli beam theory [4]. An elastic deformation of the transverse link is intentionally allowed in order to achieve an increased longitudinal elasticity of the front suspension (dashed line in Figure 3) of up to 30%. The longitudinal elasticity is therefore no longer produced by rubber bushings exclusively and it is possible to create an optimal bushing arrangement according to the loads. The unfavourable axial load of the front rubber bushing of the sickle-shaped triangular control arm concept is almost eliminated, which allows a reduction of the front rubber bushing stiffness of up to 75%. Figure 4 illustrates the force diagram of the compared concepts.

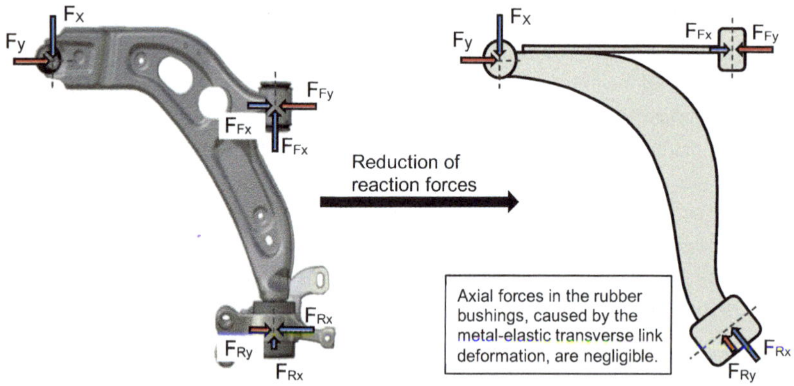

Figure 4: Force diagram of the sickle-shaped and metal-elastic triangular control arms

The concept of the metal-elastic triangular control arm increases the longitudinal elasticity of the front suspension and allows a reduced front rubber bushing stiffness at the same time. Higher driving comfort and better decoupling between road-side noise and vehicle can be achieved, resulting in a more pleasant acoustic experience inside the vehicle. In order to maintain an understeering driving behaviour of the vehicle, the lateral stiffness of the steering gear has to be adjusted if the bushing stiffness of the transverse link is reduced. For an understeering driving behaviour, the front suspension needs a toe-out tendency under lateral and longitudinal forces [4]. To generate a toe-out tendency under lateral forces, the lateral stiffness of the steering gear has to be reduced too, caused by the lever ratios at the steering knuckle. For this reason, an elastic steering gear mounting is used instead of a rigid connection between the steering gear and front subframe. By selecting the stiffness ratio between the transverse link rubber bushing and the elastic steering gear mounting, the driving behaviour can be adapted to the character of the vehicle individually.

3.2 Virtual dimensioning and validation

The virtual dimensioning and validation of the metal-elastic triangular control arm passed through several steps. In addition to the usual process of the virtual dimensioning and validation of conventional suspension components, the influence of the metal-elastic part on the structural durability and elasto-kinematic properties has to be taken into account. The geometry of the metal-elastic triangular control arm is determined by the results decisively. The following Figure shows the mentioned correlations.

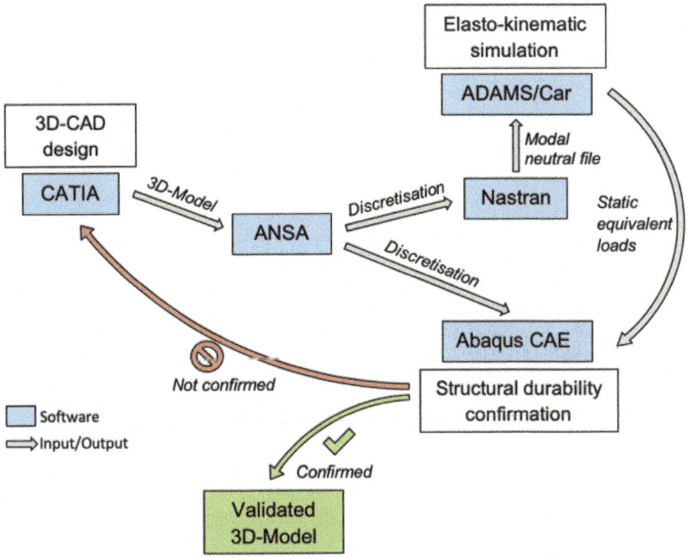

Figure 5: Procedure and correlations of the virtual dimensioning and validation

The structural durability is usually determined by a linear finite element analysis. Static equivalent loads are applied to the geometry, discretized with finite elements, to evaluate the strength of the component [2]. The maximum stress in the component should always be below the yield strength of the material.

In addition to the stress analysis, the buckling force of the metal-elastic transverse control arm has to be defined, which depends on the material properties, moment of inertia, boundary conditions and the size of the component. To prevent a plastic, irreversible deformation or failure, the buckling force needs to be higher than the loads applied to the metal-elastic transverse link. A sufficient safety factor has to be selected. The buckling force can be determined using a non-linear finite element analysis, which includes material and geometry modifications during the load application. This approach leads to more accurate results than a calculation using the Euler-Bernoulli beam formula [5].

To analyse the dynamic interaction of all suspension components, a multi-body system simulation was carried out. One element of the multi-body system simulation is an elasto-kinematic analysis, which determines the toe and chamber properties of the suspension, considering elastic connections between the components, e.g. rubber bushings. It represents a virtual suspension test rig and enables an evaluation of the expected driving behaviour of the vehicle. Usually suspension control arms are integrated as rigid bodies in the simulation. To consider the elasticity of the metal-elastic transverse link, it needs to be embedded as flexible body. This is realized by a „modal neutral file" (MNF) in the simulation program ADAMS/Car. The modal neutral file consists of condensed finite element nodes and includes information about geometry, nodal mass, inertia and mode shapes [6]. Figure 6 shows the setup of the simulation model in ADAMS/Car with the metal-elastic triangular control arm as a flexible body.

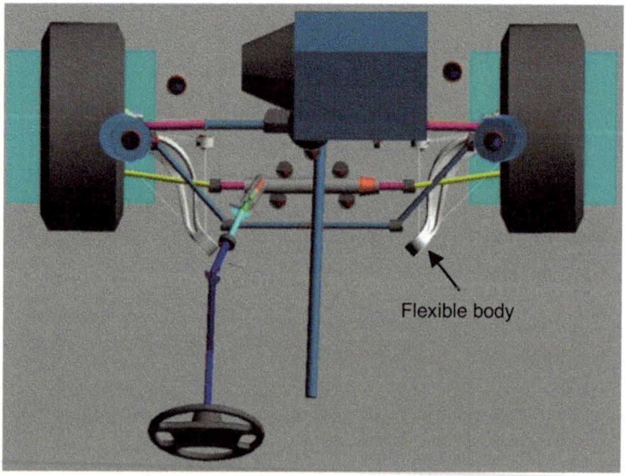

Figure 6: Multi-body system simulation of the front suspension with the metal-elastic triangular control arm implemented as a flexible body in ADAMS/Car

3.3 Prototype

To determine the function of the metal-elastic triangular control arm, prototype testing has to be performed in a test vehicle. For simplifying reasons, it was decided to separate the first metal-elastic triangular control arm prototype into two components temporarily. Therefore, it is possible to manufacture the bigger part of the concept, the compression strut region, as an aluminium milled part. In prototype manufacturing the material aluminium offers advantages in shorter production times and lower tool wear compared to the material steel [7]. To evaluate the metal-elastic properties for a later sheet metal forming part, a steel material was chosen for the transverse link. Due to the simple geometry and size it can be realized as milled part as well. The two components were connected with bushes and bolts. Based on the previously defined premises, the geometry of the metal-elastic triangular control arm was implemented with a computer-aided design program. Figure 7 shows the final geometry of the prototype.

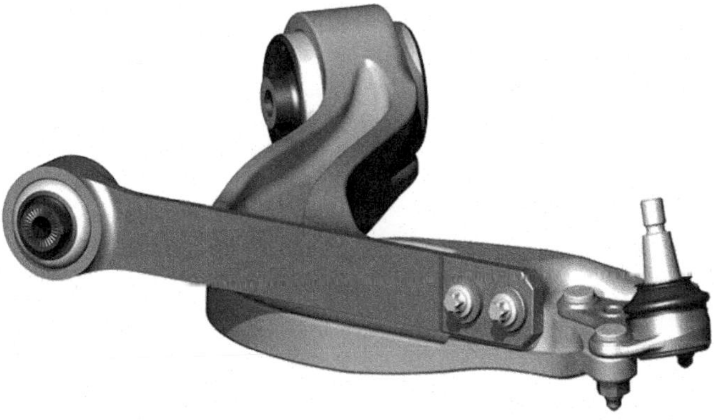

Figure 7: 3D-Model of the metal-elastic triangular control arm prototype

To prevent contact-corrosion of the material pairing of aluminium and steel, the metal-elastic transverse link was painted across its connection to the compression strut after successful production and assembly of the prototype [7].

4 Comparison of functional data

With the first prototype of the metal-elastic triangular control arm the driving behaviour as well as the acoustic and comfort properties of the test vehicle were determined and evaluated.

To evaluate the driving behaviour, the elasto-kinematic parameters of the front suspension of the test vehicle were measured on a Kinematics and Compliance (KnC) test rig. Based on the measurement it is possible to validate the associated simulation model, see chapter 3.2, and to compare the elasto-kinematic parameters with those of other vehicles. The KnC measurements of the series BMW one-joint strut front suspension with sickle-shaped triangular control arm and a series BMW two-joint strut front suspension are used to compare the driving behaviour of the test vehicle. The driving behaviour is primarily influenced by the toe characteristics, which are shown for the three compared vehicles in the following Figure.

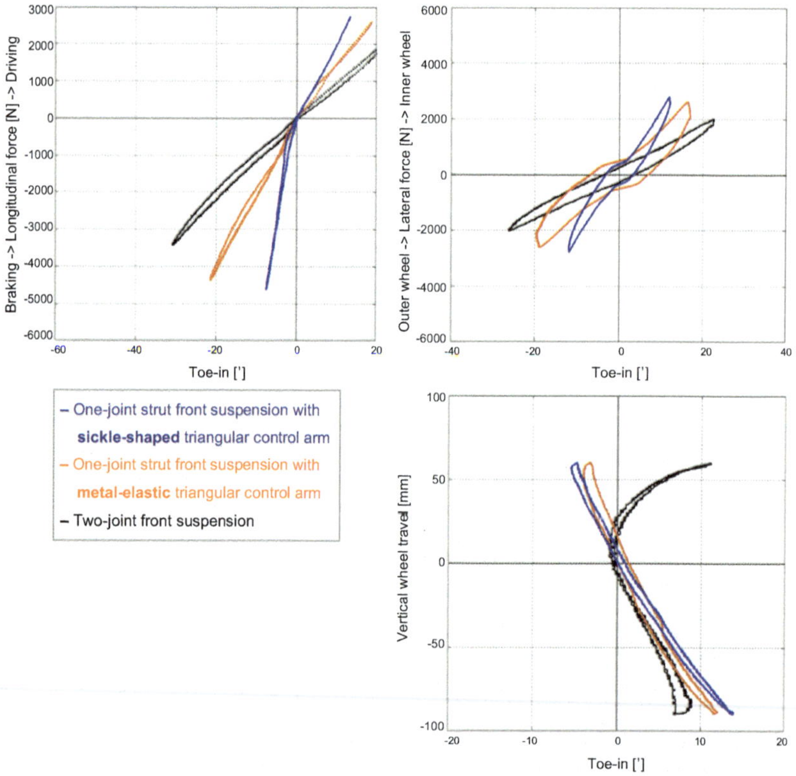

Figure 8: Comparison of the toe characteristics to evaluate the driving behaviour of the test vehicle using different lower control arm systems

All three toe curves have a toe-out tendency under braking force and outer wheel side force and contribute to understeering driving behaviour of the vehicles. A lower slope of the toe curve signifies a good-tempered, more understeering driving behaviour, a higher slope signifies a sporty, direct driving behaviour. The toe values of the strut front suspension with the metal-elastic triangular control arm are located in the target corridor, given by the two series suspensions. Under positive vertical wheel travel, the strut front suspension with metal-elastic triangular control arm is oriented to the preferred toe-out tendency of the strut front suspension with sickle-shaped control arm. The driving behaviour of the test vehicle was expected as good-tempered, not to sporty, approaching the characteristics of the vehicle with a two-joint strut front suspension, which is known for good driving performance and comfort.

To evaluate the test vehicle's interior acoustics and driving comfort, first acoustics measurements, compared to the vehicle with the sickle-shaped triangular control arm, were carried out. In the case of curb crossing, a noise reduction up to 4dB was determined, effected by the higher longitudinal elasticity and a higher acoustic decoupling potential of the reduced rubber bushings stiffness of the metal-elastic suspension concept. The interior vehicle noise level, measured on an acoustic roller dynamometer, has shown to be improved as well.

The subsequent subjective driving test confirmed a good vehicle handling at series level and noticeable improvements in driving comfort and vehicle interior acoustics.

5 Summary & future prospects

This paper presents the optimization potential of the one-joint strut front suspensions of small and mid-sized vehicles. For this reason, an innovative suspension control arm is developed, which has on the one hand functional advantages and on the other hand the potential for a cost-effective manufacturing compared to existing solutions. The simple working principle of the metal-elastic part enables a higher longitudinal elasticity of the front suspension and the application of soft rubber bushings at the same time and leads to higher driving comfort and improved acoustic properties of vehicle. By a considerable virtual dimensioning and validation process, a prototype of the metal-elastic triangular control arm concept was realized, whose functions were confirmed by prototype testing in a test vehicle. Further steps to this research work is to develop an one-shell sheet metal forming part of the metal-elastic triangular control arm, considering appropriate materials and manufacturing processes, which might be suitable for the series application in future (electric) vehicles. Based on this, weight and cost assessments will take place and lifetime limiting effects (e.g. corrosion) will be analysed.

References

[1] Matschinsky, W.: *Radführungen der Straßenfahrzeuge. Kinematik, Elasto-Kinematik und Konstruktion*. Berlin-Heidelberg: Springer Verlag, 2007.

[2] Ersoy, M.; Gies, S. (eds.): *Fahrwerkhandbuch*. Wiesbaden: Springer Vieweg, 2017.

[3] Zeller, P. (eds.): *Handbuch Fahrzeugakustik. Wiesbaden. Grundlagen, Auslegung, Berechnung, Versuch*. Wiesbaden: Vieweg+Teubner Verlag, 2012.

[4] Braess, H.-H.; Seiffert, H. (eds.): *Vieweg Handbuch Kraftfahrzeugtechnik*. Wiesbaden: Springer Vieweg, 2013.

[5] Öchsner, A.: *Theorie der Balkenbiegung. Einführung und Modellierung der statischen Verformung und Beanspruchung*. Wiesbaden: Springer Vieweg, 2016.

[6] *Getting Started Using Adams/Car*. Mechanical Dynamics, 2002.

[7] Ostermann, F.: *Anwendungstechnologie Aluminium*. Berlin-Heidelberg: Springer Vieweg, 2014.

You drive me crazy: enhance the driving ability of professional drivers to improve controllability judgement of hazardous situations

Dr. Marcus Perner, Marius Levers, Martin Gebhardt, Simon Heine
IAV GmbH, Vehicle Dynamics, Chassis Development

Introduction

Nowadays, a huge amount of research work relates to autonomous driving. Nevertheless, the degree of automation in current series-production cars is limited to the 2nd and to the 3rd level according to SAE J3016 [1]. The electrical powered steering system is an essential component of these cars because it provides controllability for the driver and assist for comfort and safety reasons. In contrast, the relating motor torque is able to harm other cars or even pedestrians in case of malfunction. Functional safety considers this issue. Within the development lifecycle, several verification activities are performed to obtain a safe and comfortable steering system. Test drivers verify the functionality and validate safety measures various times before start of production. In case of an initial status of hardware and software, only professional drivers are allowed to perform these drive tests using an early prototype on proving ground. When the component reaches a certain development progress, test persons like everyone are allowed.

This paper introduces a concept that professional drivers are able to adapt their driving skills closer to average drivers. In literature, hand wheel torque is seen as an objective criterion to judge controllability of a sudden loss of steering assist. Therefore, this paper proposes a system to display the current hand wheel torque. The assumption is that the professional driver is able to adapt the driving ability to a colored bargraph display. At first, a proposal to relate hand wheel torque with controllability is given. The controllability rating is based on the international standard ISO 26262 [2]. Based on that, the functional architecture of the system is shown. The design consists of a steering wheel with enhanced torque sensing, a microcontroller and a display for visualization. In addition, the results from vehicle tests are discussed. The initial study implies the potential of the proposed approach to get the professional driver to adapt his ability closer to the so-called every-human. The paper concludes with a plan for further development.

Verification and validation for development of steering systems

Verification and validation are two essential methods during safety lifecycle. Figure 2 illustrates a basic overview of safety activities. The overall development is based on the common V-model process. First, activities and documents related to the management phase need to be created. For instances, an impact analysis defines all necessary steps for further development based on ISO 26262 [2]. The safety plan mainly includes safety organization of the project, responsibilities of tasks, strategy for validation, confirmation reviews and information for supporting processes. During the complete lifecycle, management tasks are ongoing. Though, activities of the concept, development and release phase are performed subsequently.

To give an example, the hazard and risk analysis is performed within the concept phase. The aim is to identify and to categorize hazardous incidents. The result is a list of safety goals with corresponding automotive safety integrity level (ASIL) to prevent or disarm hazards and to avoid unreasonable risk. Harm means an injury to the life, body or health of a person. The procedure is illustrated in Figure 1. The figure shows the single steps of the analysis in the middle. Input documents and necessary engineering knowledge are related to each task. As a first step, project information is analyzed to define the field of application and to determine relevant driving scenarios.

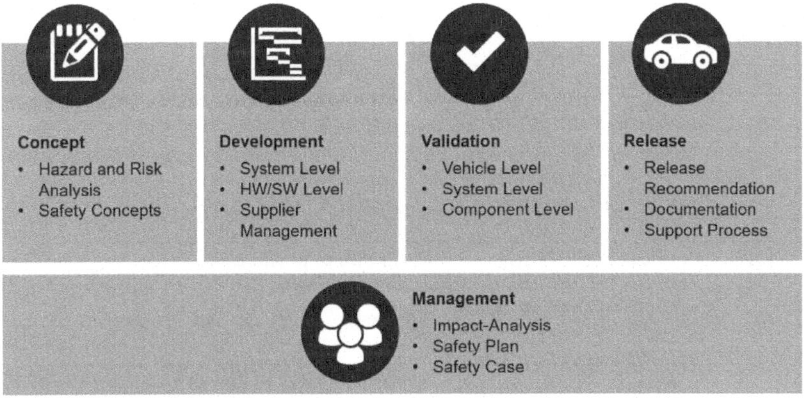

Figure 1 Generic methodology of a hazard and risk analysis

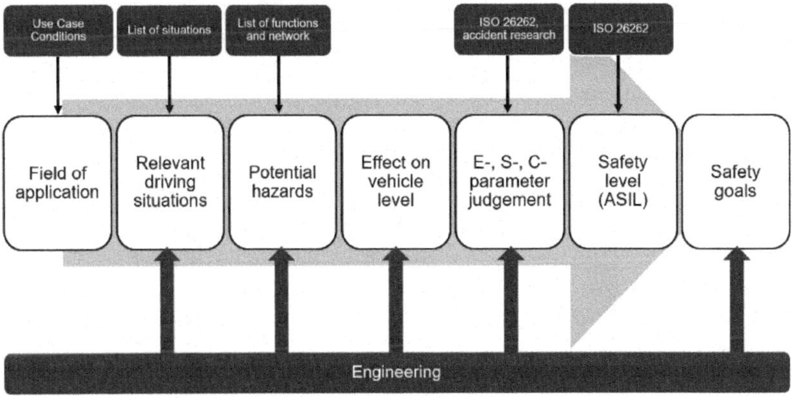

Figure 2 Overview of basic functional safety activities

Potential hazards are evaluated using the list of functions of the item and network conditions. As a next crucial step, the effect on vehicle level is assumed. The ASIL is determined by using parameters to classify the probability of the situation, the severity of the failure and the controllability of the vehicle reaction. The ASIL of the hazard indicates the remaining permitted risk. Finally, safety goals for further development of safety measures are formulated. Important information how to judge E-S-C-parameters are listed in the ISO 26262 [2]. Additional information for severity can be given by accident research. Though, assumptions of the controllability of certain hazardous situations have to be made. These assumptions need to be verified. Therefore, certain verification and validation activities are performed. ISO 26262 gives a definition of the controllability rating and the meaning (cf. Table 1). Four different ratings exist. The rating C0 means there is a noticeable reaction of the vehicle and the driver needs to compensate the reaction. The amount of compensation is usually controllable by almost every driver. The highest rating is C3 because it results in an uncontrollable situation. It defines that less than 90% of the average drivers are able to avoid harm.

Table 1 General classification of the controllability rating by means of ISO 26262 [2]

Rating	Meaning	Definition to practice
C0	Controllable in general	Noticeable, but controllable in general
C1	Simply controllable	More than 99% of the average drivers or other traffic participants are able to avoid harm
C2	Normally controllable	Between 90% and 99% of the average drivers or other traffic participants are able to avoid harm
C3	Difficult to control or uncontrollable	Less than 90% of the average drivers or other traffic participants are able to avoid harm

Proposal to classify driving ability by hand wheel torque

In literature, hand wheel torque is seen as one possible measure to classify controllability [3]. In the event of no steering assist, hand wheel torque increases suddenly. Therefore, the driver is surprised and needs to counteract. The paper of Harbluk et al. [4] investigates the impact on driver's vehicle control. During slow driving 35 % of the drivers could not keep the lane. Moreover, in case of driving a large SUV every turn was unsafe if assist failed. Pierce et al. [5] evaluate the effect of loss of assist of a steering system and put focus on the maximum physical effort under different test maneuvers. For the study, data of 182 female test drivers is analyzed. In conclusion, the limit of 68 N is recommended not to exceed at any rate.

ECE R79 requires that the maximum force while losing steering assist is limited to 150 N [6]. In contrast, the military standard MIL-STD-1472G defines force by 220 N for men and 183 N for women [7]. The standard puts focus on very well trained persons. That is why the steering force is significantly higher than in other studies. However, development for public transportation and for traffic development is independent from gender point of view. Therefore, the lowest amount of force is considered from this specific standard. As a first approach, we propose to combine the controllability of the vehicle with the maximum hand wheel force based on the named studies. Figure 3 illustrates the relation. The proposal is strictly related to the failure of a sudden loss of assist. If a loss of assist happens, the necessary hand wheel torque to steer the curve increases. If the driver cannot exert sufficient torque, the lane will be departed rapidly. Sure, dynamic driving influences the resulting torque. To give an example, an increase of lateral acceleration causes high return torque transmitted due to the contact between street and wheels. Nevertheless, the resulting torque that the driver has to manage is the key factor to keep the lane.

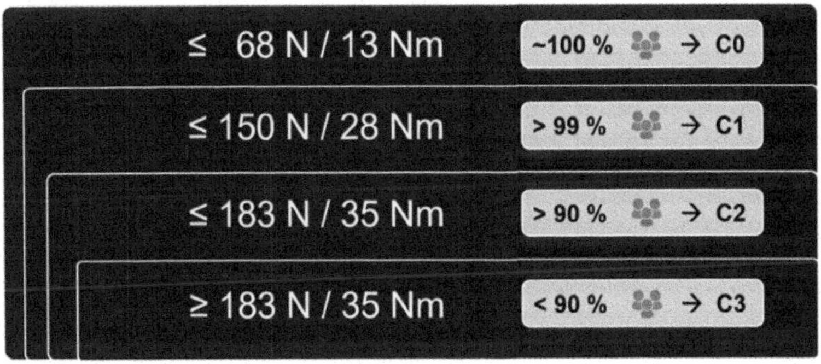

Figure 3 Proposed relation between controllability and hand wheel force

System Design

This chapter introduces the design of the proposed bargraph system. Figure 4 shows the functional design. The steering wheel is equipped with an additional sensing device. This is necessary because the measurement range of the torque sensor commonly used in series production steering systems is not sufficient in case of a loss of steering assist. The sensing device is connected to a signal converter. One purpose of this converter is to supply power for the sensor. Another purpose is to transform the sensor output to suitable CAN-signal. Another CAN-controller saves the CAN-Frames. A microcontroller reads out the memory of the CAN-controller, processes data and triggers the bargraph. Finally, the current torque is visualized to the driver.

For understanding the approach, Figure 5 shows the visualization concept of the bargraph. The bargraph consists of 24 LEDs. The upper edge of the bargraph relates to the C2/C3 border of 183 N. Therefore, each single LED represents about 8 N of hand wheel force. Based on the current hand wheel torque the microcontroller manages the coloring. Until the limit of 68 N only the first 8 LEDs light green. Above this C0/C1-limit, the next 10 LEDs light yellow if the current value equals up to 150 N. If the current value is between 150 N and 183 N, the next LEDs light red. Above 183 N the complete bargraph flashes red that equals the controllability level C3.

The proposed design is applicable to any test vehicle. That is one great benefit. The only restriction is the measurement steering wheel. In case of different versions or producers, the microcontroller offers automated detection of the connected version. Therefore, the system is easy to integrate into vehicles and ready to use if needed.

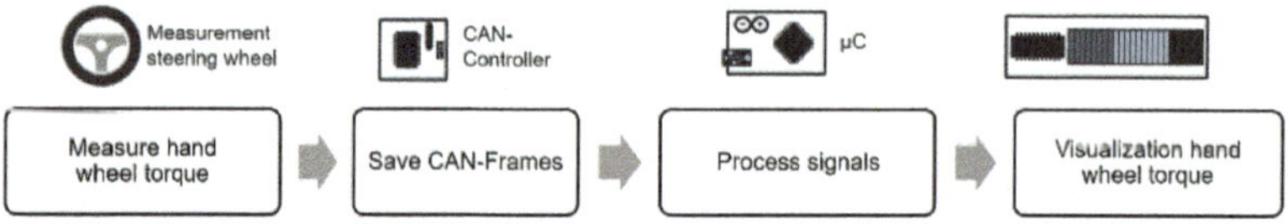

Figure 4: Functional design of the proposed bargraph system

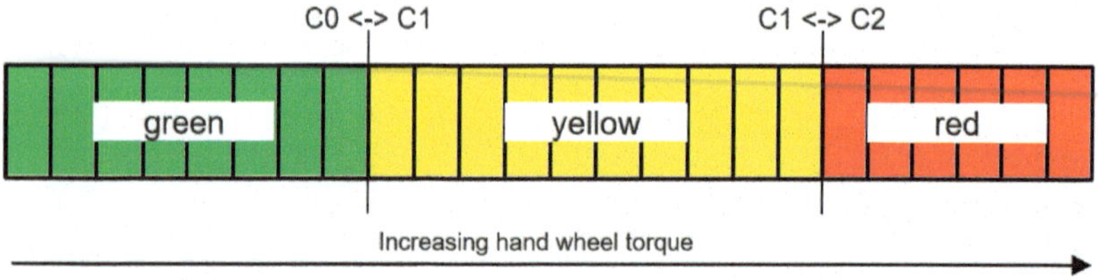

Figure 5 Visualization concept of the bargraph: green = C0, yellow = C1, red = C2, red alert = C3 (not shown)

Integration and Testing

By integration and testing, the behavior of the bargraph system is investigated. As a first step, the testing is done on a test rack to demonstrate a successfully implemented functionality. After that, the system is integrated and connected into a vehicle. During vehicle tests the performance and acceptance is evaluated.

Test rack

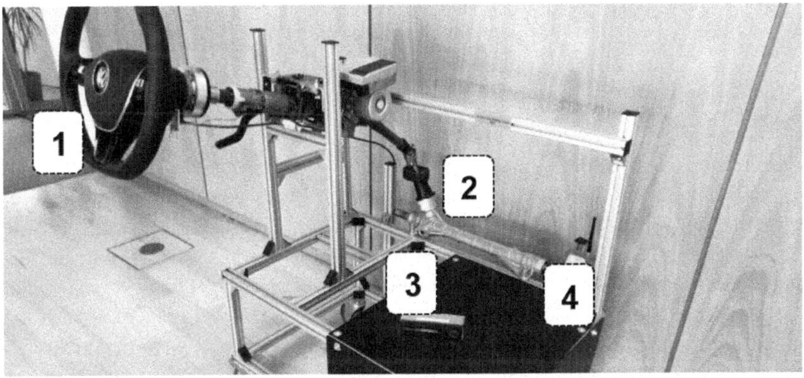

Figure 6 Test rack with 1- measurement steering wheel, 2- steering system, 3- bargraph system, 4- supply box

The test rack mainly consists of a measurement steering wheel (1), a series production electrically powered steering system (2), the bargraph system (3) and the supply management equipment (4). Figure 6 shows these components at the test rack. For instance, additional power supply and signal converter are needed for operation (cf. Figure 4). Because there are no wheels connected to the rack and there is no velocity, no counter force arises in case of steering. Finally, no force would be detectable by the torque-sensing device within steering wheel (1). Therefore, an additional mechanical end stop is attached to the rack at both sides. However, a counter force is detectable and is displayed by the bargraph system.

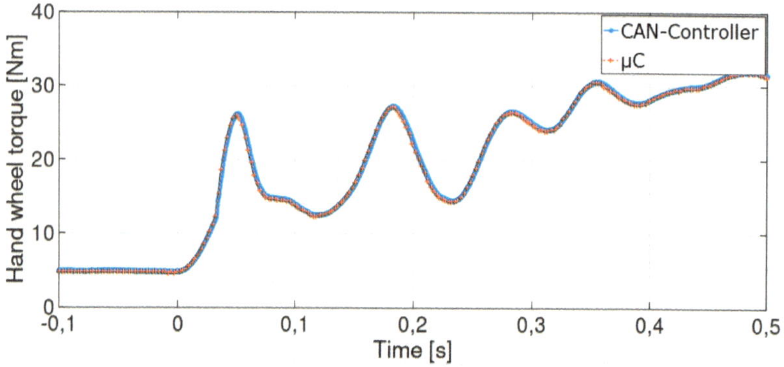

Figure 7 Simultaneous measurement of hand wheel torque using CAN-Controller (solid) and microcontroller (dash)

For functional testing, the measured signal is saved simultaneously via CAN-Tool and via microcontroller. Therefore, a signal converter is directly connected to the measurement steering wheel. It converts the voltage signal to digital signal suitable for CAN communication. At this point, an automotive CAN-to-USB-Adapter is used to connect to the measurement laptop. To see the signal that is visualized by the bargraph system, the microcontroller is connected to the measurement laptop, as well. Figure 7 shows the result of the simultaneous measurement. Both signals match very well. In conclusion, the functional design is sufficient for visualization of the hand wheel torque.

Vehicle Tests

The functionality needs to be tested in vehicle environment. Therefore, the measurement steering wheel is integrated into a test vehicle. The system design remains the same as shown in Figure 4. For power supply, the signal converter of the sensing device is connected to common 12V lead-out in the car. The bargraph system is connected to the USB-port of the infotainment system. The output of 5V DC is sufficient for the microcontroller and its parts. For start of operation, several maneuvers are performed. First, no steering assist is given by the steering device to provoke hand wheel torque at the desired maximum. Secondly, the working limits between the proposed controllability classes as explained above are tested. In detail, this is the crossing from green to yellow (C0/C1), yellow to red (C1/C2) and red to flash red (C2/C3). Since, functional testing in the car results as a success, the examination of the acceptance of the bargraph system by test drivers is the next step.

Figure 8 Bargraph system installed during vehicle testing

In this study, the event of a sudden loss of assist while steady circular driving is investigated. In practice, the amount of hand wheel torque increases rapidly and forces the vehicle into straight driving and the steering angle decreases. The current yaw rate decreases delayed due to vehicle geometry, kinematic and characteristics of the tires. Finally, lane departure is unavoidable. As soon as the driver reacts to compensate, the steering angle will increase again and the yaw rate, respectively. The driver applies the highest possible hand wheel torque to avoid or minimize lane departure.

During vehicle tests, several dynamic parameters are measured. As an addition to the hand wheel torque, the lateral acceleration and the yaw rate is measured. The test maneuver is a steady state circular driving. The velocity and radius define the lateral acceleration in steady state.

All conditions are listed in Table 2. Due to these boundaries, a resulting hand wheel torque is expected in the range of approximately 20 – 28 Nm, respectively in the field of C1-rating. The delta yaw rate is seen as another indicator to judge unstable vehicle condition [8]. It describes the change of the steady yaw rate prior to the test activation and the yaw rate due to the test case.

At first, the test maneuver are performed without the knowledge of the bargraph system. That means the task for the test driver is to guarantee lane keeping during sudden loss of assist. In a second step, the maneuvers are performed again with the additional task to use minimized hand wheel torque. Therefore, the driver should align with the visualization supported by the bargraph system (cf. Figure 8).

Table 2 Test conditions (wet road)

Velocity [km/h]	Radius [m]	Lateral acceleration [m/s^2]
30	25	3,0
70	125	3,0
80	200	2,5
85	200	2,7
90	200	3,0
110	200	4,5
110	300	3,0

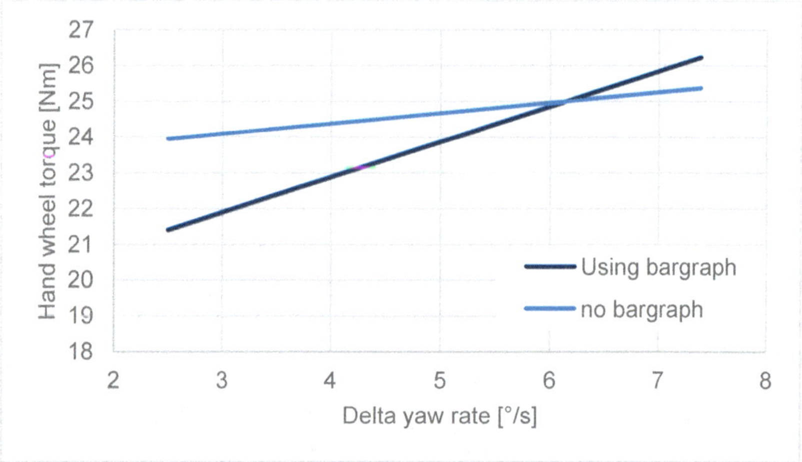

Figure 9 Mean value curve of all test measurements comparing the hand wheel torque and the delta yaw rate

The mean value curves of the corresponding data points are illustrated in Figure 9. The results are labeled with "no bargraph" and "Using bargraph". The figure shows that the test driver uses nearly the same amount of hand wheel torque over a wide range of the delta yaw rate. However, the use of the bargraph system and the additional task to keep the lane using minimized hand wheel torque results in a different steering behavior. In the area of a small delta yaw rate, less hand wheel torque is needed as the test drivers has used in prior test maneuvers. The test driver seems to be able to adapt driving ability. Finally, the diagram shows the actual required hand wheel torque to keep the lane.

Evaluation remarks

The result is based on 85 measurements done by 18 test drivers. All measurements are performed with the same test vehicle in the same configuration. Every test was performed on wet road, thus there is no influence of weather conditions. Nevertheless, the study is not representative. The goal of the study is to analyze the potential of the bargraph system to support validation activities during the development phase and to give an insight towards the acceptance by test drivers.

Conclusion and outlook

This study gave an overview of a common development cycle within the scope of functional safety. The controllability judgement as one essential step of the hazard and risk analyses was pointed out. Based on literature, a relation between the hand wheel torque and the ISO 26262 controllability rate was proposed for the specific failure of a sudden loss of assist. The challenge to judge controllability in early project progress was examined. Therefore, this paper introduced the concept of a bargraph system to visualize current hand wheel torque. The main aim is to support test drivers in their judgement of vehicle behavior in hazardous situations. The paper investigated the failure of a sudden loss of assist during steady state circular driving. The result of the driving tests shows the potential of the proposed system. In average, the test driver were able to adapt their driving ability by means of using minimized hand wheel torque to keep the lane during the failure. Nevertheless, the study is limited to 18 test drivers and 85 measurements. Consequently, the result is not representative. Therefore, more driving tests need to be performed. Another future issue is to investigate influence factors in case of other failures that result in a more dynamical vehicle behavior likewise oscillation of the steering wheel. In addition, the influence of different maneuvers will be investigated. We expect additional dependencies to the hand wheel torque in case of slalom or lane change maneuvers. However, dependencies need to be isolated to see the actual effect. Moreover, we are looking forward to achieve a mobile version of the bargraph system using a tablet application. Due to the wireless connection, we expect easier installation and an improvement of the usability.

With regard to level 2 and level 3 automation, NHTSA pointed out the need to analyze the driving ability to avoid lane departure in case of a loss of assisted or automated functionality [9]. The bargraph system is suitable to investigate this ability during takeover request in case of automated driving. We will definitely follow this issue to provide higher value for our customers.

References

[1] Society of Automotive Engineers (SAE), *J3016: Taxonomy and Definitions for Terms Related to Driving Automation,* 2018th ed.: SAE International.

[2] *ISO 26262: Road vehicles - Functional Safety - Part 3: Concept phase*, ISO 26262-3:2018 (E), 2018.

[3] Dipl.-Psych. A.Neukum, Prof. Dr. H.-P. Krüger, "Fahrerreaktionen bei Lenksystemstörungen- Untersuchungsmethodik bei Bewertungskriterien,"

[4] Joanne L. Harbluk, Peter C. Burns, Darryl Malone & Jeremy Hamilton, *Power Steering Assist Failures: Driver Behavior, Safety Impacts, and Implications for Automated Vehicles.*

[5] B. F. Pierce, *Human Force Considerations in the Failure of Power Assisted Devices. Final Report*, 1973.

[6] *Einheitliche Bedingungen für die Genehmigung der Fahrzeuge hinsichtlich der Lenkanlage*, R79, 2008.

[7] *Department of defense design criteria standard: human engineering*, 1472G, 2012.

[8] D. Simmermacher, "Objektive Beherrschbarkeit von Gierstörungen in Bremsmanövern," Dissertation, Technische Universität Darmstadt, Darmstadt, 2013.

[9] Myra Blanco et al., *Human Factors Evaluation of Level 2 and Level 3 Automated Driving Concepts*: NHTSA.

CHASSIS.TECH SECTION

CHARACTERIZATION OF ADAS DRIVING PROPERTIES

Attribute-based development of driver assistance systems

Dr.-Ing. Manuel Höfer[1], Dr.-Ing. Florian Fuhr[1], Prof. Bernhard Schick[2], Prof. Peter E. Pfeffer[3]

[1] Dr. Ing. h.c. F. Porsche AG

[2] University of Applied Sciences Kempten

[3] MdynamiX AG

1 Introduction

Advanced Driver Assistance Systems (ADAS) and Highly Automated Driving Systems (HAD) are among the most important megatrends in the automotive development. Accompanying this one big question arises: do all assisted and automated driving cars drive the same or will vehicle manufacturers be able to differentiate themselves with DNA of their own? And - especially for a sportscar-manufacturer like Porsche - how can ADAS and HAD impart typical attributes like driving fun and sportiness, Figure 1?

Figure 1: Porsche typical driving characteristics of ADAS and HAD.

In order to achieve this, clear driving characteristic goals (from a customer's point of view) must be defined and the system requirements for ADAS and HAD (including all components like sensors, ECU's and actors) shall be derived from this. However, what are driving and brand characteristics in the context of assisted and automated driving? And how can those be realized in the development? Porsche has addressed this question together with the University of Applied Sciences Kempten and MdynamiX. The main questions are: "How can an attributed-based development be achieved? How can Porsche effectively design brand-typical characteristics for ADAS and HAD? And how can Porsche establish this in the ADAS- and HAD-development?"

2 Motivation

Today, ADAS-development is mostly sensor-based compared to the classic chassis systems, which are attribute-based. The evaluation of those is commonly done individually on a subjective basis and not systematically. Various studies with over 160 test persons regarding longitudinal and lateral control assisted systems in current benchmark vehicles have shown that the functional and driving characteristics of the actual systems still offer a great potential upward. The customer acceptance is – in all benchmark vehicles – unsatisfactory [1] [2] [3]. The challenge for vehicle manufacturers is also the optimization of the systems and the difficulty of differentiating themselves regarding the brand-DNA. The brand-specific characteristics and the brand position of the vehicle have hardly been taken into account in the ADAS/HAD development so far. For Porsche with its sports- and luxury DNA it is very important to transfer the product and the brand to the age of ADAS and HAD, so that they can be experienced by their customers. Those should experience special emotions, sportiness and everyday practicability, distinguishable from other brands and products.

Using a lane keeping system (LKS) as a first example, a generic procedure model should be developed to translate subjective customer experiences into subjective expert evaluations and finally into objective measurable key performance indicators (KPIs) with defined manoeuvres [4]. Afterwards this method is applied to other systems – for example the Adaptive Cruise Control (ACC). Using those KPIs it is possible to define objective attribute targets for a Porsche typical characteristic and to validate them at any time in all phases of the development – from simulation, software-, hardware- and vehice-in-the-loop tests up to road tests. Furthermore it can be used for an objective benchmark comparison, especially in vehicles without BUS-access. As the next step this procedure model can be transferred to evaluate other ADAS and functions of higher automation levels.

3 Porsche typical characteristics of Advanced Driver Assistance Systems

As a sports car manufacturer, Porsche offered its customers only a few safety functions and a cruise control system when the first Panamera was launched in 2009. In the following years, Porsche pursued a late-follower strategy in the expansion of ADAS, focusing primarily on the Panamera and the SUV models. Since the introduction of the 2nd generation of the Panamera 2016, a trend reversal has been initiated. With predictive longitudinal control functions such as Porsche InnoDrive, for the first time Porsche introduced an in-house developed ADAS function and optimizes existing functions by means of Porsche typical extensions, such as sportiness recognition [5]. With the 3rd generation of the Cayenne, Porsche launched its first level 2-function with the Adaptive Cruise Assist, a combined lateral- and longitudinal function, Figure 2.

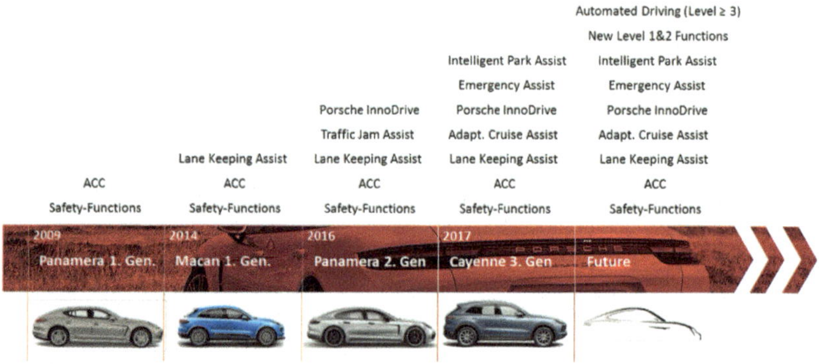

Figure 2: Evaluation of ADAS at Porsche.

Porsche pursues the approach of offering driver assistance functions with their own DNA by complying brand-typical attributes such as reliability, sovereignty, performance, intelligence and trustworthiness. In order to address this claim for future assisted and highly automated driving functions, professional methods are required for an attribute-based development.

4 Method

By applying principles and approaches of classic vehicle dynamics and chassis development [6], a generic procedure model is developed using the example of the LKS. This procedure model is afterwards transferred to the ACC-system.

4.1 Evaluation Level Model

In numerous expert workshops, benchmark tests, subjects-studies and measurement campaigns, the relevant evaluation-attributes were systematically developed and verified. These subjective characteristics are transferred to a so-called level model.

Using a suitable linkage between subjective criteria and measurement signals (e.g. lateral acceleration) the objective level results, Figure 3 [4].

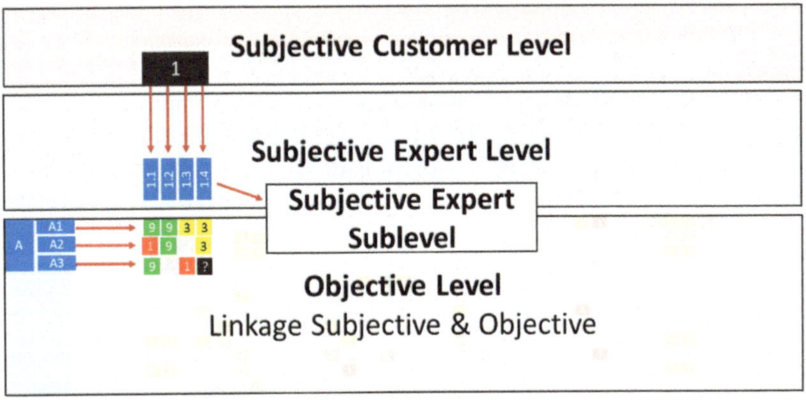

Figure 3: Evaluation level model with customer-, expert and objective level.

The model consists of the subjective customer evaluation, subjective expert evaluation, measurement signals and objective characteristic values (KPI) which are collected in defined driving manoeuvres/driving scenarios. At the highest customer level, there are key criteria such as track guidance quality, vehicle reaction, driver-vehicle interaction, availability, degree of relief, sense of safety and HMI (operation, display, monitoring and warning). At the expert level, the main criteria are broken down into sub-criteria. In the next step, all relevant and measurable vehicle signals are identified in expert workshops, in which the subjective expert criteria are expected to be clearly visible. Based on the expert knowledge, subjective and measurable are linked afterwards. The experts rate the degree of visibility to be expected as high (9), moderate (3), low (1), none (0) or unknown (?). Therefrom, KPIs for the relevant signals are developed according to the individual expert criteria in analogy to characteristic values of vehicle dynamics. It is possible to subdivide the second level into one more sub-level to consider much more detailed facets, like steering feel. In this case, steering feel can be evaluated during assistance guiding with common subjective steering feel criteria and special, new criteria such as the steering torque gradient.

4.2 Ground Truth measurement method

For the objective evaluation of the driving characteristics of ADAS and HAD a new measurement and test method need to be developed. This method must be unattached of BUS-access. In the assessment of driving dynamics, it is a common procedure that the driver – vehicle – environment control loop should be considered in the global vehicle evaluation. Therefore, driver input as well as road and traffic input, control intervention and the resulting vehicle reaction/movement should be evaluated in its six degrees of freedom. Derived from the automated and assisted lateral control, it is necessary to acquire a high level of knowledge of the road excitation, essentially road markings, surface geometry etc. as well as the driver input in order to be able to evaluate the resulting vehicle reaction caused by the system. In the case of ACC it is necessary to know all about the surrounding traffic, especially the preceding car.

Like all sensors, ADAS- and HAD sensors (camera, radar, lidar) are faulty and not available or sufficiently accurate at all time and in all situations. This can have a significant impact on the system behaviour and its driving characteristics. For example, the camera may not be able to reproduce the curvature of the road accurately, which can cause difficulties for the lane-keeping controller. This repeatedly leads to uncertainties if the experienced driving characteristics are a result of the poor performance of sensors, trajectories, controllers, actuators or the poor response of the vehicle influenced by steering, axles, tires and chassis control systems.

In order to investigate this cause and effect chain, a much more accurate reference measurement should be used as a "Ground Truth". For the LKS an approach was chosen which integrates a highly accurate measured vehicle position and movement into highly accurate Ground Truth maps (Figure 4). For this maps the company Atlatec has generated a digital high accurate 3D-map of tracks around the cities of Weissach and Kempten, which can also be used as a digital twin in simulation environment. Using an Inertia Measurement Unit (IMU) with fiber optic gyroscope, Kalman filter, RTK-DGPS and SAOPS correction service the vehicle position and motion was measured (Figure 5a) [7]. Great efforts have been made to achieve the objective of a relative accuracy between both absolute measurements (digital map and vehicle position) of less than +/-5cm. For measuring the steering wheel torque and angle a measurement steering wheel from Kistler has been used. It is important to keep the original steering wheel installed in order not to influence the haptics, control functions, hands-off detection and airbag function. The LKS state-icon in the head unit was recorded by a special camera (AVAD) and converted into a measurement signal (Figure 5b).

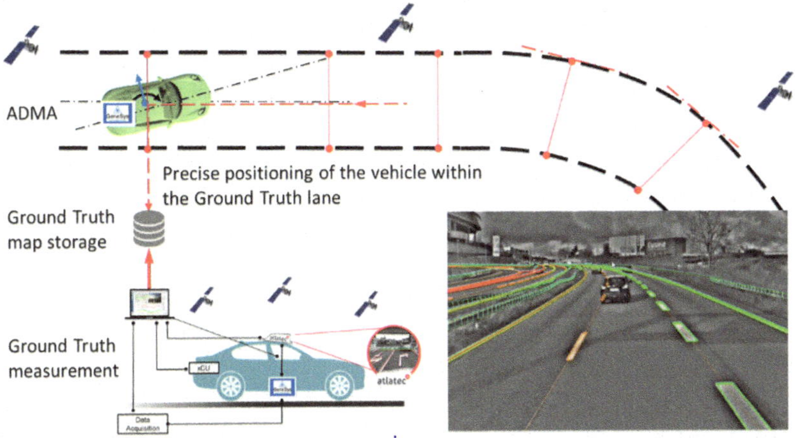

Figure 4: Measuring Ground Truth maps.

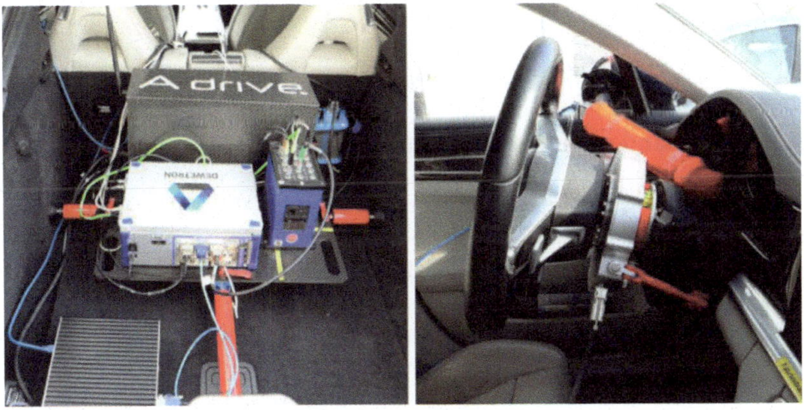

Figure 5a: Measurement Setup: IMU; 5b: measurement steering wheel and AVAD.

With this measurement setup it is possible to identify driver-interventions, to measure all relevant signals for the objective evaluation and to evaluate the on board sensors. The setup is designed to be used in all benchmark vehicles, however it is also possible to use the on board signals from the vehicle BUS. This allows a quality evaluation of the sensors, trajectory, controller and the vehicle response. It is also an enabler to investigate the effect chain of the system and the requirements of all individual components.

For the ACC measurement a second DGPS-unit and IMU on the preceding car are used for an exact measurement of distance, differential velocity and transverse offset instead of the Ground Truth map, Figure 6. All other measurement equipment remains identical.

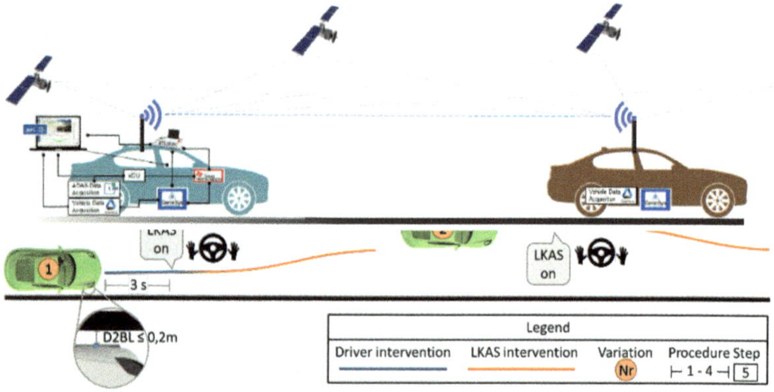

Figure 6: Measurement Setup for ACC with two cars.

4.3 Route and maneuver catalogue

For a comparable subjective and objective evaluation of ADAS and HAD it is important to test it in identical situations. To be able to evaluate the driving characteristics in comprehensive driving situations, the previous standard procedures such as EuroNCAP [8] are not sufficient. The variance in possible real road events (road types, curvature, road markings, cross slopes etc. for LKS and distance, velocity, acceleration, driving behaviour etc. for ACC) is far too great. For this purpose, a comprehensive route and manoeuvre catalogue with detailed driving manoeuvres was developed. In this catalogue, the routes, sections, areas and events are listed as well as the relevant measurement signals and the resulting objective criteria. Each manoeuvre is precisely defined and described with a trace of the driving situation (figure 7 & 8).

For LKS the manoeuvres are for example a so called free ride, lane change test, transient test, drop-off-tests, step-steer-tests, on-center handling test, etc. The free ride test includes test drives with different drivers and at different times and serves the recording of lane accuracy and availability. Most of the manoeuvres should be driven on open roads. For special KPIs like the sublevel of steering feel it is much easier to drive on a test track.

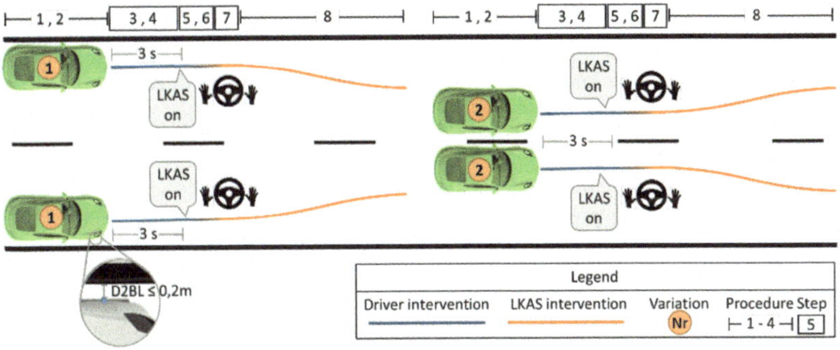

Figure 7: Driving manoeuvre for LKS.

In case of ACC the manoeuvres are for example braking and accelerating target vehicles, cut-in and cut-out situations, lane-change manoeuvres and stop-and-go-situations. In comparison to the LKS a free run is not intended and it is favourable to perform the tests on a proving ground.

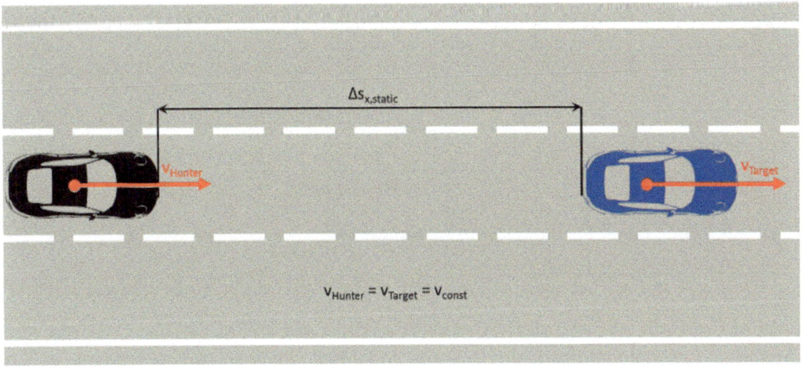

Figure 8: Driving manoeuvre for ACC.

4.4 Objective evaluation of driving characteristics using KPI's

By using suitable algorithms, the KPI's can be generated automatically from the measurement data. For this purpose e.g. reference signals of yaw rate and lateral acceleration, based on the "Ground Truth" curvature, are generated as target and the deviation from the actual measurement is evaluated. For free travel, statistical distributions or counting methods are used, such as the availability measurement, tracking precision measurement or jerk measurement, as well as finding specific states and events using an event finder. For example the stationary states can be identified by comparing the stationary lateral position to the lateral acceleration. The following chart, Figure 9, provides information on how the vehicle is carried outwards (negative curve cutting gradient) or on how it cuts curves slightly (positive curve cutting gradient). In addition, the chart shows the center position when driving straight ahead (offset at ay=0).

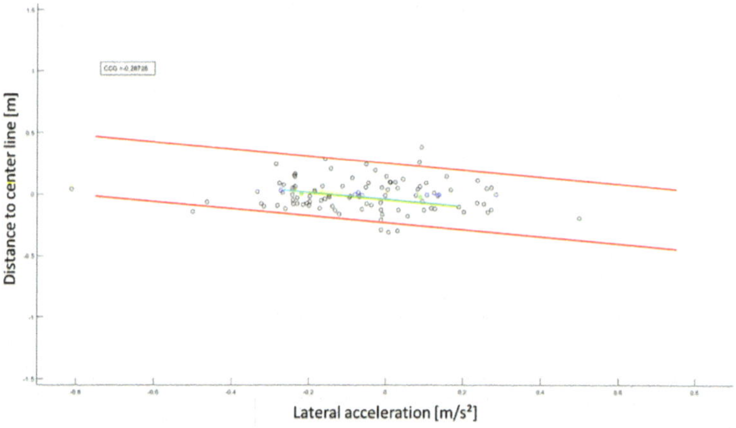

Figure 9: Side offset for curved and straight driving.

Furthermore, the lateral acceleration limits, steering torque limits, dropout limits, response, lock-in and lock-in times, steering torque gradients, steering hysteresis and drift speed are determined. In this scheme, over 80% of the subjective expert evaluations can be objectified, Figure 10.

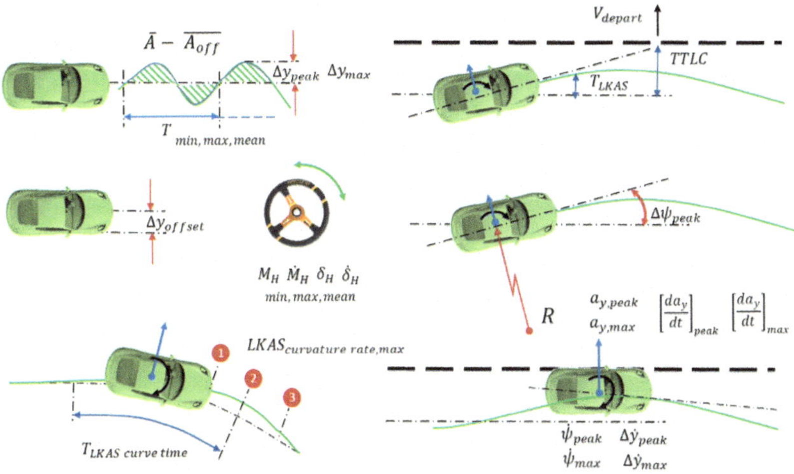

Figure 10: Concept of objective criteria.

4.5 On-Board evaluation

With the presented method it is possible to evaluate and benchmark LKS and ACC systems with a high accuracy, but it is associated with a high expenditure. For the usage during vehicle development in test- and calibration-drives, it is required to have a quick and easy evaluation system. For this special use case a so-called online-evaluation-system is developed, which evaluates for example the last calibration parameter set or two different vehicles with the same KPIs. The evaluation system uses on on-board signals including the track recognition of the vehicle camera. As a result, the score is not as precise as using the Ground Truth method, but it helps to compare different calibration sets quickly and easily during test drives.

5 Driving characteristic evaluation and KPI usage in development

The ADAS and HAD evaluation method addresses various steps of development. In the beginning KPIs help to define system requirements, based on earlier projects, knowledge and benchmark studies. Then the evaluation algorithms can be used in simulation and during the system verification and calibration process. At last, the subjective evaluation on expert level and the objective evaluation are an important part of the approval process of ADAS and HAD.

5.1 Benchmark studies and target definition

Especially in the highly innovative environment of ADAS/HAD it is necessary to observe the performance and solution approaches of competitor vehicles to learn from the good and to avoid poor solutions. Even in the age of ADAS/HAD, it is important for Porsche to design the product in a way that it can be experienced as brand-typical and to differentiate itself from others. This requires clearly recognizable driving characteristics that are associated with the Porsche brand and can be compared to the benchmark. Familiar brand attributes such as driving pleasure, performance, precision, driver feedback, transparency and reliability are to be addressed here as well, with a high degree of suitability for everyday use. Customers would expect a Porsche, for example, to drive precisely in lane and always provide the driver with pleasant but not disturbing feedback on the driving condition. A Porsche typical ACC should accelerate and brake like a sports car. For this purpose, the desired brand attributes are linked to the criteria in the level model, Figure 3. This makes it possible to define objective targets for a typical Porsche characteristic and validate them at all times in all phases of development. Figure 11 shows an example of the benchmark evaluation of three vehicles for track precision, as sub-criteria of the track guidance quality based on the calculated distance to centreline measured in the free ride test.

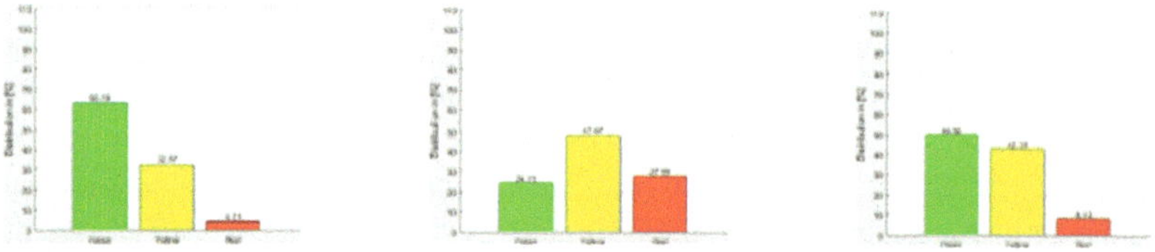

Figure 11: Lane precision distribution of three benchmark vehicles.

The green area is defined with ±20 cm, the yellow area with ±50 cm and the red area is more than 50 cm. For example a KPI-target can be more than 95% in the green corridor and 0% in the red area. Figure 12 shows the curve cutting coefficients of the three benchmark vehicles. The left vehicle possesses a relatively strong drift to the outside of the curve. In contrast the right car slightly cuts the curve, which would be desirable for a sports car. The bandwidth of the red lines represents the lane precision of the systems – for this aspect the left car posses the best properties.

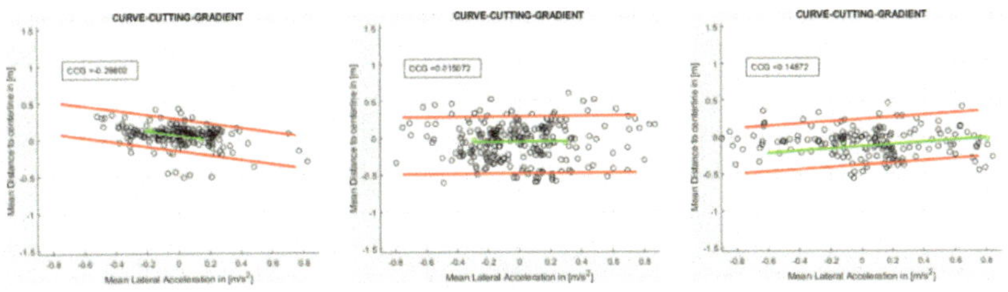

Figure 12: Curve Cutting Gradient of three benchmark vehicles.

5.2 Simulation-based development process

For using the objective evaluation in simulation (including model, software, hardware and vehicle in the loop) a modular simulation environment based on Vires VTD, Porsche driving dynamics model and a control system network including a LKS- and ACC-controller is developed. A real car model with a good steering, braking and acceleration behaviour is the premise for correct results. For this purpose, the Pfeffer steering model of MdynamiX was integrated into the Porsche driving dynamics model. Furthermore, the Ground Truth maps and the defined driving manoeuvres were implemented analogously to the real road tests, Figure 13.

Figure 13: Implementation of Ground Truth maps and manoeuvre catalogue.

6 Conclusion and outlook

ADAS and HAD functions are important to Porsche and should be developed with typical Porsche characteristics. Based on the examples Lane Keeping Systems (LKS) and Adaptive Cruise Control (ACC) a procedure model was successfully established to show how to evaluate different systems or calibrations and how typical Porsche characteristics can be effectively achieved in an attribute-based development. The procedure model can now be transferred to other assistance and automative driving systems. It also allows the definition of specific calibration goals for different regions e.g. China. In the future, Porsche customers will be able to experience brand-typical automated driving functions and the associated emotional driving pleasure.

Bibliography

[1] B. Schick, C. Seidler, S. Aydogdu, Y.-J. Kuo: "Driving Experience vs. Mental Stress with Automated Lateral Control from the Customer's Point of View", München, ATZ, 2018.

[2] C. Seidler, B. Schick: "Stress And Workload When Using The Lane Keeping Assistant - Driving Experience With Advanced Driver Assistance Systems", in 27th Aachen Colloquium Automobile and Engine Technology 2018, Aachen, 2018.

[3] S. Aydogdu, B. Schick, M. Wolf: "Claim And Reality? Lane Keeping Assistant - The Conflict Between Expectation And Customer Experience", in 27th Aachen Colloquium Automobile and Engine Technology 2018, Aachen, 2018.

[4] B. Schick, S. Resch, M. Yamamoto, I. Kushiro, N. Hagiwara: "Optimization of steering behavior through systematic implementation of customer requirements in technical targets on the basis of quality function deployment", Yokohama/Japan: FISITA, 2006.

[5] M. Höfer: "Fahrerzustandsadaptive Assistenzfunktionen", Dissertation, Stuttgart: Fraunofer Verlag, 2015.

[6] P. Pfeffer, M. Harrer: "Steering Hand Book", Springer Vieweg, 2011.

[7] D. Schneider, B. Huber, H. Lategahn, B. Schick: "Measuring method for function and quality of automated lateral control based on high-precision digital "Ground Truth" maps", in VDI Tagung Fahrerassistenzsysteme und automatisches Fahren, Wolfsburg, 2018.

[8] European New Car Assessment Programme (EuroNCAP): "Test Protocol – Lane Support Systems Version 2.0.2", European new car assessment program (EuroNCAP), 2018.

Assisted/autonomous vs. human driving assessment on the DiM driving simulator using objective/subjective characterization

A. Affanni[(1)], P. Zontone[(1)], R. Fenici[(2)], D. Brisinda[(2)], D. Bacchin[(3)], L. Gamberini[(3)], P. Pluchino[(3)], M. Bruschetta[(4)], C. Savorgnan[(5)], F. Formaggia[(5)], M. Minen[(5)], D. Minen[(5)]

(1) DPIA, University of Udine, Italy

(2) BACPIC, UCSC Roma, Italy

(3) HIT University Padova, Italy

(4) DEI University of Padova, Italy

(5) VI-grade

Abstract

More and more sophisticated assisted/autonomous vehicles are becoming available in the market. Automation levels 2 and 3 have been given for settled just a couple of years ago, and the path to fully autonomous car seemed to have no obstacles. In reality, OEMs started recently realizing that the impact of semi and/or fully robotized cars on drivers as well as on passengers is all but predictable. VI-grade has more than ten years experience with developing turn-key solution driving simulators, and has been working for more than five years on a research project to collect meaningful bio-signals from the driver during simulator sessions, in collaboration with the BACPIC of the Catholic University of Sacred Heart and the DPIA of the University of Udine. Recently, a collaboration with the Human Inspired Technology Research Center of University of Padova allowed the extension of the assessment at physio-emotional level.

Introduction

There is increasing interest from major OEMs for technologies providing real-time monitoring of the driver psycho-physiological reactions to vehicle dynamics under direct or robotized control: a) to study the robotic acceptance; b) to evaluate the difference between self and assisted/autonomous driving in identical scenarios; c) to evaluate the impact of on-the-fly setup variations of the vehicle when autonomously driven. The DiM150 driving simulator is ideal to replicate, in a fully immersive and deterministic environment, any kind of real-time vehicle dynamics under any kind of driving conditions. A novel wearable device to record bio-signals and a special software to analyse and monitor in real-time multiple physiological parameters have been developed. All the bio-signals are synchronized with vehicle's ones. Driving-induced transient variations of autonomic nervous system (ASN) modulation derived from heart rate variability (HRV) and skin potential response (SPR) are collected letting the driver on the simulator drive through a series of tasks (separated by recovery intervals) on a variable highway scenario. Aim of this study was: a) to identify even minor human reactions differences induced by tasks during active and passive driving; b) to investigate possible correlations between the analysed bio-signals and the driver's subjective experience, including the perception of danger, the sense of presence, the in-vehicle perceived comfort and the quality of the driving experience, the level of engagement.

The present work represents a qualitatively investigation of a methodology, based on Driving Simulator session, which could give interesting indications to the vehicle development teams for autonomous and assisted driving about the efficiency/comfortability of the driving intelligence in realistic highway driving scenarios. To prove the methodology, we have preliminary performed a test with 13 participants. The number is still not statistically significative, and the result of the present work are to be intended for a) investigation and tuning of the methodology and b) for the researchers to understand if, applying signal processing developed in previous works [1 – 5] and given a bigger number of participants, the approach could scientifically provide innovative and quantitative indexes for classifying autonomous driving algorithms.

Sensor description

The instrument block diagram is shown in Figure. 1. The ECG is acquired on three channels by amplifying and conditioning the differential voltages V_5-V_1, V_4-V_3 and V_3-V_2. These voltages are picked using a commercial vest having wet electrodes positioned accordingly to Figure 1. The SPR signal is acquired by amplifying and conditioning the differential voltage picked between the palm and the back of each hand using Ag/AgCl electrodes. The reference voltage V_{REF} is applied on the chest and on the wrists.

As shown in Figure 1 each differential voltage (3 ECG and 2 SPR channels) is properly amplified and filtered; a Digital Signal Processor (DSP) acquires the amplified analog voltage and transmits through Universal Asynchronous Receiver Transmitter (UART) the acquired data to the Wi-Fi module that converts UART data into IEEE 802.11 wlan protocol. The device is supplied with one Li-Ion cell: a 3.7 V 1000 mAh lithium polymer battery is used. The device current consumption during transmission is 200 mA, allowing 5 hours of continuous data acquisition. A power supply section of the circuit reduces the battery voltage to +3.3 V using a buck DC-DC converter; the voltage V_{REF} = 1.65 V for the reference electrodes and for the instrument is provided by a linear voltage reference.

SPR circuit design

The voltages on the palm and back electrodes are high-pass filtered by a couple of passive first order filters having cutoff frequency of 0.08 Hz, with the aim of removing the common mode DC voltage that may be present on the skin; the input impedance of the instrument is 100 MΩ, in order to assure a load error lower than 1%, assuming the skin impedance in the order of 1 MΩ. The filters are connected to the instrumentation amplifier which amplifies the differential voltage between palm and back of the hand with a gain G = 160, since the expected maximum SPR pulse is in the range of ±10 mV and must be amplified into 3.3 Vpp. Finally, the anti-alias filter has been designed as a third order low-pass filter with cutoff frequency of 40 Hz.

ECG circuit design

The circuit topology of each channel of the ECG circuit is similar to the SPR with different gain and bandwidth. As in previous subsection at each channel input there are two first order high pass passive filters. The input impedance is, as in previous subsection, 100 MΩ in the pass band. The filters outputs are connected to the instrumentation amplifier, which amplifies the differential voltage relative to the channel (i.e. V_5-V_1, V_4-V_3 or V_3-V_2). The maximum range of the input differential voltage is supposed to be ±4.5 mV, this yields to a gain G = 370 if it must be amplified into a signal with maximum amplitude 3.3 Vpp. The frequency behavior of the sensor is a band pass system with gain 370 in band center, a lower cutoff frequency at 0.022 Hz with a slope of 40 dB/dec and an upper cutoff frequency at 170 Hz with a slope of -60 dB/dec.

DSP and WiFi data transfer

The conditioned analog signals are sent to an analog input of the DSP. The chosen DSP has an on board 12 bit A/D converter operating at 8 Mega Instructions per Second; the sample rate has been set to 1 kSa/s. The converted data are sent via UART protocol to the WiFi module. The Baud Rate for data transfer has been set at 115.2 kbps in order to allow the data flow without crowding the channel.

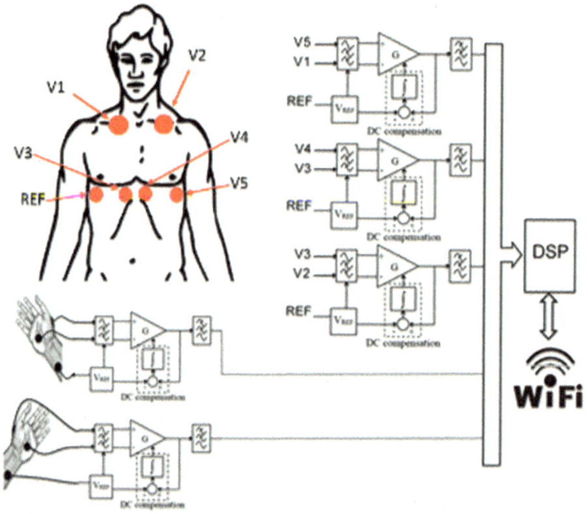

Figure 1: Sensor block diagram

Acquisition software description

For data acquisition, the WINTAX 4 PRO software by Magneti Marelli has been used. It is a suite of data analysis tools developed for motorsport and it provides highly advanced real time analysis functions as well a standard interface to team's proprietary software applications. Specific characteristics of the program can be found in [6]. The advantage of this choice is the capability of monitoring, in real time and synchronized both the vehicle telemetry and the biotelemetry data. Moreover, we can add real time processing of the data channels implementing complex functions like the features used in heart rate variability analysis. In Figure 2 an example is shown of real time monitoring and analysis of ECG, SPR and HRV parameters dynamic time-varying changes.

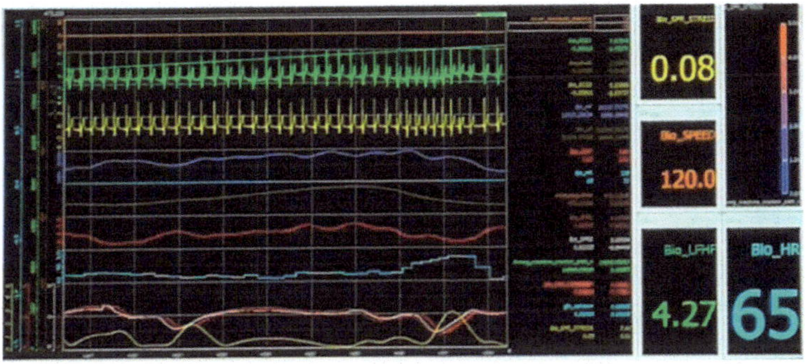

Figure 2: Real time monitoring of multiple physiological parameters using WINTAX 4 PRO

Motion artifact removal on SPR signals

The proposed algorithm is based on the assumption that the interferences due to motion cause an increase of the energy of the measured signals, because a first energy contribution is provided by EDA pulses and a second energy contribution is related to hands movement. On the basis of this assumption, we first evaluate the RMS_1 and RMS_2 values of the spr_1 and spr_2 signals on a moving time window of duration of 1 s. At the ith sample, $RMS_{1,2}$ results:

$$RMS_{1,2}(i) = \sqrt{\frac{\sum_{n=i-N+1}^{i} spr(n)_{1,2}^{2}}{N}} \quad (1)$$

We consider the function:

$$f(x) = \frac{1}{1 + e^{-2(x-1)}} \tag{2}$$

$\alpha(i)$ used to combine spr_1 and spr_2. The ith sample of $\alpha(i)$ is obtained as

$$\alpha(i) = \begin{cases} f\left(\frac{RMS_1(i)}{RMS_2(i)}\right) & if\ RMS_2 \neq 0 \\ 1 & if\ RMS_2 = 0 \end{cases} \tag{3}$$

From (2) and (3), it is clear that $\alpha(i) \rightarrow 0$ when $RMS_1 \ll RMS_2$ and $\alpha(i) \rightarrow 1$ when $RMS_1 \gg RMS_2$. The output is finally obtained as a linear combination of spr_1 and spr_2:

$$OUT(i) = \alpha(i) \cdot spr_2(i) + [1 - \alpha(i)] \cdot spr_1(i) \tag{4}$$

Intuitively, the input signal (whether spr_1 or spr_2) with lower energy content (within the moving window of duration 1 s) is considered more reliable for the output in (4).

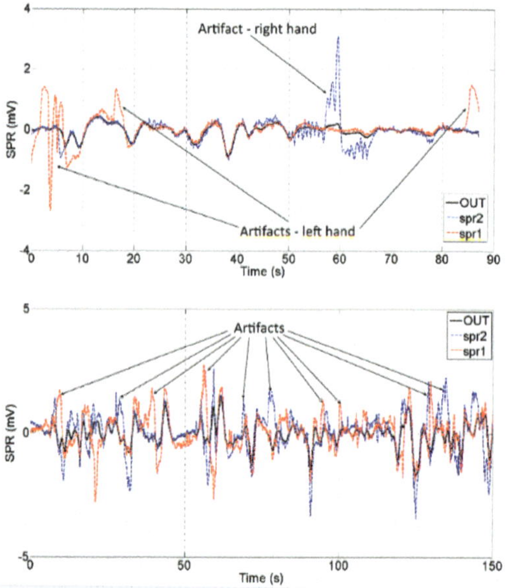

Figure 3: motion artifact removal: in lab experiments (top) and on simulator (bottom)

Figure 3 shows the results obtained from the motion artifact removal algorithm. In top graph the experiments are initially conducted in laboratory, just moving the hands and disturbing the electrodes during acquisition; in bottom graph we acquired the signals during a lap driven on a driving simulator. The simulated circuit is Jerez de la Frontera, the driver never drove on it before. It is evident that the algorithm follows the input with minimum energy when there is discordance between inputs. In both charts, it is possible to see that the motion artifacts are perfectly removed by the proposed algorithm.

Test description

A group composed by 13 healthy volunteers (age 31.4 ± 9, 8 males and 5 females) drove on simulator in a highway scenario in three different phases. The first phase had the aim of training people and getting familiar with the simulator driving with very low traffic in a highway for 5 minutes, while their baseline bio-signals were acquired. In the second phase participants had to manually drive in a highway for a distance of 40 km (corresponding approximately to 20 minutes). On the path, at specified positions, where posed four tasks: 1) **overtaking** another car which is undecided on which lane to keep; 2) brake maneuver because two **trucks** ahead are overtaking each other; 3) lane **narrowing** and mandatory shift due to road works and 4) unexpected lateral **wind** gusts. Figure 4 shows the screenshots of the tasks.

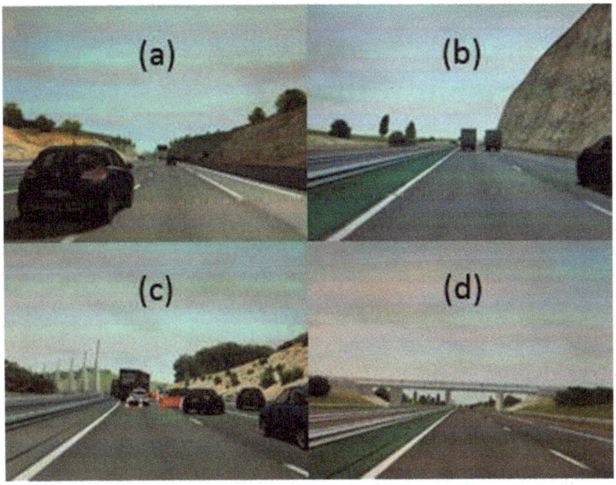

Figure 4: the tasks positioned on the road. (a) overtake of undecided car; (b) trucks overtaking; (c) lane narrowing; (d) wind gusts.

During the third phase, the subjects were on the same highway with an autonomous car which had to cope with the same tasks. In order to avoid data bias, one half of the volunteers did the manual drive before the autonomous, and the other half vice versa.

As an example, Figure 5 shows the typical behavior of heart rate and SPR signal during the tasks.

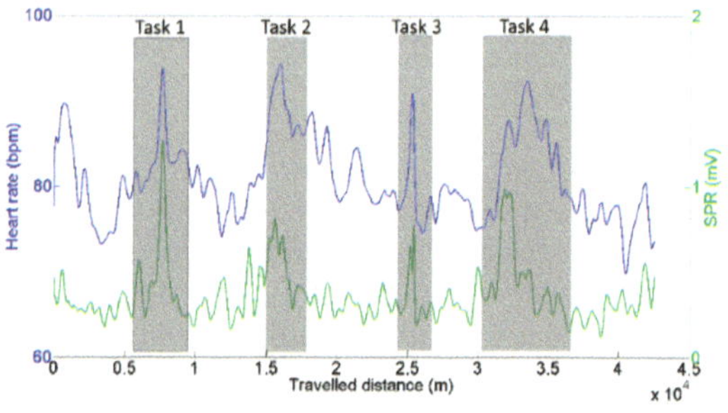

Figure 5: heart rate and SPR signal versus the travelled distance. The effect of the tasks is clearly visible in both traces

As it can be seen, during the tasks there is a significant increase of heart rate and SPR signal.

Heart Rate Variability Analysis

To validate real-time assessment obtained with the VI-grade experimental set-up in the Driving Simulator , the same sensor system was tested in the electrophysiology laboratory in the BACPIC, where the quantitative HRV analysis was also performed with the Kubios software (version 3.0.2), with time-varying algorithms and, according to the European Society of Cardiology guidelines [7], in the time domain (TD), in the frequency domain (FD), and with non-linear (NL) methods. In the FD, the LF/HF ratio, calculated from LF and HF in normalized units, was accepted as an index of sympathovagal interaction adequate to explore autonomic modulation [8]. For comparison, 12-lead ECG was continuously recorded also with Mortara Surveyor/X-Scribe

The validation protocol consisted of five phases: implying: 1) 10-minutes baseline supine; 2) 10-minutes Head-up tilting 70^0 (HUTT); 3) 10-minutes supine recovery); 4) 20-minutes mental stress with "Mensa" preliminary Tests; 5) exercise test, at bicycle-ergometer, until muscle exhaustion.

Experimental results

SPR Activity

As a first analysis, we want to investigate if there is a significant difference in SPR activity from a task to another among all the subjects. To do this, since every person has different electrodermal activity, we evaluated the RMS value of the SPR over the tasks normalized with respect to the RMS evaluated on the entire trace. Figure 6 shows the graph with the normalized SPR for manual (red line) and autonomous (blue line) driving.

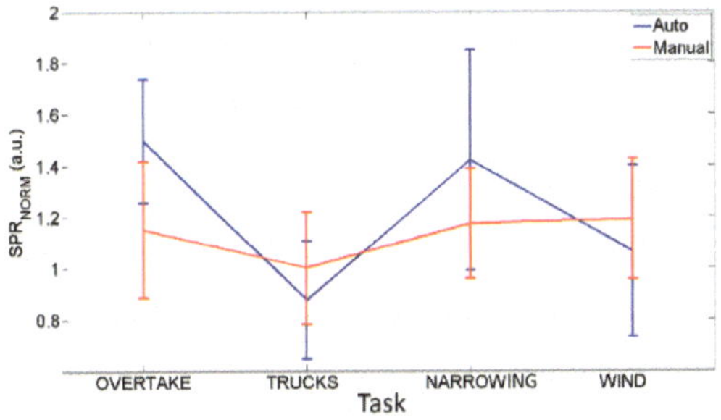

Figure 6: Normalized SPR over the tasks for autonomous (blue line) and manual (red line). Error bars represent standard deviation among the subjects

Referring to manual driving (red line), we observe that "wind" is slightly more stressful than the other tasks and that "trucks" is slightly less stressful than the others, although SPR activity is comparable in all tasks when the subject is driving manually ($p>5\%$). Instead during autonomous driving (blue line), "overtake" is perceived as more dangerous compared with "trucks" and with "wind" ($p<5\%$). "Trucks" are perceived as less dangerous than lane narrowing, which is more dangerous than "wind".

As a second analysis, we wanted to evaluate if the tasks are perceived more dangerous in autonomous or manual drive. Performing the t-test for each obstacle, we obtain that "overtake" is perceived significantly more dangerous ($p=0.2\%$) in autonomous. Although non-significant the lane narrowing is perceived slightly more stressful in autonomous drive ($p=8\%$), whereas "trucks" and "wind" are perceived in the same way ($p=18\%$ and 28%, respectively).

As a third analysis, we want to investigate if the four tasks (in general) are perceived more stressful in autonomous or manual drive by each participant. Figure 7 shows the results.

Before providing any comment, we must point out that two different autonomous algorithms were tested: an aggressive autonomous drive algorithm (subjects 1 – 6) versus a smooth autonomous algorithm (subjects 7 – 13).

As it can be seen in Figure 7, there is a huge variability among the subjects. However, increasing the confidence of t-test to 20%, subjects 1 and 3 perceived autonomous drive as more dangerous than manual, subjects 7, 9, 10, 12 and 13 had the opposite reaction (autonomous drive was perceived less stressful than manual), whereas no difference was appreciable in the remaining ones.

Particularly interesting is the fact that, for most participants the autonomous drive was significantly less stressful than manual when the smooth algorithm was used. Instead, with the aggressive autonomous algorithm, two people over six perceived autonomous drive as significantly more dangerous than manual.

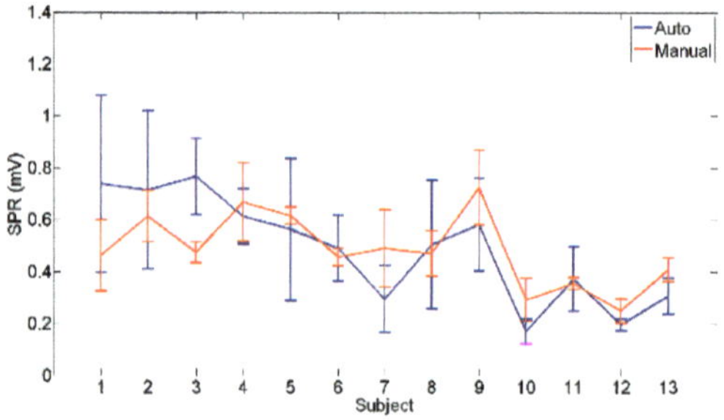

Figure 7: SPR averaged on the tasks for each subject in autonomous (blue line) and manual (red line). Error bars represent standard deviation among the tasks.

Finally, we evaluated if, over the entire distance, the autonomous drive is perceived as less stressful than manual. The autonomous drive resulted significantly ($p<5\%$) less stressful than manual for subjects 7 – 13 (i.e. when smooth autonomous algorithm was used. (Figure 8).

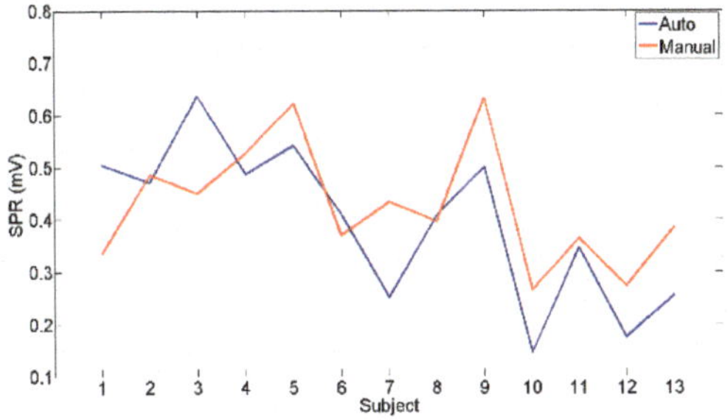

Figure 8: SPR averaged on the entire distance for each subject in autonomous (blue line) and manual (red line).

Real- time HRV-Sensor validation in the physiology laboratory.

The signals quality obtained with the VI-grade sensor system was optimal for reproducible real-time calculation and monitoring of time-variant HRV parameters (Figure 9).

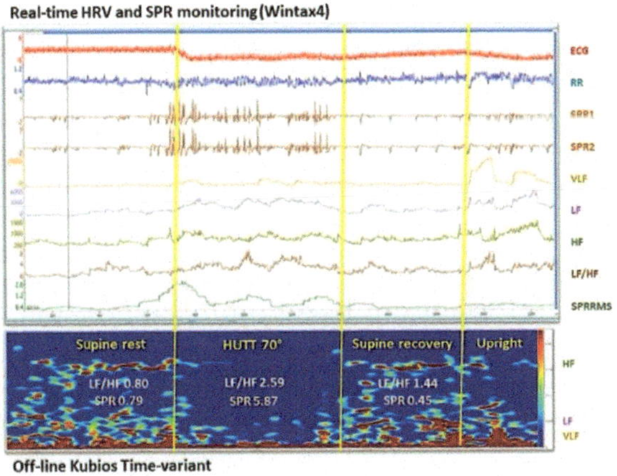

Figure 9: example of comparison between real-time monitoring of HRV and SPR parameters with WINTAX 4 and and off-line time-variant HRV analysis performed with Kubios from ECG simultaneously recorded with the BACPIC Mortara Surveyor/X-Scribe system.

Baseline SPR RMS activity ranged between 0.01 and 0.8 mV, with wide inter-individual variability. A significant ($p<0.05$) increment (up to 2.6 mV) was induced by HUTT, but not by physical exercise. SPRRMS and LF/HF had similar trend along test session. Good agreement was found between HRV parameters calculated in real-time, their off-line recalculation with Kubios software and those obtained from Mortara telemetry.

Questionnaire method and results

We have considered several subjective metrics to fully understand the overall users' experience in described context [9 – 13]. The sense of presence, defined as the sense of being in a virtual environment, it is a crucial variable to assess whenever considering a virtual environment (VE). The better a VE is designed the higher sense of presence is experienced. In different studies various parameters were manipulated in order to comprehend their influence on the perceive sense of realism of the VEs like field of view increment or presence of autonomous traffic. A second relevant self-reported measure is the User Experience (UX), namely, "the perception and reactions of a user that derive from the use, or from the prediction of use of a product, system or service" (ISO, 2009). In the automotive research field, UX questionnaires are used to assess the interacting experience regarding the in-Vehicle Infotainment Systems (IVIS) or the driving simulator itself. A third aspect important to evaluate is the systems usability insofar as it definitely affects the drivers' experience. It is defined by Shackel (2009) as "the capability to be used by humans easily and effectively" and, like UX, its applicable both to IVIS and to simulators. Several questionnaires were considered. A demographic questionnaire was used to gather background information (e.g., age, gender, experience with virtual reality devices, driving videogames, large screens). The NASA Task Load Index (NASA-TLX) is a multidimensional scale index employed to assess subjective workload. It comprises six subscales: mental, physical demands, and temporal demands, frustration, effort, and performance. The response scale ranged from low (0) to high (100). The performance is the only dimension that present different labels good (0) and poor (100). The System Usability Scale (SUS) allows evaluating the perceived usability of the system. The responses can be provided on a 5-point Likert scale. Sum scores can range between 0 and 100. A Presence questionnaire was also administered. Finally, three ad hoc questionnaires were administered to assess respectively: perceived difficulty and danger, as well as UX. The Perceived Danger and Difficulty questionnaires were considered to assess the perception of drivers, in terms of and danger and difficulty, considering the obstacles faced during the driving tasks. Both questionnaires comprised 4 items on a 5-point scale (from 1-low to 5-high). The UX questionnaire aimed at evaluating three dimensions: pleasantness, engagement, and utilization and time flow. It comprises a total of 12 items. The response could be provided utilizing a

5-point Likert scale. The Presence questionnaire shows a statistic difference in the possibility to act dimension, higher for the manual condition (p-value 0.023), due to the control exercised by participants over the vehicle. In both conditions, the realism and the quality of interface dimensions were close to the maximum scale. This information shows the high fidelity of the driving simulator used. System Usability Scale doesn't show statistical differences among conditions but the results show a general very high usability of the driving simulator. Moreover, also the User Experience questionnaire shows the high quality of the virtual environment used, with the three scales, Pleasantness, Engagement and use&Time flow, with results near the maximum for all of them. One of the main limitations of this experiment was the restricted sample size. Nevertheless, on the first half of the participants, we wanted to verify the effect of an aggressive behavior of the robot driver: sudden brake actions (instead of smooth) during the execution of some tasks. With the second half of the participants, the robot driver acted much more smoothly for the same tasks. The analysis of the questionnaires on the total population shows some important trends that are indicated in the form of percentual increment between the two conditions. For a preliminary comparison with bio-signal data, we selected perceived Difficulty from questionnaires. Figure 10 shows the results of questionnaires relative to the perceived difficulty.

Finally, NASA-TLX shows, as expected, that manual driving is more demanding in terms of physical demand (20%), mental demand (5%) and effort (20%). On the contrary, the frustration and performance dimensions show the impact of the AI aggressiveness on participants. In fact, in the manual condition, the performance was evaluated lower by 10%. This can be explained by the perceived difficulty results in the 7-13 pool that shows that the Wind obstacle is the most challenging in manual condition. This information gives us an explanation of the performance higher value for autonomous driving. The frustration dimensions, higher in the autonomous condition, is due to the errors in the behaviors of AI vehicles, that cannot be avoided without control over the vehicle, producing a higher value in the participants.

Conclusions

Our experimental set-up has proven reliable for real-time monitoring and quantitative estimate of driver's instantaneous stress reaction induced by driving demands. The research performed has shown some promising results for the methodology used to define a subjective/objective metrics to classify the effect of various robotic driving styles in highway realistic driving on a professional Driving Simulator, which offers a a seamless real/virtual human reaction. This means that, when a statistically meaningful number of participant data would be available, the methodology applied on the Driving Simulator could be extended to a real-life case. While with the former the external traffic environment is provided as deterministic, with the latter a random variation is unavoidable, thus the approach is very useful to isolate the most influencing parameters. The research team is planning to repeat the experiment to reach at least a total of 30-40 participant in the upcoming months.

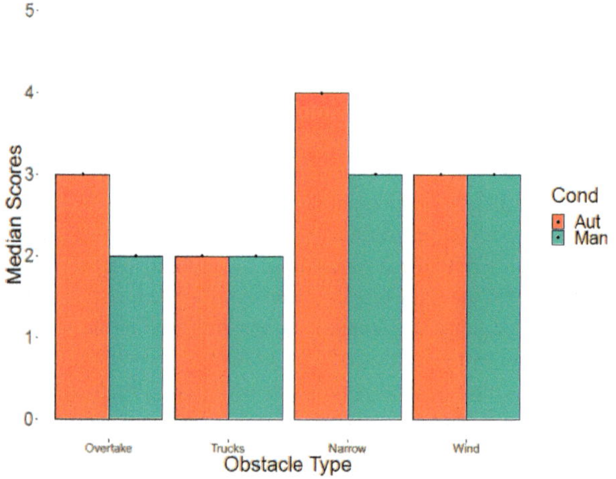

Figure 10: Median scores of the questionnaires on the perceived difficulty.

References

[1] A. Affanni, "Wearable instrument to measure simultaneously cardiac and electrodermal activities", IEEE International Symposium on Medical Measurements and Applications, MeMeA 2016 - Proceedings, 2016.

[2] A. Affanni and G. Chiorboli, "Wearable instrument for skin potential response analysis in AAL applications", 20th IMEKO TC4 Symposiu-m on Measurements of Electrical Quantities: Research on Electrical and Electronic Measurement for the Economic Upturn, Together with 18th TC4 International Workshop on ADC and DCA Modeling and Testing, IWADC 2014, 2014.

[3] A. Affanni and G. Chiorboli, "Design and characterization of a real-time, wearable, endosomatic electrodermal system", Measurement, vol. 75, pp.111-121, 2015.

[4] A. Affanni, G. Chiorboli, and D. Minen, "Motion artifact removal in stress sensors used in driver in motion simulators ", Medical Measurements and Applications (MeMeA), 2016 IEEE International Symposium on, 2016.

[5] A. Affanni, R. Bernardini, A. Piras, R. Rinaldo, and P. Zontone, "Drivers stress detection using skin potential response signals", Measurement, vol. 122, pp. 264-274, 2018.

[6] http://docs.gestionaweb.cat/1662/catalogo-magneti-marelli-2017.pdf

[7] Task force of the European Society of Cardiology and the North American Society of pacing and Electrophysiology.Heart rate variability. Standards measurement, physiological interpretation, and clinical use. Eur Heart J. 1996;17:354-81.

[8]. Lombardi, F., & Stein, P. K. (2011). Origin of heart rate variability and turbulence: an appraisal of autonomic modulation of cardiovascular function.Front Physiol, 2, 95.

[9] Wright, S., Ward, N. J., & Cohn, A. G. (2002). Enhanced presence in driving simulators using autonomous traffic with virtual personalities. Presence: Teleoperators & Virtual Environments, 11(6), 578-590.

[10] Lin, J. W., Duh, H. B. L., Parker, D. E., Abi-Rached, H., & Furness, T. A. (2002). Effects of field of view on presence, enjoyment, memory, and simulator sickness in a virtual environment. In Proceedings ieee virtual reality 2002 (pp. 164-171). IEEE.

[11] Hess, A., Jung, J., Maier, A., Taib, R., Yu, K., & Itzstein, B. (2013, June). Elicitation of mental states and user experience factors in a driving simulator. In 2013 IEEE Intelligent Vehicles Symposium (IV) (pp. 43-48). IEEE.

[12] Schultheis, M. T., Rebimbas, J., Mourant, R., & Millis, S. R. (2007). Examining the usability of a virtual reality driving simulator. Assistive Technology, 19(1), 1-10.

[13] Robillard, G., Bouchard, S., Fournier, T., & Renaud, P. (2003). Anxiety and presence during VR immersion: A comparative study of the reactions of phobic and non-phobic participants in therapeutic virtual environments derived from computer games. CyberPsychology & Behavior, 6(5), 467-476.

Development of model predictive motion planning and control for autonomous vehicles

Guido Tosolin, Senior Manager, Chassis Development
Jaume Cartró, Project Engineer
Vishwas Sharma, Project Engineer

1 Abstract

Full autonomous vehicles for the general public are getting closer every year. Among all the challenges to overcome, one of them is the acceptance of this technology which translates to make the passengers enjoy being driven. To achieve this objective, automated vehicles will have to focus on performance attributes such as comfort, stability or efficiency and vehicles dynamics development will take care of it.

At Idiada, research about this topic is being carried out and strategies about motion planning and control (path follower) will be proposed in this paper. The development is based on optimal control method like Model Predictive Control (MPC).

Among all the possibilities to face the problem, MPC was chosen for several reasons. MPC allows setting constraints in our control inputs like maximum steering wheel angle or vehicle states like accelerations. However, the main reason to use MPC is its way of planning in advance control actions that behaves very similar to how a human driver would do. This feature is key in our understanding to make autonomous vehicles be accepted by all passengers.

Also, our contribution yields in finding the correct vehicle dynamics metrics to design and adjust all the cost functions.

Finally, thanks to our new acquisition, the DIM 250 VI-Grade Simulator which is able to reproduce up to 2.5G accelerations, all our development will be evaluated in a fast and secure testing environment.

2 Automated driving overview

Figure 1 provides an overview of how automated driving can be structured:

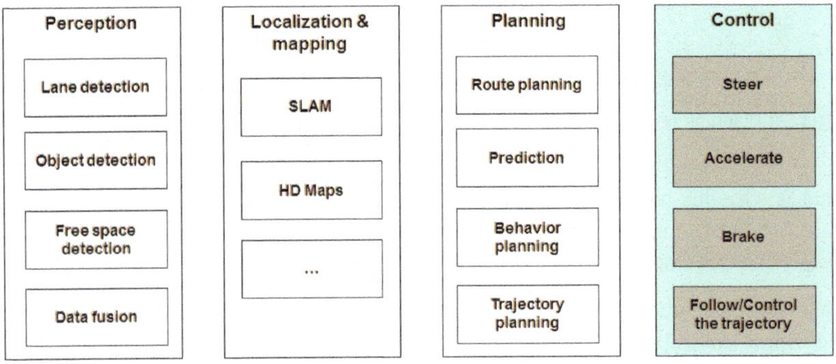

Figure 1: Automated driving blocks

The first layer of perception will get all data available from the sensors to capture the environment around the vehicle. Then a second layer will locate the vehicle and build the surroundings with data of the first layer. Then planning algorithms will find a trajectory to follow and finally the control of the vehicle is managed to get the desired vehicle motion.

3 Development

In this paper a high-level controller will be presented. The outputs of this module are the steering wheel angle and pedals reference positions to be followed by other low-level controllers and as a result, the controller is able to follow the path described by planning accurately.

On the planning side, this block can be divided into several layers very related to the frequency of operation as described in [1]. The first layer is the route planning which ideally runs once per trip. Then a local motion planner decides which maneuver to do considering all obstacles, dynamic and static. This last layer normally considers a horizon based on what the sensors can see which can be of hundreds of meters and with very combinatorial solutions which makes it very hard to use with a realistic dynamic model. Then the most common strategies such as A* search algorithms [1], or RRT algorithms [2], use a very simplified vehicle dynamic model or directly a kinematic model.

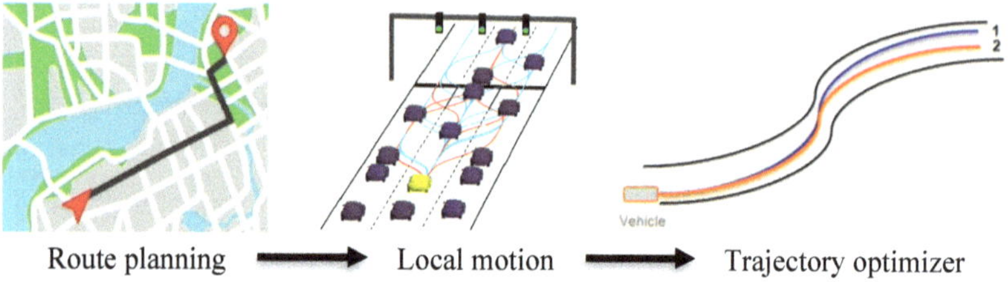

Figure 2: Planning layers workflow

That is why this research work proposes to add a new layer that optimizes the final trajectory using a more detailed dynamic model. In the end, this is going to be the trajectory that passengers would feel and to keep them in good shape and conscious.

Then the contribution of this paper lies on the dynamic behavior that can be referred as a dynamic controller which includes the dynamic planner, that adapts the trajectory references for the controller, and the controller that follows the optimized path.

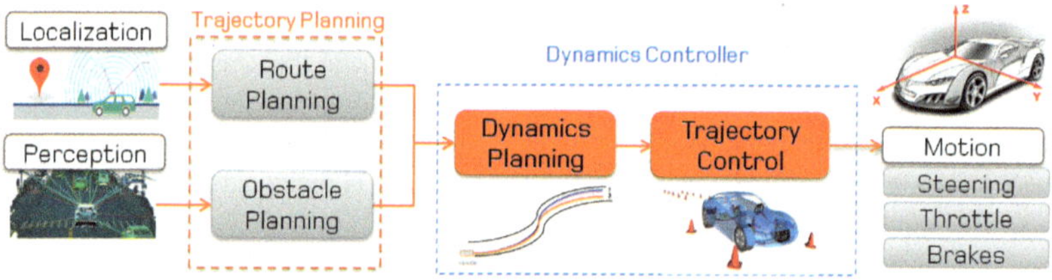

Figure 3: Dynamics controller coupling with the rest of layers

Then, two different approaches for trajectory optimization are discussed in the paper. On one hand a planner which works off-line and uses a nonlinear model to iterate till final solution and, on the other hand, one that works online with linear model constraints.

3.1 MPC Controller

The controller receives the position of the discretized trajectory to follow and predict the control inputs. The number of prediction points chosen is in the range of 15 to 30 and are separated depending on the sample time and speed at each point. The sample time is fixed and around tens of milliseconds so in the end, the total distance predicted will vary depending on the speed reference profile.

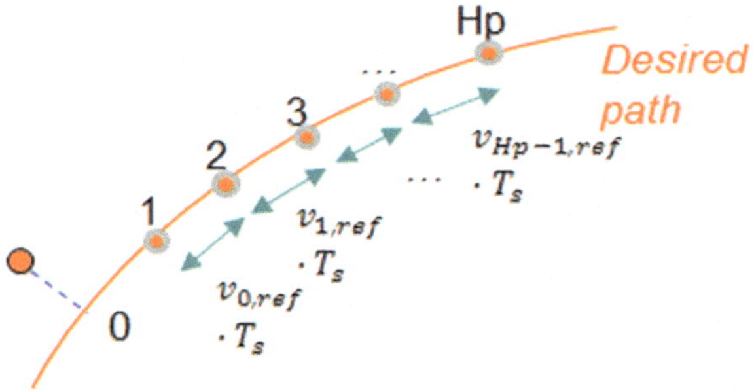

Figure 4: Controller references from planning

The controller uses a 3DOF bicycle model (Figure 5), which is very efficient computationally and at the same time represents all the essential dynamics of the vehicle. For lateral tire forces, the linear region is considered which will be suitable for most scenarios. Only in some emergency maneuvers, the vehicle could drive at its limits and then there are some solutions to consider this effect that are effective (explained later on).

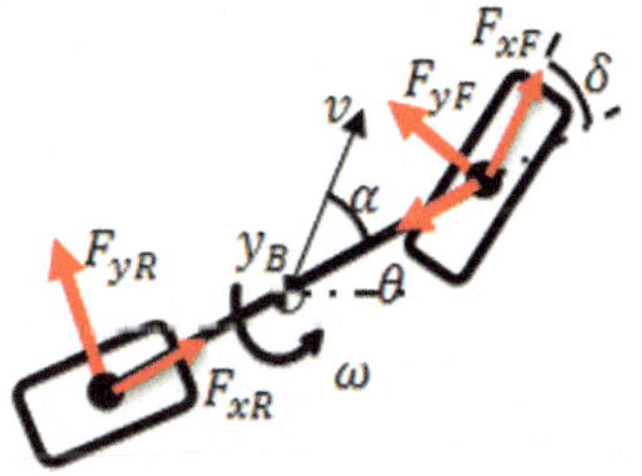

Figure 5: 3DOF bicycle model

Before showing the MPC problem, say that quadratic programming solvers are used to find the optimal solution. This means that the cost function is restricted to be quadratic and the restrictions must be linear. By using these kinds of solvers instead of working with the nonlinear vehicle model, a linear vehicle model is predicted for each prediction and thus the space of solution that the solver works with becomes convex. Then the solver converges to a unique solution very close to the one that a nonlinear solver would found but in a much faster rate and enables us to work in real time. This approximation is correct when working in high frequencies of sampling like the controller is doing.

The cost function used in the controller looks like (Equation 1):

$$\boldsymbol{min}\ (\omega_e \cdot Je + \omega_u \cdot Ju + \omega_c \cdot Jc) \qquad \text{(Equation 1)}$$

- Je: error in trajectory tracking (lateral and angular deviation)
- Ju: control effort → smooth actions (steering wheel rate and longitudinal force rate)
- Jc: stability/comfort parameters (lateral/longitudinal accelerations/jerks)
- ω_i : associated weight

Notice that MPC works as a filter so to really follow correctly the reference the controller emphasizes error tracking. For what concerns restrictions, they are provided by the bicycle model, one for each prediction, and control actions operational limits.

So far, these are the typical constraints and cost function. However, more features have been added to have a more stable and comfortable controller.

For example, not only accelerations are minimized but also a limit acceleration ellipse is set, Figure 6, to always try to be inside of this behavior and allowing to be out of it (for example in an emergency maneuver) but penalizing more in the cost function. It must be remembered though, that constraints must be linear so the implementation is done by multiple straights.

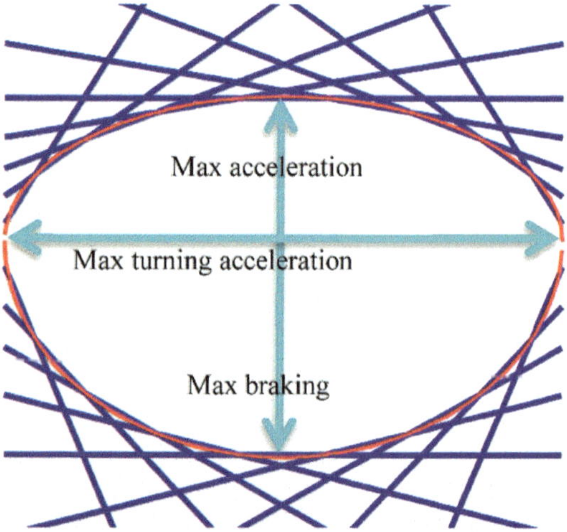

Figure 6: Acceleration ellipse limit

Also, to face the problem of tire limits commented before, the following strategy is adopted.

For understeering behavior, looking at Figure 7, for front tire lateral force the limit is reached for different vertical loads at a specific front sideslip angle. Then a new constraint is added (Equation 2), with a new variable, ε_1, which is added in the cost function with a high weight. Then this variable will be zero with no penalizing effect in the cost function unless the front sideslip angle is overpassed and then solver will try to minimize front sideslip.

$$-Sideslip_{front\ limit} - \varepsilon_1 \leq Sideslip_{front} \leq Sideslip_{front\ limit} + \varepsilon_1$$

$$\varepsilon_1 \geq 0 \qquad \text{(Equation 2)}$$

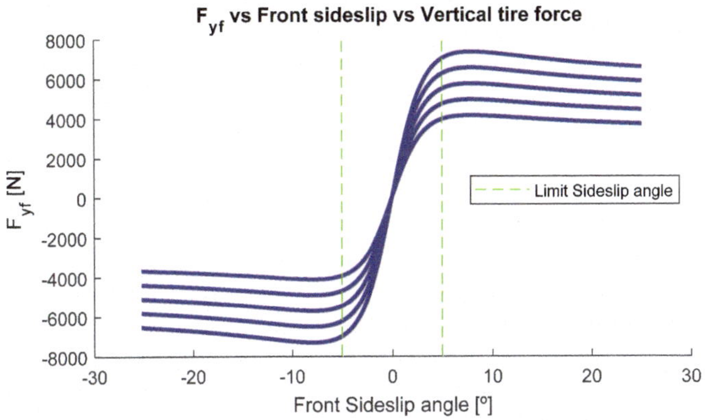

Figure 7: Pacejka tire model for lateral force

Finally, for oversteering behavior, the rear sideslip angle is penalized and thus adding a stabilizing effect in the whole driving experience.

To implement the controller Simulink was used, scheme in Figure 8, as it is a good testing environment to integrate simulation software like CarMaker or CarRealTime to have a realistic vehicle behavior. Then inputs are: in yellow the position and velocity references plus current vehicles states. Outputs are vehicle control inputs.

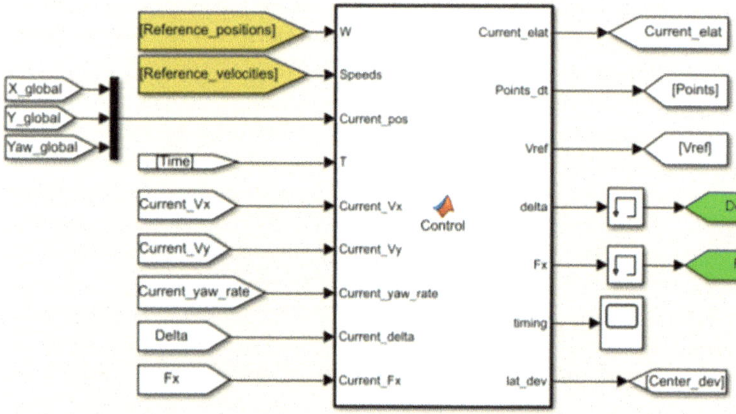

Figure 8: Simulink implementation

3.2 MPC Off-line Planner optimizer

This first approach for planning solves an optimization problem like the controller explained in section 3.1 but with some changes. As an input, it gets a trajectory and optimizes it making more feasible for driving taking into account dynamic metrics.

The first change is that the horizon of prediction considered is bigger than the controller prediction. The horizon selected can vary depending on the purpose and can cover as much distance as desired taking into account that the solution time will increase.

Then the cost function no longer penalizes so much the deviation errors but it focuses on other metrics:

$$\min\ (\omega\_c \cdot \mathrm{Jcomf} + \omega\_s \cdot \mathrm{Jstab} + \omega\_t \cdot \mathrm{Jtime}) \qquad \text{(Equation 3)}$$

- Jcomf: comfort parameters (vertical acceleration, lateral and longitudinal jerk)
- Jstab: stability parameters (sideslip and sideslip rate)
- Jtime: time minimization
- ω_i : associated weight

In this case, the planner uses the same 3DOF of the controller for the handling dynamics plus a quarter car suspension model for the vertical dynamics. The planner no longer linearizes the model for each prediction and uses the nonlinear model directly.

The workflow of the planner is detailed as follows. The input trajectory is discretized in points which the solver is able to move in the normal direction of the trajectory, always respecting boundary limits. Then, in one iteration, it moves all points and runs a simulation where the vehicle follows the trajectory (a controller is also incorporated in the algorithm in order to track the optimized trajectory and provide accurate states). When the iteration is finished using the metrics commented before it defines the trajectory points for the next iteration and also the speed to follow.

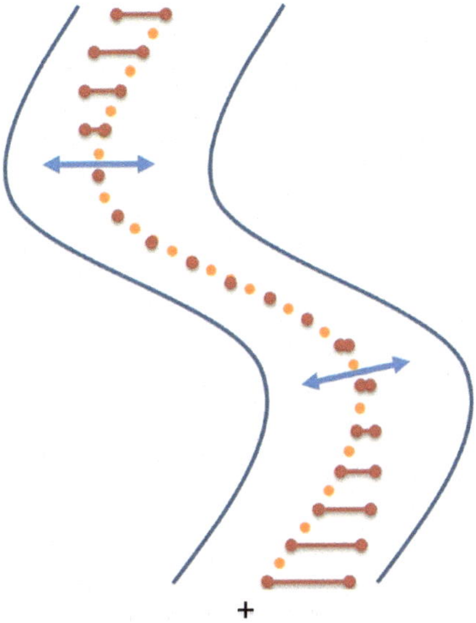

Figure 9: Orange: input trajectory; Red: optimized trajectory

The planner optimizer is following a sectional planning approach to decrease the solving time. However, due to its nonlinear nature is not able to run in real time. Nevertheless, it serves a relevant purpose, because it can be used to validate online optimizers that assume more simplifications.

3.3 MPC On-line Planner optimizer

The planner of section 3.2 provides good results with the only problem of real-time feasibility. That is why Idiada started developing a planner optimizer capable of working in real time.

In collaboration with the 'Institut de Robòtica Industrial (IRI)' of Polytechnic University of Catalonia (UPC), an online planner is being developed. The idea of this planner is very similar to the one of the controller: use MPC structure giving linear vehicle model for each prediction. Again, as the objective is to change the trajectory, the cost function will not focus on tracking errors but on comfort and efficiency metrics. The most difficult aspect is that, unlike the controller model, that considers a relatively short horizon and for which it is easy to get future states, the planner optimizer considers a longer horizon and the future states are changing at each sample time since the trajectory optimized is being updated continuously. The best way to deal with this problem is to use the predicted states in the previous trajectory found by MPC solver to get future vehicle models. This implementation is possible due to the rate of sampling used to update the trajectory which is about hundreds of milliseconds.

The way of computing the linear vehicle model was changed [3]. Typically for linearizing the vehicle model and in order to get a State Space model, Taylor derivative methods are used. On the other side, now a novel technique is used which no longer uses derivate but instead gets a Linear Parameter Varying (LPV) model that translates to reduce the calculations and help with the real-time feasibility.

4 Results

The following section contains the optimization results of the controller and offline planner.

4.1 MPC Controller

For evaluation of the controller, two scenarios are presented. In the first scenario, the trajectory selected is the one in Figure 10. The scenario created emulates an urban scenario where the speed limit is at 40 km/h and in Figure 11 it can be seen how the vehicle is able to follow the reference trajectory.

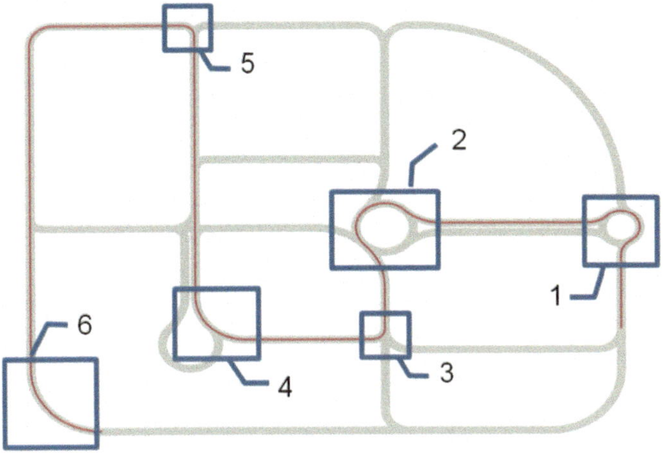

Figure 10: Urban track

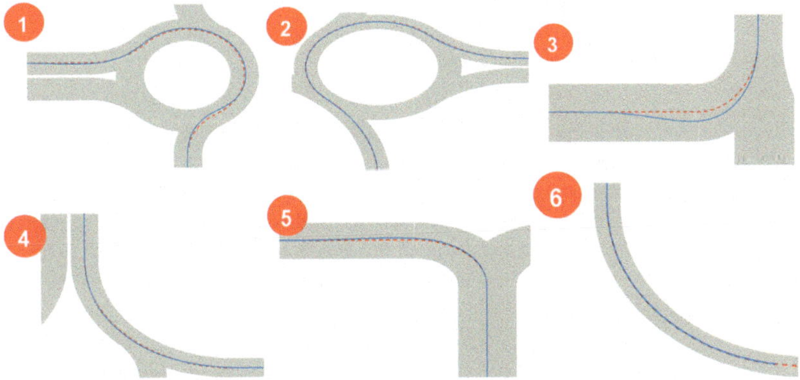

Figure 11: Urban track, 6 trajectory curves detail (red: reference; blue: real trajectory)

In Figure 12 some metrics are displayed like speed reference or real, lateral deviation, acceleration or wheel toe angle. It shall be remarked the fact that longitudinal and lateral tracking is done simultaneously as both dynamics are dependent on each other. That is an important feature because it allows avoiding if for instance, our planner tells the vehicle to do a curve at a speed which is not safe, the MPC solver will reduce the speed in order to maintain the safety parameters like maximum lateral acceleration.

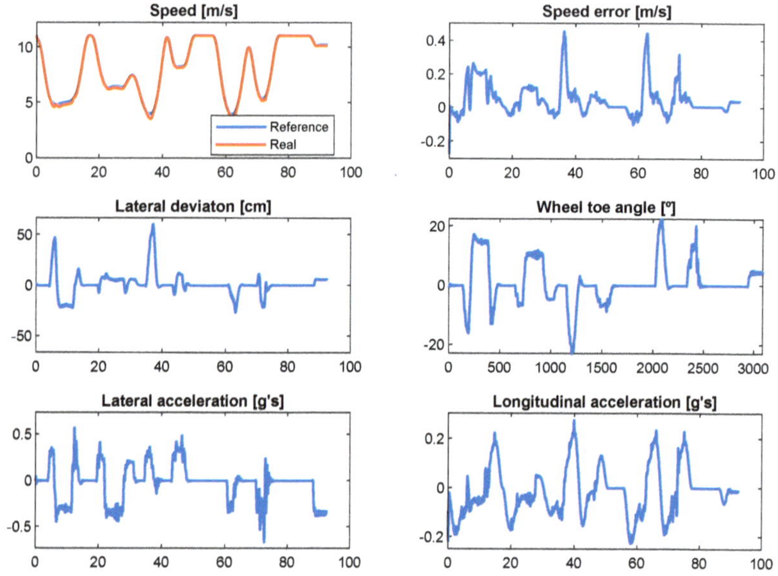

Figure 12: Analytical results at urban scenario

The second scenario is one of Idiada's testing tracks. The idea of this test was to adjust the controller for very demanding situations. Also, the aim of the test was to quickly evaluate the cost function weights without a trajectory planner module. For this, it was decided to use a driving simulator (Figure 13) to generate the trajectories., An experienced driver executed track laps at different levels of intensity and telemetry was transmitted in real time to give references to the controller. An extreme race car was used for this experiment.

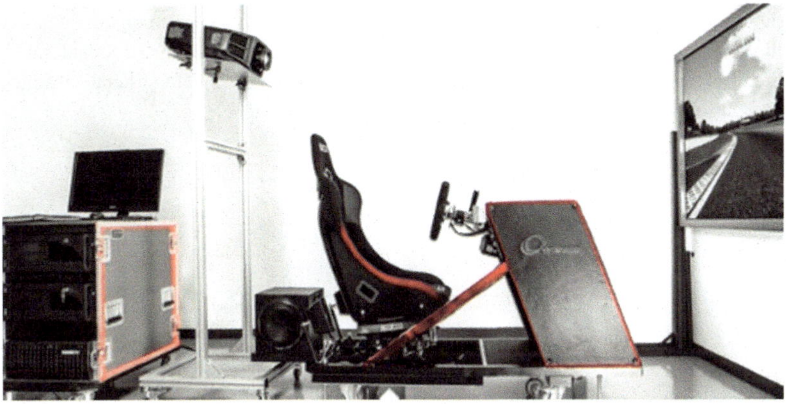

Figure 13: VI-Grade Compact Simulator

The results in Table 1 show that the controller is able to follow the trajectory even at higher levels of lateral accelerations. Lateral deviation increases but in an acceptable range considering that that vehicle was driven at the limit reaching drifting behavior in some curves.

Table 1: Results from following a real driver at different intensity

	Reference time	Lap time	Average lateral deviation	Max lateral acceleration
Relax lap	62.9 sec	62.52 sec	13 cm	1.41 g
High intensity	48.3 sec	48.81 sec	36.31 cm	2.22 g
Limit lap	44.6 sec	45.3 sec	59.39 cm	2.75

4.2 MPC Off-line Planner optimizer

For showing the effectiveness of the planner three different scenarios are chosen to be analyzed: a double lane change maneuver, again Idiada's racing track 'Dry Handling' and an S track.

Double lane change is chosen as it is a demanding maneuver which is very useful to evaluate the lateral dynamics behavior. Starting at a speed of 80 km/h, the trajectory in blue in Figure 14, based on splines considering geometric kinematics, is given as an input to the planner optimizer. In orange, the optimized trajectory based on the dynamic model is obtained. It can be seen in Figure 15 that the jerk and sideslip are considerably reduced as compared to the simulation with the not optimized trajectory, hence producing a comfortable trajectory for the passengers.

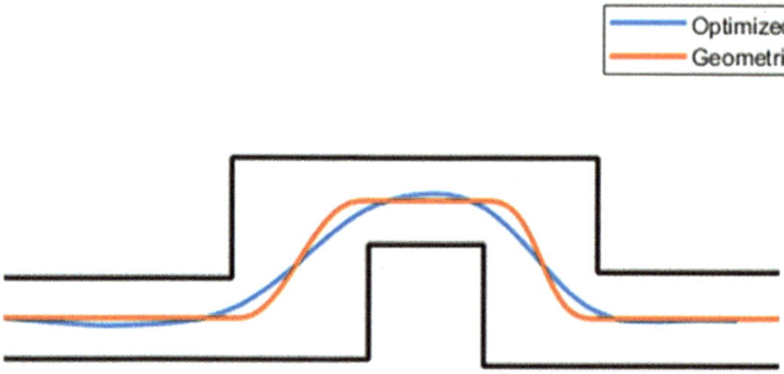

Figure 14: Trajectory comparison between input and optimized

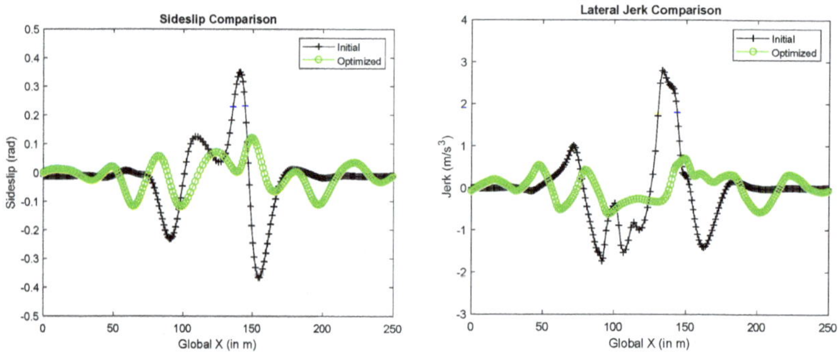

Figure 15: Double lane change trajectory optimization metrics

On the second analysis, Dry Handling track is used for the optimisation. With this kind of scenario, the planner not only has to optimize the next curve, but it also has to take into account the following ones.

As an input centre line was given and the result is in Figure 17.

Figure 16: Dry Handling testing track in Idiada

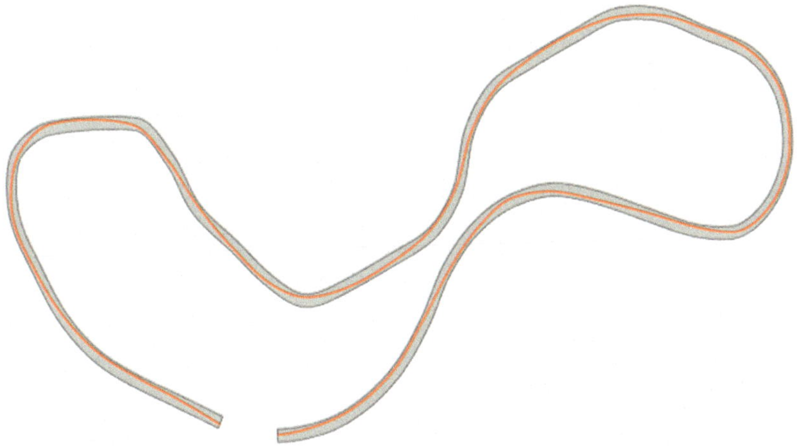

Figure 17: Dry handling optimized

Finally, the third scenario was used to demonstrate the possibility to have different optimised trajectories, depending on the adjustment of the cost function of the dynamic planner. This feature shall be used to define different driving behaviours that shall be tuned in order to fit the driving scenarios or the preference of the people inside the vehicle.

In this case, an S track is optimized for two different sets of weights in the cost function of the planner: one where sideslip is minimized and the other one where lateral jerk is the metric of focus. In Figure 18 differences can be observed.

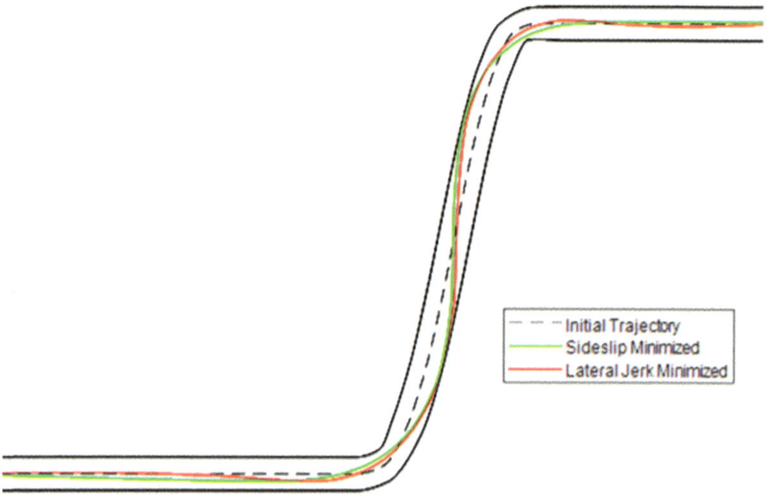

Figure 18: S track optimization

5 Future developments

The future work will focus on the following points:

1. Real time feasibility and integration of the on-line planner based on LPV.
2. Validation and testing of the dynamic controller with subjective feedback.

For the second point, the plan is to use a dynamic simulator. Idiada has recently acquired a DiM250 dynamic simulator from VI-Grade, which will be ready to use in June of 2019.

The simulator will be used in order to refine the cost functions in terms of the metrics that are considered and in ters of the balance between different weight factors. In fact, the simulator allows to receive a direct subjective feedback about how the vehicle motion is perceived by the people in the vehicle.

The big difference with tests done so far will be the capability of evaluating the strategies with one or two passengers inside the car that will be able to do objective and subjective analysis. Compared to tests that can be done in a real car, the dynamic simulator is a safe environment and also very adjustable and easy to change from one situation or scenario to another one. In this line of research, Idiada has integrated the AV Scaner Studio software, to facilitate the task of generating scenarios.

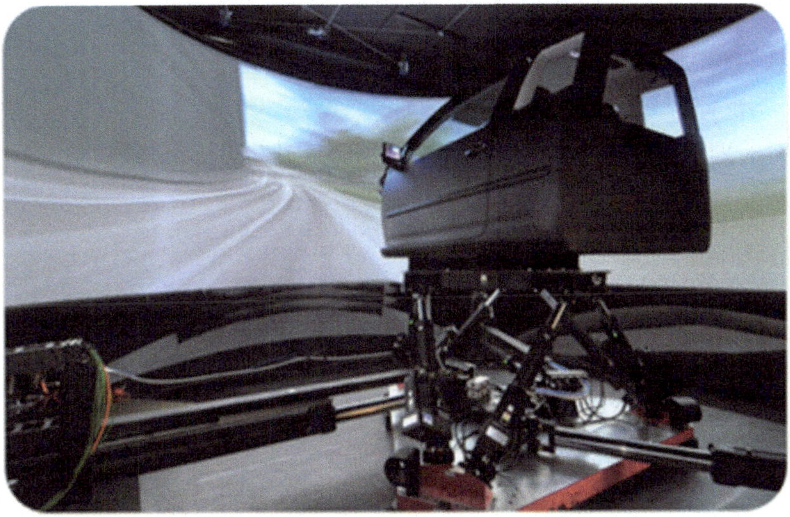

Figure 20: Driving simulator DIM 250

6 Summary and conclusions

In this paper route optimization and control strategies based on Model Predictive techniques is presented for autonomous vehicles. Both strategies have been already tested in a simulation environment and will soon be tested in a dynamic simulator. MPC has proven to be suitable for autonomous driving applications as behaves very similar to how human drivers act.

The authors believe that vehicle motion and will be an important aspect to consider for achieving the best driving experience for the passengers of the future. For this, autonomous driving controllers shall incorporate vehicle dynamics aspects in order to achieve a high level of acceptance in the transition from human to self-driving driving vehicles.

7 References

[1] 'Search-based optimal motion planning for automated driving', Zlatan Ajanovic1, Bakir Lacevic, Barys Shyrokau, Michael Stolz and Martin Horn, August 2018

[2] 'Real-time Motion Planning with Applications to Autonomous Urban Driving', Yoshiaki Kuwata, Gaston Fiore, Emilio Frazzoli, September 2009

[3] 'Autonomous racing using Linear Parameter Varying -Model Predictive Control (LPV-MPC)', Eugenio Alcalá, Vicenç Puig, Joseba Quevedo, Ugo Rosolia

DEVELOPMENT METHODS

Challenges and solutions of converting conventional vehicles to hybrid electric or battery electric vehicles

Matthias Wilmes

Rob Kraaijeveld

Marvin Schrage

Frank Schummers

Introduction

Legal and regulatory pressures in Europe and China to reduce emissions and fuel consumption are increasing the demand for hybrid and electric vehicles. This growing demand is the main driver for technological improvements in electric powertrains and batteries. Nearly every car manufacturer is developing hybrid or electric vehicles to fulfil the fleet consumption targets in order to avoid penalties.

The car manufacturers developed several powertrain options, to improve fuel economy and reduce emissions to meet legal regulations. An overview is given in the following figure.

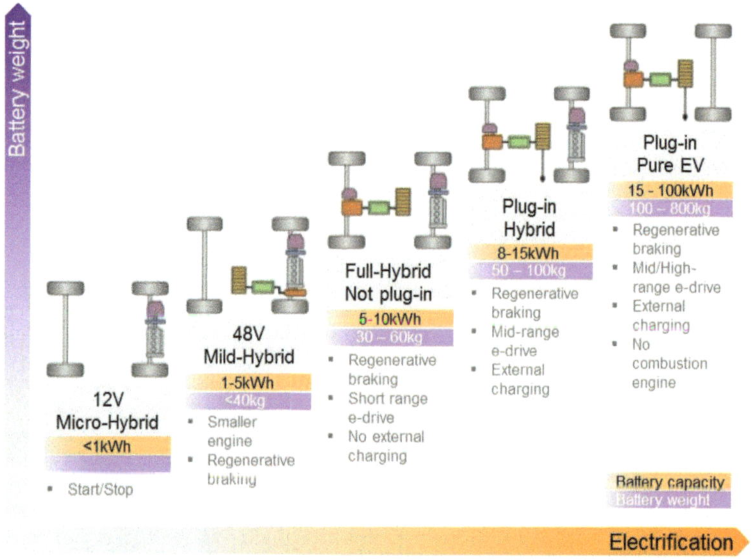

Figure 1: Different Levels of Vehicle Electrification

The most effective way to avoid emissions caused by internal combustion engines is to totally remove the conventional powertrain by a fully electric powertrain.

This case study describes a time and cost-efficient approach to convert a conventional vehicle with internal combustion engine (ICE) to a full electric vehicle. The major challenge is to integrate an existing battery into the body in white with only minor changes of both systems to reduce development time and cost. This article describes how this is realized by chassis system modifications.

Challenges of converting a conventional vehicle to a hybrid vehicle / BEV

A big advantage of converting a vehicle with a conventional internal combustion engine into a full electric vehicle is the fast time-to-market compared to a new development. On the other hand, the shortened development time implies only minor changes at the major systems body in white and battery.

The original drivetrain is removed, making space for the battery and the electric drivetrain. Vehicles with a smaller range can place separate battery packs in the trunk and in the original engine bay. If a higher range is desired, a common way is to use a battery pack which is placed underfloor. This means that the body has to be lifted in order to retain the same ground clearance as before. The main requirement for the new chassis system is to ensure that the vehicle dynamics behaviour of the electric vehicle and the original vehicle are the same.

Methodology

The conversion of a conventional vehicle to a battery electric vehicle is done completely in a virtual development process. Based on the CAD-data of the conventional vehicle and the existing e-powertrain including the battery, the first step is to start with the package analysis and the first layout for the new chassis system. The resulting design and engineering data serves as input to build-up first ADAMS- and CarSim-Models which are used to simulate the chassis system and the vehicle dynamics behaviour.

The CAD concept provides the hardpoints of the suspension system. This information is used in ADAMS simulations to determine the Kinematics and Compliance (K&C) behaviour of the modified suspension. The new torsional stiffness of the twist beam rear axle is defined with the help of additional simulations.

The results from the K&C simulations are used to support the CAD-Design and the FE-Calculations afterwards. If necessary, iteration loops between K&C simulation and CAD-Design are performed to meet the defined goals and requirements.

Finally, the results are used in the CarSim model to validate the vehicle dynamics behaviour with regard to the conventional vehicle. The described method is shown in the following figure.

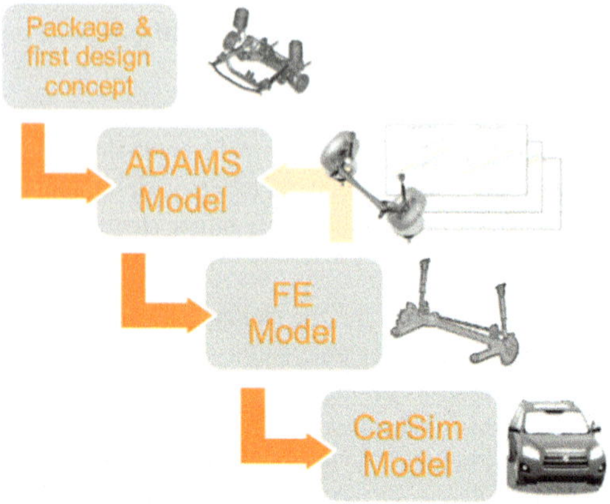

Figure 2: Methodology of the virtual development process

Impact to Vehicle Mass

One important market demand for an electric vehicle is great driving range. In this case study, the vehicle has achieved a driving range of more than 400 km with a battery energy of ~50 kWh and a battery weight of ~450 kg.

Although the conventional powertrain was removed, the new electric powertrain and the battery system increase the curb weight by 250 kg in comparison to the original vehicle. The significant weight of the battery itself and the positioning underfloor leads to a different overall weight distribution. The Centre of Gravity (CoG) is lowered by ~30 mm and the weight distribution between front and rear axle is shifted towards the rear axle. The new distribution between front / rear axle is close to 50 % / 50 % with battery compared to approx. 60 % / 40 % of the conventional car. The repositioning of the CoG is shown in the next figure.

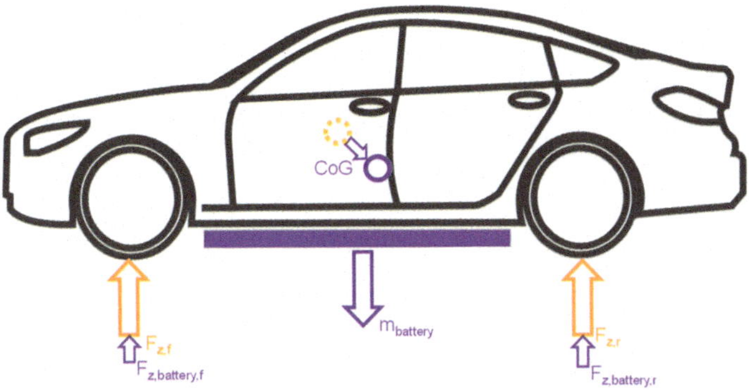

Figure 3: Effects on CoG and weight distribution

Another result of the higher mass and the different weight distribution is an increase of Moments of Inertia (MoI), which will also influence the vehicle dynamics behaviour.

The increase of the overall vehicle mass and the Moments of Inertia will affect the vehicle dynamics behaviour, especially to the response time of the vehicle in lateral manoeuvres and the transient behaviour.

Package

Due to the size and shape of the existing battery the only position left is an underfloor solution. This leads of course to a decreased ground clearance, which has to be compensated by lifting the body of the car in z-direction of 115 mm to regain the original ground clearance. The following figure explains this fact.

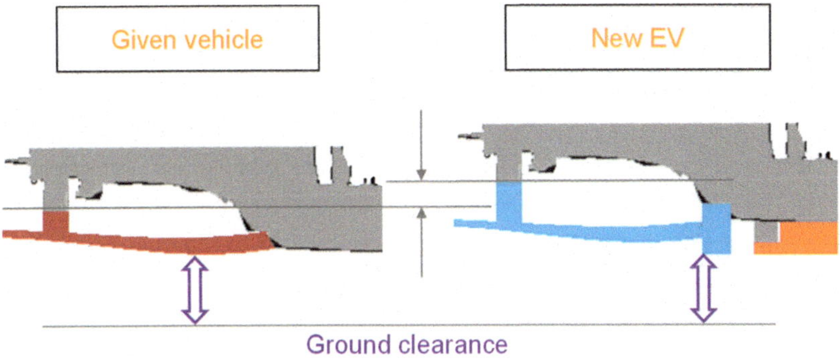

Figure 4: Ground clearance of conventional and electric vehicle

Naturally, this lifting of the body by 115 mm has an impact on the chassis system. The main requirements for the new chassis system is to achieve the same vehicle dynamics behaviour and the same wheel travel as the vehicle with the conventional powertrain. Additionally, an increased tire size in comparison to the original vehicle has to be considered in the package of the system.

To investigate the package situation with the new chassis concept, a tire envelope with maximum steer and maximum bump conditions was created. The envelope for the front axle is shown in the following figure.

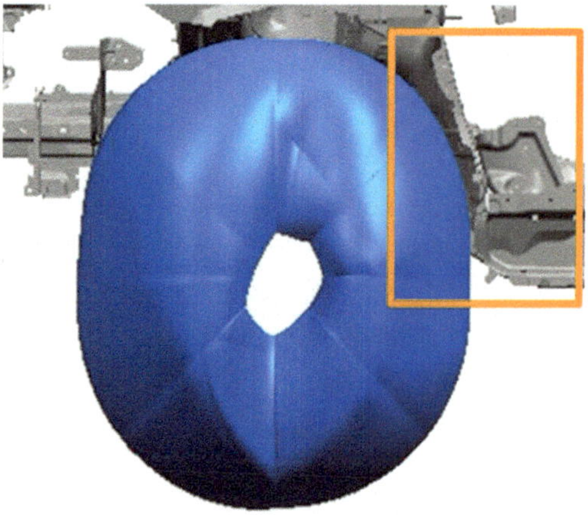

Figure 5: Tire envelope with interference

The investigation on the front axle indicates that there is an interference with the body, which is mainly caused by the bigger tire size. The interference has a size of app. 11 mm.

One mandatory requirement for the front axle is that no body modifications are allowed to integrate the new suspension system. Therefore, a new subframe and a new knuckle were designed to move the wheel centre forwards in order to clear the interference.

The same investigation method was applied at the rear axle. The result is presented in the following figure.

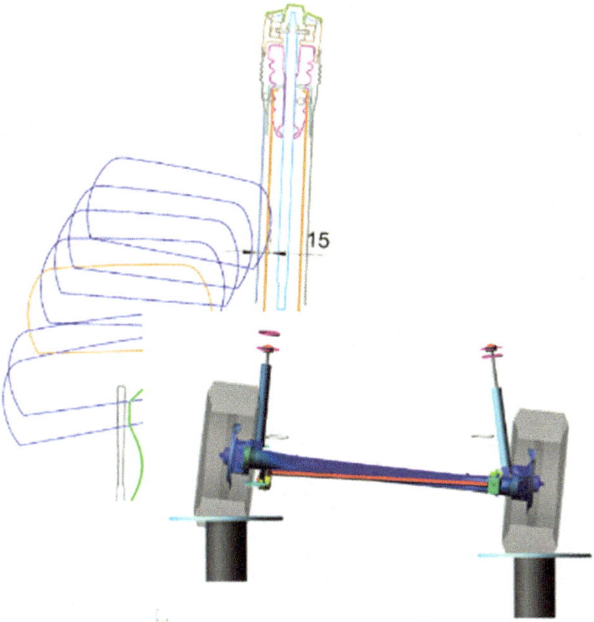

Figure 6: Interference at rear axle

An extension of the damper according to the body lift of 115 mm results in an interference between the tire and the damper. As was the case for the front axle, the requirement that the body top mount of the damper has to stay the same applies to the rear axle as well.

In addition to the interference between rear axle and body, there is a more general conflict, as well. The conventional car has a twist beam rear axle. In combination with the new underfloor battery, the cross beam of the twist beam interferes with the battery. As a result, the cross member has to be moved rearwards and nearly the complete rear axle has to be redesigned.

New Chassis Design

As mentioned in the previous section, new designs for both axles are necessary due to the interferences and the conflict at the rear axle.

For the front axle a new subframe and a new knuckle were designed. The new subframe is shown in the following figure.

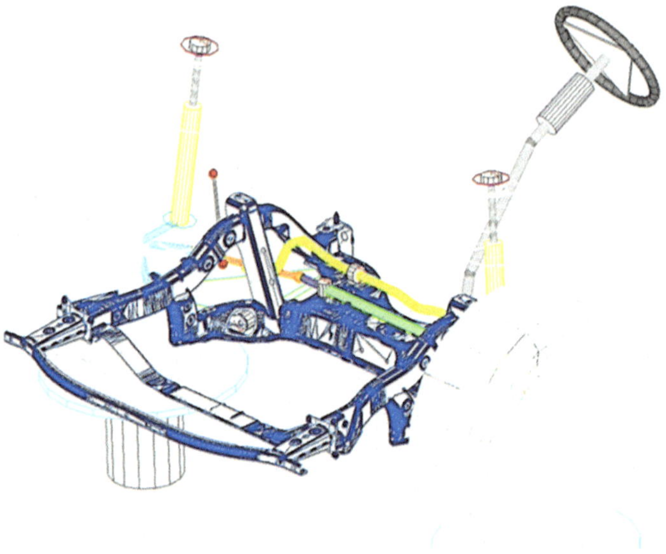

Figure 7: New subframe design

This new subframe and extended struts accommodate the body lift by 115 mm without any changes regarding the topmount, the attachment points to the body or the linkages of the axle. The axis of the strut / damper remains the same, so that there are nearly no changes regarding the kingpin angle and caster angle. As a result of the extension of the strut, the new front axle has an increased track width of ~7 mm.

In addition to the changes to the subframe, the knuckle was redesigned as well. The main reason is the interference at the front axle between the tire and the body, which leads to a forward movement of the wheel centre by ~11 mm. Concurrently, the caster offset is increased by 11 mm leading to an increase of the wheel base of 11 mm. All other hardpoints remain the same, so that there are no changes to the steering kinematics or the linkage of the axle.

Because of the conflict regarding the cross member of the twist beam axle and the new battery system, a redesign of the rear axle is necessary. To solve this matter, the cross member was moved backwards. This leads to longitudinal arms with increased length. Additionally, the attachment points of the original rear axle also interfere with the battery. Therefore, new brackets for the mounts of the rear axle are needed and are placed beside the battery. The required body changes are kept to a minimum. The original and the new rear axle design are shown in the following figure.

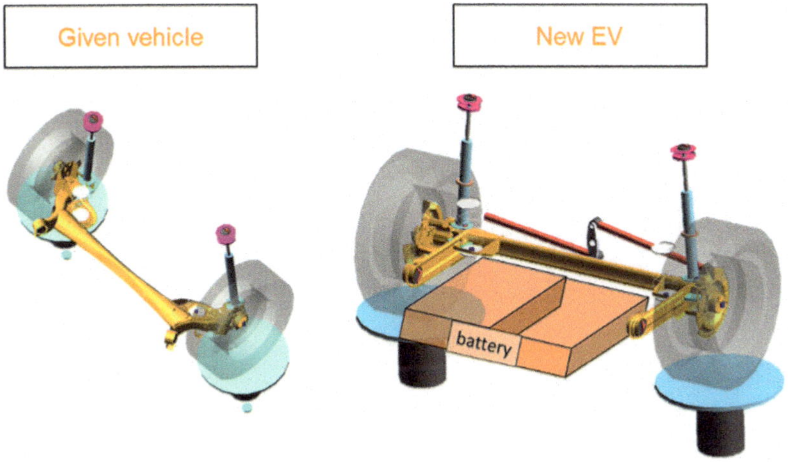

Figure 8: Original and new rear axle design

Another interference at the rear axle occurs between the tire and the damper. To solve this the lower attachment point of the damper to the twist beam was moved towards the vehicle center. In addition to this modification, the camber and the toe angle of the wheel were reduced.

Another major difference to the original vehicle is the weight distribution. The main reason for the increase of the static wheel loads at the rear axle is the additional battery mass. This additional weight leads to higher tire forces in dynamic manoeuvres, especially in lateral manoeuvres. A watt linkage was implemented to optimize the vehicle dynamics behaviour, by achieving a higher lateral stiffness of the rear axle.

Simulation

A CarSim model is used to ensure that the vehicle dynamics behaviour of the new electric vehicle is similar to the original vehicle, several manoeuvres are performed to investigate the behaviour in longitudinal, lateral and transient conditions:

- Steady state circle test
- Lateral transient response test (step steer test)
- Single sine steer test
- Double lane change test
- Braking test

Figure 9: Steady sate circle test simulation

The results of the electric vehicle are compared to the results of the conventional vehicle. In case of deviations between the conventional and the electric vehicle, the cause is investigated and actions are derived out of this. Afterwards, the simulations are performed again to get the improved results. The modified parameters are e.g. spring stifnesses, damper characteristics, cross member stiffness, etc. An example of the results is shown in the following figure:

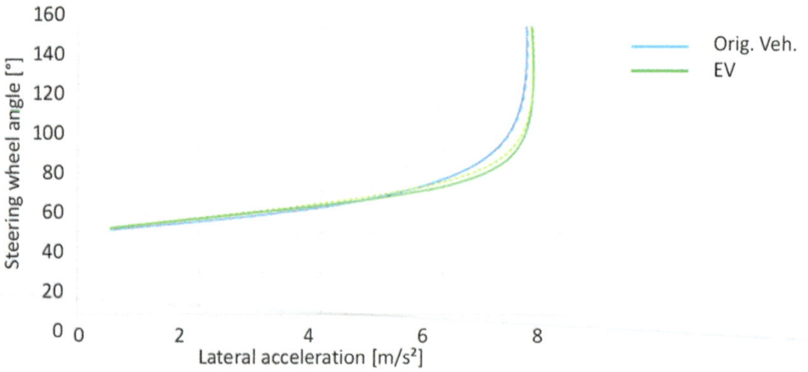

Figure 10: Results of steady state circle test

Conclusions

The short time-to-market makes it very appealing to convert an existing conventional car into a battery electric vehicle. This case study shows a fast and cost-efficient approach. In addition to the required electrical and body engineering work, the front and rear axle systems were modified as well. For the front axle a new subframe and a new knuckle have been designed to solve the chassis-body interferences, which occurred during the package investigations. The kinematics at the front axle remained nearly the same.

The main driver for the redesign of the rear axle is package conflict with the battery system. The cross member of the twist beam was moved backwards, which caused an increase of the length of the longitudinal arms. To gain a better vehicle dynamics behaviour, the lateral stiffness was increase by implementing a watt linkage.

All these design changes were evaluated in vehicle dynamics simulation with CarSim to assure an identical vehicle dynamics behaviour as the conventional vehicle. Therefore, several manoeuvres have been performed to prove that requirement.

Automatically reconfigurable actuator control for reliable autonomous driving functions (AutoKonf)

Dr. Sergey Orlov (HELLA GmbH & Co. KGaA)

Dr. Matthias Korte (Intedis GmbH & Co KGaA)

Florian Oszwald (BMW Group)

Pascal Vollmer (ITK Engineering GmbH)

Introduction and motivation

The introduction of highly automated driving features will lead to major changes in the vehicle's E/E architecture: among other things, a reliable infrastructure will be indispensable. Currently, the driver is responsible for the vehicle at any time and serves as a backup for possible functional failures. With the introduction of highly automated driving functions, this will no longer be the case in the future, as the driver can have other activities in the car like reading books, working or even sleeping. Psychologists have determined how long a driver needs to finish his job and regain control of the vehicle: about 15 seconds are needed for this change [1], [2], [3]. Therefore, the vehicle must work safely for at least 40 seconds up to several minutes after a system error occurs to give the driver time to take back control of the vehicle or, if the driver is not able to do so, to automatically bring himself into a safe state. Currently, the E/E architectures in any vehicle are not reliable enough to meet this challenge yet. New approaches are needed to detect, isolate and respond to errors. Examples of such failure scenarios include false, delayed or missing information, component failure or loss of electrical power. Therefore, it is necessary to design a failure-tolerant system.

For example, the approach currently established for automatic driving functions is to incorporate enough sensors with overlapping functions and different physical principles (e.g. cameras, radars, lasers, ultrasonic sensors) to minimize the risk of misinterpreting the environment. Such an approach is not possible for the braking or steering function, where mechanical components are physically bound to motors and associated electronic control devices. It has already been shown that a partial overlap of brakes and steering functions is achievable [4], [5]. The steering function can be realized by means of adaptive braking force distribution with the disadvantage of reduced brake capacity reserves, which is unacceptable from a safety point of view. For braking, however, functional redundancy is impossible. Therefore, current approaches are intended to duplicate the actuators completely, as shown below. So far this solution has been transferred from aviation and adapted to drive-by-wire technology [6]: two paths are used in parallel and in the event of an error, the faulty one is deactivated (Figure 1c).

Already known approaches face a huge challenge in terms of cost and space requirements. In lieu of this, the Autokonf[1] project attempts to limit the number of redundant components to be introduced. For this purpose, the overall number of redundant devices should be reduced by being reconfigurable during operation. If a component could be used for two different purposes (e.g. braking or steering), then this single two-function component could replace two single-function components. Thus, a component that has to be developed in addition to the regular control devices takes over the function of the faulty component. To address this challenge, this project will look at how to separate the ECUs from the actuators, how to create generic chassis components, what kind of fault-tolerant infrastructure is required and how to utilize an infrastructure with reconfigurable connectors.

---

[1] "Autokonf" stands for "AUTOmatische reKONFigurierbare Aktorikansteuerungen für ausfallsichere automatisierte Fahrfunktionen" (English: Automatically reconfigurable actuator control for reliable autonomous driving functions).

Overview state-of-the-art systems

For the better understanding of the Autokonf architecture proposal and to differentiate it from the "state-of-the-art" steering and braking systems, a short overview regarding current developments is presented below.

Steering system

An electrical power system (EPS) generates the power assist by means of an electric motor whose torque is fed into the steering rack column by a servo gear unit (Figure 1a). This motor is controlled by power electronics integrated into the ECU: Electrical Control Unit [7]. This ECU is designed in a way to assist the driver in steering the car in the desired direction and to reduce the amount of force necessary to be applied. The safety mechanics of this system prevents causing torque in the wrong direction or generating torque if it is not expected by the driver. In this way, a fail-safe operation is guaranteed (Figure 1b). If, nevertheless, any kind of failure is observed by the system, the integrated safety mechanism on a software and hardware level will deactivate the power supply to the 3-phase motor so prevent the undesired behavior. Since it is ultimately intended that the driver controls the car, this situation remains safe. Electrical support to turn the wheels is deactivated, but this is not the issue in most driving situations.

With the introduction of autonomous driving, the driver will be at least partially relieved from vehicle control responsibility. Other possibilities are needed to realize a fail-operational functionality. As already mentioned, the simplest way would be to duplicate the needed electrical components (see Figure 1c). Because mechanics are designed to be "unbreakable", the complete duplication of these components is not necessary. This architecture allows to control a 6-phase EPS motor from two similar ECUs independently and simultaneously. In case of failure of one or another ECU, the malfunctioning one will be deactivated, as this is a case for a 3-phase operation. Reduced torque can still be provided with a remaining ECU. This means that the system is still operational at the cost of more necessary components.

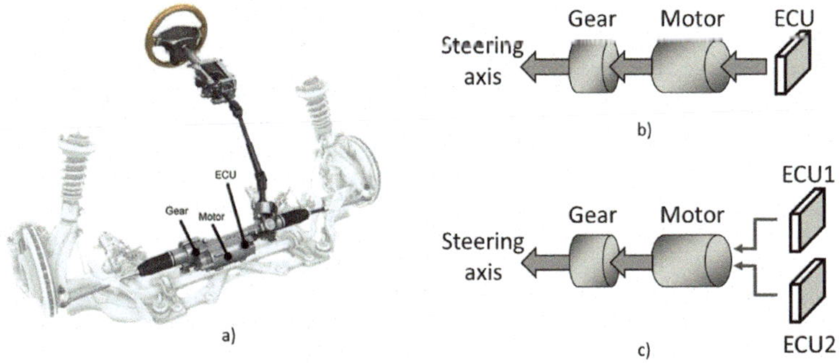

Figure 1: Electrical power steering (a) [12]; different architectures considering "driver in the loop" (b) and "driverless" (c) redundancy realization.

Nissan Infiniti Q50

Currently, the only known Steer-by-Wire (SbW) system in a series vehicle is the Nissan Infiniti Q50 (Figure 2). It has a double redundancy on the ECU level and an additional mechanical backup which closes the clutch on the steering rod during system malfunction. Two steering angle actuators generate the requested steering torque for the vehicle control. When the SbW-mode is on, a steering force actuator at the steering wheel provides feedback about the road to the driver. The three ECU´s control the steering angle actuators for an optimal, electrical support of the driver. Moreover, they are monitoring each other by sharing the result.

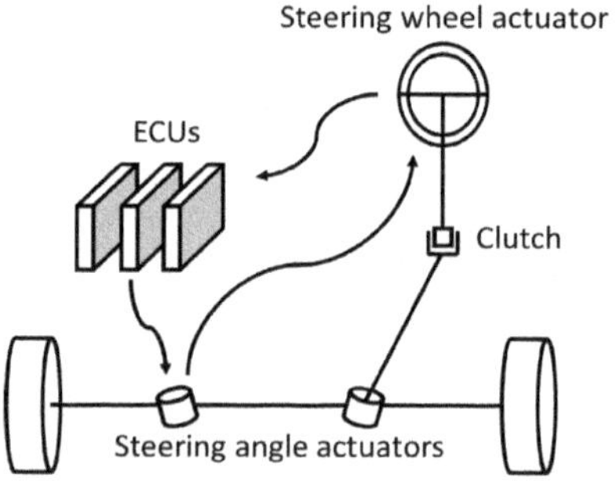

Figure 2: Steer-by-Wire concept of Nissan Infiniti Q50 [15].

Braking system

There is currently a variety of different braking systems on the market. Having the same safety approach as the steering system, braking actuators will still rely on the driver in case of any failures. So called "one-box" solutions (e.g. Continental MK C1 [8], see Figure 3a) are "all-in-one" systems including braking and vehicle stability functionality in one component. If failures are detected, the system will be deactivated in a time less than failure tolerance time interval (FTTI). In going the direction of autonomous driving, a combination of an MK-C1 unit plus an additional unit e.g. MK 100 is needed as a fallback solution. In the case of a two-box topology (e.g. Bosch iBooster + ESPhev [9]), a malfunction in one component can be at least partially covered by the remaining component and, in this way, deceleration can take place (Figure 3b). That being said, the ESP box is mainly used to correct errors made by the driver (e.g. acceleration, braking or steerability is too aggressive in critical situations). For automated driving, where driver's behavior in controlling the car will be reduced if not eliminated, the purpose of the ESP needs to be reconsidered. Nevertheless, a second redundant braking unit will still be required, otherwise realization of a fail-operational functionality is not possible.

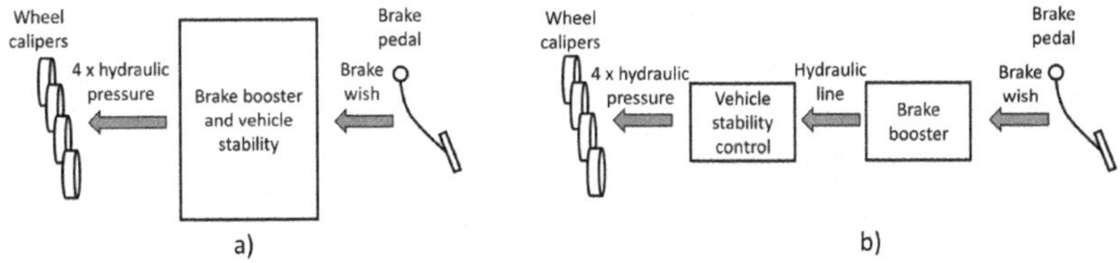

Figure 3: Principal braking system architecture based on a "one-box" (a) and "two-box" (b) solution.

Summarizing the current steering and braking technologies towards autonomous driving, component or functionality doubling is observed all the time even if different systems or approaches are used.

Safety goals and system requirements

The target of the AutoKonf project is to develop a redundant control unit being able to replace either a nominal braking control unit or a steering control unit in case of a failure without any functional degradation. Automatic reconfiguration is an important asset for highly automated vehicles (SAE Automation Level 4 [10]). At this level, a failure of the braking and the steering ECU must not prevent the autonomous vehicle from operating. This paper considers a single failure; simultaneous malfunction of the braking and the steering ECU is not covered.

Failure event	Failure scenario	Conditions	Fault detection time [ms]	Fault condition	Effect	FTTI [ms]
Loss of nominal braking function	Living street	Speed = 5 km/h	30	When stepping on the pedal	Brake travel distance = 0,24 m	160
	Highway	Speed = 130 km/h		When stepping on the pedal	Brake travel distance = 140 m	4260
Loss of nominal steering function	Traffic circle	Speed = 40 km/h		Start leaving the driving trajectory	Leaving the lane by max 0,5 m	570
	Highway entry	Speed = 100 km/h		Start leaving the driving trajectory	Leaving the lane by max 0,5 m	720
	VDA lane change	Speed = 50 km/h		Start leaving the driving trajectory	First pylon contact	90

Table 1: Summary of analyzed critical scenarios

To set up the system architecture, the following steps are necessary: investigate the most critical driving scenarios that are supported by brake and steering systems of highly and fully automated driving; conduct hazard and risk assessments for the identification of failures according to ISO 26262 and derive timing requirements for the reconfiguration mechanism.

As a first step, different driving scenarios were considered (see Table 1). Numerical simulations helped to substantiate the requirements, especially in order to quantify the time between the incidence of a fault and its effect. For instance, in the VDA double lane change scenario, a maximum lateral offset of 0,5 meters was set, preventing the car from hitting the traffic cones. By simulation, the corresponding worst case start time of malfunction was determined in this way.

The Hazard and Risk Analysis is the next step. It starts from an item definition and a temporal specification and defines safety goals with respect to unintended behavior and loss of the nominal function.

According to ISO 26262, each safety goal comes with a Fault Tolerant Time Interval (FTTI). The Fault Reaction Time should be shorter than FTTI (see Figure 4). The Fault Reaction Time consists of the Failure Detection Time and the Reconfiguration Time. The most critical driving scenario is the VDA lane change with a FTTI of 90 ms (see Table 1). Assuming a Failure Detection Time of 30 ms, a Reconfiguration Time of max. 60 ms remains. Within this time, the fail operational system deactivates the faulty ECU, reads out motor control parameters and guarantees safe actuator control by transferring control to a newly reconfigured redundant ECU.

System description

In this chapter, the proposed system will be presented in detail. In Figure 5, the Autokonf architecture is shown. It is mainly divided into three different levels:

- The architecture level takes the role of the vehicle high performance ECU including trajectory calculation, communication with other vehicle systems and having a high-ranking command to initiate a reconfiguration if needed.
- Actuator ECUs (original and redundant), including switches, are located at component level. Their role is to generate control signals and to send them to the corresponding actuators.

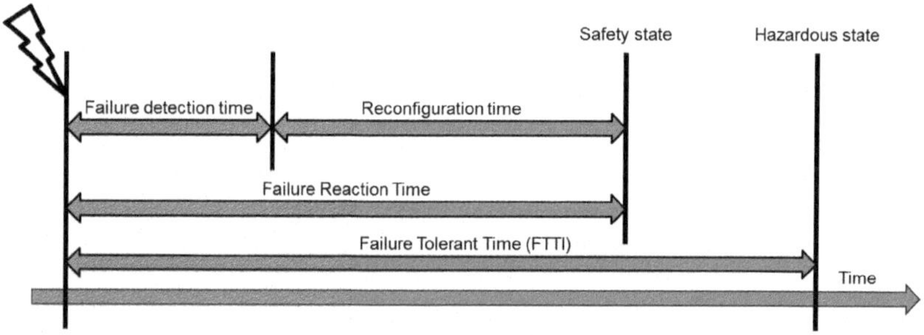

Figure 4: Definition of reconfiguration time

- The actuator level is a level where electric motors including mechanical components are located. There is no microcontroller logic integrated here.

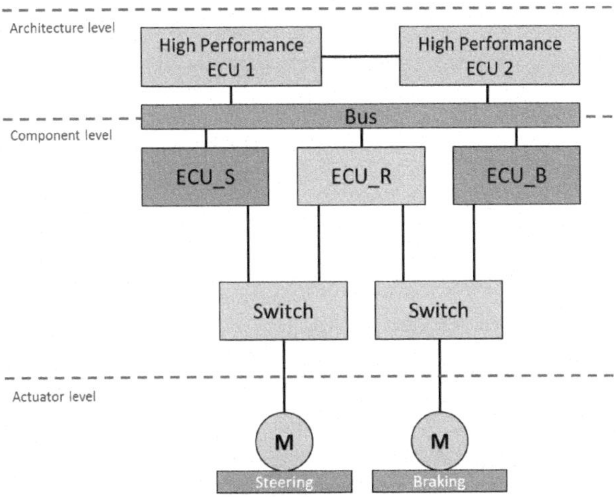

Figure 5: Autokonf system architecture

Architecture level

For the high-performance ECUs at this level, Xilinx ZCU 102 boards have been selected on which the Autosar architecture has been implemented [11]. The data connections between the ECUs on the top layer are Ethernet based, considering that the goal was to implement a service-oriented middleware in the form of the scalable service-oriented middleware over IP (SOME/IP) which is in line with the POSIX-based AUTOSAR adaptive operating system.

Between the ECUs on the top and mid layer, a CAN bus was implemented because it is the easiest option for this setup and achieves the necessary requirements regarding timing.

Component level

Both braking and steering systems have been selected for this project to show the principle of a fail-operational system without complete components duplication. There are several reasons to select especially these two safety related sub-systems:

- Both systems have the highest safety requirement (ASIL-D);
- Fail-operational functionality of both systems is mandatory for the future of autonomous driving;
- Both systems have a similar architecture topology: power ECU needs to control a brushless DC motor;

- Both motors are of the same power class (in the range of equal or less than 1kW);
- In the past, both systems have shown the same evolutional, developmental steps: starting with a complete mechanical system, then moving on to an electro-hydraulic support and lastly to the driver towards complete electro-mechanical "x-by-wire" solutions.

In the Autokonf project, component level was defined and limited in the following way: for each steering and braking system, one main ECU is used to control the actuator motor.

Whereas for the steering system, this definition can be used without any limits and is the mainly used solution currently, there are some remarks to be mentioned regarding the braking system.

Here, the brake booster control ECU is considered only (e.g. iBooster). Vehicle stability function is "beyond the project scope", since this function is not active most of the time and is activated only in some special cases where vehicle dynamics behavior may become unstable.

For the steering system, an EPS ECU is used to control the steering motor on the actuator level. For the braking system, the same ECU can be used based on above-mentioned assumptions. Both ECU_S and ECU_B are (see Figure 5) responsible for controlling their own actuators and in case of a failure, they will be deactivated based on their own safety strategies.

Between them, a redundant ECU_R can be found, which will be put on hot standby during the normal operation. This ECU_R is equipped with both steering and braking software and will start a corresponding software block in the case that one of the original ECUs will have some malfunctions. All three ECUs have the same topology, interface and HW components, so that cost saving potential is possible.

All ECUs are equipped with a redundant power supply and have a redundant communication bus to the vehicle high performance ECUs (not shown in the figure).

On the other side, all of them are connected to the actuators via so called switches. The role of these switches is to connect the corresponding power electronics to the outputs of the driving motors. Based on the safety analysis, switches have the highest ASIL-D safety level and are the key components in the proposed architecture.

Actuator level

As the actuator, a typical EPS steering motor or a brake booster motor can be used. Both motors are equipped with a motor position sensor which are necessary to apply a field-oriented control and read out its position.

System specific torque or pressure sensors are not considered here and are therefore out of the scope.

Alternative system concepts and state-of-the-art architectures

The system architecture, presented in Figure 5, is one possible solution, but not the only one. During the system requirements definition phase, several criteria have been set up to validate our approach with other possible architectures. Their summary is found in Table 2.

Criteria	Space	EMC[2]	Safety	Weight	Costs	Total
Influence factor	0,5	0,2	1,0	0,3	0,6	
Concept						
4 motors equipped with their own power electronics and switches	1,00	0,33	0,14	1,00	1,00	**1,61**
6-phase motors connected directly to the ECUs without opening switches	0,57	1,00	0,57	0,95	0,78	**1,81**
6-phase motors with opening switches	0,71	1,00	0,29	0,97	0,85	**1,64**
Redundant ECU without redundant power electronics	0,14	0,67	1,00	0,53	0,28	**1,53**
Redundant ECU including redundant power electronics and switches	0,57	0,33	0,43	0,68	0,66	**1,38**
Redundant ECU and power electronics without opening switches	0,43	1,00	0,86	0,63	0,44	**1,72**
Autokonf proposal	0,43	1,00	0,43	0,65	0,51	**1,34**

Table 2: Alternative architectures comparison

Different combinations (3-phase and 6-phase motors, with and without switches, double-ended or single-ended, etc.) have been analyzed. Every concept has been validated based on the pre-defined criteria. Importance of every criteria was described by the influence factor. Thereafter, every own concept value was multiplied by the influence factor and all the values were added together providing the final score value. The lower the total value is, the better the current concept is in regard to the costs, weight, safety, etc.

Based on this comparison, the Autokonf architecture, with its best score, has proven to be a promising solution. Nevertheless, according to the scoring table, there are some other possible alternative concepts available, which can be investigated later.

Project partners and roles

The "Autokonf" project group consists of one OEM, one Tier-1 supplier, two engineering system suppliers and one research institute.

[2] EMC: electro-magnetic compatibility

BMW AG

BWM, with its large system expertise, is responsible for the complete vehicle architecture and communication between different systems in the car. BMW AG is mainly working on defining and developing high performance ECUs and their communication with each other and with other components within the predefined architecture. The focus of the BMW development is to investigate the dynamic reconfiguration for fail-operational systems.

HELLA GmbH and Co. KGaA

As a component supplier for EPS ECUs, Hella is mainly active at “component level”, according to the predefined architecture, and responsible for development of all three ECUs and their communication with actuators and high performance ECUs. Safety and self-protection mechanisms are integrated at this level as well.

Intedis GmbH & Co KGaA

Intedis is a joint-venture consisting of two suppliers “Hella” and “Leoni” and has its focus on the building of architecture for the development of electrical and electronic devices intended for the future automotive market. In the Autokonf project Intedis, the project leader is also responsible for the preparation of the test demonstrator.

ITK Engineering GmbH

ITK Engineering GmbH develops steering and chassis systems since many years. Their main role in the AutoKonf project is to define system requirements and to derive the system architecture and the safety concept.

Fraunhofer Institute EMFT

Fraunhofer Institute EMFT has a high level of expertise in the field of packaging of electric elements, especially when relying on a very small amount of space. Due to their special requirements with high load and external protection, connectors in particular have been the focus of research for several years. In the Autokonf project, Fraunhofer EMFT will be responsible for the development of the switches.

Experimental results

The predefined reconfiguration time of 60 ms must be validated experimentally.

For this purpose, a test setup has been prepared (see Figure 6). It consists of one original ECU_x being configured as a braking ECU_B or steering ECU_S, one redundant ECU_R including both steering and braking software blocks, one opening switch, one closing switch combined in the “switch” and one motor actuator, connected to a load motor, simulating the feedback from the real system.

Since the steering or braking-specific profiles can be loaded into the system, the purpose was not to build the complete setup, but only to validate the needed time and measurement-corresponding parameters. The whole system, including all components, will be used for demonstration purposes. For the technical validation, a part of the overall architecture is sufficient.

Original driving scenario (VDA double lane change), with its most critical reconfiguration time, has been used to generate steering wheel profiles for the validation test. This trajectory profile was transmitted by the

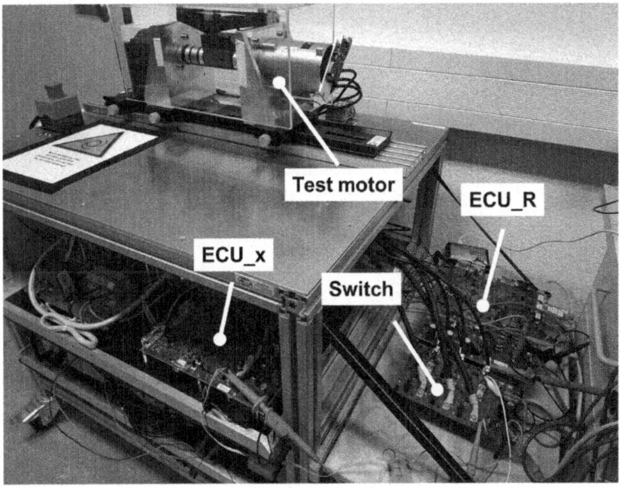

Figure 6: Experimental setup

high performance ECUs to the actuator in the form of desired values. A pre-defined failure has been injected into the system with a goal to trigger an internal diagnostics and fail-safe mechanics of the original ECU_L. Based on this failure detection, ECU_L has generated a signal for the switch to connect motor phases in a passive way without active motor control to the ECU_R (see "closing switch_U" edge in Figure 7). This allows for avoidance of any current spikes, coming from the rotating motor, that could damage the hardware electronics during the following "disconnect ECU_L" phase. This way, the possible electronics damage is prevented. As far as the motor is safely connected to the ECU_R, disconnection of the original malfunctioning ECU_L has been initiated by the switch. A time delay between these two phases is about 1 ms.

Parallel to the connection phase, ECU_R gets the signal that the ECU_L has a malfunction and needs to shut down. To be prepared to take over the motor control, ECU_R loads the corresponding "steering" section from the SW block and neglects any "braking"-related activities in the future.

Motor position data is available for both original and redundant ECUs all the time. Therefore, as far as the motor is disconnected from the original ECU_L, ECU_R will start its own motor control and bring the motor to its original profile. This happens due to an increasing deviation between the desired and current profiles during the loss of control by the ECU_L.

From the VDA test profile a most critical motor parameter has been selected. The motor turned with a speed of 1000 rpm and generating a torque of 1 Nm. A one possible failure scenario with a malfunctioning gate driver unit has been injected in the same point of time. As it is shown in Figure 7, a time of 12 ms is required until the actuator motor has reached the steady state control again.

With this test, a required reconfiguration time of less than or equal to 60 ms has been proven and the project safety goals have been achieved.

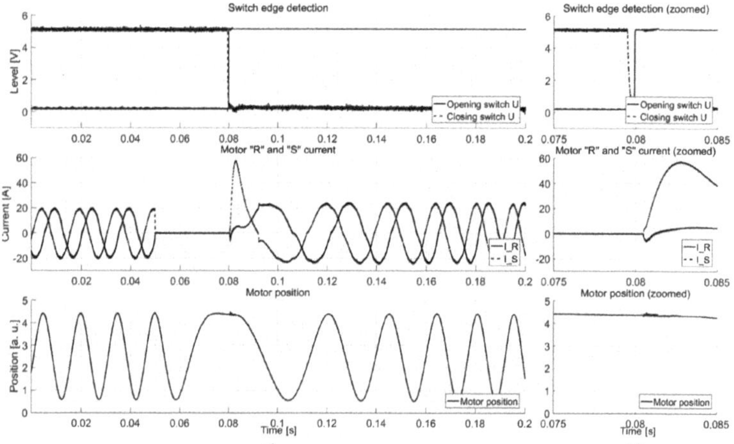

a) b)

Figure 7: Experimental results: overview (a) and zoomed around the reconfiguration event (b).

Outlook and perspectives

Open points

There were some limitations and simplifications used to set up the current architecture. The goal was to have a first step in the direction of a reconfigurable system. The detection of a fault is not described in detail. It was simplified, since the focus of this project is on the investigation of the reconfiguration behavior and not on the reason of this reconfiguration. Multiple and cascaded faults were not considered due to the same reason. Power supply and communication buses have been set as already redundant components providing their services and not considering their faults in more detail, since it is out of the scope of this project. Their redundancy architecture for a safety critical application was already investigated for example in the funded project "HiBord" [12].

Once the system was reconfigured, there is no mechanisms implemented in case a new failure can appear. A strategy how to handle the initial or any further failures has to be considered in the future activities.

Proposed vehicle integration project

After achieving the goals of the AutoKonf project, only the first step into vehicle integration and mass market has been achieved. Based on the preliminary results, the project partners decided to commit a sketch for a follow-up project. The aim of this project is to develop a new reconfigurable system with close to production components and their integration into a demonstrator vehicle. The main purpose is to guarantee vehicle stability in all demanding driving situations and, if necessary, to transfer the vehicle to a safe state. In the event of a fault in individual components, the affected function will be transferred to the error-free component within a short handover time by reconfiguring a reserve control device. Future drive-by-wire technologies will be the focus of the project, as they are particularly suitable for highly automated vehicles to meet the changed customer requirements. The new Re2Ko[3] system will contain reconfigurable control units, actuator control changeover switches and a central driving computer.

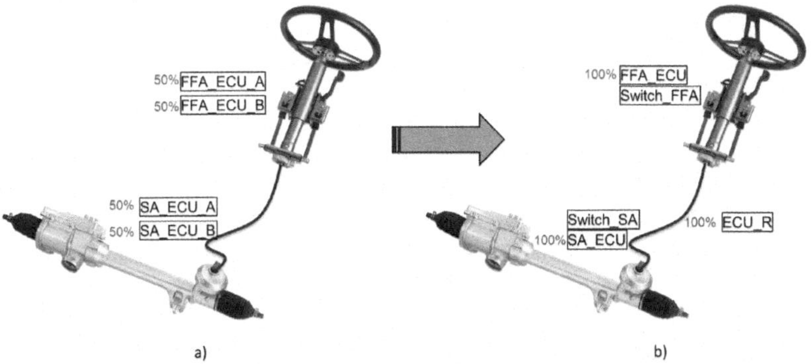

Figure 8: Current development of a steer-by-wire systems (a) and proposed system design based on Autokonf architecture (b).

In Figure 8 possible SbW architecture concepts are shown. In contrast to the doubled ECUs for rack and steering wheel with 50% power (Figure 8a), the reconfigurable system is equipped with two single 100% ECUs for steering wheel and rack and one generic, redundant 100% ECU which can take over a defective function control (Figure 8b).

Conclusion

In this paper, a proposal for the new system architecture based on the safety related braking and steering systems with a reconfigurable redundant ECU has been described. This system allows to reduce the number of components, if an interaction between different vehicle sub-systems is considered, which was mainly out of scope in the past, but will play an important role in the future. Different driving scenarios with their critical time have been discussed. A test bench was prepared with a goal to confirm theoretical and simulated results. Experimental results have shown the successful switching of a running actuator from one

[3] Re2Ko = REal REconfiguration: proposed project name for the follow-up project.

control ECU to another one and in this way to prove a fail-operation functionality. The next possible step in a following project would be an integration of this system architecture in a test vehicle and to conduct a validation test "in the field".

Acknowledgment

Authors acknowledge the financial support by the Federal Ministry of Education and Research of Germany in the project "AutoKonf" [13].

Bibliography

[1] N. Merat, Transition to manual: Driver behaiviour when resuming control from a highly automated vehicle., Elsevier Verlag, 2014.

[2] D. Damböck, Automationseffekte im Fahrzeug – von der Reaktion zur Übernahme., München, 2013.

[3] Christian Gold et al., "Take over!" How long does it take to get the driver back into the loop?, 2013.

[4] D. P. Grabs, E/E architecture proposals for automated driving and their failure robustness., Bad Boll: EEHE, 2015.

[5] W. Kaewkerd, „Steer by Braking – Redundancy concept for EPS," Swiss federal institute of technology, Lausanne, 2014.

[6] S. Armruster, Synergies in the development of powernets for X-by-wire-Systems and Aircraft Concepts – Realisation in the European Project SPARC., Hannover: Forum Electrical Energy Distribution Systems, 2005.

[7] M. Harrer, P. Pfeffer, Steering Handbook, Munich: Springer, 2016.

[8] „Continental Engineering Services GmbH," [Online]. Available: www.conti-engineering.com.

[9] „iBooster – Intelligent verstärkte Bremskraft," Bosch, [Online]. Available: https://www.bosch-presse.de/pressportal/de/de/ibooster-%E2%80%93-intelligent-verstaerkte-bremskraft-103365.html.

[10] „SAE J3016 Levels of driving automation," SAE Org, 11 2018. [Online]. Available: www.sae.org.

[11] F. Oszwald; J. Becker, P. Obergfell, M. Traub, „Dynamic Reconfiguration for Real-Time Automotive Embedded Systems in Fail-Operational Context," 2018.

[12] B. Projekte, „HiBord: Hochzuverlässige und intelligente Bordnetztopologien für automatisierte Fahrzeuge," BMBF, 2018. [Online]. Available: https://www.elektronikforschung.de/projekte/hibord.

[13] D. P. Grabs, „Autokonf," 2018. [Online]. Available: www.autokonf.de.

[14] M. Fernie, „How Electric Power Assisted Steering Works, And Why It's Better Than Hydraulic," 2016. [Online]. Available: https://www.carthrottle.com/post/electronic-power-assisted-steering-how-does-it-work/.

[15] infinity, „Infinity Techinfo," Infinity, 2018. [Online]. Available: https://www.infiniti-techinfo.com.

Nonlinear brake control of a Formula 1 racing car and the small step to nonlinear vehicle dynamic control

MSc. Michele Sigilló
Founder and Head of Technology, SIGICONTROL s.r.o.

MSc. Marc Zurbuchen
Control Systems Engineer, Alfa Romeo Sauber F1 Team

BSc. Davide Spagnol
Head of Systems Engineering, Alfa Romeo Sauber F1 Team

Abstract

Brake system dynamics are governed by highly nonlinear effects when required to propagate high hydraulic pressure within a short amount of time to the brake calipers. Due to substantially more demanding conditions and requirements in a Formula 1 car's brake system do not only these effects have a bigger impact, but make it also more difficult to robustly control the brake caliper pressure when brake-by-wire is adopted. The control system presented in this paper originates from Alfa Romeo Sauber F1 Team's desire (today Alfa Romeo Racing) for an alternative algorithm with enhanced robustness and similar performance when compared to the existing proportional-integral based one. To this purpose, an innovative nonlinear controller based on back-stepping methods and adapted robust control has been developed. Its ability to cope for model uncertainties and unknown disturbances, while guaranteeing stability over the full nonlinear operational range and asymptotic error convergence, are the focus of this work alongside the illustration of the controller's working principles and the small step to reach nonlinear vehicle dynamic control.

1. Introduction

In 2014 Formula 1 entered a new era by introducing hybrid technology for the vehicle's drive train, i.e. a combustion engine supported by an electrical motor whose electrical power is deployed by means of rechargeable batteries. The consequential energy recovery system paved the way for brake-by-wire (BbW) also in F1. In addition to the energy recovery system, which causes an electrically induced deceleration of the vehicle, BbW consists also of conventional friction brakes that provide mechanical deceleration. To comply with the desired overall deceleration request from the driver and to achieve a defined braking balance, a target rear brake caliper pressure is calculated [5]. The underlying BbW controller is responsible of actuating the electro-hydraulic system such that the target pressure is met as accurately as possible.

Due to the highly nonlinear and complex nature of hydro-dynamic effects from the electro-hydraulic brake actuator to the brake pads, modelling and accurate control of the brake caliper pressure becomes very challenging. Poor and inconsistent tracking of the desired rear caliper pressure reflect negatively on safety and overall vehicle performance. It's thus state of the art to use conventional control systems with varying control gains, with values originating from empirically identified look-up tables and maps. Though these methods can result in robust and precise control performance, tuning, stability margin, maintenance and user friendliness are restricted and the creation of empirical data sets lengthy and costly.

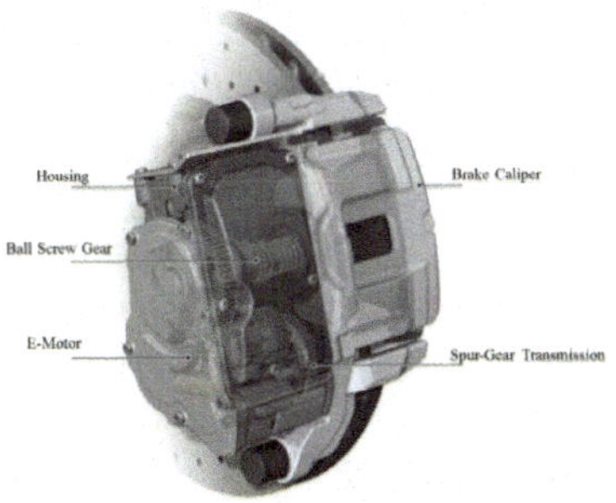

Figure 1: Electro-mechanical brake

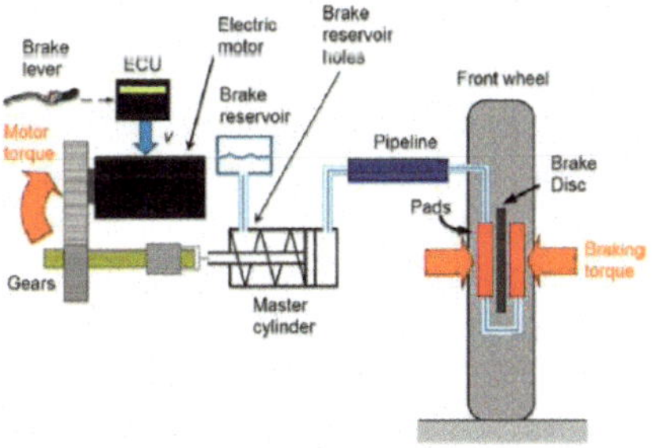

Figure 2: Electro-hydraulic brake

Albeit the conditions under which Formula1 brake systems work are tougher, it's worthwhile looking at BbW solutions for road vehicles. There exist two classes of systems as shown in figure 1 and 2: electro-mechanical and electro-hydraulic. Both have an electric motor as actuator, but in former the torque is transferred to the brake pads through a direct mechanical path, whereas in the latter the motor either pushes a piston as in figure 2 or regulates a flow. Nonlinear approaches like the PID - Fuzzy-logic control in [1] and [2], as well as the sliding mode control (SMC) presented in [3], cannot be applied to the F1 BbW system under discussion, as it belongs to the electro-hydraulic class.

For this class of systems [7] proposes a switched controller solution. A first controller moves the hydraulic piston past the actuation dead zone and then switches to a second PID controller responsible for the brake caliper pressure control. The nonlinear, time-varying characteristic of the brake system requires a set of transfer functions for different working points, an adaptation algorithm to enhance robustness, position-pressure maps and a state-machine. Using AMESim allows for meticulous modeling and simulation of the BbW system in [8]. Although it leads to careful tuning of the proposed PID controller, its limited bandwidth causes jittering around the target pressure, while neglected nonlinear effects put a question mark on the performance in presence of uncertainties in the real system. AMESim is used similarly in [9], but the applied sliding mode controller is able to overcome the deficiencies of [8]. Despite good simulation results, the developed control input contains the time derivative of a sign function, which makes a practical application of the controller questionable. The classical SMC developed in [13] shows better results than [9]. To reduce the jittering effects, caused by the discrete controller, a saturation function has been implemented. Reduced jittering however comes at costs of slower system response. The system architecture presented in [10] comes close to the one built in the F1 vehicle of this paper, i.e. by varying the hydraulic flow, the pressure in the brake system is influenced. Generalized predictive control (GPC) is applied for controlling the motor speed with good pressure tracking performance for nominal braking scenarios. As for any predictive control, the tracking performance depends also on the prediction horizon. For high dynamic systems a shorter control horizon is required, which in turn increases computational effort. The underlying linear nature of GPC calls for linearization during run time of the highly nonlinear braking system, needing accurate parameter identification and requiring bigger control effort in presence of system uncertainties.

The controller developed by SIGICONTROL s.r.o. for Alfa Romeo Sauber F1 Team's BbW system is an adapted robust controller based on nonlinear Lyapunov back-stepping methods. Similarly to Bosch's Integrated Traction Control [12], it starts from an exact model knowledge design and is augmented during the development of the overall back-stepping control with concepts used in the Lyapunov-based nonlinear control of [13]. Back-stepping allows for improved dynamic controller response, while the adapted nonlinear robust part of the algorithm guarantees stability over the full nonlinear working range, good tracking performance and asymptotic error convergence despite the presence of model uncertainties and unknown disturbances. It's a single controller with a single tuning set purely based on 1st order physical principles and doesn't make use of any gain scheduling, empirical data or maps in contrast to most BbW controller in literature and the currently PI-based algorithm. It is evident that the same nonlinear approach can be applied to a variety of vehicle dynamics control problems to improve performance, safety and robust behavior, especially in view of the increasingly automated and autonomous driving scenarios.

The following section of this paper is dedicated to the BbW system description of Alfa Romeo Sauber F1 Team's racing car, followed by the third section describing the system dynamics. In section four and five the control objective and development are explained before the stability analysis is carried out in section six. In section seven measurement results are presented followed by the close relationship to nonlinear vehicle dynamic control and closing comments of section eight.

2. Brake-by-Wire System

2.1 System Layout

The installed mechanical braking system consists of an ECU (4), an electrically actuated hydraulic servo valve from the company Moog (1), a hydraulic double and single piston combined unit provided by the company Brembo (2), brake pipes, calipers,

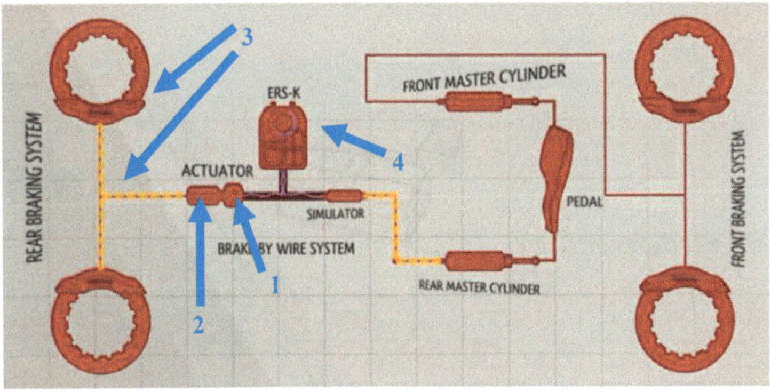

Figure 3: Formula 1 brake by wire simplified system

pads and discs (3). The simplified system [6] is represented in figure 3.

In brake-by-wire mode the servo valve is electrically actuated and the hydraulic flow through it moves the double piston in the combined unit. This hydraulic sub-circuit is part of the high pressure system circuit and the above described flow is the real control input for the BbW system. The double piston is mechanically connected to a single piston which, forced by the movement of the double piston, builds up a pressure in the brake fluid. This pressure is then forwarded through the brake pipes to the calipers and finally causes the frictional braking between brake pads and discs.

2.2 Components

2.2.1 Servo Valve

The used servo valve, depicted in figure 4, is produced by the company Moog Inc. and shows a dual gain flow characteristic achieved by a T-shaped flow slot.

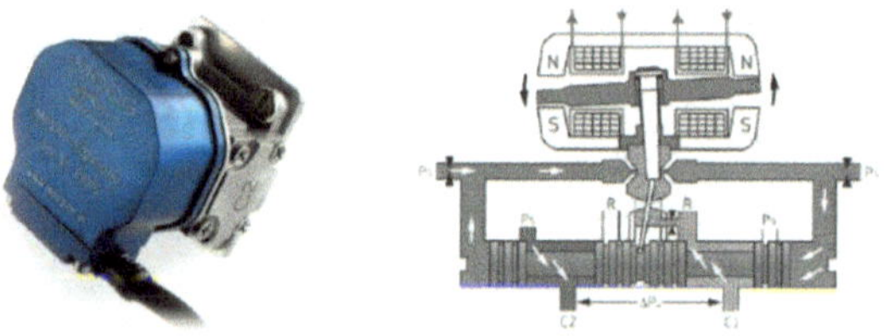

Figure 4: Hydraulic servo valve - Moog

This special shape causes an additional nonlinearity due to the nonlinear relationship between valve current and flow as shown in figure 5.

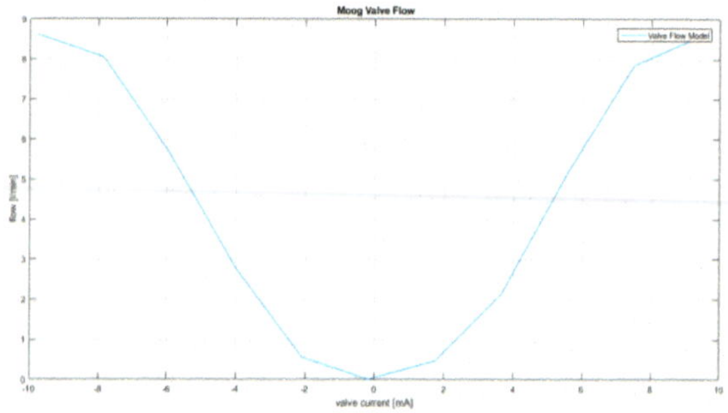

Figure 5: Nonlinear current-flow characteristic

2.2.2. Brembo Combined Unit

The main actuator of the BbW system is provided as a combined unit by the company Brembo S.p.A. It consists of a double piston cylinder mechanically connected to a single piston, which is responsible for building up the pressure in the brake pipe. Depending on the pressure differential across the two chambers of the double piston unit, caused by the flow of hydraulic fluid directed through the servo valve, the piston position changes. The single piston unit is constrained to move by the same amount, so that the pressure in the brake pipes and calipers can be influenced through the flow rate into the two separated chambers.

2.2.3 Brake Pipes

Standard aluminum pipes are used as brake pipes. Given the position of the hydraulic brake units in the vehicle, the brake pipes have a high length - diameter ratio of about 800. This ratio points out that nonlinear turbulent fluid dynamic effects in the pipe can occur for quick pressure requests and cause an additional challenge to the brake controller.

2.2.4 Friction Brake – Discs, Pads, Calipers

Discs

Carbon discs are used in Formula 1. This discs weigh a magnitude of order less than commercial car's cast iron discs (1,2kg vs. 15kg) and can manage brake heat temperatures in the range of 350-1'000°C.

Pads

Also brake pads are made of carbon material, though with different properties. Their weight is around 200g with a friction coefficient between 0.7 and 0.9 but similar temperature range.

Calipers

Fixed type calipers with a higher braking performance are used. They are more suitable for high speed vehicles and show an even distribution of braking forces on the disc during braking.

2.3 Working principle

During the braking process the desired amount of caliper brake pressure is calculated. This value serves as reference value for the BbW control which calculates the current for the servo valve [4]. The current influences the spool position and thus the hydrau lic flow entering and exiting the chambers of the double piston unit. This flow is responsible for the pressures in the single chambers and therefore also for the differential pressure p_L across the piston. Accordingly to the pressure differential, the double piston position x_P changes and thus also the position of the rigidly connected single piston in the unit moves by the same amount. The movement of the single piston in positive direction compresses hydraulic brake fluid and presses it through a pipe to the brake calipers. It's this pressure, arriving at the calipers, that is responsible to push the brake pads against the brake discs and create so the friction for mechanical deceleration.

From the described working principle it should be apparent that the brake caliper pressure achievement undergoes several dynamic stages, starting with the valve current and ending with the caliper pressure. Control input (servo current) and control goal (caliper pressure) are hence dynamically not directly related and lead to consider the presented back-stepping control method.

3. System Dynamics

The above mentioned cascaded nature of the entire BbW system inspires the creation of a set of dynamic equations describing the single systems. Schematically the brake system can be illustrated as in figure 6 and used as reference for the single dynamic component equations.

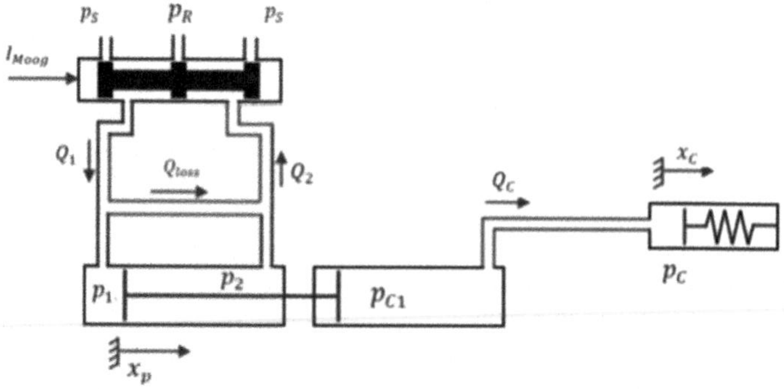

Figure 6: Brake-by-wire schematic first order physical principles

3.1 Equations of Motion

For each subsystem the equation of motion is derived from first order physical principles, starting from the caliper pressure dynamics and working backwards to the servo valve, hence the control method's name "back-stepping control".

The caliper pressure dynamic $\dot{p}_C$ is given by

$$\dot{p}_C = \left(\underbrace{\frac{\pi r^4}{8L\mu_{bf}} (p_{C_1} - p_C)}_{Q_C} - A_C \dot{x}_C \right) \frac{\beta_{bf}}{V_{C_0} + A_C x_C} \tag{1}$$

With A_C as caliper piston area, x_C as caliper piston travel, L and r as pipe length and radius respectively, p_{C_1} as output pressure of the Brembo actuator, V_{Co} as initial volume of the brake caliper chamber and μ_{bf} and β_{bf} as absolute viscosity and bulk modulus of the brake fluid respectively. Q_C describes the flow rate through the brake pipe.

From eq.(1) follows that the caliper pressure dynamics can be controlled through the pressure p_{C_1} at the beginning of the pipe, which itself is influenced by the piston velocity of the combined Brembo unit $\dot{x}_P$, as eq.(2) shows

$$\dot{p}_{C_1} = \left(A_{C_1} \dot{x}_p + \underbrace{\frac{\pi r^4}{8L\mu_{bf}} (p_C - p_{C_1})}_{Q_C} \right) \frac{\beta_{bf}}{V_{C_{10}} - A_{C_1} x_p} \tag{2}$$

Here A_{C1} and x_P represent the area of the single piston in the actuator and its displacement, while the initial volume is denoted by $V_{C_{10}}$.

Newtonian physics governs the dynamic behavior of the piston velocity $\dot{x}_P$ by the pressure differential $(p_1 - p_2)$ in the Brembo combined unit together with the friction force F_F, the piston mass m_P and the elastic properties of the fluid k_P.

$$\ddot{x}_p = \frac{1}{m_p}\left[\underbrace{(p_1 - p_2)}_{p_L}A_P - p_{C_1}A_{C_1} - F_F - k_p x_p\right] \tag{3}$$

Given that the flow into and out of the Brembo unit is assumed to be identical,the differential pressure dynamic $\dot{p}_L$ can be described as in eq.(4).

$$\dot{p}_L = (Q_1 - Q_{loss} - A_p\dot{x}_p)\,\beta_{hyd}\left[\frac{1}{V_{10} + A_p x_p} + \frac{1}{V_{20} - A_p x_p}\right] \tag{4}$$

With A_P being the double piston area, Q_1, the hydraulic flow into the Brembo unit, Q_{loss} the flow losses from one chamber to the other, V_{10} and V_{20} the initial volumes of the chambers in the double piston unit and β_{hyd} the bulk modulus of the hydraulic system fluid of the race car.

There is obviously a further nonlinear dynamic relationship between the servo valve and the flow Q_1. Preliminary experimental results however showed that for the vast majority of braking requests the transition through the nonlinear part of the valve characteristic is fast enough that a fairly linear relationship between the current and the flow can be established. Deviations from this linear description will be handled by the robust controller. It has therefore been decided to cut off the model at this stage and define the flow Q_1 as the real control input, which weighted by the proportionality factor becomes the valve current I_{Moog}, real physical control input entity of the BbW system.

4. Control Objective

The control objective is the robust control of a rear break caliper pressure p_C that tracks a desired caliper pressure p_{C_d} asymptotically.

5. Control Development

Similar to cascaded control, in back stepping control the system of equations has a nested structure, i.e. the output of one dynamic equation is described by a differential equation which simultaneously represents the control input for the subsequent differential equation whose output is the control input of the following equation and so on.

For simplicity, the back stepping control development will be carried out on the following simple system of equations where the final, true control input is denoted by u.

$$\dot{x}_1 = ax_1^2 + x_2 \tag{5}$$

$$\dot{x}_2 = x_3 \tag{6}$$

$$\dot{x}_3 = u \tag{7}$$

Let's assume the goal is to control the state x_1 with the system dynamics described in eq.(5). Let's further assume that the nonlinear differential equation can be controlled by another state x_2 that is physically not directly available, e.g. the torque of an electric motor, but is accessible through a further state, e.g. the motor's electric current that generates torque. The control input x_2 becomes so a virtual control input of eq.(5), but can be influenced by the state x_3 through the dynamic relation of eq.(6). Assuming that the state x_3 is itself not accessible, thus becoming a virtual control input, one can see that by varying the real control input u the state x_1 will be influenced indirectly through the different levels of equations of motion

To this purpose eq.(5) is extended by a desired control input x_{2d} that controls the individual dynamic, leading to the following augmented systems of equation

$$\dot{x}_1 = ax_1^2 + \underbrace{x_2 - x_{2d}}_{\eta_1} + x_{2d} = ax_1^2 + \eta_1 + x_{2d}$$

$$\dot{\eta}_1 = \dot{x}_2 - \dot{x}_{2d} = \underbrace{x_3 - x_{3d}}_{\eta_2} + x_{3d} - \dot{x}_{2d} = \eta_2 + x_{3d} - \dot{x}_{2d}$$

$$\dot{\eta}_2 = \dot{x}_3 - \dot{x}_{3d} = u - \dot{x}_{3d}$$

The back stepping errors η_i allow to tie the three dynamic equations eq.(5) – eq.(7) and define the virtual control inputs x_{2d}, x_{3d} that solve and stabilize the individual control problems. Details on the design of the real control input u follow in the next section, however the control input that controls the state x_1 to the desired state x_{1d}, forming the error $e_1 = x_{1d} - x_1$, has the following form:

$$u(t) = -k_{\eta_2}\eta_2(t) + \dot{x}_{3d}(t) - \eta_1(t) \tag{8}$$

with

$$\int \dot{x}_{3d} = x_{3d} = \dot{x}_{2d} - k_{\eta_1}\eta_1 + e_1 \tag{9}$$

$$\int \dot{x}_{2d} = x_{2d} = \dot{x}_{1d} - ax_1^2 + k_1 e_1 \tag{10}$$

6. Stability Analysis

6.1 Lyapunov Stability Theorem

Lyapunov based control systems find their roots in the equally named stability analysis of systems. Starting from the notion of a system's energy the stability behavior over time can be analyzed. Along the same line of thought, Lyapunov based control system developments take an error-energy like term (with underlying system dynamics) and develop during the stability analysis a control input that guarantees the system stability as defined by Lyapunov's theory.

A system that continuously dissipates energy is stable as its kinetic energy will eventually fade (or at least not increase) over time. Therefore, if an energy-like function based on the tracking error is defined, an appropriate control input can force the energy's time derivative to be always negative (or equal zero) and reduces so the tracking error to zero. To show how the virtual and real control inputs in eq.(8) – eq.(10) guarantee asymptotic tracking and full range stability, the stability analysis for the system in section 5 shall be analyzed.

6.2. Lyapunov Stability Analysis

Since the elimination of the single errors η_i in the individual augmented dynamic equations leads to the annulment of the tracking error e_1, the energy-like function $V(t)$ is based on those errors

$$V(t) = \frac{1}{2}e_1^2 + \frac{1}{2}\eta_1^2 + \frac{1}{2}\eta_2^2 \tag{11}$$

As per Lyapunov's theorem, the time derivative of eq.(11) should always be negative for a stable system

$$\dot{V} = e_1\dot{e}_1 + \eta_1\dot{\eta}_1 + \eta_2\dot{\eta}_2$$

$$\dot{V} = e_1\left(\underbrace{\dot{x}_{1d} - \dot{x}_1}_{\dot{e}_1}\right) + \eta_1\left(\underbrace{\dot{x}_2 - \dot{x}_{2d}}_{\dot{\eta}_1}\right) + \eta_2\left(\underbrace{\dot{x}_3 - \dot{x}_{3d}}_{\dot{\eta}_2}\right) \tag{12}$$

Replacing the brackets in eq.(12) with the augmented dynamic equations containing the virtual inputs leads to

$$\dot{V} = e_1\left(\dot{x}_{1d} - ax_1^2 - \eta_1 - x_{2d}\right) + \eta_1\left(\eta_2 + x_{3d} - \dot{x}_{2d}\right) + \eta_2\left(u - \dot{x}_{3d}\right) \tag{13}$$

Inserting the control inputs defined in eq.(8)-eq.(10) in eq.(13) results in

$$\dot{V} = -k_1 e_1^2 - k_{\eta_1}\eta_1^2 - k_{\eta_2}\eta_2^2 \tag{14}$$

with k_1 , $k_{\eta 1}$, $k_{\eta 2}$ as tunable control gains. Clearly eq.(14) is always negative for positive k_i's and for any state of the error variables. In other words, the control system is always stable and the errors, in particular the tracking error e_1, asymptotically converge to zero. The controller in this example was designed using exact model knowledge (EMK), which in reality is not available. Applying adaptive and robust control terms to the developed equations eq.(8) – eq.(10) however, allow to handle model uncertainties, varying parameters and a priori unknow disturbances while still achieving asymptotic convergence of the tracking error and full range stability.

6.3 Brake-by-Wire Controller

The high dynamic stiffness and pressures in a F1braking system don't give enough time to parameter adaptation algorithms. Nevertheless, any model uncertainty can be seen as a disturbance. Therefore the developed BbW control algorithm has been designed with particularly adapted nonlinear robust terms to guarantee asymptotic tracking and overall stability fo the system, as shown in the course of this section.

The control algorithm for the BbW system was developed analogous to the example in section 5. Eq.(12) reveals that the tracking error e_1, the back stepping errors η_i and their derivatives are key to the Lyapunov stability analysis and control development. While e_1 is simply the mismatch between the desired caliper pressure and the measured pressure, the BbW back stepping errors are

$$e_1 = p_{C_d} - p_C$$

$$\eta_1 = p_{C_1} - p_{C_{1d}}$$

$$\eta_2 = \dot{x}_p - \dot{x}_{pd}$$

$$\eta_3 = p_L - p_{L_d}$$

and consequently the Lyapunov candidate, who's time derivative needs to be kept negative for all working conditions of the BbW system by the controller, is defined as

$$V(t) = \frac{1}{2}e_1^2 + \frac{1}{2}\eta_1^2 + \frac{1}{2}\eta_2^2 + \frac{1}{2}\eta_3^2 \qquad (15)$$

Carrying out the time derivative of eq.(15) and replacing the emerging terms with the set of equations of motion found in section 3.1, the virtual control inputs $p_{C_{1d}}$, $\dot{x}_{pd}$, p_{L_d} and the real control input Q_1 can be designed as the following functions

$$p_{C_{1d}} = f_1(A_C, x_C, V_{C0}, p_C, \dot{p}_{C_d}, L, r, \beta_{bf}, \mu_{bf}, k_1 e_1) \qquad (16)$$

$$\dot{x}_{pd} = f_2(A_{C_1}, x_P, A_C, x_C, V_{C_{10}}, V_{C0}, e_1, \dot{p}_{C_{1d}}, p_C, p_{C_1}, L, r, \beta_{bf}, \mu_{bf,} k_{\eta 1}\eta_1) \tag{17}$$

$$\begin{aligned} p_{L_d} = f_3(A_{C_1}, p_{C_1}, x_P, V_{C_{10}}, \beta_{bf}, p_{C_1}, F_F, k_p, m_p, \dot{x}_{pd}, A_p, \eta_2 k_{\eta 2}) \\ + v_{p_{Ld}}(\eta_2, k_{\eta 2}, sgn(\eta_2), \alpha_{p_{Ld}}, \beta_{p_{Ld}}) \end{aligned} \tag{18}$$

$$\begin{aligned} Q_1 = (Q_{loss}, \dot{x}_p, A_p, \beta_{hyd,} V_{10}, V_{20}, \dot{p}_{L_d}, \eta_2, m_p, \eta_3 k_{\eta 3}) \\ + v_{Q_1}(\eta_3, k_{\eta 3}, sgn(\eta_3), \alpha_{Q_1}, \beta_{Q_1}) \end{aligned} \tag{19}$$

The particularly adapted nonlinear robust functions $v_{p_{Ld}}$ and v_{Q_1}, that guarantee asymtotic tracking tracking performance and stability over the full working range, are injected in the last two stages of the control desing in order to counteract on cumulative modelling uncerteainties.

The original PI-based controller is supported by several maps and look-up tables for its control gains. This ad-hoc parametrization leads to very good tracking performance, however the empirical method comes at high costs in terms of time and flexibility as a slightly different system architecture would require a retuning of all maps and tables. In addition, proving stability of the controller is almost impossible, meaning that the stability boundaries are often unclear and therefore force the tuning to be very conservative. The proposed controller on the other hand, has just eight tuning parameters and no look up tables whatsoever. Since it's based on first order physical principles, a change in system architecture is easily applicable in the controller and a new stability proof established. Out of the eight parameters however, just six have a major impact on the control performance, such that tuning time can be significantly reduced as opposed to the standard controller.

7. Measurement results

The control performance was evaluated with measurements acquired with the recunstructed real brake system testbench. Figure 7 shows two brake maneuvers picked from the Canadian Grandprix of 2017, one of the hardest for the brakes. In fact, in addition to the electrically induced deceleration, the desired brake caliper pressure has to reach almost 70 bar to achive the total brake request.

The measurements show that despite highly nonlinear characteristics of the brake system, model uncertainties coming from the neglected T-shpe of the servo valve, flow in the brake pipe continuously switching its state from linear to turbulent and despite parameter variations like hydraulic temperature and mechanical friction, a good tracking performance can be achieved.

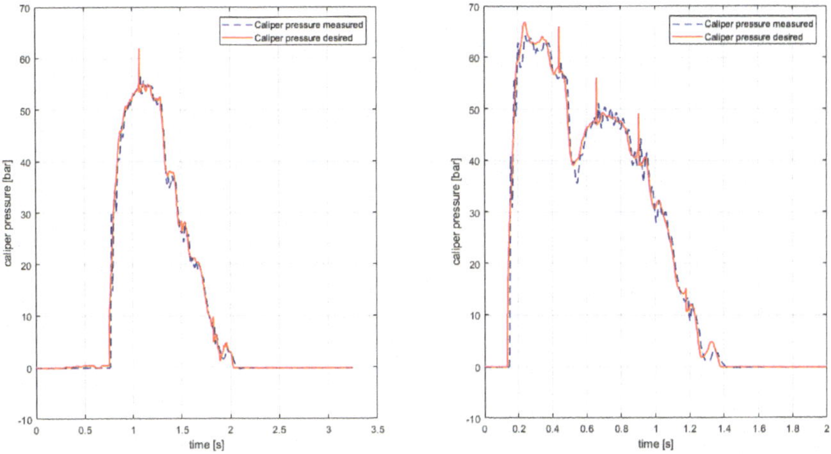

Figure 7: Brake maneuver test data from Canadian Grandprix 2017

8. Little Step to Vehicl Dynamic Control and Conclusion

Obviously the presented controller has an even wider range of possible applications in road cars than on F1 cars. Longitudinal and lateral control as well as traction , brake and engine control are the primary candidates for nonlinear back-stepping control methods and in general for Lyapunov-based nonlinear adapted robust control. For instance automated steering in Trucks, where the high steering forces still require an hydraulic powersteering but the input troque could be provided by an electric motor. Or electro-mechanical BbW, or ABS and Traction control where between the traction torque and the ground tire dynamics significantly influence the vehicles behavior. With today's level of automation in road vehicles, good sensors are important but not sufficient to ensure vehicle stability, increasing performance requests and not least safety. Smart sensors need to be supported by smart and innovative control techniques of which Lyapunov-based nonlinear control is part of.

In the previous sections the system layout of Alfa Rome Sauber F1 Team's BbW has been described and its dynamic equations analyzed. Using a simplified example, the development of a Lyapunov-based backstepping nonlinear robust controller has been carried out. The subsequent Lyapunov stability analyisis proved asymptotic error convergence and stability over the whole working range also in presence of model uncertainties and a priori unknown disturbances with one controller and eight tuning parameters. The good results are underlined by testbench measurements on the real system.

References

[1] Ameer Sider, Mohd Khair Hassan „Worked Example of X-by-Wire Technology in ElectricVehicle: Braking and Steering“, IntechOpen, 2018

[2] Weidong Xiang et al. „Automobile Brake-by-Wire Control System Design and Analysis“, IEEE Transactions on vehicular technology, 2008

[3] Mostafa R. A. Atia et al.“ Enhanced Electro-Mechanical Brake-By-Wire System Using Sliding Mode Controller“, Journal of Dynamic Systems, Measurement and Control, 2015

[4] Brembo S.p.A „The secrets of Brembo brake system for Formula One“, Youtube, https://www.youtube.com/watch?v=1Gv_nOI5T4U

[5] Sam Collins „2014 F1 explained: What is brake by wire?“, Racecar Engineering, 2014, https://www.racecar-engineering.com/articles/2014-f1-explained-what-is-brake-by-wire/

[6] Giorgio Piola „How F1 deals with its toughest braking challenge“, Motorsport.com, 2018, https://www.motorsport.com/f1/news/canadian-gp-brakes-tech-piola-1044399/3115789/

[7] FabioTodeschini „Adaptive position–pressure control of a brake by wire actuator for sport motorcycles“, European Journal of Control, 2013

[8] Yang Yang et al. „A Pressure-Coordinated Control for Vehicle Electro-Hydraulic Braking Systems“, 2018

[9] Qiping Chen et al. „On the Control of the Master Cylinder Hydraulic Pressure for Electro-Hydraulic Brake (EHB) Systems with the Sliding Mode Design Methodology“, FDMP, 2018

[10] Zhe Xiong et al. „Modeling and pressure tracking control of a novel electro-hydraulic braking system“, Advances in Mechanical Engineering, 2018

[11] Yong Jia-wang et al. „Pressure-tracking control of a novel electro-hydraulic braking system considering friction compensation“, Central South University Press and Springer-Verlag GmbH Germany 2017

[12] Lars König et al. „ITC – model-based feed forward traction control“, 9th International Munich Chassis Symposium, 2018, 2018

[13] M. Sigilló et. al. "Implementation and Testing of Different Control Strategies on a Steer-by-Wire Research platform", 6th International Munich Chassis Symposium, 2015

RIDE COMFORT

Characterization and potential analysis of passive and (semi-)active suspension systems by means of equivalent suspension parameters

Florian Klinger, Johannes Edelmann, and Manfred Plöchl

TU Wien, Institute of Mechanics and Mechatronics

1 Introduction

Providing adequate ride comfort at premium class passenger cars is essential to meet the costumers' expectations. A key feature of "good ride comfort" is the capability of the vehicle's suspension system to isolate the vehicle body – and thus the passengers – from road imperfections and undulations, which otherwise cause unpleasant oscillations and accelerations. The respective low-frequency range, often referred to as primary ride, [1], is of main interest within this paper. Passive suspension systems have evolved to a high level of maturity, and due to the particular subjective nature of the human perception of ride comfort, this achievement is largely based on the experience of generations of test engineers and their efforts on vehicle testing on proving grounds and/or public roads. No further substantial improvements are to be expected for passive spring and damper, [2, 3].

Nonetheless, to fulfil rising customer expectations w.r.t. both handling dynamics and ride comfort – which are likely to increase even further for the "drivers" of future autonomous vehicles – active and related semi-active suspension systems have proven to be a remedy and became popular in premium class cars as well as for lower segments. Different strategies with varying levels of complexity have been proposed to control the vertical suspension forces; respective detailed information can be found in [4–6]. However, a common understanding about advantages, limitations and drawbacks of these strategies still needs to be obtained. In addition, finding appropriate setups for active and semi-active suspensions and dampers is even more demanding than for passive ones, and less (practical) experience and less (theoretical) guidelines for tuning the control parameters are available.

It would thus be practical to transfer findings on passive suspension and damper design to active and semi-active suspension systems. For this purpose, the concept of *equivalent suspension parameters* of active suspension systems has been elaborated in [7] and presented in [8] from a rather theoretical point of view. Here, additional practical benefits of the concept will be highlighted. The concept is originally based on a quarter-car model with random road excitation and Linear-Quadratic (LQ) optimal control strategy. The popular skyhook and/or groundhook strategies are a sub-set thereof and are considered as well. Equivalent suspension parameters relate tuning parameters of the controller to suspension design parameters (of active and passive systems) with a clear physical meaning. This is in particular helpful to apply knowledge gained from theoretical and practical experience with passive suspension and damper design also to the active systems. In [7], this concept is further extended to a half-car model, focusing on coupling effects between the front and rear axle, and on ride comfort related effects of heave and pitch, respectively. It is shown that front–rear combined control strategies are superior to front–rear independent designs, in particular if pitch moment of inertia and body mass of a vehicle deviate considerably from "usual" values (ratios), which might easily be the case for electrically driven vehicles with heavy batteries, packed between the axles.

The paper is organised as follows. In the following section, some main results from a non-linear quarter-car model and (virtual) passive suspension damper design are re-capped, which also serve as a reference for the subsequent analysis of differently controlled suspension setups. The performance of passive, semi-active and active suspensions is compared by means of a multi-objective optimisation w.r.t. performance criteria related to ride comfort and road holding, section three. The present results extend and confirm findings from literature on the one hand; on the other hand, the complexity to extract individual setup candidates from such studies becomes obvious. To overcome these difficulties, findings with regard to the LQ-optimal control strategy for active suspensions from [9, 10] are extended further in section four to the concept of equivalent suspension parameters, which will be discussed qualitatively and quantitatively based on illustrative examples in section five. Finally, a brief summary is given.

2 Passive Suspension Damper Design

Passive suspensions include remarkably non-linear features. Spring characteristics are typically progressive with increasing stiffness over wheel-travel to ensure a proper body posture for (dynamic) load transfers or variable (static) payloads. The damper characteristics typically flatten out for higher damper stroke velocities to provide good ride comfort on rough roads, while being firm and markedly asymmetric at lower stroke velocities; the latter is necessary to reduce the rattle space demand on long wavelength road disturbances, as well as to ensure a good vehicle handling on (rather) flat terrain.

As stated above, (passive) suspension design is largely driven from subjective evaluation of ride comfort, and the tuning of the parameters of interacting spring, damper, bump-stop etc. is well-coordinated by experienced engineers and test drivers with test vehicles and components. In addition, a basic theoretical understanding thereof has been obtained using basic vehicle models, such as the quarter-car model. Simplifications are applied for good reason: the model is detailed enough to capture main dynamic effects w.r.t. road induced disturbances, yet it is simple enough (in terms of less parameters) to understand the basic interactions between those few parameters and their impact on the overall behaviour of the system.

Figure 1(a) shows a full non-linear multibody system (MBS) simulation model of a front axle, loaded with proportional weight of the respective vehicle, on a virtual two-post test-rig. Since excitations at the left and right post are equal, this model is referred to as "extended quarter-car model". See the studies [11, 12] on preliminary

Figure 1(a): MBS model of front axle on a two-post test-rig, [11, 12].

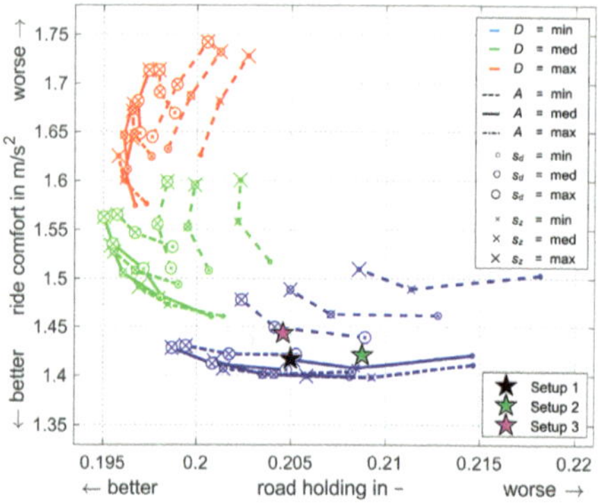

Figure 1(b): Variation of passive damper setups w.r.t. standard deviations of sprung mass acceleration and normalised wheel load variations, i.e. ride comfort and road holding, [11, 12].

virtual chassis and damper tuning for a more detailed description hereof. Related Fig. 1(b) shows results from random/stochastic "road" excitation and variations of the shape of the non-linear passive damper spline. Therein, different colours represent a different "amount" of damping, while different symbols specify the distribution of the respective damping over damper stroke velocity. Well-established performance criteria for this type of excitation are the standard deviations of vertical sprung mass acceleration and (normalised) wheel load variations to account for ride comfort and road holding, respectively.

The conflicting nature of these criteria is well known from basic linear investigations, see e.g. [13–15], and holds for the present non-linear model as well. Roughly summarising the significance of Fig. 1(b), analyses based on stochastic road excitations confirmed to be a good indicator to determine an adequate amount of overall damping for a given axle, while additional performance criteria and excitations, such as single obstacles of different wavelengths, are considered necessary to assess the shape of respective damper splines. The intensities of both types of excitations need to fit to the vehicle (model) and should be selected carefully.

Improving ride comfort, i.e. reducing the standard deviation of the vertical sprung mass acceleration below the blue setups in Fig. 1(b), cannot be achieved by adapting passive damper characteristics only. Reducing the stiffness of spring and/or bump-stop seems to be an option, but is inherently restricted by the need for proper attitude control and rattle space/packaging constraints. Controlled suspension forces allow to relax these limitations.

In case of semi-active suspensions, instead of dampers with fixed force–stroke velocity characteristics, (continuously) variable dampers allow for an instant modulation of the actual damping force. However, such systems also require a passive spring to support the weight of the vehicle body, hence above restrictions apply for them as well. Bearing this in mind, the task of respective control strategies is to use the available suspension working space more effectively compared to suspensions with passive dampers. Several methods to control the damping force have been proposed in literature; two representative strategies are discussed below. In case of active suspensions, most freedom to balance requirements w.r.t. ride comfort, road holding, and rattle space usage can be achieved.

3 Potential Analysis of (Semi-)Active Suspensions

Since automobiles are exposed to stochastic road excitations in every day driving, it represents an integral part in (virtual) passive suspension analyses and tuning approaches, and seems to be representative to compare the performance of active and semi-active suspension systems as well. For the comparison, a quarter-car model, allowing only vertical motions of sprung and unsprung mass, is used. It includes main nonlinearities of the extended quarter-car model from Fig. 1(a), such as non-linear spring – and thus restricted suspension working space –, non-linear passive damper and a potential lift-off of the tyre from the road. Additionally in the semi-active case, a major nonlinearity results from the semi-active damper itself: the device is considered to provide basically any demanded force (from the controller), as long as it acts against the damper's stroke velocity; otherwise, no force is applied. Unrealistically low or high damper forces are though prevented, which also counteracts possible jerk induced due to switching on and off the damper force. In the case of the active suspension system, spring and damper are replaced by an active, "force-producing" element. However, elastic bump- and rebound-stops remain to limit the working space of the suspension. The specification of the stochastic road excitation is adopted from [13], and the following parameters are applied: input displacement spectral density $\Omega_0 = 20\,\mathrm{cm}^3$, waviness $w = 2$, vehicle forward speed $v = 80\,\mathrm{km/h}$.

Figure 2 shows a conflict diagram for passive, semi-active, and active suspension systems. For the passive suspension, the markers of different damper setups similar to Fig. 1(b) are depicted; different values for the standard deviations of sprung mass acceleration (ordinate) and normalised wheel load variations (abscissa) are due to different model parameters. Note that the coloured pattern fully represents the attainable tuning range for passive dampers, from comfort oriented soft damping to a road holding orientated firm damping. Apparently, the respective range for the performance criteria may serve as a reference for the comparison with the controlled suspension systems. Each controlled system is represented by a coloured line, which has been calculated by means of numerical multi-objective optimisations of the particular control parameters w.r.t. the (conflicting) performance criteria. Thus, each line (i.e. Pareto-front) constitutes the best possible performances for the given conditions. Results are summarised as follows.

The popular "skyhook" control strategy, illustrated by the brown line in Fig. 2, is considered first. In contrast to a conventional vehicle damper, whose force depends on the relative velocity of sprung and unsprung mass, respective force at skyhook damping is proportional to the absolute vertical velocity of the sprung mass only. Hence, the vehicle body is virtually damped against the inertial system (fixed to the sky). Application of the respective force by the semi-active damper between sprung and unsprung mass leads *"to a well-damped body mode with no negative transmissibility-related side-effects."* (Hrovat, [9]). However, a pure skyhook damping system lacks damping for the unsprung mass mode (wheel-hop) and results in insufficient road holding abilities. Therefore, skyhook damping and passive damping is combined here, and optimal composition of both damping forces follows from the optimisation procedure.

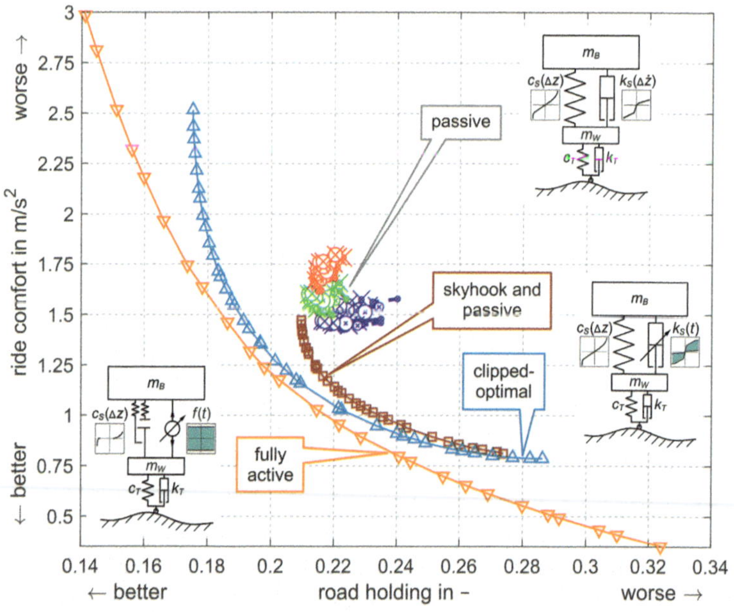

Figure 2: Comparison of passive, semi-active and fully active suspension systems w.r.t. standard deviations of sprung mass acceleration and normalised wheel load variations, i.e. ride comfort and road holding.

The setup on the rightward end of the brown Pareto-front consists of minimum passive and highest skyhook damping, which results in best ride comfort at the expense of road holding. Proceeding leftwards on the line, passive damping increases and skyhook damping decreases steadily. For comfort oriented setups with a standard deviation of wheel load variations similar to comfort orientated passive setups, improvements w.r.t. ride comfort are still substantial and at least double the span of the tuning range of (reasonable) passive dampers in terms of standard deviation of sprung mass acceleration. Notably in this range, optimal passive damping of the skyhook system is similar to respective (blue) passive setups, and corresponding optimal skyhook damping is approx. three times the (linearised) passive damping coefficient. Later in this paper, this relationship is found again when discussing LQ-optimal suspension forces of a respectively tuned active system.

The left end of the brown line reveals no potential of the skyhook strategy to improve road holding compared to a respectively tuned passive damper setup. More complex control strategies are required in this respect, see [4, 7] and references therein for details. As a representative control approach, which potentially allows improvements of both ride comfort and road holding, the clipped-optimal strategy is discussed next.

Linear-Quadratic (LQ) optimal control theory is regularly applied to various vehicle dynamics control problems, [16], and in particular to active and semi-active suspension systems, [9, 10]. The semi-active implementation of the LQ-optimal control is commonly referred to as "clipped-optimal" control. In contrast to the skyhook strategy, which basically relies on the absolute velocity of the sprung mass only, all (motion) states of the vehicle model are taken into account now. In practise, this involves effort for appropriate measurements/estimations of the motion of sprung and unsprung mass, which may only be justified when a superior performance is to be expected. Comparing the Pareto-fronts of the skyhook (brown) and the clipped-optimal (light-blue) strategy in Fig. 2 gives ambivalent evidence to this expectation: if ride comfort is of interest only, similar performance is found, and the increased complexity of the clipped-optimal strategy most likely does not seem to be worthwhile. However, if road holding abilities and more balanced setups are of interest as well, advantages of the latter control strategy increase from right to left, and show remarkable improvements compared to skyhook and passive suspension setups.

The orange Pareto-front finally represents a fully active system. It shares the control strategy with the light-blue semi-active system. Differences are consequently related to the implementation/actuation of the control force: in contrast to the former controllable damper, the actuator is now able to apply the demanded force (limited only to reasonably min/max values) independent from the relative motion of sprung and unsprung mass. Note, that the suspension working space is restricted here as well. Expected advances over the semi-active strategies w.r.t. both performance criteria can be observed in Fig. 2, especially as long as the limits of ride comfort or road holding are of interest. For balanced setups in between, advantages of the active suspension system appear to be rather limited, which is in accordance to prior findings in [17]. However, when considering these results, the ability of active suspension systems to additionally adapt to different operation conditions and to (fully) control the position of vehicle body in all kind of driving scenarios still needs to be kept in mind.

To exploit the theoretical capabilities of each control system in practical applications as well, a proper tuning of the control parameters is crucial. Simulation based approaches as above may be helpful to derive initial setup proposals. For a final setup however, corrections thereof corresponding to the experience of test drivers are decisive. For the skyhook control strategy, the "tuning knobs", i.e. skyhook and passive damping coefficients, are obviously directly related to the desired behaviour of the suspension system. For the LQ-based strategies, this is no longer the case. Respective tuning knobs are related to the (mathematical) control design procedure rather than to physical suspension effects. Yet to combine benefits of the LQ control strategy (optimality w.r.t. well-defined criteria, robustness, stability) with the physical clearness of the skyhook strategy, the concept of *equivalent suspension parameters* has been developed.

4 Derivation of Equivalent Suspension Parameters

A linear representation of the quarter-car model is required to design the LQ controller. The respective model is shown in Fig. 3(a), its mathematical state space description can be found in [7, 8]. The state vector $\underline{x}(t)$ consists of tyre deflection $x_1 = \zeta - z_W$, unsprung mass velocity $x_2 = \dot{z}_W$, suspension deflection $x_3 = z_W - z_B$, and sprung mass velocity $x_4 = \dot{z}_B$. With this representation, the cost function to be minimized by the LQ controller,

$$J = \lim_{T \to \infty} \frac{1}{2T} \int_0^T \left(r_1 x_1^2 + r_3 x_3^2 + \ddot{z}_B^2 \right) \mathrm{d}t, \qquad (1)$$

considers mean-square values of tyre and suspension deflection to account for road holding and rattle space usage, as well as of sprung mass acceleration to account for ride comfort. Weighting factor r_1 on tyre deflection, x_1, together with weighting factor r_3 on suspension deflection, x_3, allow for different control setups. Effects of these weighting factors on performance criteria for ride comfort, road holding and suspension rattle space have been thoroughly studied by Hrovat in [9, 10] for the fully linear case. Roughly summarising, the weighing factor r_3 on suspension deflection adjusts rattle space demands, while the weighing factor r_1 on tyre deflection can be used to balance a compromise between ride comfort and road holding. The lower r_3, the more efficiently this compromise is resolved.

The latter no longer holds for the present non-linear vehicle model, as can be seen in Fig. 3(b). The figure shows the optimised control parameters of the LQ-based systems from Fig. 2 (light-blue: clipped-optimal semi-active system; orange: fully active system). According to the lower lines in Fig. 3(b), (very) low values of r_3 are obviously prevented by the optimisation, in particular for the semi-active case. Otherwise, for the active case, bump- and rebound stops were hit too often with adverse effects, especially on ride comfort. For the semi-active case, an illustrative explanation of involved effects is given at the end of the next section. For weighting factor r_1, a similar behaviour of active and semi-active system is observed. In each case, lower values relate to better ride comfort, higher values to better road holding (but degrade ride comfort).

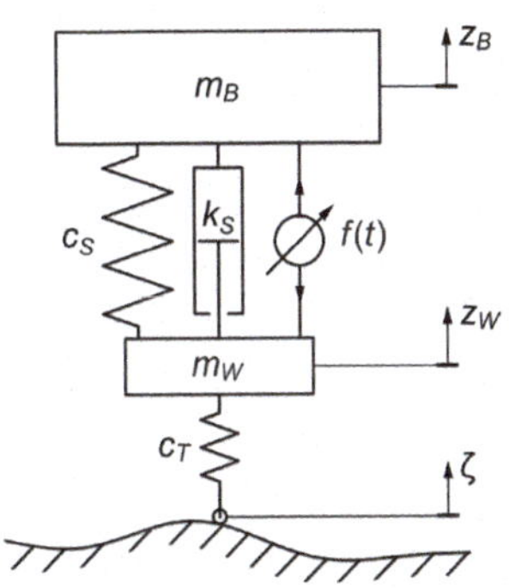

Figure 3(a): Linear quarter-car model with active and passive suspension.

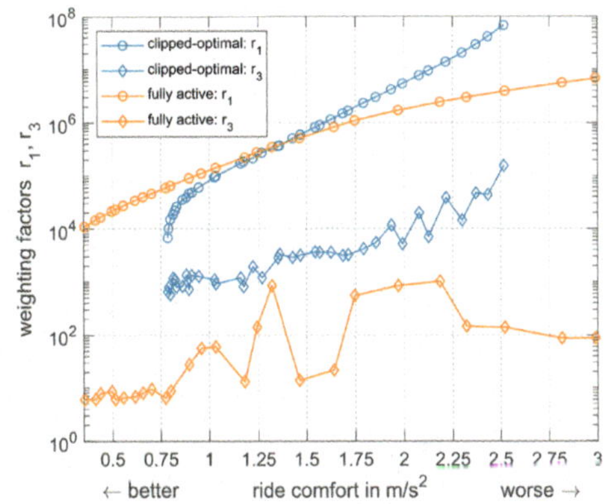

Figure 3(b): Optimised weighting factors for LQ-based control strategies over standard deviation of sprung mass acceleration according to Fig. 2.

The actuation force output from the LQ controller is now analysed in more detail. It results from minimizing the cost function (1) and full state vector feedback:

$$f(t) = -\underline{k}^{\mathrm{T}} \underline{x}(t) = -[k_1, k_2, k_3, k_4]^{\mathrm{T}} \underline{x}(t). \tag{2}$$

The constant controller gains k_{1-4} depend on the selected weighting factors r_1 and r_3, as well as on the model parameters of the quarter-car. Since passive spring and damper are accounted for in the control design procedure, the total suspension force F between sprung and unsprung mass is optimal w.r.t. (1) and involves the force of passive spring and damper and the actuation force:

$$\begin{aligned} F = {} & c_S x_3 + k_S (x_2 - x_4) \\ & - k_1 x_1 - k_2 x_2 - k_3 x_3 - k_4 x_4. \end{aligned} \tag{3}$$

Rearranging terms in (3) and considering two cases

1. if $(k_2 + k_4) > 0$:

$$F = -k_1 x_1 + \overbrace{(c_S - k_3) x_3}^{\text{equiv. passive spring}} \\ \underbrace{+(k_S - k_2)(x_2 - x_4)}_{\text{equiv. passive damping}} \underbrace{-(k_2 + k_4) x_4}_{\text{skyhook damping}}, \quad (4)$$

2. if $(k_2 + k_4) < 0$:

$$F = -k_1 x_1 + \overbrace{(c_S - k_3) x_3}^{\text{equiv. passive spring}} \\ \underbrace{+(k_S + k_4)(x_2 - x_4)}_{\text{equiv. passive damping}} \underbrace{-(k_2 + k_4) x_2}_{\text{groundhook damping}}, \quad (5)$$

illustrates, that LQ-optimal suspension forces combine forces of: (i) an equivalent passive spring, (ii) an equivalent passive damping, (iii) an inertial damping, either of sprung mass (skyhook) or of unsprung mass (groundhook), and (iv) a force proportional to tyre deflection. Terms in front brackets in Eq. (4) und (5) are thus referred to as *equivalent suspension parameters*. Additionally, k_1 contributes to an equivalent tyre stiffness $(c_T + k_1)$, see [7, 8]. Different control setups r_1 and r_3 imply different controller gains k_{1-4}, and consequently result in different equivalent suspension parameters. In practical application, forces (i) and (ii) may be applied by passive spring and damper with appropriate parameters; forces (iii) and (iv) require an active element, c.f. [18].

5 Analysis of Equivalent Suspension Parameters

Figure 4(a–b) shows interrelations of weighting factors, performance criteria and equivalent suspension parameters, resulting from the linear quarter-car model; model parameters are listed in Table 1. Weighting factor r_3 on suspension deflection is kept constant in each figure, with $r_3 = 10^{2.5}$ in Fig. 4(a) and $r_3 = 10^{3.56}$ in Fig. 4(b). Weighting factor r_1 on tire deflection is varied over a broad range in both cases and can be read off the top abscissa. On the ordinate, equivalent suspension parameters are depicted, each of them normalized by the passive reference counterpart. Skyhook and groundhook damping coefficients are normalized by the passive damping coefficient k_S^* as well. Additionally, variations of the damping coefficient of the passive system, grey line, are illustrated. Coefficient k_S^*, which gives best ride comfort (indicated by grey star in Fig. 4) has been selected as reference coefficient in Table 1. On the bottom abscissa, resulting standard deviations of sprung mass acceleration are shown, again normalised w.r.t. the passive reference system. Active setups further to the left improve ride comfort, setups further to the right enhance road holding.

The equivalent passive stiffness, $(c_S - k_3)$, green line, turns out to be independent of r_1 and is thus directly linked to weighing factor r_3 on suspension deflection. The higher r_3, the larger the equivalent passive stiffness. In Fig. 4(a), the equivalent passive stiffness is approximately 30% of the passive reference; in Fig. 4(b), passive and equivalent

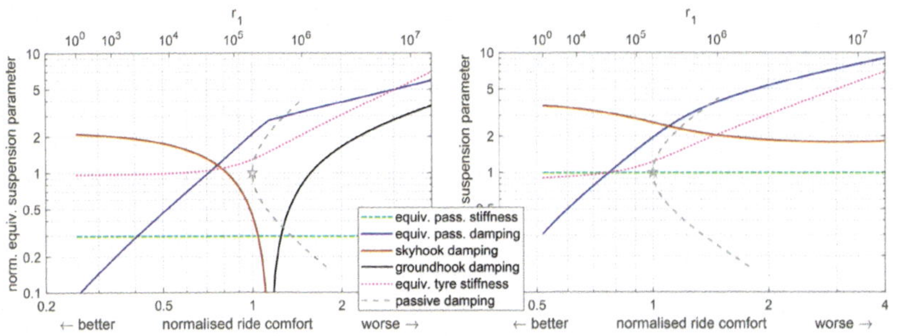

Figure 4(a–b): Equivalent suspension parameters of the fully active linear quarter-car model for variations of weighting factors r_1 and r_3 over performance criterion for ride comfort; (a): variation of r_1 and $r_3 = 10^{2.5}$; (b): variation of r_1 and $r_3 = 10^{3.56}$.

passive stiffness are equal. Other parameters depend on r_1. Considering the equivalent passive damping coefficient, $(k_S - k_2)$ resp. $(k_S + k_4)$, blue line, small values relate to lower sprung mass accelerations. Very low values suggest best ride comfort, but may be impractical due to little wheel-hop damping and (too) poor road holding. More balanced setups comprise equivalent damping similar to the passive reference. Finally, for pronounced road holding, rather high equivalent passive damping is observed.

Contributions from inertial damping are of particular interest. For small r_1, $(k_2 + k_4) > 0$ holds in both cases, and the respective skyhook damping, brown line, is considerably larger than the equivalent passive damping. Skyhook damping tends to decrease with increasing r_1. For higher r_1 and Fig. 4(a), $(k_2 + k_4) < 0$, and skyhook damping is replaced by groundhook damping, black line. However, in Fig. 4(b) skyhook remains effective also for road holding orientated setups. Then, in both cases, impacts from the equivalent tyre stiffness, $(c_T + k_1)$, magenta line, are essential. For improved ride comfort, in contrast, the equivalent tyre stiffness is close to the original tyre stiffness. In the latter case, the contribution from tyre deflection feedback to the actuation force can be neglected without much consequence, [7].

parameter	symbol	value	unit
sprung mass	m_B	465.7	kg
unsprung mass	m_W	50.4	kg
suspension stiffness	c_S	27922.4	N/m
suspension damping coeff.	k_S^*	1238.8	Ns/m
tyre stiffness	c_T	262200	N/m

Table 1: Parameters of the passive linear quarter-car model from Fig. 3(a).

Obviously, analysing setups of active suspensions by means of equivalent suspension parameters helps to achieve a basic understanding of the relations between respective weighting factors, composition of the suspension forces, and system behaviour. Several beneficial applications may be concluded therefrom; some ideas are briefly addressed in the remainder of the paper.

To begin with, above findings are used to complete the investigations w.r.t. semi-active and active systems from the previous sections, c.f. Fig. 2 and Fig. 3(b). Considering the clipped-optimal strategy, optimised weighting factors r_3 are found from Fig. 3(b) to be mainly within $10^3 < r_3 < 10^4$. Hence, stiffness values of originally installed suspension spring and equivalent passive stiffness are not very different. In case of a respective lower (or higher) equivalent stiffness, the force of the semi-active damper counteracts (or boosts) the force of the passive spring. Apparently, this cannot be entirely achieved by the semi-active device at any time, and more or less aligned stiffnesses appear to be most effective. To put it the other way round: specifying r_3 such that equivalent and actual passive suspension stiffness are equal may serve as a reasonable initial setup; the respective value can easily be derived from above considerations. Additionally, by inspection of Fig. 4(b), a characteristic setup can be found, at which equivalent passive damping and passive reference damping are equal and skyhook damping is of approx. three times the magnitude. For lower values of r_1, clipped-optimal setups behave more or less like skyhook systems, which actually explains the small differences between skyhook and clipped-optimal strategy in Fig. 2 w.r.t. comfort orientated setups. For higher r_1 and improved road holding, effects of increasing equivalent tyre stiffness are obvious.

Considering the fully active system, rather small values of r_3 appear to be favourable, at least in the context of the optimisation w.r.t. standard deviations of sprung mass acceleration and wheel load variations. Following above considerations, small r_3 result in low equivalent passive stiffness, and consequently in (very) low heave natural frequencies. Additional investigations and/or field tests need to show whether such setups are desirable in practical application. Raising r_3 moderately does not cause much harm to the performance under discussion, see [7–10]. Then, on one hand, weighting factor r_1 balances requirements w.r.t. road holding or ride comfort. On the other hand and in more detail, r_1 decides on the active and passive damping of the suspension system. Resulting effects especially influence sprung mass acceleration amplitudes at certain frequencies. Please refer to [7, 8] for a detailed discussion.

If, for instance, a proper suspension setup has been found for the front axle of a vehicle, the task to adjust the parameters of the suspension of respective rear axle can be substantially simplified by using equivalent suspension parameters. From passive suspension design, advantages of a slightly stiffer rear suspension to reduce pitch reactions are well-known, c.f. [15], and can easily by employed for (semi-)active systems as well, just by specifying equivalent passive stiffnesses properly. Active and passive damping can then be balanced accordingly.

6 Conclusions

A non-linear quarter-car model was utilised to compare passive, semi-active and active suspension setups under random/stochastic road excitation. Nonlinearities result from restrictions of the suspension working space (elastic bump- and rebound stops), as well as from passive and semi-active damper models. Pareto-optimal setups w.r.t. criteria for ride comfort and road holding have been analysed, considering both, achievable performance as well as underlying control parameters.

The semi-active skyhook strategy is known for substantial improvements of ride comfort compared to reasonable passive suspension setups. If combined with appropriate passive damping, road holding qualities can be guaranteed; however, the higher the passive damping, the less the advantage w.r.t. ride comfort, and a setup, which combines the damping of a passive system for best ride comfort, and a skyhook damping of approximately three times the magnitude, may provide a reasonable compromise. Except for pure ride comfort, the semi-active system based on the clipped-optimal strategy outperforms the skyhook system in any case and maintains advantages over passive suspensions even for improved road holding. Further benefits due to a fully active system are revealed for either pronounced ride comfort or pronounced road holding. Less improvement for balanced setups may rather be understood as a merit of the principle of semi-active dampers than as a shortcoming of active suspensions; for other disturbance scenarios, advantages of active systems are evident.

Although the clipped-optimal (semi-active) system and the fully active system share the same control strategy, i.e. Linear-Quadratic (LQ) optimal control, optimised control parameters differ systematically. Differences were found to arise from the semi-active system and the interaction of the forces of controllable damper and passive spring. Proposals for a respectively adapted tuning of the control parameters of both active and semi-active systems are facilitated by the concept of equivalent suspension parameters. The concept has been elaborated to reveal a better understanding of the interrelations of control parameters, composition of the actuation force, and resulting effects on system behaviour. From a research point of view, it can not only be applied to identify above effects, but may allow to better adopt control strategy and design parameters for a desired behaviour of the suspension system, rather than by considering just a cost function to minimise respective performance criteria.

From a practical point of view, considering equivalent suspension parameters allows to apply available broad knowledge from passive suspension tuning to active systems. Equivalent suspension parameters relate the (fictitious) control parameters to vehicle (model) parameters with a clear physical meaning, and can thus be used as a practical way to choose appropriate control parameters based on experiences with (passive) suspension design. Recently, active suspension systems have been presented, which consist of passive (air-) spring and damper and additionally allow for actively superposed, dynamic vertical forces. Whenever a LQ strategy is of interest to control these forces, equivalent suspension parameters straightforwardly indicate well-coordinated contributions/designs of active and passive suspension forces. In addition to suspension performance considering ride and handling, highlighting synergy effects of the (functionally) separated forces may also be of interest from an energy efficiency perspective.

References

[1] P. Zeller: *Handbuch Fahrzeugakustik: Grundlagen, Auslegung, Berechnung, Versuch.* 3rd ed., Wiesbaden: Springer Fachmedien, 2018.

[2] R. S. Sharp and D. A. Crolla: *Road Vehicle Suspension System Design – a review.* Vehicle System Dynamics 16(3), 1987, 167–192.

[3] D. Cao, X. Song and M. Ahmadian: *Editors' perspectives: road vehicle suspension design, dynamics, and control.* Vehicle System Dynamics 49(1–2), 2011, 3–28.

[4] M. Valášek and W. Kortüm: *Semi-Active Suspension Systems II.* In: The mechanical systems design handbook: modeling, measurement, and control, O. D. I. Nwokah (ed.), Boca Raton, Fla.: CRC Press, 2014, Chap. 31, 221–238.

[5] E. Guglielmino: *Semi-active suspension control: improved vehicle ride and road friendliness.* London: Springer, 2008.

[6] G. Mastinu and M. Plöchl: *Road and off-road vehicle system dynamics handbook.* Boca Raton, Fla.: CRC Press, 2014, Chap. 22 and 31.

[7] F. Klinger: *Kombinierte Betrachtung passiver und (semi-)aktiver Radaufhängungen mittels äquivalenter Fahrwerksparameter.* Dissertation, TU Wien, 2018.

[8] F. Klinger, J. Edelmann, S. Jeindl and M. Plöchl: *Equivalent suspension parameters of active suspension systems.* Proceedings of AVEC'18, Peking, 2018, 6 p.

[9] D. Hrovat: *Survey of Advanced Suspension Developments and Related Optimal Control Applications.* Automatica 33(10), 1997, 1781–1817.

[10] H. E. Tseng and D. Hrovat: *State of the art survey: active and semi-active suspension control.* Vehicle System Dynamics 53(7), 2015, 1034–1062.

[11] F. Klinger et al.: *Virtual chassis tuning with emphasis on the damper characteristics.* In: 6th International Munich Chassis Symposium. Munich, 2015, 325–346.

[12] F. Klinger J. Edelmann, M. Plöchl and S. Jeindl: *Effects of different excitation spectra on virtual suspension damper design.* In: Proceedings of the IAVSD 2015. Graz, 2015, 115–124.

[13] B. Richter: *Schwerpunkte der Fahrzeugdynamik.* Fahrzeugtechnische Schriftenreihe, Köln: TÜV Rheinland, 1990.

[14] M. Mitschke and H. Wallentowitz: *Dynamik der Kraftfahrzeuge*. 5th ed., Wiesbaden: Springer Vieweg, 2014.

[15] T.D. Gillespie: *Fundamentals of vehicle dynamics*. Warrendale, Pa.: Society of Automotive Engineers, 1992.

[16] R.S. Sharp and H.Peng: *Vehicle dynamics applications of optimal control theory.* Vehicle System Dynamics 49(7), 2011, 1073–1111.

[17] R.S. Sharp and S.A. Hassan: *The Relative Performance Capabilities of Passive, Active and Semi-Active Car Suspension Systems*. Proc. Instn. Mech. Eng. D 200(3), 1986, 219–228.

[18] M.C. Smith and G.W. Walker: *Performance Limitations and Constraints for Active and Passive Suspensions: a Mechanical Multi-port Approach.* Vehicle System Dynamics 33(3), 2000, 137–168.

Damper HiL-tests and damper characterizations with focus on real-time application

M.Sc. David Benz*
Prof. Dr. Peter E. Pfeffer*

* Munich University of Applied Sciences

Abstract

Testing and evaluating of dampers, top mounts and its assemblies, can be challenging when it comes to HiL-applications and characterization in the small signal range with real-time capability. Especially the real-time environment has to cope with signal input, signal recording, vehicle model simulation and the control of actuators at the same time to meet the specification of hard real-time capability. Another challenge is the low velocity identification of small damper forces which are important for adequate friction modelling. In this paper a successfully implemented concept of a chassis HiL test bench will be presented. This chassis HiL test bench is designed especially for ride comfort tests in the small signal range. Although there are some limitations due to actuator dynamics, it is possible to carry out characterization and HiL tests with small displacement and velocity amplitudes. A fast actuator control is possible by implementing a fiber optic based communication system which guarantees 125 µs latency. Furthermore, test scenarios based on advanced characterization and HiL tests are presented. These tests scenarios were developed for tests and ride comfort investigations in the small signal range. Using the example of a rear axle twin-tube damper assembled with a top mount, these test scenarios show very good results. With this chassis HiL test bench and the tests scenarios more accurately test results for the product development process of passive, semi-active and active dampers, top mounts and its assemblies are possible.

Introduction

During development phase, the shock absorbers (hereinafter called dampers) and the so called top mounts (TM) undergo a product development process until they are released for series production. The focus often lies on the specifications of vehicle manufacturers. In this case, the damping behavior of the damper by means of the so-called VDA characterization consists of large amplitudes e.g. stroke of 100 mm and therefore high velocities (Figure 1).

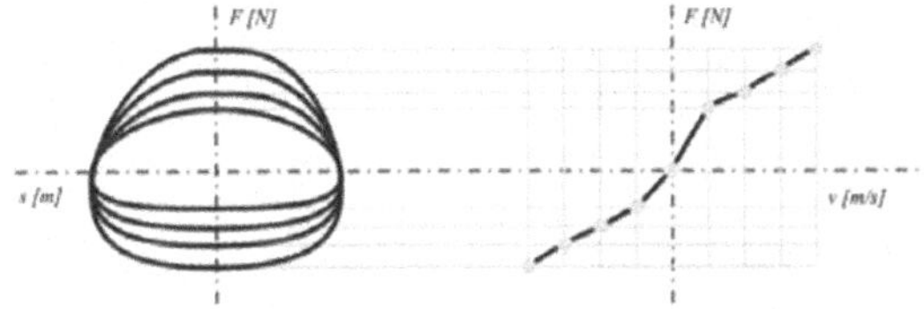

Figure 1: VDA characterization with large excitations and velocities [1]

The static and dynamic properties of the damper and the TM are often optimized separately from each other. In the sense of global function in the vehicle, the static and dynamic small-signal behavior of the damper, TM and damper-top-mount-assembly (hereinafter called DTA), which is largely responsible for the noticeable ride comfort range up to 30 Hz, is not in the focus [2]. To perform DTA characterization and DTA HiL tests (hereinafter called DTA tests) for product development and adequate DTA models for virtual development a test bench must be designed accordingly. A special focus was on generating and measuring low level of forces and displacements to map the effects in the so-called small signal range. This is very important for the ride comfort. As part of this paper, a successfully implemented concept of a chassis HiL test bench (hereinafter called CHiL) with the capability for extensive DTA tests will be presented (Figure 2).

Figure 2: CHiL with testing and real-time environment

The development included the design of the CHiL actuator system, as well as the passive components such as the stiff test frame for DTA mounting. Furthermore, the focus of the development was on a fast 8 kHz actuator control and the real-time operating system for vehicle model calculation to make real-time applications possible. All components implemented for DTA tests are described in more detail below.

CHiL Environment

To visualize how the components of the CHiL are build up and communicate with each other, a schematic diagram is shown in Figure 3.

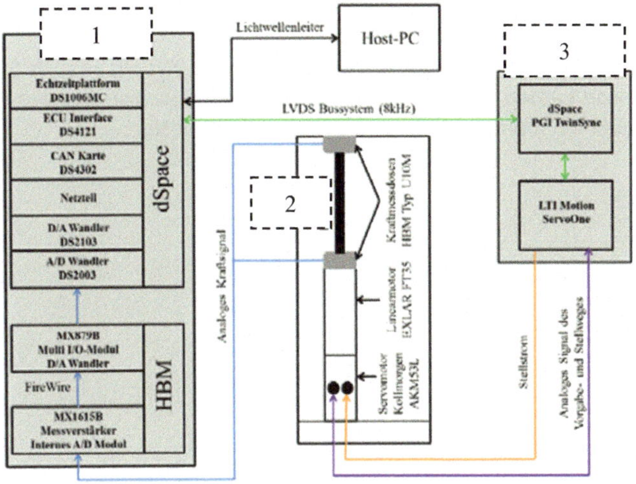

Figure 3: Schematic diagram of the CHiL with 1) Control cabinet with real-time environment, 2) DTA test set-up and actuator system and 3) Power cabinet [3]

The main components such as the control cabinet with the real-time environment, the DTA test set-up with sensors and actuators, as well as the power cabinet are shown, which represent the CHiL as a whole. The system has been modularly designed so that both hardware and software components can be expanded or replaced at any time. The DTA test procedure and measurement data acquisition were also standardized so far that the creation of the test vectors, the post-processing and the evaluation of the measurement data are almost fully automated.

CHiL Hardware

Especially for DTA tests it is necessary to uncouple all external interference as far as possible from the measuring equipment of the CHiL. Even the smallest disturbances from the CHiL environment are undesirable for the application in the small signal range. For this reason, the rigid test plate used was largely decoupled from the environment.

DTA Test-Setup

The DTA test-setup shown in Figure 4 exemplifies the test options with the CHiL, which can be carried out with both front axle and rear axle passive or semi-active dampers in mono-tube as well as in twin-tube design.

Figure 4: Typical DTA test-setup on CHiL with 1) TM and 2) twin-tube rear axle damper

Due to the CHiL concept, the connection via the TM is modular and a variety of different TM types can be used. The rear axle DTA shown here, consisting of the TM and a two-pipe rear axle damper, was connected to the load cell via an adapter. The upper load cell is used to measure the reaction force of the DTA while the lower load cell can be used for force control of the actuator. The adaptation to the test frame and the test frame itself have been designed to be very stiff in order to keep the natural frequencies outside the ride comfort range to be considered, which minimizes the error. The damper parts involved in the damper friction are heavily dependent on the lateral force occurring in the damper. Due to the tight toleration of all CHiL components, it was possible to optimally design the entire tolerance chain of the linear actuator and the linear guide. The central axis of the DTA is located on the upper and lower vertical axis of the adapters and load cells.

CHiL Actuators

Here the principle of the characterization of DTA is based on imprinting an excitation at the damper base. The deviation of the excitation represents a corresponding velocity and thus a reaction force (whole damper force) on the TM. Neglecting the stiffness of the damper bottom elastomer bearing this excitation is the differential excitation between the wheel carrier and the TM. The servomotor linear actuator (hereinafter called actuator) consists of several subcomponents shown in Figure 5.

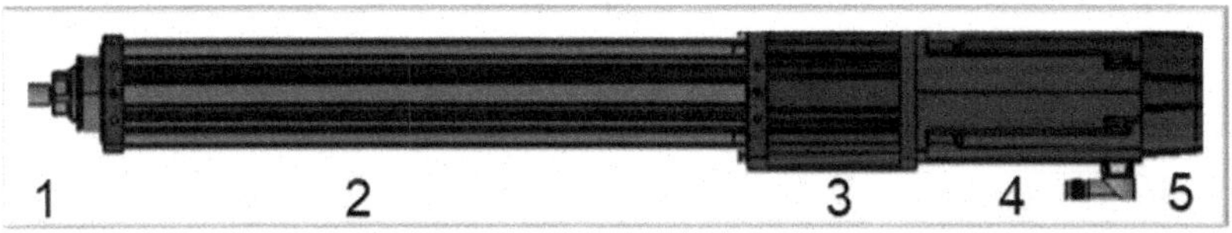

Figure 5: Actuator consisting of 1) linearly guided piston, 2) linear actuator FT35-1210, 3) clutch, 4) AC servomotor and 5) brake

The slip-free linear module FT35-1210 with planetary roller screw drive from Exlar and the AC servomotor AKM53L from Kollmorgen are installed via an in-line bellows coupling. This results in a low overall clearance and an increase in the dynamics of the actuator, since in the case of a parallel connection, the connection between the servomotor and the linear module would be implemented by means of a belt. A pairing between belt and disc can be difficult to control because of vibrations and possible slip. The actuator system is current-controlled, because on the one hand the periphery, such as hydraulic units and tubes were not desirable, but also the controllability of servomotors is considerably more dynamic and efficient. The requirements on small signal excitations and a low overall clearance, maximum forces, high positioning accuracy for the small signal range, etc. have been included in the selection of the actuator. This actuator has the following performance specifications shown in table 1

Tab 1: Performance specifications of the actuator

Actuator type	Linear guided actuator with break
Max. Force	18 kN
Nom. Force	8 kN
Max. Excitation	305 mm
Max. Velocity	0.75 m/s
Max. Acceleration	10 m/s^2
Spindle Accuracy	0.025/300 mm/mm
System Clearance	<0.06 mm
Electric Motor	AC Servomotor
Nom. Current	11.8 A
Nom. Moment	11.6 Nm
Actuator Control	8 kHz Cascade current control / Position control
Position Feedback	High resolution Sin-Cos-Encoder

A tightly tolerated linear guide compensates any possible transverse forces on the linear module. With this actuator, it was possible to implement all requirements placed on the overall system. Only the maximum achievable frequencies at higher sinusoidal excitation amplitudes over 2 mm must be limited to 15 Hz, since the dynamic limits of the actuator with respect to maximum velocity and acceleration.

CHiL Sensors

The actual actuator displacement required for the position control can be determined indirectly due to the slip-free linear module and the low clearance in the actuator via an in the servomotor absolute integrated encoder. Considering the translation from the rotary to the translatory actuator part of 10 mm per revolution, the Heidenhain high-resolution multiturn sine-cosine rotary encoder enables a rotational positioning accuracy of 20 arc seconds. This means a theoretical translational positioning accuracy of 15 µm. Since the linear module has a positioning accuracy of 25 µm per 300 mm travel as shown in Table 1, which can be assigned to the large signal range, the overall positioning accuracy in the small signal range could be set to 15 µm.

For highly accurate detection of the resulting DTA reaction forces, load cells from Hottinger Baldwin Messtechnik (HBM) were used. These have a six-conductor technology to compensate voltage drops, a high fundamental resonance frequency and can electrically compensate bending moments. Depending on the DTA test scenario a 5 kN load cell or for the determination of very small forces such as friction forces a 500 N load cell can be used. If semi-active dampers are tested, damper forces above 5 kN can occur. In this case an additional 25 kN load cells could be used. However, this is not scope of this paper. The measured forces are amplified via a strain gage bridge measuring amplifier and routed to a multi-I/O module where it can be read as an analogy voltage signal with +-10 V on the A/D card DS2003 in the real-time environment.

Due to the filtering process of the measured signal an adequate trade-off between latency and interference free force signal had to be made depending on the DTA test scenario. For real-time applications with a vehicle model and a real DTA (DTA HiL) low latency is needed to provide the measured damping force to the vehicle model within an adequate time range. The latency behaviour for a filter setting is constant, thus it can be corrected by suitable time shifting in post-processing for non-real-time DTA tests without HiL application.

In DTA testing a differential displacement between the damper piston rod and the TM arises. Due to the new degree of freedom, the piston rod can move relatively between the test frame and the damper housing as can be seen in the schematic drawing in Figure 6.

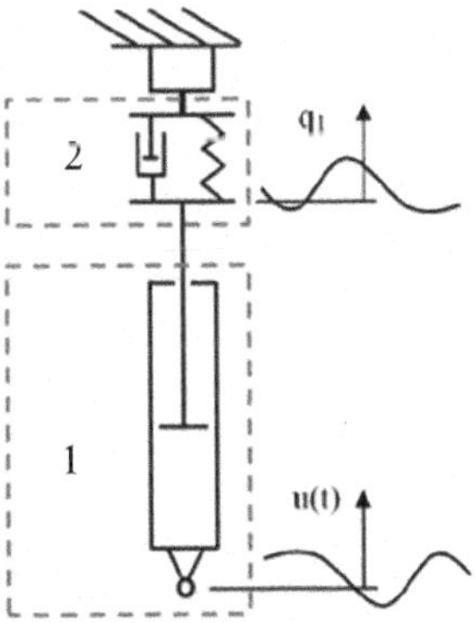

Figure 6: Schematic drawing with 1) damper bottom excitations u(t) and 2) TM with additional degree of freedom q_1 of the damper piston rod

The measurement of this difference was carried out with a Heidenhain high-resolution excitation sensor. The excitation sensor has a high measuring accuracy due to the measuring method and the preload of the measuring tip by means of a spring. The take-off frequency of the measuring tip is far outside the considered frequency range while its spring stiffness is kept low to minimize the influence of the measuring errors. Thus, displacement measurements up to a measuring accuracy of +-1 µm in the small signal range are possible.

During the DTA tests in the small signal range, no high temperature changes of the damper oil are to be expected due to the low energy dissipation in the damper. However, to document the current temperatures, a CAN-based temperature measurement system was installed to record the temperature at the damper tube.

CHiL Software

All automatization and control software developed for the CHiL is based on inhouse developments. On the one hand, due to the modular structure, software modules could be reused from existing test benches, on the other hand, new modular programmed test automation and CHiL control concepts were also created modular. The predominantly Matlab/Simulink, ControlDesk Next Generation from dSPACE (CDNG) and Python-based CHiL operation allows standardization of the interface between the hardware and software by using the test platform-independent so-called XIL API from dSPACE. Thus, it is possible to use programmed test automation platform independent for other test benches.

CHiL Real-time Environment and Control

For the CHiL control, a Simulink-based control was implemented. After compiling to real-time code it runs on the real-time DS1006MC processor board and can be accessed via CDNG which runs on the host PC. The high computing potential of the processor board allowed splitting the CHiL control into two parts. The first part runs on the processor board with a discrete frequency of 1 kHz and includes all non-time-critical control processes of the CHiL. Especially the control of the actuator and the vehicle model for DTA HiL tests as well as the force measurement are summarized in an extra block and run with a discrete frequency of 8 kHz. This high-speed block allows control of the actuator and the calculation of other time-critical processes, e.g. the vehicle model, with 125 µs cycle time. Via the high-speed block and a fiber optic communication between the control request from the CHiL controller and the converter of the actuator a bidirectional cycle time of 125 µs is possible.

During the development process, the cascade control that ran on the inverter could be replaced on the Simulink CHiL environment. This offered the advantage of high control speed and access to all control parameters at all times and thus could be optimally designed. Only the current loop has not been implemented on the CHiL control yet. The position controller of the CHiL directly specifies an equivalent actuating current which is send via the high-speed block to the converter. This causes a rotation of the actuator which is guided as the actual displacement back into the CHiL control and thus closes the control loop.

In an offline optimization process the best controller parameters for the DTA tests have been found. This optimization based on a fully developed CHiL model which includes the mechanical and electrical actuator parts as well as a damper model, a quarter vehicle model and the data communication.

DTA Test Scenarios

The DTA test scenarios are split in characterization and HiL tests. The characterization tests are designed to extract the model parameters for an advanced damper model (ADM) [7], a TM model and there assembly, the DTA model. This ADM uses also parts of a sub-force model approach described in [2] and required corresponding characterization and parameterization procedures for the parametrization. The basic structure of the not in detail described DTA model is shown in Figure 7.

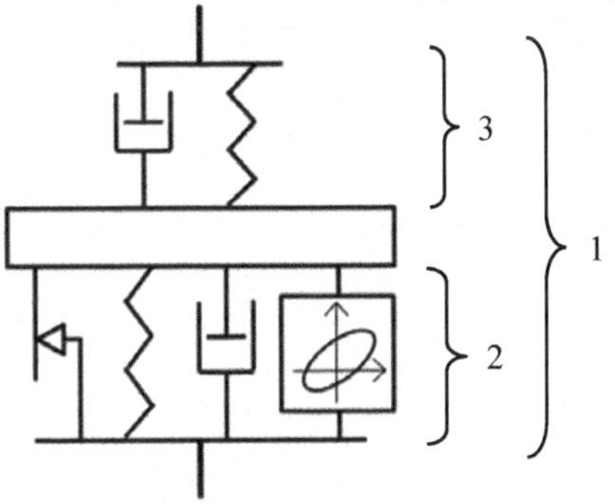

Figure 7: Basic structure of 1) DTA model [7] with 2) ADM sub force models [2] and 3) TM model

The model parameter extraction included the following test procedures:

1) DTA model

- 2) ADM parameters
  - Friction force
  - Gas force
  - Damper characteristic forces
  - Dynamic force
- 3) TM parameters
  - (Quasi-)static stiffness
  - TM Damping force
- Piston rod mass

The HiL tests were made with the DTA and a two mass quarter vehicle model (QVM). In Figure 8 the HiL test structure based on [2] is shown.

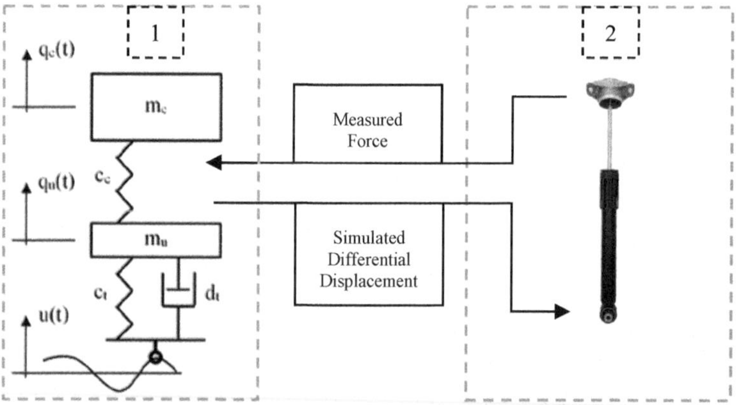

Figure 8: HiL test structure based on [2] with 1) QVM on the real-time system and 2) DTA hardware test-setup on CHiL

The QVM computes the differential displacement between the unsprung mass m_u and the body mass m_c which is needed for position control of the CHiL actuator. The measured DTA reaction force is traced back into the QVM and closes the loop between DTA (hardware) and QVM (software).

As part of the development of the CHiL so-called test vectors were needed to create the test scenarios developed for the parameter evaluation. The code generation for these test vectors based on own developed code and programming work based on [3, 4 and 5]. In addition to meta data such, as test naming, amplitudes and frequencies they also contain the signals required for the actuator controller setting and so force. For an efficient evaluation of the measurement data, an additional signal was implemented in the test vectors, which is used as a start and stop trigger for each measurement to be evaluated. In Figure 9 an example of a sinusoidal displacement signal is shown.

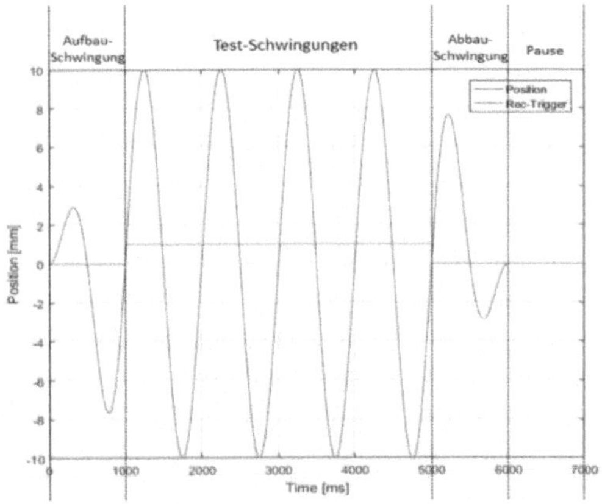

Figure 9: Example of a test vector with 10 mm amplitude, 1 Hz frequency with 4 repeats and start and stop trigger signal [4]

To get comparable damping behaviour during all DTA tests equal test conditions have to be ensured. For this purpose, a preconditioning of the DTA is necessary. This can be achieved by 3 conditioning strokes which are performed over the full available damper stroke at a velocity of 60 mm/s prior to each test. The centre of the DTA stroke is determined in a special test.

The DTA test scenarios have been designed so that they can be carried out for passive and semi-active front axle and rear axle dampers as well as for DTA. In this paper, e. g. the rear damper and its DTA will be discussed.

DTA Standard and Advanced Characterization

The challenging part with **friction force** measurement is the method for obtaining the required parameterization values. They are generally determined with relatively small displacements and velocities and very small frictional force values are to be expected. The force measurement system as well as the actuator control must be able to handle with these small signals to allow a parameterization of an adequate friction model.

The friction model is based on a two state friction model with elasto-plasticity [6] and can be found in [7]. It has five parameters to be parameterized and is able to describe the damper friction force in two different regimes, the presliding regime and the transition regime in quasi-static condition. The parameter extractions for the two regimes were made with velocities smaller than 2 mm/s and displacement smaller than 0.5 mm which doesn't trigger break away between friction planes. To evaluate measurement data and validate the friction model, small displacement amplitudes, small velocities and accelerations are needed to reduce the errors due to system dynamic forces to a minimum.

As also shown in [2], the **gas force** has little influence on the total force of the damper in the small signal range. For the sake of completeness, the generation of the measured values by means of test vectors is described here, although not in detail. The gas force model which consists of a quadratic interpolation method is only used to correct the measurements for determining the parameters of the damper characteristic model. It also represents the constant gas force offset which is established in the centre position of the damper.

The damper attached to the DTA HiL is traversed with the largest possible damper stroke at a constant velocity of 4 mm/s. This makes it possible to determine the gas force characteristics in the rebound and compressive directions. By quadratic interpolation over the entire measuring stroke, three parameters described in [2] can then be obtained.

The damper characteristic model is based on an extended so-called look up table [2]. The normally used standard VDA test consists of velocities higher than 52 mm/s. In order to get an adequate **damper characteristic model** valid for the small signal range, the determination of the forces in the low velocity range has been considerably expanded. Since the determinations of the break points as well as the VDA test from the measured values of the force-displacement-hysteresis were obtained, this method was adapted here.

The basis of the VDA test is a sinusoidal displacement signal with fixed amplitude of 40 mm. In each case a test vector with 6 amplitude repeats with an equivalent maximum velocity is given. The following break points stored in table 2 have proven to be a good compromise between measurement effort and break point density.

Tab 2: Test procedure for damper characteristic model parameter extraction with advanced VDA test

Velocity range in mm/s	**Velocities in mm/s**
Velocity $\leq$ 52	2, 4, 6, 8, 10, 12, 14, 16, 18, 20, 40
Velocity $\geq$ 52	60, 80, 100, 200, 300, 400, 500, 600, 700

If the VDA test procedure for a damper has been passed through the test automation, the raw data can be corrected using the results of the gas-force measurement.

The dynamic model is able to depict amplitude and frequency dependent **dynamic forces** of the ADM. To extract the corresponding parameters for the dynamic model, the following test procedure with focus on the small signal range described in Table 3 was carried out.

Tab 3: Test procedure for dynamic model parameter extraction with dynamic sinusoidal displacement test

Amplitude in mm	**Frequencies in Hz**
0.2, 0.5	1, 2, …, 30
1.0	1, 2, …, 22
2.0	1, 2, …, 15

The actuator is specified by physical limits with respect to the maximum speed and acceleration. Thus, tests with amplitudes over 1 mm. can no longer be done with the complete frequency band, which leads to a limitation.

The extraction of the parameters for the TM model consists of a (quasi-)static measurement in which the TM is deformed by a constant speed of 10 mm/min in pull and compression direction (Figure 10).

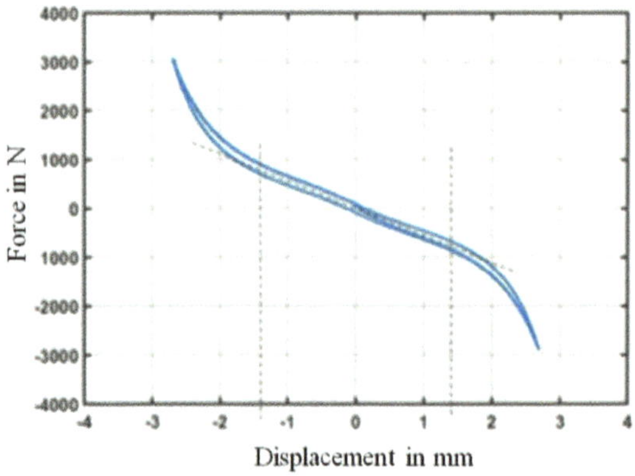

Figure 10: Measured (quasi-)static nonlinear spring characteristic of the TM

The resulting spring characteristic then serves as break points for the TM model look up table representing the **(quasi-)static stiffness** by means of a partly non-linear stiffness characteristic.

The parameter extraction from which the **TM damping force** of the TM model is derived is not described in detail. However, the area of the hysteresis surface obtained from the dynamic tests which have been also tested according to table 3 gives information of the TM damping.

The DTA model consists of the combination of the ADM and the series-connected TM model. Thus, it is possible to estimate the unknown model parts of the DTA and validate the DTA Model with additional DTA measurements. One of the most important parameter to be determined is the damper piston mass which represents the additional degree of freedom described in figure 6. It can be extracted by means of dynamic measurements with amplitudes 0.2 and 2 mm at 10 and 1 Hz. An optimization algorithm approximates adequate DTA model parameters for a good predictive power of the DTA model.

DTA HiL Tests

At the CHiL DTA HiL tests have been conducted successfully. These tests consisted of DTA HiL tests and validation tests for DTA models. HiL and validation tests with semi-active DTA are also possible.

In [2] DTA HiL tests with road profile irregularities were presented. These tests based on excitations measured from real road surfaces which could be classified as very good [1] and thus can be categorized in the small signal range. In [7] HiL tests with synthetic road surface irregularities resulting from a filter-based road profile generator [5] were made. With this generator any road classification with its correlating power spectrum density, road length and tire filtering effect [8] is possible. In figure 11 an example of a generated road surface used for DTA HiL tests in [7] is shown.

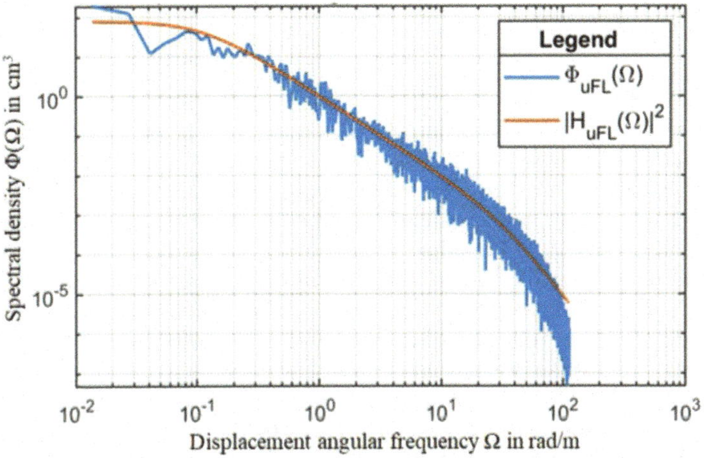

Figure 11: Spectral density of a synthetic road surface for DTA HiL tests

For DTA Model validation tests DTA HiL tests with additional (quasi-)static and dynamic test procedures were made accordingly. These tests mainly consisted of excitations and frequencies in the small signal range.

For both, the HiL tests with the measured and the synthetic road surface and the validation tests, the actuator control and the real-time environment of the CHiL show very good HiL performance.

Conclusion

In this paper a successfully implemented concept of a chassis HiL test bench was presented. The focus lies on characterizing and measuring in the small signal range as well as execute real-time capable HiL tests with dampers, top mounts and their assemblies. For this purpose the hard- and software components have been developed to fulfill real-time requirements. The actuators and the DTA mounting were developed with small clearance to ensure best actuator control behavior. In addition, the actuator controller communicates via a fiber optic system which guarantees a 125 µs latency. For all tests a Python-based automatization with a platform-independent so-called XIL API has been developed, which allows to use the programmed test automation platform independent with other test benches.

The DTA test scenarios, which can be performed with passive and active dampers, are split in characterization and HiL tests. The presented DTA characterization scenarios are based on both standardized and advanced characterizing methods for expanded damper, top mount assembly models. Especially the friction model, the damper characteristic model and the dynamic model have high impact on predictive power of the DTA model in the small signal range. They can be performed by means of very small excitations and velocities. For DTA HiL test scenarios measured and synthetic road surfaces can be used. The synthetic road surface with any power spectral density and classification can be created via a generator which also takes tire filtering into account. The actuator controllers perform very well for all (quasi-)static and dynamic DTA tests. With the developed CHIL and the test scenarios, the passive, the semi-active and the active damper and top mount product development process can be accelerated and enhanced.

Literatur

[1] Heißing B, Ersoy M. (2011) Chassis Handbook. Fundamentals, Driving Dynamics, Components,Mechatronics, Perspectives, 1. Aufl. ATZ / MTZ-Fachbuch. Vieweg+Teubner Verlag / Springer Fachmedien Wiesbaden GmbH Wiesbaden, Wiesbaden

[2] Benz, David, Pfeffer, Peter E. (2018) A NEW METHOD FOR DAMPER CHARACTERIZATION AND REAL-TIME CAPABLE DAMPER MODELING FOR RIDE COMFORT. FISITA World Automotive Congress 2018

[3] Rona, Patrick. (2018) Entwicklung und Optimierung eines Hardware-in-the-Loop Testverfahrens für Dämpfer. Bachelorthesis, Munich University of Applied Sciences

[4] Hofmann, Patrick. (2017) Entwicklung einer durchgängigen Tool-Kette zur Testvektorerstellung, Testautomatisierung, Testauswertung und Reportgenerierung. Bachelorthesis, Munich University of Applied Sciences

[5] Sigg, Stefan. (2019) Entwicklung eines Modells für semiaktive Dämpfer für Fahrkomfortuntersuchungen mit Fahrbahnunebenheitsprofilen. Bachelorthesis, Munich University of Applied Sciences

[6] Rudermann, Michael. (2012) Zur Modellierung und Kompensation dynamischer Reibung in Aktuatorsystemen. Dissertation, TU Dortmund

[7] Benz, David. (expected 2019) Analyse und Modellierung des dynamischen und nichtlinearen Verhaltens des Stützlager-Dämpfer-Fußlager-Verbundes und dessen HiL-Tests im Kleinsignalbereich. Dissertation, TU Wien

[8] Ammon, Dieter. (1997) Modellbildung und Systementwicklung in der Fahrzeugdynamik. Zugl.: Karlsruhe, Univ., Habil.-Schr., 1996. Stuttgart. Teubner, 1997. – ISBN 3-519-02378-4

Road to virtual tuning: support for damper tuning in Hyundai Motor Europe Technical Center with new lumped model and testing protocol

A. Salgarello, D. Hahn, A. Pizzuto

Engineering Design Department
Hyundai Motor Europe Technical Center GmbH

Abstract

Vehicle dynamics is a fundamental part of vehicle performance. It combines functional requirements (i.e. road safety) with emotional content ("fun to drive", "comfort") [1, 2]. The balance of Ride & Handling is what often characterizes the car manufacturer (OEM) driving DNA. The level reached nowadays by most of the OEMs on Ride & Handling is very high. This has been possible combining experience of years of development with virtual vehicle simulation. Vehicle simulation allows finding the best solutions yet in the pre-development phase, in purely virtual stage. Simulation tools can reduce significantly development costs, fundamental from OEM prospective, and they can still be useful to support the physical development of the vehicle (i.e. tuning).

One of the most important chassis components, contributing to define the character of the vehicle, is the damper [3]. Despite of its importance, most of the OEMs are mainly relying on supplier's "know-how" about design and on experienced drivers about tuning. Usually 1D lookup tables Force vs. Velocity, generated from tests like the standard VDA, are normally used by OEMs to describe the damper behavior. This approach is not representative of the damper's construction. This modelling is not able to describe how the damper is actually developing the force, mainly dependent on valve topology and tuning strategy. Different dampers display the same Force vs. Velocity curve but they can give different feeling to the driver. Consequently, the capability to represent the full damper behavior, in testing and numerical simulation, is fundamental.

To do that, an advanced CAE damper model (starting from Duym research [4]) and a new testing protocol have been developed in collaboration between Hyundai Motor Group and Politecnico di Torino. The virtual model has been developed in Matlab/Simulink® to be easily integrated in co-simulation with the CAE process used in HMETC (e.g. Driving Simulator).

It represents the damper behavior by the physical properties of its components (such as rod, valves components, oil properties [5], friction [6], etc.). Most of the parameters are sourced directly from damper drawings, Bill of Material or by measurements. The model has been tuned and verified against the output of the testing protocol, showing a good level of correlation up to 30 Hz. This is very important to correlate the vehicle ride up to high frequencies. Because of that, it is possible to use the driving simulator not only for Handling but also for Ride vehicle evaluation.

The testing protocol has been defined in order to have more details on how the damper is developing the force on selected range of component speeds & frequencies, under quasi-static and dynamic conditions. Moreover, a new way to analyze results in frequency domain has been proposed, to better understand, describe and correlate the damper performance to whole vehicle behavior.

1 Introduction

Vehicle dynamics is a fundamental part of vehicle performance development. Chassis engineering is usually responsible to study the vehicle behavior in terms of Ride Comfort (i.e. vertical dynamic) and Handling (i.e. lateral dynamic).

Once the vehicle mission is defined, vehicle's chassis setup is studied. Suspension kinematic and compliance, together with springs, anti-roll bars (when available) and dampers, are tailored to fulfil the target mission.

Differently from most of the chassis components, dampers are the only ones which have a (component-)speed-dependent performance. They are used to dissipate the energy stored by the suspension by transforming it into heat due to friction [1, 2, 7]. The suspension stores energy that comes from ground irregularity or from body motion.

Setting the damper means on one side defining how the energy is dissipated by the suspension and, as a consequence, how vibrations are transmitted to the passengers, main scope of ride studies; on the other side, dampers are used to limit the body motion and the tire oscillation during transient handling maneuvers, to ensure vehicle safety.

These two requirements cannot be obtained at their optimum using only one component's characteristic: trade-off is needed. Finding the right trade-off is called "damper tuning" phase.

Hyundai Motor Group (HMG) has improved a lot the dynamic behavior of their cars in the past few years. One reason of this improvement is the daily effort of the teams working on vehicle dynamics in Hyundai Motor Europe Technical Center (HMETC).

To improve the usage of a component, is important to have a physical understanding of its behavior. In HMETC we focused to better understand how a damper is working, how to define its performance, how to judge different dampers (i.e. different suppliers or tunings) and as well how to obtain a realistic damper behavior in CAE.

This paper will show the development and validation of a new advanced damper model and the application of a new testing protocol. The knowledge coming from modeling the component has been used to improve the testing and the testing has been used to improve the model itself.

We decided to put effort in developing a new CAE model to better predict the component behavior, especially at high frequency. Being able to represent not only the frictional hysteresis, but as well the hydraulic one is very important when the mission is predicting the R&H behavior of the vehicle via simulation. Thanks to the full-physical parametrization of the model, this tool enables us to virtually tune the damper, already in pre-development phase when no prototype vehicle is available and only simulation can give an indication of the new vehicle behavior.

2 Advanced Testing Protocol

As the advanced damper model is trying to represent the physical behavior of the component in its overall utilization range, a testing protocol able to provide high detail on how the damper is developing the force is needed.

The testing has been divided in two main parts:

1. Quasi-static test for very low-speed behavior and friction
2. Dynamic test for dynamic damping performance

Quasi-static Tests

The test performed with the component in quasi-static condition are focusing on friction caused mainly by the sealings, on gas behaving as a non-linear spring and on the initial behavior of the hydraulic.

Prior of making any kind of test in a machine and to be able to have an internal common testing language, it is important to define testing nomenclature and testing points: this is done in the so-called travel check test. During this test, performed manually by the operator, the maximum available stroke is identified, including 10 mm of safety margin as well, in both directions, in order to make the tests in safe conditions. Once this manual test has been performed, all the testing points for all the tests (quasi-static & dynamics) are already in place.

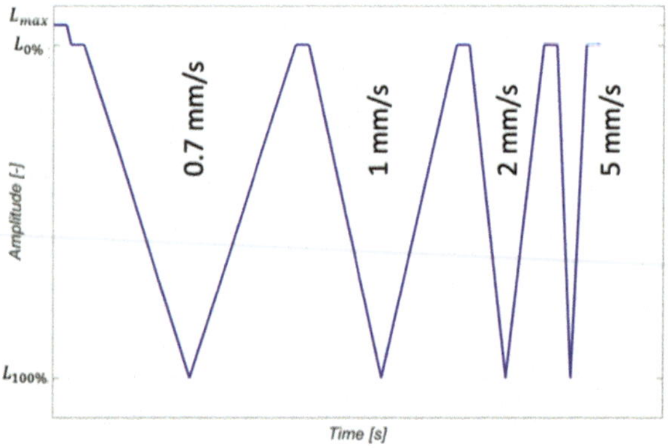

Figure 1. GFBA Test Input

The most important quasi-static test is the so-called GFBA (Gas-, Friction and Break Away Force) Test. It is able to provide information about friction in the damper, initial damper pressurization, gas charge behavior and initial hydraulic behavior of the valves. This test is performed three times using the full component stroke, in both rebound and compression, at four different speeds 0.7, 1, 2, 5 mm/s. At the end of each cycle, 30 seconds of component holding is performed to evaluate eventual creep of the system. An example of the input is shown in figure 1.

Dynamic Tests

The standard VDA test is acceptable when, given a defined damper characteristic curve (Force vs. Speed), the tester wants to understand if the component is fitting the requirements.

To understand how the damper is actually developing the force, the standard VDA has not enough resolution at low speed (speed < 0.3 m/s) where most of R&H events happen [13] and it is not testing any amplitude/frequency damping behavior. The post-processing of the results is as well a critical point. In the standard VDA the focus is to obtain only the force developed at a defined speed, in order to build the characteristic curve shown in figure 2.

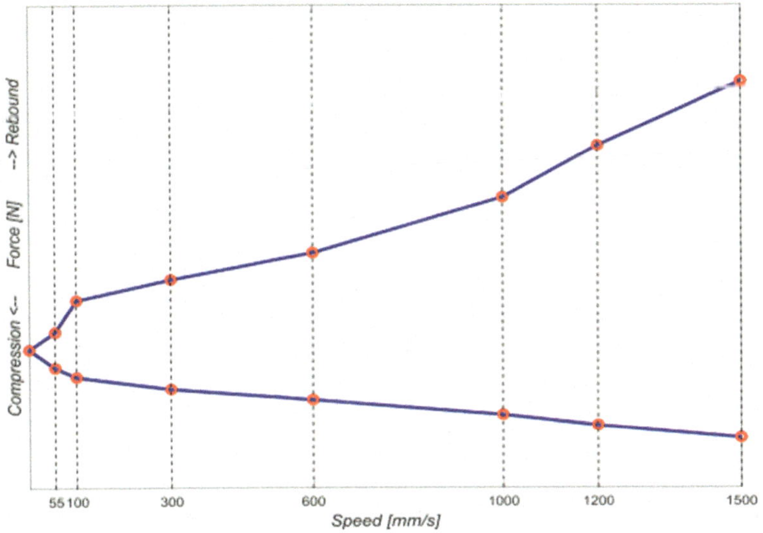

Figure 2. Example of Damper Characteristic Curve from Standard VDA Test

The new tests are capable to fulfil the weak points of the standard test, to test the component as well to conditions closer to real usage. This is also important in order to connect the component behavior to the vehicle one. As well the post-processing is not focused only on creating a more detailed characteristic curve, but as well to understand how that force is developed during the excitement. To do that, is fundamental to read the work diagram (Force vs. Stroke) together with the continuous evolution of the force in speed domain during the test. This enables the test engineer to have a picture on how the hydraulic is working and to be able to understand where to apply countermeasures if needed.

The new introduced tests are called **Advanced-VDA** test and **Sinusoidal Sweep** test.

The **Advanced-VDA** is considered as an extension of the actual standard-VDA test. The input is a sinusoidal signal with fixed amplitude and exponentially increased speed. The testing amplitudes are different, from ±2 mm to ±40 mm. The ±40 mm amplitude and the seven speeds of the standard-VDA are maintained to compare the results of the standard test with the advanced one.

The second introduced test is the Sinusoidal Sweep test. Instead of incrementing exponentially the excitation speed, this test is incrementing the excitation frequency, still maintaining the amplitude constant. The relation between the damper and the vehicle performance can be better understood, linking the damping performance at vehicle level with a new damping performance representation: force in frequency domain.

The tests are limited to 30Hz as excitation frequency, as this is considered the threshold between vehicle dynamic and NVH frequencies.

Depending on the needs, two way of post-processing can be applied. The most common and used algorithm is evaluating the force at the maximum peak only: this kind of algorithm is commonly called peak velocity plot (PVP). The other way of post-processing damper test is evaluating the full force development by the component: this this kind of algorithm is commonly called continuous velocity plot (CVP).

As we introduced tests which are testing the damper in speed and frequency, we are able to define a new representation of the damping performance in 3D with the PVP algorithm: Force vs. Frequency vs. Speed.

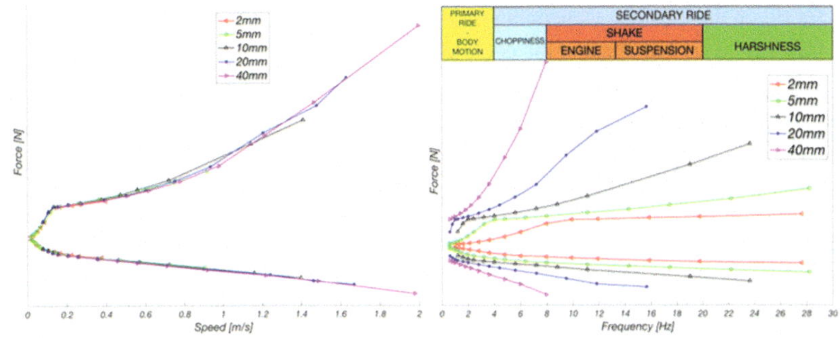

Figure 3. Damper Characteristic Curves in Speed and Frequency Domains

If we look the damper in speed domain (figure 3) we can notice that all the curves are overlapping one on the other, creating a single, uniform, damper characteristic curve: this is why this representation is very useful for damper tuning. The high detail on low speed enables better evaluation of important parts of the curve effecting vehicle ride and handling. As all the curves are overlapping one on the other, without any frequency-dependency, is difficult to relate the damper performance to the vehicle performance.

To better understand the damper-vehicle connection, the frequency domain is better. Figure 5 is showing an example with just an indicative frequency division taken from HMG Ride & Handling evaluation method. It is possible to highlight where and how the damper is acting on the vehicle in different ride and handling usage conditions.

The presented curves are useful to study characteristics like the concavity of the initial part of the curve, together with the curve's gradient after the knee, the roundness of the knee-point as well as the rebound to compression ratio characterizing the solid-feel of the vehicle. In frequency domain instead is interesting to study the shifting of the curves over frequency and the linearity at high frequency.

These curves are used as well for model correlation as every change in the curve's gradient correspond to a specific valve component.

Using the CVP algorithm instead, it is possible to study more hydraulic-related effects which are impacting the damping behavior, related to valve design, hydraulic hysteresis or cavitation.

Once a certain frequency is overcome, oil inertia, compressibility and temperature are playing fundamental roles in the damping behavior: most of these effects are not visible with a PVP analysis, so the detail given by CVP is needed. At high frequency the valve is acting as a wall for the fluid, giving main responsibility for the damper performance to the oil characteristics. Depending on these characteristics, a certain amount of hydraulic hysteresis is present in the damper.

As for the hysteresis, as well for cavitation phenomenon is important to use the CVP representation. When cavitation is present, a high increase of hysteresis at the beginning of the excitement is present, giving a high delay for the requested damping and affecting the oil characteristics as well. Cavitation is related to the gas present in the oil at ambient pressure that, when a high excitation is performed, provokes a formation of gas bubbles into the fluid [3].

3 Advanced Damper Model

When this research activity started, an extensive literature study has been performed. We wanted to develop a model fully in Matlab/Simulink® environment, as most of other HMETC models. As well we were searching for an accurate model, fast enough to be usable in real-time: one of the next steps of this activity is to use the developed model in a driving simulator to give more realistic feedback to the driver.

From HMETC point of view, the model developed by S. Duym was able to meet most of the requirements: it is representing the damper behavior also at high frequency, it can be easily adapted to measurements and it is fast to be correlated with the usage of a DOE [4, 8]. The valve modeling needed an important re-work, as it is made using estimated parameters together with others deriving from sub-component's properties.

The valve is the major damper sub-component: it is defining the complete damper performance. We started thinking about a valve modeled using only parameter derived from the part's properties (i.e. the diameter or the thickness of a disc), available in the drawings. This is enabling us to have a model not only capable to represent the damper behavior, but as well to use it to estimate different valve constructions, which give different damping characteristics.

For an OEM company like Hyundai, the possibility to tune virtually the damper (together with some of the other complex components in the suspension) gives a big advantage in terms of final product quality, development time and cost.

We started analyzing deeply the valve characteristics in order to have a deep understanding of how all the components present in the valve are working and are influencing each other. Only with this deep investigation we were able to understand how to represent the most important component's properties.

The developed damper model is representing a twin-tube rear damper with full-displaced piston valve and a rod-displaced base valve. The equations used to model the damper have been implemented in Simulink, using parameters defined in a Matlab script and derived from the technical drawings and the damper BOM.

We used the new testing protocol to test the damper that we used as prototype for our model. With the new tests we are able to have a damper performance representation up to 30 Hz. We used these data to validate the model as well up to this frequency, which is usually defined as threshold between ride and handling frequencies (0-30 Hz) and noise, vibration and harshness (NVH) frequencies (higher than 30Hz). From now on this paper will refer to the developed model as the "Advanced Damper Model" or "ADM".

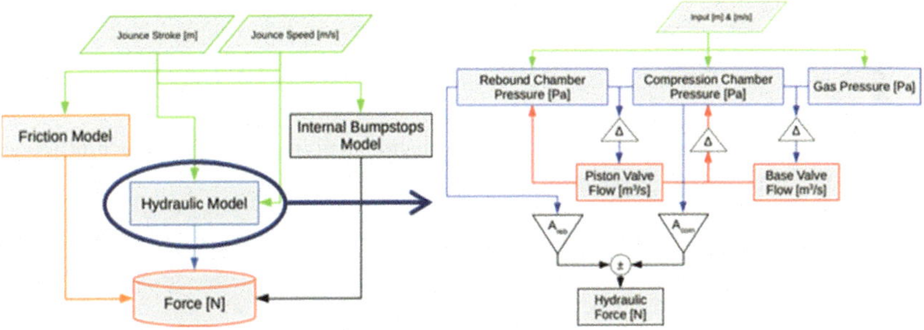

Figure 4. Advanced Damper Model and Hydraulic Model Structures

The model, as represented in figure 4, is made by three sub-models to have the three main source of damping force:

1. Model of the hydraulic ($F_{hydraulic}$);
2. Model of friction via the LuGre model ($F_{friction}$);
3. Model of the internal rebound stop ($F_{internal\ stop}$).

The sum of each modeled force is giving the damping force made by the modeled damper ($F_{damping}$):

$$F_{damping} = F_{hydraulic} + F_{friction} + F_{internal\ stop} \quad (1)$$

Model of the Hydraulic

How the energy stored by the suspension is dissipated depends mainly on the damper. How the damper is dissipating this energy depends mainly on its hydraulic characteristic. This is the reason why the model of the hydraulic, shown in figure 1, is the core of the entire model. It takes as input the relative stroke of the rod with respect to the tube and it delivers as output the force caused by the pressure difference between the chambers inside the damper (figure 5). The delta in pressure is determined by the oil flowing through the valves present inside the damper: higher is the restriction for the flow to pass from one chamber to the other, higher is the delta pressure.

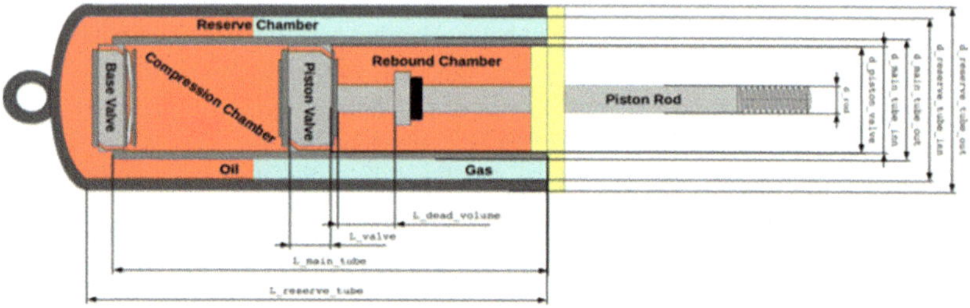

Figure 5. Defined Damper Model Layout and Nomenclature

The difference between the pressures of two chambers divided by one valve, multiplied by their effective areas, gives the hydraulic force:

$$F_{hydraulic} = F_{rebound\ chamber} - F_{compression\ chamber} \qquad (2)$$

Where:

$$F_{rebound\ chamber} = (A_{pt} - A_{rod}) * p_{reb} \qquad (3)$$

And:

$$F_{compression\ chamber} = A_{pt} * p_{com} \qquad (4)$$

The valve is restricting the fluid to pass from one chamber to the other. The restriction depends on the components which are present in the valve. This is the reason why is important to model each component individually, in order to have representation of the full topology of the system.

In our case we had to model a shim valve, with a stack of discs in both rebound and compression side. A shim valve is usually composed by a combination of:

- Discs with bleeding spots for low-speed behavior;
- Discs with holes (called orifices) for high-speed behavior;
- Flat discs for mid-speed behavior;
- Check valve made by Bellville or conical springs.

The main equation used to represent the behavior of the fluid passing through the valve derives from the Bernoulli's equation for turbulent flow, connecting the flow rate and the delta in pressure:

$$Q = \sqrt{\frac{2(\Delta p_{1-2} - \Delta p_0)}{\rho}} C_D A_{eff} \qquad (5)$$

In the equation we find:

1. Δp_0 which is the preload applied to the valve;
2. C_D which is the "discharge coefficient" that takes into account the ratio between the actual and the theoretical flow;
3. A_{eff} which is the "effective area" constricting the flow.

For high frequency behavior, the oil compressibility is playing an important role. As no thermal model is yet implemented, we used a model for isothermal oil compressibility, which account for "the relative decrease of the volume of a system with increasing pressure in an isothermal process" [9].

Two possibilities are given to the ADM user:

1. Constant oil compressibility (α_{oil}) over pressure;
2. Oil compressibility variable over pressure ($\alpha_{oil}(p)$).

To define α_{oil} (which is the reciprocal of the oil bulk modulus), we analyzed different models and we selected the one developed by Boes, which is valid of low pressure system (pressure lower than 100 bar). All the models give similar results at high pressure but they can have big difference on the behavior at low pressure. The Boes's model [10] is calculating the bulk modulus with the equation:

$$\beta_{oil}(p) = 0.5 * \beta_{oil,ref} * \log\left(99 * \frac{p}{p_{ref}} + 1\right) \qquad (6)$$

Model of the Friction

The main friction present in a twin-tube damper is the one caused by the steel-to-rubber contacts between the piston rod and the top guide on one hand and between the valve sealing and the piston tube on the other hand. In case of mono-tube damper with floating piston, friction between the floating piston and the piston tube needs to be taken as well in account. Most of the mathematical formulations of friction are dependent of the sign (i.e. direction) of the speed. This kind of modeling can lead to numerical instability which wants to be avoided. One of the most complete friction models, conceived to avoid discontinuities, is the LuGre Friction Model [11, 12].

This model is a memory-model and it can represent the friction hysteresis as well with a defined rise time (figure 6). It is an empirical dynamic friction model based on the elasticity in the contact: being empirical, a test focused on friction evaluation is needed.

The equation governing the LuGre friction model is:

$$F_{friction} = \sigma_0 z(t) + \sigma_1 \frac{dz(t)}{dt} + F_{viscous} v(t) \qquad (7)$$

Where:

$$\frac{dz(t)}{dt} = v(t) - \frac{\sigma_0}{g(v(t))} z(t)|v(t)| \qquad (8)$$

And:

$$g(v(t)) = F_{coulomb} + (F_{stiction} - F_{coulomb})e^{-\left|\frac{v}{v_s}\right|^{\eta}} \qquad (9)$$

The variable $z(t)$ represents the friction state and can be interpreted as the mean deflection of the junctions between two sliding surfaces. The variable σ_0 describes the stiffness for a spring-like behavior for small displacements while σ_1 gives the micro-damping. The state variable *g(v)* gives the modeling of the Stribeck effect [6].

The variables$F_{coulomb}$, $F_{stiction}$ and $F_{viscous}$ represent respectively the Coulomb, Stiction and Viscous friction terms. The variable vs determines how fast is *g(v)* in approaching $F_{coulomb}$.

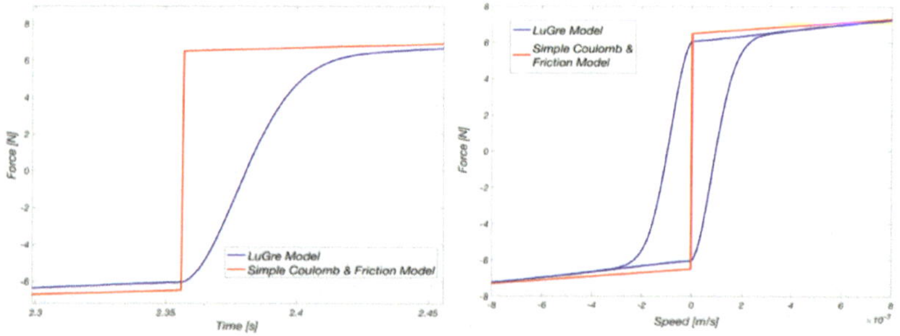

Figure 6. Coulomb vs. LuGre Friction Models

4 Validation of the Damper Model

The inputs of the new testing protocol have been used as input for the damper model. Because of that, the data coming from the new testing protocol have been used to validate the model ability to represent the real damper behavior. Same input has been given in simulation to a 1D lookup table filled with the data coming from the standard VDA, offset by the gas force value.

In the next figures the results from the lookup table simulation are presented in green, the one coming from the new model in red and the one coming from the Dynamometer Test Bench in blue. The force evolution in amplitude domain rotates clockwise while the one in speed domain rotates anti-clockwise.

In the next tables the standard deviation (Std. Dev.) of the results and the root mean square (RMS) of the residuals, in percentage with respect to the test results, are as well presented.

Simulation of GFBA Test – Quasi-static Behavior

As can be seen from figure 7, the new damper model is able to represent the real damper behavior without any significant difference. Only during the rest phase the model is a bit overestimating the force recorded during the test. In amplitude domain instead is possible to highlight a difference in compression, caused by the adiabatic law used to model the gas charge behavior.

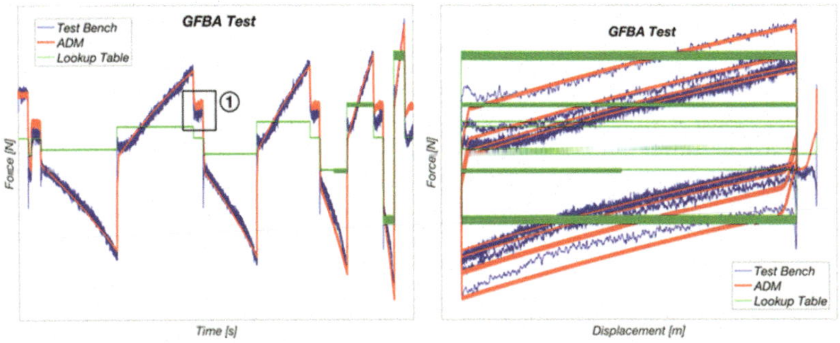

Figure 7. GFBA ADM vs. Lookup Table vs. Test Results

The improvement of the ADM with respect to the actual 1D lookup table is clear and visible. The level of correlation is shown in table 1.

Table 1. Correlation on GFBA Test

GFBA Test	**ADM**	**1D Lookup Table**
Std. Dev. [%]	95	43
RMS Residuals [%]	98	91

Simulation of Sinusoidal Sweep Test – Dynamic Behavior

Post-processing the data with the CVP algorithm, we noticed that the hysteresis at very-low speed is mainly dependent to the friction and the valve setup. These are represented by the new damper model, while they are not by the 1D lookup table.

The model is well predicting the absolute force of the damper in both speed and frequency domain but starting from 4Hz the capability to estimate the right amount of hydraulic hysteresis is limited, as shown in figures 8 and 9.

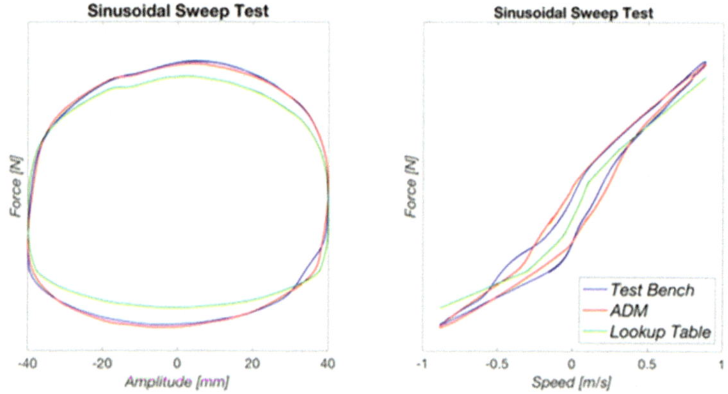

Figure 8. Sinusoidal Sweep Test (f < 4Hz): ADM vs. Lookup Table vs. Test Results

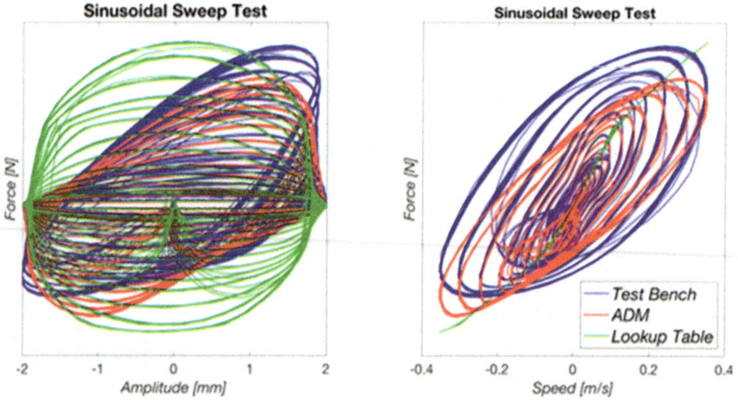

Figure 9. Sinusoidal Sweep Test (f > 4Hz): ADM vs. Lookup Table vs. Test Results

The level of correlation is shown in tables 2 and 3.

Table 2. Correlation on Sinusoidal Sweep Test – Low Frequency – CVP

Sweep Test – Low Freq.	**ADM**	**1D Lookup Table**
Std. Dev. [%]	96	88
RMS Residuals [%]	71	61

Table 3. Correlation on Sinusoidal Sweep Test – Low Frequency – CVP

Sweep Test – High Freq.	**ADM**	**1D Lookup Table**
Std. Dev. [%]	95	89
RMS Residuals [%]	56	17

Post-processing the data with the PVP algorithm, how the force is developed is then not considered and the characteristic curve is the main target.

The level of correlation is shown in table 4.

Table 4. Correlation on Sinusoidal Sweep Test – PVP

Sweep Test – High Freq.	**ADM**	**1D Lookup Table**
Std. Dev. [%]	96	82
RMS Residuals [%]	98	83

5 Main Results

Virtual Damper Tuning with Advanced Damper Model

As the parametrization of the new model is done mainly using the components specifications and properties, it should be able to represent the effect of any change as well in the component as well.

To show it, a sensitivity study on this capability has been performed, testing the model with the sinusoidal sweep test. In the pictures are shown a change in the rebound bleeding disc in terms of number of holes (±50%) and a change in the rebound flat discs thickness (±50%). The results are shown in figure 10.

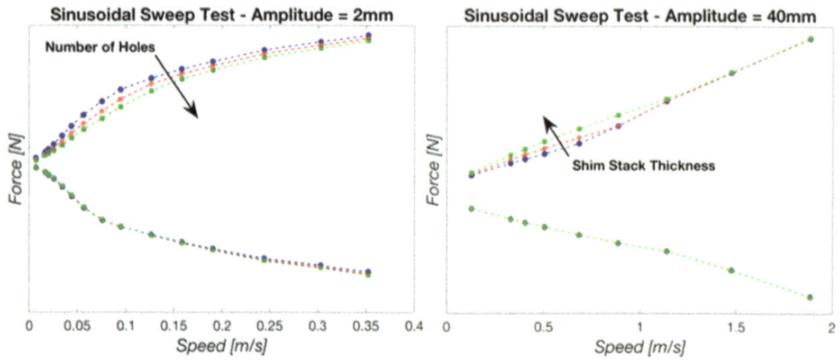

Figure 10. Sensitivity Study Made Using the New Advanced Damper Model

As damper tuning is one of the most time-consuming activities in vehicle development and it is usually very close to final vehicle release, this new tool can guide and help the damper engineer to obtain the targeted vehicle tuning.

As well the model could be used in full virtual environment or at the driving simulator to enhance the capability of the models to give right feedback to the driver and to use it as a tool for virtual damper tuning.

The model is reducing the spread between virtual and real world, thanks to its capability on representing the real behavior of the damper connected to its valves construction.

Vehicle Damping Performance Improvement

The new testing protocol enables HMETC to increment the amount of information about damper performance and damper tuning coming from testing, as shown in figure 11. The tests are giving a more in-deep picture on how the damper is really working and not only on the force is delivering.

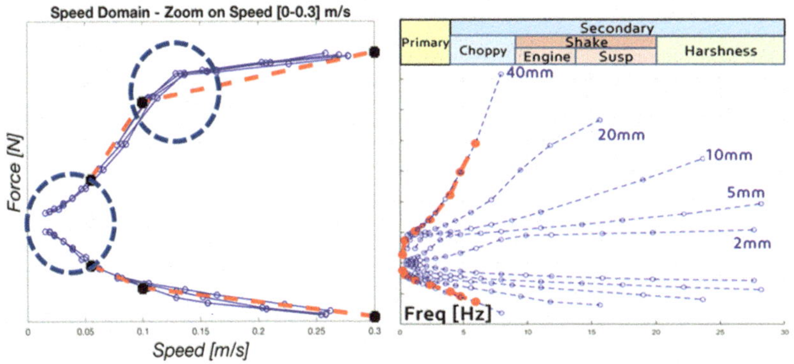

Figure 11. Standard VDA (red) vs. New Testing Protocol (blue)

First of all the fundamental area of the characteristic curve in speed domain for ride comfort and body control is now represented in high detail.

Then a new representation in frequency domain to connect the damper performance and the vehicle performance (often studied in frequency) has been proposed.

6 Conclusion

To understand how a component is physically working, it is very important to be able to model it very detailed and to test it in very different conditions. To improve both side of damper world (simulation and testing) a new testing protocol and a new damper model have been proposed.

The new testing protocol is providing much higher details on how the force is developed by the damper thanks to the possibility as well to choose between two different post-processing algorithms. In the meantime, it is as well enabling the test engineer to better connect the component performance to the vehicle performance, thanks also to the proposed vision in frequency domain.

On the other side, the new Advanced Damper Model, thanks to its parametrization connected to the physical characteristics and properties of the various component in a damper, is able to represent the damper from low to high vehicle dynamics related frequencies.

Both are touching one of the most time-consuming activity in vehicle development: damper tuning. Both will enable HMG to reduce development time and costs but increasing at the same time the final quality of the products.

This is fundamental from in OEM point of view.

In the next future HMETC will continue to improve the testing protocol, focusing on testing time reduction and frequency-dependent behavior of the damper. At the same time the damper model will improved as well, in order to be able to model different damper and valve technologies, as well to try it in real-time applications.

Acknowledgments

Thanks to the complete Chassis Team in Engineering Design

of Hyundai Motor Europe Technical Center that supported this activity, from technical and human point of view.

Hyundai Motor Europe Technical Center is the European R&D technical center for Hyundai Motor Group: more information are available at **www.hmetc.com.**

References

[1] Gillespie, Thomas D. "Vehicle dynamics." *Warren dale* (1997).

[2] Guiggiani, Massimo. "The science of vehicle dynamics." *Pisa, Italy: Springer Netherlands* (2014).

[3] Dixon, John C. *The shock absorber handbook.* John Wiley & Sons, 2008.

[4] Duym, Stefaan WR. "Simulation tools, modelling and identification, for an automotive shock absorber in the context of vehicle dynamics." *Vehicle System Dynamics* 33.4 (2000): 261-285.

[5] Boes, Christoph. *Hydraulische Achsantriebe im digitalen Regelkreis*. Diss. Verlag nicht ermittelbar, 1995.

[6] Armstrong-Hélouvry, Brian, Pierre Dupont, and Carlos Canudas De Wit. "A survey of models, analysis tools and compensation methods for the control of machines with friction." Automatica30.7 (1994): 1083-1138

[7] Genta, G., & Morello, L., "The automotive chassis", Vol. 1-2, Springer, 2009.

[8] Duym S., Stiens R., Baron V., Reybrouck K.G., " Physical modelling of the hysteretic behavior of automotive shock absorbers", SAE Technical paper, 1997.

[9] American Meteorological Society, "Glossary of Meterorology – Coefficient of Compressibility", 2012.

[10] Boes C., "Hydraulische Achsantiebe im digitalen Regelkreis, 1995.

[11] De Wit, C.C, Olsson H., Astrom K.J., Lischinsky P., "A new model for control of systems with friction", IEEE Transaction on automatic control 40 (3), pp. 419-425, 1995.

[12] Do N.B., Ferri A.A., BAuchau O.A., "Efficient simulation of a dynamic system with lugre friction", Journal of Computational and Nonlinear Dynamics, 2(4), pp. 281-289, 2007

[13] Fukushima N., Hidaka K., Iwata K., "Optimum characteristics of automotive shock absorbers under various driving conditions and road surfaces", International Journal of Vehicle Design, 4 (5), pp. 463-472, 1983.

STEERING.TECH SECTION

DEVELOPMENT METHODOLOGY AND AI

Learned steering feel by a neural network for a steer-by-wire system

Patrick Krupka, M. Sc.,
Doctoral Student, Steering Electronics – Future Development
Volkswagen Group Components, Braunschweig

Prof. Dr. Paul Lukowicz,
Scientific Director and Head
of the Research Unit Embedded Intelligence
German Research Center for Artificial Intelligence (DFKI),
Kaiserslautern

Dipl.-Ing. Christopher Kreis,
Head of Sub-Department Steering Electronics – Future Development
Volkswagen Group Components, Braunschweig

Dr.-Ing. Bastian Boßdorf-Zimmer,
Head of Department Steering Electronics
Volkswagen Group Components, Braunschweig

Abstract

High available steering systems for autonomous driving enable the development of Steer-by-Wire systems. Due to the missing mechanical linkage the steering wheel torque which depends on several inputs with different influences needs to be artificially generated by a Force-Feedback-Actuator.

The present publication describes how machine learning methods and especially artificial neural networks can be used to provide a steering feel for a Steer-by-Wire system. Therefore training data consisting of synthetic driving maneuvers is recorded to train a feedforward neural network. Measurement signals of the driver input and the vehicle reaction were selected to be used as inputs for estimating the steering wheel torque as the output of the model.

Networks of different sizes are trained and evaluated on the basis of their training and test error to examine how complex the model must be to calculate the output sufficiently. To extract more information from the training data sliding window features are used in addition to the current signal values.

A trained network has been integrated into the software of a Steer-by-Wire system in a prototype vehicle to provide the steering wheel torque to the Force-Feedback-Actuator. In this vehicle the steering feel generated by the model could be subjectively evaluated on a test site.

1 Introduction

In conventional vehicles the forces on the steered wheels are transmitted via the chassis and the steering gear to the steering wheel. In electromechanical steering systems, the supporting electric motor generates an additional torque in the steering gear. This torque can support the driver or let him feel more counter forces. The resulting steering wheel torque perceived by the driver is an essential part of the steering feel, which is an important aspect for the customer. Therefore, many steering functions are developed and optimized in the software of the steering system in order to achieve a pleasant steering feel for the customer. While the steering feel in a stricter sense mainly takes into account the angle and the torque on the steering wheel the steering feel in a wider sense also includes the vehicle dynamics perceived by the driver. This paper deals with the steering feel in the stricter sense.

1.1 Motivation

The developments for automated driving lead to changed requirements for steering systems. This favors the development of a Steer-by-Wire system and requires it over the long term. In these systems, there is no mechanical connection between the steering wheel and the steered wheels of the vehicle. So the steering wheel torque must be artificially generated by a Force-Feedback-Actuator. An example of what a Steer-by-Wire system might look like is shown in Figure 1.

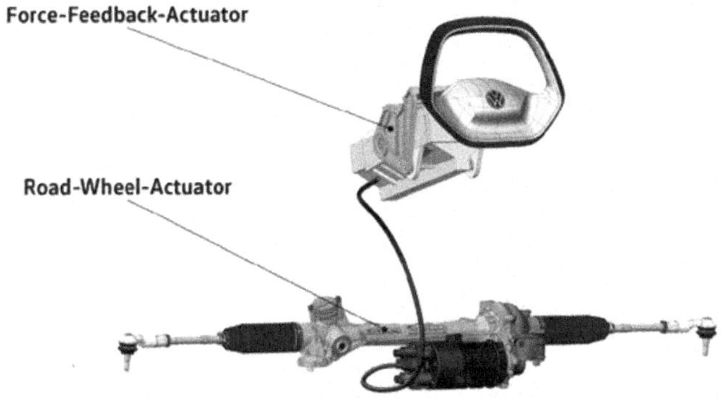

Figure 1: Example of a Steer-by-Wire system with a Road-Wheel-Actuator and a Force-Feedback-Actuator

1.2 Objective of this work

Many variables affect steering torque and the dependencies are partial complex. Steering systems must therefore be tuned extensive to achieve a good steering feel. The modeling of the steering feel for Steer-by-Wire systems is based on the steering feel of conventional systems. The dependencies between the input variables of driver and vehicle and the resulting steering wheel torque should be the same, so that today's steering feel is transferred to Steer-by-Wire vehicles. Learning these dependencies can be seen as a problem that can be solved by machine learning methods.

An approach to generate a steering feel for a Steer-by-Wire vehicle is to learn it from a conventional car. The behavior should be summarized in a model which calculates the steering torque from the given inputs. A model for learning and reproducing this behavior could be an artificial neural network. This approach will be examined in the following article.

1.3 Artificial neural networks

Artificial neural networks are mathematical models that are capable of changing their function by adapting their parameters. In supervised learning, neural networks can be used for both classification and regression. The adaptation of their function is performed by means of examples with given input and output quantities. A schematic presentation of a neural network is shown in Figure 2.

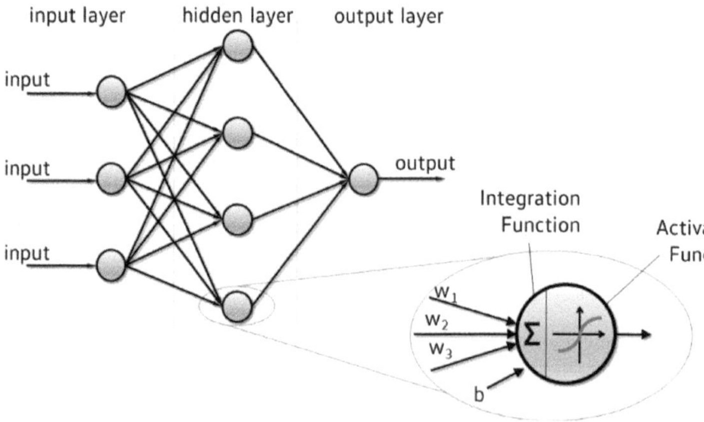

Figure 2: Schematic presentation of an artificial neural network with its neurons and weighted connections

Neural networks are composed of neurons which can be ordered in layers and are connected to each other. How the neurons of the layers are connected depends on the type of network. In feedforward neural networks, the neurons of one layer are connected exclusively to the neurons of the previous layer, so that the outputs of one layer are used as inputs of the next layer. The first layer of the network, in which each neuron is assigned to an input value of the model, is called the input layer. Accordingly, the last layer in which each neuron corresponds to an output value of the model is called the output layer. The layers in between are called the hidden layers. The number and the size of these layers affects the function of the inputs and outputs, which the model is able to learn.

Apart from the input neurons, each neuron can have multiple inputs, but only generates one output value. The inputs of each neuron are weighted (w) and summarized together with a bias value (b) in an integration function. The result is used to calculate the output via the activation function. By adapting the weights of the connections the function can be adjusted to minimize the error between the generated and given output in the training examples (Figure 3).

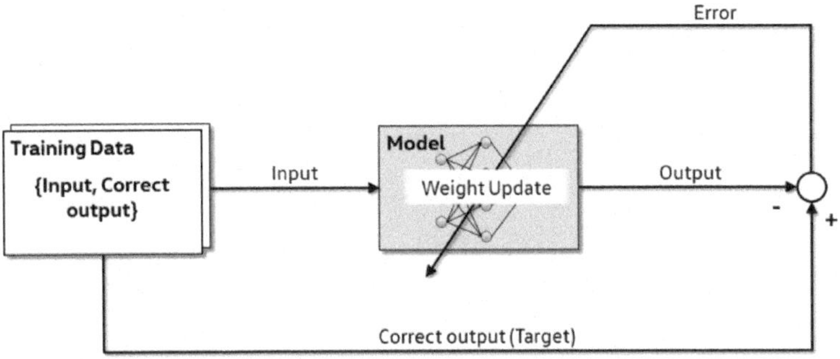

Figure 3: Updating the weights of the network on basis of the error (Kim, 2017)

2 Learning the steering feel

The procedure for learning the steering feel in this paper will be explained by means of the schematic presentation in Figure 4. The model should estimate the steering torque as the output from the driver input and the vehicle reaction (shown in Table 1) as the input variables. For this purpose input signals are to be used that are available on the bus system of the vehicle.

Table 1: Inputs and output variables of the model

Input / Output	Measurement variable	Unit
Input	Steering wheel angle	°
Input	Steering wheel angle speed	°/s
Input	Vehicle velocity	km/h
Input	Longitudinal acceleration	m/s2
Input	Lateral acceleration	m/s2
Input	Yaw rate	rad/s
Output	Steering wheel torque	Nm

To train the model, training data must be recorded and a suitable model structure and learning rule has to be chosen. The trained model can then be deployed for testing its performance and the application in a test vehicle.

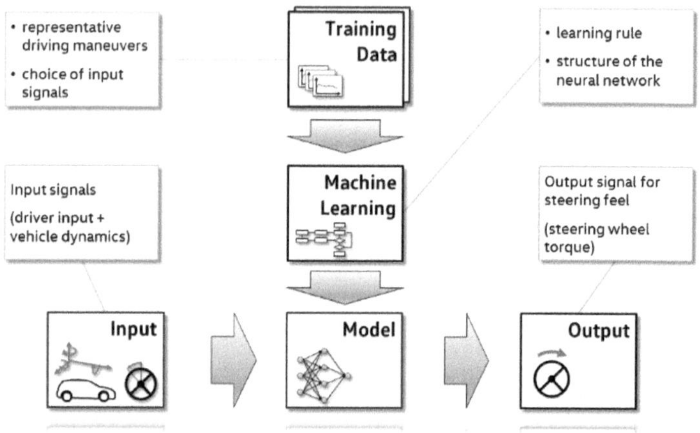

Figure 4: Methodology for learning the steering feel from training data

2.1 Training data

Training data needs to be recorded to have examples for the model to adapt its function to. For the training data, synthetic maneuvers are performed while relevant measures are recorded. In this way, it should be ensured that the data contains a large part of the driving dynamics area and is distributed as evenly as possible. As driving maneuvers, those are selected which represent variations of steering angle, steering angle velocity and vehicle speed, such as weave test, step inputs and frequency response. Each one is repeated several times to include variations in the maneuver.

In addition to the training data, another dataset is recorded which consists of a cross-section of the maneuvers used. This will be later used as a test dataset and will be therefore not used for training. Thus, the performance of the model can be tested on previously unseen data, which is also referred to as the generalization.

2.2 Training algorithm

As a model a feedforward neural network will be used, which connect the inputs to the output by weighted connections in several layers. These connection weights will be adapted during the training. The goal of the training is to get a model that will compute an output from the given inputs that is as close as possible to the given (correct) output. A simplified presentation of the learning process in the form of a flowchart can be found in Figure 5.

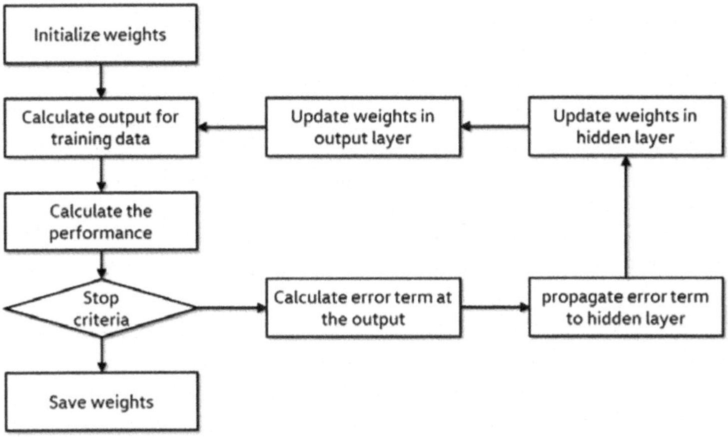

Figure 5: Flowchart of the training process for an artificial neural network

As a first step the connection weights of the network are initialized. The randomly initialized model is then executed to calculate the output of the training examples. Next the performance of the current model will be calculated which is defined as the mean squared error (MSE) in this case. At this point it is checked whether one of the stop criteria as an minimum error has been reached. If not the next step would be to calculate the error term for the output layer and propagate it back to the error term of the hidden layer. With that the weights in the hidden and the output layer can be updated. After this again the output for the training data will be calculated with the updated model and the procedure is repeated until a stop criteria is reached. Besides the minimum error other criteria can be a maximum number of training iterations or if not only the training error but also the error for a validation set is decreasing which is not used for training. At the end of the training the weights for the best iteration will be saved, so the model can be deployed for further application.

2.3 Variation of the model

In general machine learning models work best when their complexity is similar to the complexity of the problem they have to solve. Therefore, we will now investigate how complex a model must be in order to map a steering feel. The complexity of the feed-forward neural network will be varied by changing the number and size of the hidden layers.

The different networks will be trained as described before with the same dataset. Subsequently, the trained models are applied to the test set, which was not used for training, to evaluate their suitability. In the following bar graph in Figure 6 the achieved training and test errors of the differently sized networks are shown. As a comparative measure on the ordinate, the root mean square error (RMSE) was chosen so that the values are plotted in Newton meters.

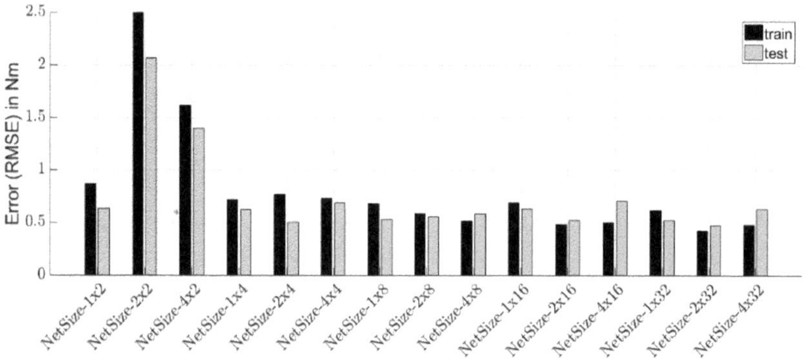

Figure 6: Training and test performance (RMSE) of trained networks with different net sizes (numbers of layers x size of layers)

The number of hidden layers was varied between one and four layers, the size of the hidden layers is varied between two and 32 neurons per layer. From the presentation, it can be guessed that the larger networks lead to better results, however, the achieved errors of most of the networks are not fundamentally different from each other. The best reached error is 0.4255 Nm (RMSE) and belongs to the network with two hidden layers with 32 neurons per hidden layer.

The steering wheel angle and the vehicle speed for the test set are plotted in the upper part of Figure 7. The test set included the steering in the standstill of the vehicle (0-10 seconds) with large angles and later smaller steering angles while increasing the vehicle speed. The lower part of Figure 7 shows the output of this network for the test set in a time series compared to the measured torque (Target). The first part of the test that represents the steering at standstill of the vehicle leads to the highest errors (0.99 Nm RMSE) in the estimation. As the vehicle speed increases in the rest of the test set, the error becomes smaller (0.23 Nm RMSE). Since this average deviation is still relatively high for a steering wheel torque, further possibilities for improving the performance are to be investigated here.

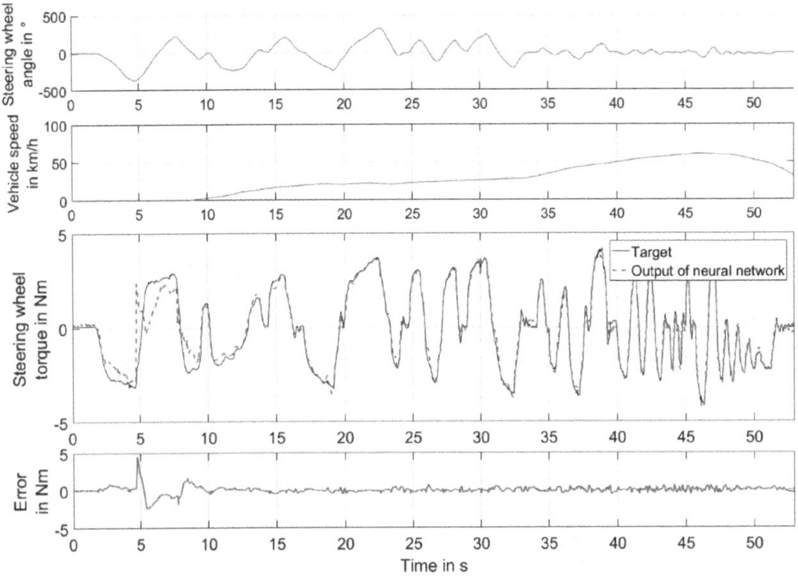

Figure 7: Time series of the test set - Steering wheel torque recorded and estimated by the trained neural network with two hidden layer with 32 neurons per hidden layer

The next step is now to test whether better results can be achieved by taking into account past measured values. By considering past time steps, the dynamics of the vehicle, chassis and steering system should be captured. Since the use of individual time steps at a low sampling rate would result in a drastic enlargement of the network, information from the past time steps should be summarized. For such a consideration of the signals in a time window, the use of sliding window features seems reasonable.

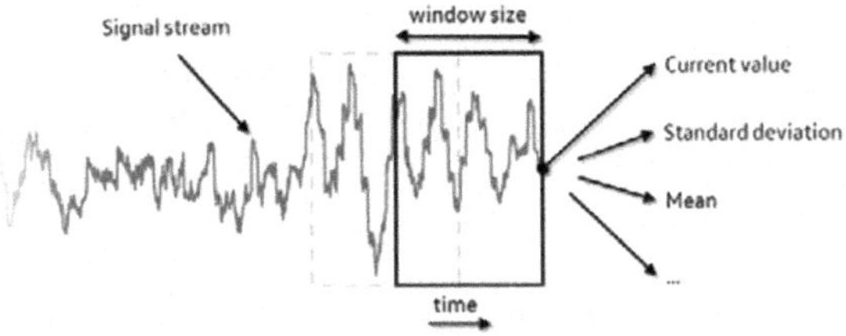

Figure 8: Principle of extracting further features as sliding window features

Sliding window features are extracted features from a time series measure that compress information from a considered time window. Sliding window features may be, for example, dispersion parameters such as the arithmetic mean, median and variance. The size of the considered time window must be selected so that relevant information is contained in it.

Now sliding window features will be extracted from the training data. Besides the current value of each input signal the mean and the standard deviation of each signal within the time window in the past will be used in the network. As more inputs are now used, the model will likely need to be more complex. Therefore, networks of different sizes are trained as before. The results of the training as well as the error of the model for the test data are shown in the bar graph in Figure 9.

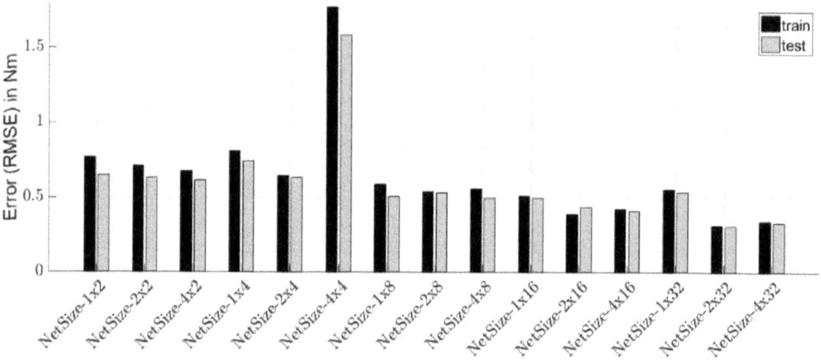

Figure 9: Training and test performance (RMSE) of trained networks with different net sizes (numbers of layers x size of layers) with sliding window features

In this presentation, it can be seen that the total error achieved is lower than with the trained networks without the use of the additional features. The network with two hidden layers with 32 neurons per hidden layer reached the lowest error of 0.3178 Nm (RMSE). Figure 10 shows the output of this network for the test set in a time series compared to the measured torque (Target). The error at the standstill of the vehicle in the first part of the test set decreased to 0.62 Nm (RMSE) and the error for the rest of the test set is now 0.18 Nm (RMSE).

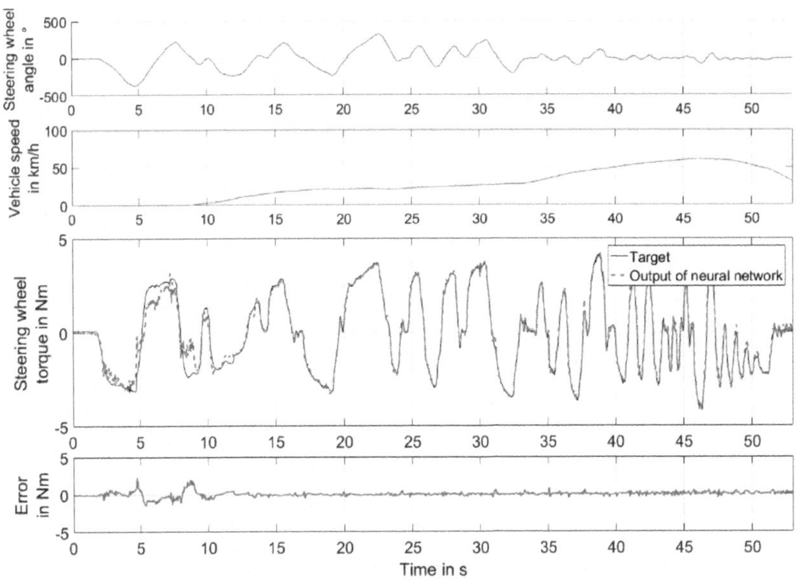

Figure 10: Time series of the test set - Steering wheel torque recorded and estimated by the trained neural network with two hidden layer with 32 neurons per hidden layer and additional sliding window features as inputs

3 Application of the trained model in the vehicle

A network from the former investigation which has a suitable performance will be used in this chapter to integrate it into a test vehicle to get a subjective impression of steering feel calculated by a neural network. As a suitable model for this the network with sliding window features which reached the lowest training and test error is chosen. It will be integrated on the software for a Steer-by-Wire system in a prototype vehicle. Subsequently, the suitability of the model in real time can be subjectively evaluated.

3.1 Test vehicle and integrated model

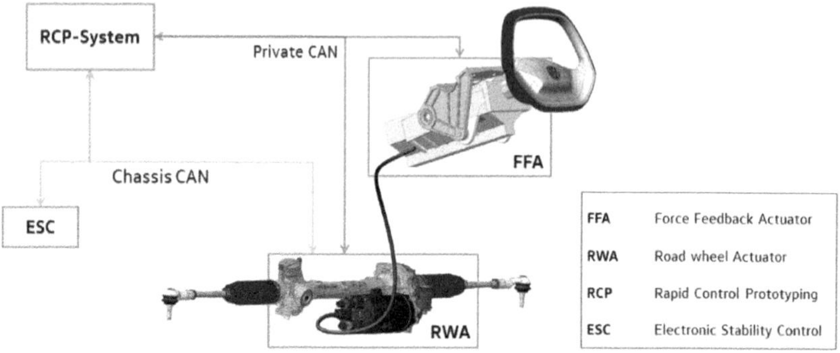

Figure 11: Overview of the Steer-by-Wire system in the prototype vehicle

As a test vehicle a vehicle is used in which the steering system was converted to a prototype Steer-by-Wire system (Figure 11). For this purpose, an electrically switchable coupling was installed at the connection of the steering intermediate shaft to the steering gear. In the active open state of the switchable coupling, the movement of the steering wheel is decoupled from the movement of the steering gear. As it is a prototype system the coupling represents a fallback level by restoring the mechanical connection in case of a system failure. To give the driver counter forces in the active open state there is a Force-Feedback-Actuator installed in the steering column. It also has an angle sensor integrated in it, which measures the steering input of the driver on the steering wheel. The steering gear software has been adjusted to control the steering angle of the front wheels depending on the driver input, thus acting as a Road-Wheel-Actuator. As an advantage of these systems the ratio of steering wheel angle to steering angle of the wheels can be adjusted.

The Force-Feedback-Actuator and the Road-Wheel-Actuator communicate with each other via a private CAN, which also includes a rapid control prototyping system. This is also connected to the chassis CAN of the vehicle and can use further messages of the vehicle over it.

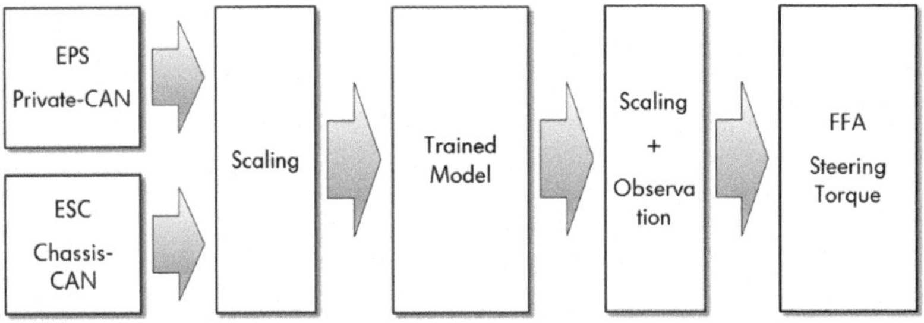

Figure 12: Integration of the steering torque calculation by the trained model

The RCP-system can request a desired steering wheel torque which is transmitted to the force feedback actuator via the private CAN. This is used to provide a steering wheel torque calculated by the trained model to the driver. As input parameters of the driver input and vehicle dynamics, the messages of the EPS and ESC control unit available on the private CAN and chassis CAN are used.

The function is built up in the model of the prototype controller as shown in Figure 12. The messages used as input to the model are first scaled to the same range of values used for the training. Also the output of the model is scaled according to the training to the original value range of the steering wheel torque. The output is limited in its extreme values and an exceeding of the limit values is recognized.

3.2 Subjective evaluation

The suitability of the trained neural network should be assessed within a test drive. For this purpose, some expert drivers were invited to test the network's function on a test site and to describe their impressions of the steering feel. During the driving task they should pay attention to the typical criteria for steering system evaluation. These were among other the steering torque level and curve, the center feeling, responsiveness and the feedback of utility information. The objective of this subjective evaluation is to make a statement as to whether the results achieved with this procedure can in principle be used to provide the steering feel and should be further optimized for this purpose.

The overall impression of the steering feel calculated by the model was described as fitting in most situations. The center feeling was good and the steering wheel torque was as expected to be. At large steering wheel angles, the system was described as a bit subdued, which was probably due to the fast falling torque when steering back to the center.

Differences from the expected behavior were noted particularly when driving the vehicle at very low vehicle speed and at a standstill. The return torque was considered as very synthetic at low vehicle speeds and led in a slow return to the center. At standstill of the vehicle, the effect of the friction and deformation forces of the wheels on the road surface could not be sufficiently reproduced. The reasons for this can be, on the one hand, that the input data used does not contain the necessary information for this state. On the other hand, the model may not be able to take these effects into account and needs to be adapted. Measures for both reasons should be investigated. Inputs representing the load variables of the steering gear should be included, and further model architectures should be tested which allow for internal feedback.

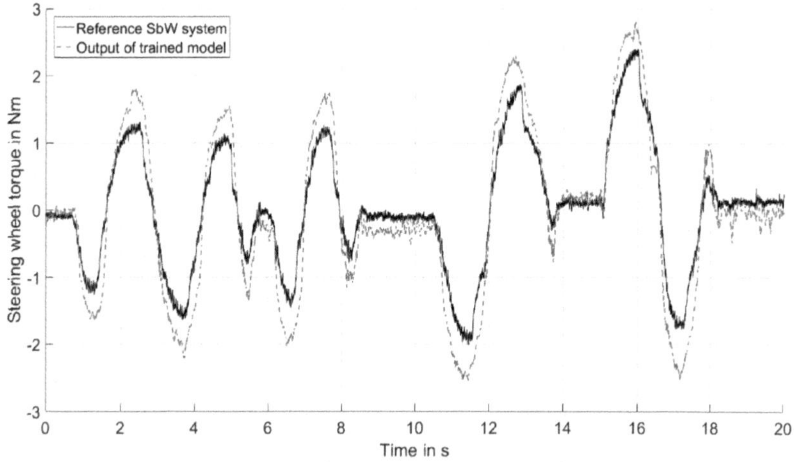

Figure 13: Measured steering wheel torque generated by the trained model compared to a reference function of a Steer-by-Wire system

Figure 13 shows measured data of the test vehicle for the sinusoidal steering at the beginning and two lane changes at the end. The steering feel of the trained model is here compared to a Steer-by-Wire system as a reference which is based on functional developed steering feel generation. Due to its construction, the reference system is able to show the effects of the forces acting in the steering gear caused by the friction, deformation and movement of the wheels on the surface. This makes it less synthetic compared to the neural network which is more focused on the vehicle dynamics. This is especially noticeable at very low speeds below 10 km/h and when the vehicle is at a standstill.

4 Conclusion and outlook

The present publication describes how a model was developed to provide a steering feel for a Steer-by-Wire system by using machine learning methods and especially artificial neural networks. Training data consisting of synthetic driving maneuvers was recorded to train a feedforward neural network. Measurement signals of the driver input and the vehicle reaction were selected to be used as inputs for estimating the steering wheel torque as the output of the model.

Several networks with different sizes have been trained and evaluated on the basis of their training and test error to examine how complex the model must be to calculate the output sufficiently. Better results were achieved by using sliding window features in addition to the current values of the signals extracting information from the time window in the past.

A trained network has been integrated into the software of a Steer-by-Wire system in a prototype vehicle to provide the steering wheel torque to the Force-Feedback-Actuator. In this vehicle the steering feel generated by the model could be subjectively evaluated by some expert drivers on a test site.

The trained neural network could be evaluated in the test drives as suitable in principle to create a steering feel for a Steer-by-Wire vehicle. In its basic form, it has fulfilled the expected function well, but it also shows areas where it does not meet the expectations of the steering feel.

Further investigations are planned by applying the training data to models with a different network architecture, e.g. convolutional neural networks (CNN) or recurrent neural networks (RNN), especially long short-term memory (LSTM) models. With these types of networks, a better consideration of inputs in the time domain or the consideration of the state in the immediate past could be possible.

Another investigation is to check the additional use of inputs that represent the load variables of the steering gear. This should improve the model in detecting the actual forces on the steered wheels. However, it should be noted that the loads of the steering gear in the case of conventional steering systems and Steer-by-Wire systems are not directly comparable and therefore appropriate measures must be determined.

The steering torques while driving and at a standstill are due to different underlying effects. It may therefore be useful to consider this in the model. On the one hand, the distinction between these states could be supported. On the other hand, different models could be trained for these states, and their results could then be included proportionately in the calculation.

Finally options and methods should be examined for how a Steer-by-Wire system which has a learned steering feel can be verified and validated. It must be ensured, for example, that the trained model does not lead to unexpected or undesirable results for previously unseen input combinations.

5 References

Garimella, G., Funke, J., Wang, C., & Kobilarov, M. (2017). Neural Network Modeling for Steering Control of an Autonomous Vehicle. *International Conference on Intelligent Robots and Systems (IROS).* Vancouver, BC, Canada: IEEE/RSJ.

Goodfellow, I., Bengio , Y., & Courville, A. (2016). *Deep Learning.* Cambridge, Massachusetts, London, England: The MIT Press.

Jayachandran , J., & Ashok, D. (2016). Neural Network Based Approach for the Generation of Road Feel in a Steer-By-Wire System. *Eng. J., vol. 20, no. 5,* 49-66.

Kim, P. (2017). *MATLAB deep learning : with machine learning, neural networks and artificial intelligence.* Berkeley, CA: Apress.

Koch, T. (2010). *Untersuchungen zum Lenkgefühl von Steer-by-Wire Lenksystemen.* München.

Pfeffer, P. (2013). *Lenkungshandbuch.* Wiesbaden: Springer Vieweg.

van Ende, K., Schaare, D., Kaste, J., Kücükay, F., Henze, R., & Kallmeyer, F. (2016, 07 03). Practicability study on the suitability of artificial, neural networks for the approximation of unknown steering torques. *Vehicle System Dynamics,* pp. 1362-1383 .

Robust control for electric power steering with unconsidered modes and parameter uncertainties

Marcus Irmer,
CLM – Cologne Laboratory of Mechatronics, TH Köln;

Michael Haßenberg, Hermann Briese,
DMecS GmbH & Co. KG;

Prof. Dr.-Ing. Hermann Henrichfreise,
CLM – Cologne Laboratory of Mechatronics, TH Köln

Abstract

For the control of electric steering systems, there is still a need for a robust implementation due to unconsidered degrees of freedom, nonlinear spring stiffnesses and gear ratios. Therefore, this article describes the modeling, control design and system analysis of an electric power steering system. It is focused on the degree of robustness of the possible controls. Two LQG-controllers are designed which are able to perform active vibration damping and disturbance compensation. This allows a high control bandwidth. Both resulting control systems are robust against unconsidered degrees of freedom, nonlinear system behavior and variations of plant parameters. Thus, the presented controllers fulfill the requirements of a modern steering system and allow the adaptation of the steering feel to the current driving situation.

1 Introduction

An LQG/LTR-design to control the driver's steering torque for electromechanical power steering (EPS) systems was presented in [1,2]. There, a simplified plant model for the steering system with two degrees of freedom, which only considers the elasticity of the torsion bar, is used for the controller resp. observer design. However, several other elastic components exist in a real steering mechanism. These elasticities were considered in [3,4] in a new detailed model of the steering system.

Disregarding degrees of freedom (eigenmodes) of the plant model for the controller resp. observer design can lead to stability problems (spillover instability) in the closed-loop system [5]. The effects of unconsidered (residual) eigenmodes on the behavior of the feedback control of a steering system were investigated in [3,4]. In this article, further reduced (simplified) design models are developed using a model-based approach, and a systematic analysis in the time and frequency domain is performed. For this, in section 2 the detailed model developed in [3,4] is briefly described. It contains all relevant elasticities occurring in a steering mechanism. Based on this detailed model, in section 3 reduced design models for the controller resp. observer design are derived. In section 4 the closed-loop systems resulting from these different design models are analyzed and compared with respect to their dynamic behavior as well as their robustness against unconsidered eigenmodes and parameter uncertainties in the plant model. Finally, further measures for improvement are described and a summary is given in section 5.

All investigations are done based on an axially parallel EPS. The methods can also be applied to other configurations and systems.

2 Detailed Model of the Steering Mechanism

Fig. 1 depicts the physical model of the detailed steering mechanism with eight degrees of freedom (8DOF). The individual rigid bodies are labeled with the indices S (steering), P (pinion), R (rack), N (nut), M (motor), C (casing), V (vehicle), WL (left wheel) and WR (right wheel). The elasticities in the model are considered by the torsion bar stiffness c_{tb} (torque sensor), stiffness c_{NR} of the ball screw drive, belt drive stiffness c_{MN}, stiffness c_{NC} of the axial nut bearing, stiffness c_{CV} of the casing attachment and the stiffnesses c_{RWL} and c_{RWR} of the attachments of the left and right wheel. The equations of motion of this model have been derived, linearized and converted into state space representation. Based on this model with eight degrees of freedom, further simplified models are generated by model order reduction.

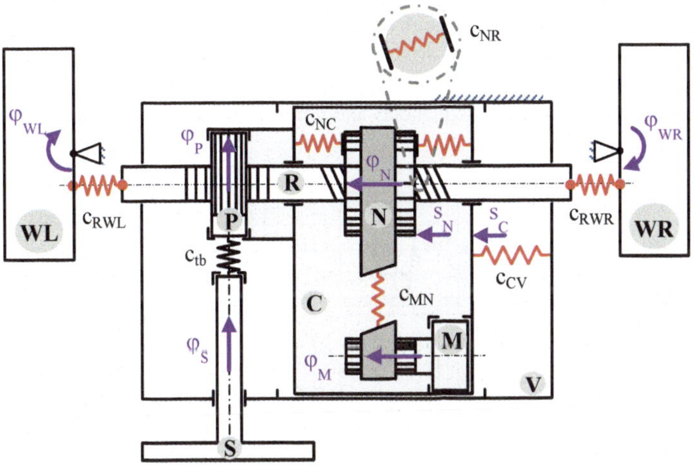

Fig. 1: Physical Model of the Steering Mechanism with eight Degrees of Freedom

Each of these different models for the steering mechanism is extended by a simplified, linear model of a current-controlled EPS motor in form of a first order lag system. Combined, they form the plant model. The plant model consisting of the current-controlled EPS motor and the detailed model of the steering mechanism with eight degrees of freedom will be called "8DOF" in the subsequent sections.

The frequency response of the control transfer path of the detailed plant model "8DOF" from the reference current i_{ref} to the torsion bar torque T_{tb} is shown in Fig. 2. There, the frequency response of the undamped model is illustrated to highlight the locations of the eigenfrequencies (black circled digits) and the notch frequencies (grey circled digits). The stiffness c_{tb} of the elastic torsion bar dominantly affects the oscillation with the eigenfrequency ① of about 51 rad/s (8 Hz) [1,2]. The eigenfrequencies ② and ③ as well as the notch frequencies ① and ② of the control transfer path of the plant model are characterized by the in-phase and counter-phase displacement of the wheels [6]. Thus, the locations of these eigenfrequencies and notch frequencies depend essentially on the parameters of the wheels and the wheel attachments. A stiffer connection c_{RW} of the wheels to the rack increases the eigenfrequencies and notch frequencies, whereas a higher moment of inertia J_W of the wheels reduces them. A change in the gear ratio i_{RW} between rack and wheel also mainly affects these eigenfrequencies and notch frequencies.

Furthermore, the stiffness c_{NC} of the axial nut bearing influences the eigenfrequencies ①, ②, ④ and ⑤ as well as the notch frequency ④, whereas the stiffness c_{CV} of the casing attachment affects the eigenfrequencies ②, ④ and ⑤ as well as the notch frequencies ① and ③. The belt drive stiffness c_{MN} influences the eigenfrequency ⑥ and the stiffness c_{NR} of the ball screw drive the eigenfrequency ⑦. As a stiffness increases, the respective eigenfrequencies and notch frequencies increase and vice versa.

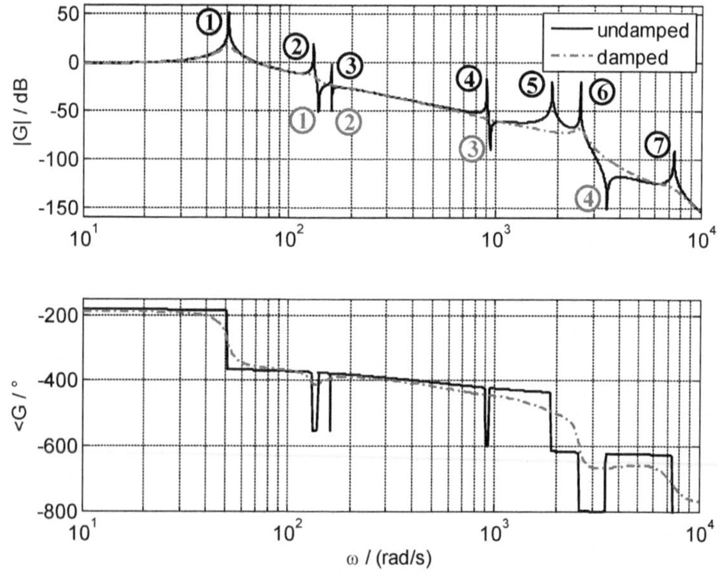

Fig 2: Frequency Response of the Control Transfer Path of the detailed Plant Model

3 Control Design

Fig. 3 shows the block diagram of the closed-loop system consisting of the detailed plant model "8DOF" and the dynamic LQG-compensator. The compensator contains the linear optimal static state space controller (LQR) and the linear optimal state space observer (LQE). The controller and the observer are designed with reduced plant models. These are based on the detailed plant model "8DOF" described in section 2.

For a constant steering wheel angle, the torsion bar torque T_{tb} is equal to the steering torque T_S induced by the driver at the steering wheel. Therefore, the task of the compensator is to adjust the torsion bar torque T_{tb} instead of the steering torque T_S to a given externally generated reference steering torque T_{req}. By setting the desired steering torque T_{req} as the reference variable, it is possible to generate a defined steering feel for the driver with an additional outer control loop [2] which should not too much affect the dynamic behavior and stability characteristics of the steering control. The steering torque T_S itself and the rack force F_R resulting from the contact between the tire and the road as well as from the friction at the rack represent disturbance variables for the control system [1]. The effect of these disturbance variables is compensated by a disturbance feedforward. Since it is not possible to measure all state and disturbance variables, an optimal observer is used to calculate the reconstructed state vector $\underline{\hat{x}}_p$ and the estimated disturbance vector $\underline{\hat{x}}_d$. For this, the observer uses measurements of the torsion bar torque T_{tb}, motor angular velocity Ω_M and motor current i_a as well as the reference current i_{ref} for the inner control loop of the EPS motor. [1,2]

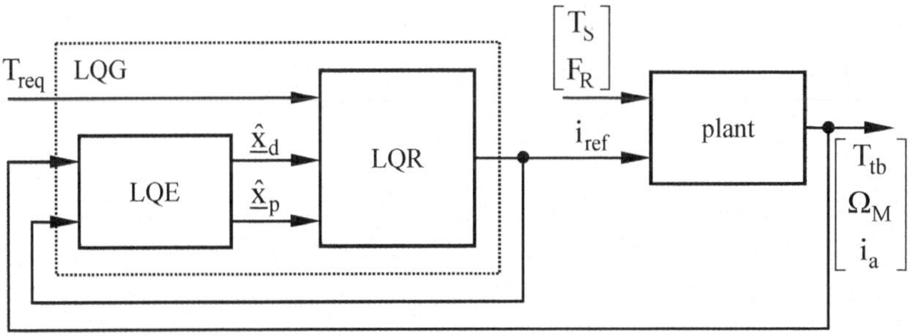

Fig. 3: Block diagram of the Closed-Loop System with dynamic LQG-Compensator

The control system must satisfy the requirements of a good control and disturbance behavior as well as a high degree of robustness against unconsidered eigenmodes and parameter uncertainties in the plant model. A precondition for a good dynamic behavior is an active vibration damping of the oscillating modes of the mechanical system. For this purpose, an LQG-compensator is designed based on [1] and [7] which considers the natural limitations of the real system so that no bounds are exceeded during normal operation. This leads to a control bandwidth of 80 rad/s (13 Hz).

However, the use of a high order model for the compensator (controller and observer) design causes problems, and a compensator of the lowest possible order should always be implemented. For example, the high-frequency eigenfrequencies change under load due to the nonlinear spring characteristics. This is detrimental to the control, since it does not match the corresponding eigenmodes sufficiently. Therefore, two methods are proposed in the subsequent sections that can be used to reduce (simplify) the detailed model without excessively impairing its characteristics. Both methods maintain stability and provide a high approximation quality in the relevant frequency range for the control. Starting point in each case is the state space representation

$$\begin{aligned} \dot{\underline{x}} &= \underline{A}\underline{x} + \underline{B}\underline{u} \\ \underline{y} &= \underline{C}\underline{x} + \underline{D}\underline{u} \end{aligned} \tag{1}$$

of the detailed plant model "8DOF" with the (n)-vector $\underline{x}$ of the state variables. The aim is to determine a well-suited reduced plant model

$$\begin{aligned} \dot{\underline{x}}_r &= \underline{A}_r\underline{x}_r + \underline{B}_r\underline{u} \\ \hat{\underline{y}} &= \underline{C}_r\underline{x}_r + \underline{D}_r\underline{u} \end{aligned} \tag{2}$$

with a (n_r)-vector $\underline{x}_r$ and $n_r < n$ for the design of a robust control system. [3,4]

Design Model based on Physically Motivated Order Reduction

By considering just the torsion bar stiffness c_{tb}, while the other components are assumed to be rigid, the detailed model of the steering mechanism with eight degrees of freedom is reduced to a model with two degrees of freedom [1,2]. However, the control system resulting from this reduced design model is not robust against unconsidered eigenmodes. To resolve this problem, similar assumptions can be used to develop a reduced model of the steering mechanism with the three degrees of freedom φ_s, φ_M and φ_W by physically motivated order reduction [3,4]. Its physical model is shown in Fig. 4. This model results from the detailed model of the steering mechanism with eight degrees of freedom by introducing constraints for the assumptions that the toothed belt is rigid and the wheels only deflect in the same direction. In addition, individual masses and moments of inertia are redistributed to the three bodies S, $\tilde{M}$ and W, and the resulting model is statically condensed. The reduced plant model consisting of this reduced model of the steering mechanism and the current-controlled EPS motor will henceforth be denoted by "3DOF physical".

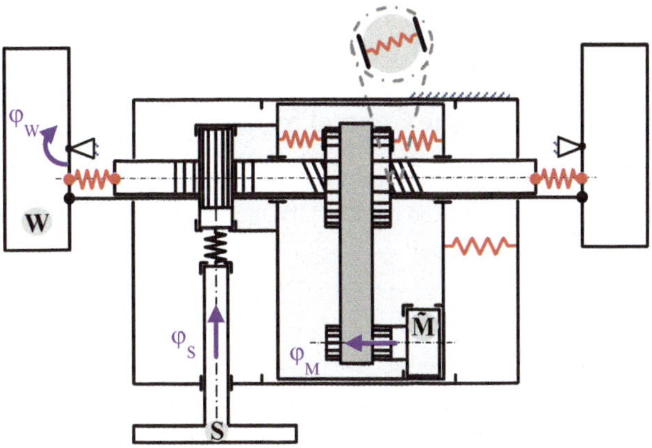

Fig. 4: Physical Model of the Steering Mechanism with three Degrees of Freedom

Design Model based on Modal Order Reduction

Alternatively, a reduced model of the steering mechanism with three degrees of freedom can be determined by modal order reduction. For this purpose, the detailed model of the steering mechanism with eight degrees of freedom is transformed into (real) modal form, and only the most dominant eigenvalues are considered. The reduced plant model resulting from this reduced model of the steering mechanism is called "3DOF modal" in the subsequent sections.

A design model which has been modally reduced in this way provides excellent robustness characteristics if it is designed with the exact parameters of the plant model. This will be shown in the next chapter. However, the modal order reduction cannot guarantee the precise reproduction of the stationary gains of individual transfer paths of the detailed model. This could be corrected by a balanced reduction, which is not used because it negatively influences the robustness behavior of the control system. Moreover, it must be ensured that no control and observation spillover occur. Control spillover means that unconsidered eigenmodes are excited, whereas observation spillover implies that these residual modes are visible in the measured variables. If both occur, the control system can become unstable (spillover instability). [5]

4 System Analysis

The behavior of the control system is reviewed by an analysis with linear models in the time and frequency domain. In each case, the closed-loop system consists of the detailed plant model “8DOF” and a compensator designed with one of the previously described reduced models “3DOF physical” or “3DOF modal”.

Dynamic Behavior of the Control Systems

Fig. 5 depicts the reference step responses of the closed-loop systems for the design of the compensator with the reduced models “3DOF physical” and “3DOF modal”.

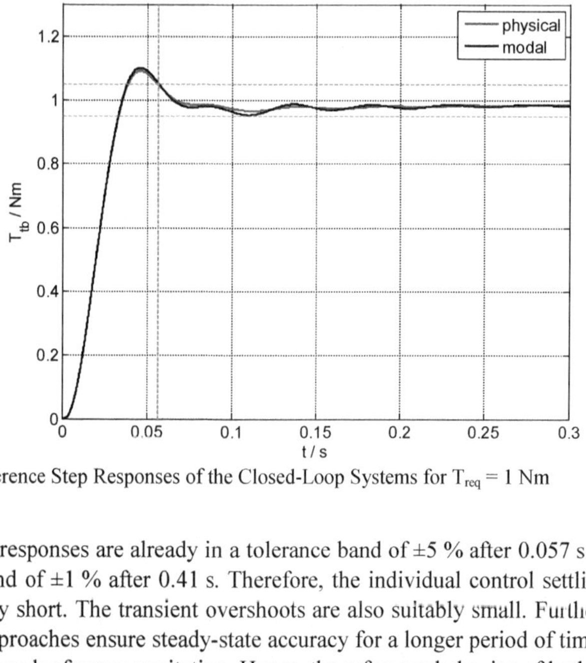

Fig. 5: Reference Step Responses of the Closed-Loop Systems for T_{req} = 1 Nm

Both step responses are already in a tolerance band of ±5 % after 0.057 s and in a tolerance band of ±1 % after 0.41 s. Therefore, the individual control settling times are sufficiently short. The transient overshoots are also suitably small. Furthermore, both control approaches ensure steady-state accuracy for a longer period of time in the case of step-shaped reference excitation. Hence, the reference behavior of both control systems is similarly good.

In comparison, Fig. 6 illustrates the disturbance step responses of the closed-loop systems for a rack force F_R of 1 kN. The disturbance step response of the closed-loop system with compensator for the design model “3DOF physical” shows smaller values than that of the closed-loop system with compensator for the design model “3DOF modal”. This can be explained by the fact that the design model resulting from modal order reduction does not accurately reproduce the stationary gain of this disturbance transfer path. Nevertheless, steady-state accuracy for step-shaped disturbance excitation is achieved with both design models, and both control systems behave well under the influence of disturbances.

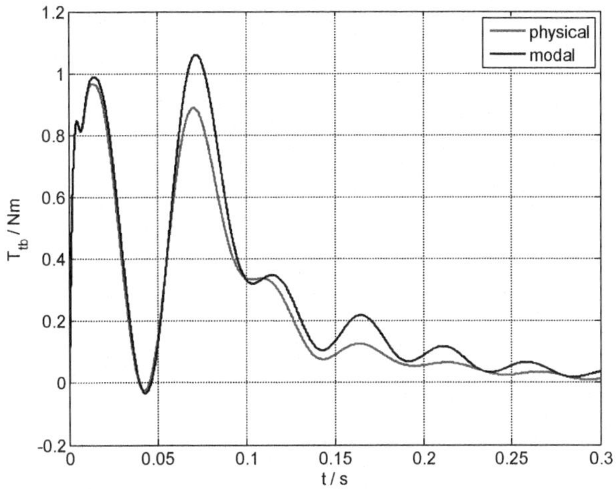

Fig. 6: Disturbance Step Responses of the Closed-Loop Systems for $F_R = 1$ kN

In an additional outer control loop [2], the estimated rack force $\hat{F}_R$ is used to generate a reference steering torque T_{req}, which creates a steering feel for the driver. Thus, the driver is given feedback about the forces and torques acting on the tires and hence about the road conditions.

Robustness of the Control Systems

In this section, the robustness of both control systems will be analyzed. The robustness characteristics can be derived from the Nyquist plots of the open-loop systems which are attained by cutting the closed-loop systems from Fig. 3 at the control input of the plant model. For the given system with only one control input variable i_{ref}, the gain and phase margin can be used to assess the robustness [1]. Fig. 7 shows the Nyquist plots of the open-loop systems from the control input of the plant to the output of the compensator. These open-loop systems consist of the detailed plant model "8DOF" and the state space controllers (LQR) resp. of the plant model "8DOF" and the compensators (LQG) for the design models "3DOF physical" and "3DOF modal".

Based on the frequency response of the open-loop system consisting of the plant model "8DOF" and an LQR-controller, the guaranteed robustness characteristics of an ideal static state space controller can be seen. It has a phase margin of at least 60° and an infinite[1] gain margin. The proximity of the curves of the frequency responses of the open-loop systems consisting of the plant model "8DOF" and either the state space controller or the compensator verifies a sufficient degree of Loop Transfer Recovery (LTR). Thus, the good robustness characteristics of the static state space controller are well approximated with the corresponding dynamic compensator. [7]

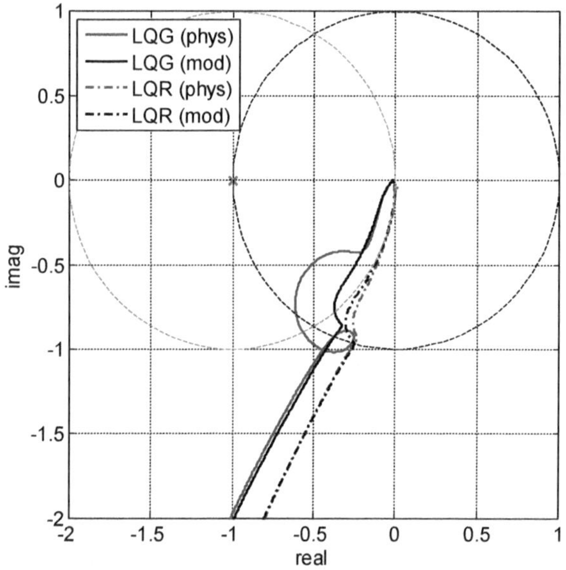

Fig. 7: Nyquist Plots of the Open-Loop Systems

However, the robustness characteristics of the control system with compensator for the design model "3DOF modal" are better than those of the control system with compensator for the design model "3DOF physical". This can also be seen from the sensitivity frequency response of the open-loop systems in Fig. 8. There, the eigenmodes of the plant which are not considered in the design models as well as the parameter uncertainties in the plant model are described as deviations from the design model (stable nominal plant) by a multiplicative perturbation $P_i(j\omega)$. The perturbation $P_i(j\omega)$ can be interpreted as a relative change of a variation of the plant compared to the nominal plant. In addition, an upper limit of the maximum allowed perturbation $P_{i,max}(j\omega)$ can be calculated up to which the closed-loop system is still guaranteed to be stable.

1 The Nyquist plot has no intersection with the negative real axis.

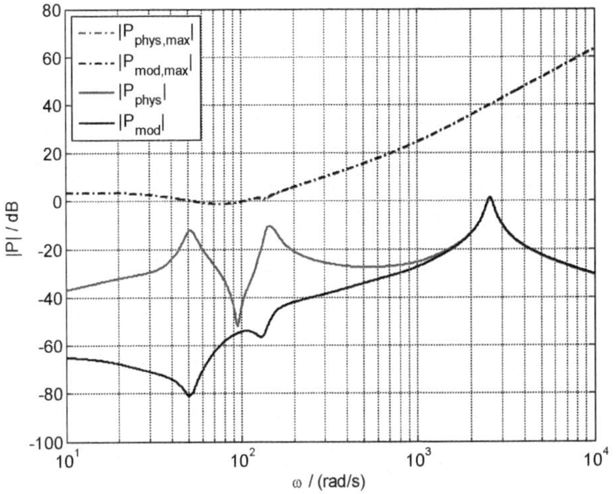

Fig. 8: Sensitivity Frequency Response of the Open-Loop Systems

The absolute value of the multiplicative perturbation $P_i(j\omega)$ is calculated according to [8] and depends on the frequency responses of the open-loop systems

$$|P_i(j\omega)| = \frac{|F_{o,i}(j\omega) - F_{o,i,0}(j\omega)|}{|F_{o,i,0}(j\omega)|} \tag{3}$$

where $i \in \{phys, mod\}$. $F_{o,i,0}(j\omega)$ describes the nominal frequency response of the open-loop system consisting of the nominal plant model "3DOF physical" or "3DOF modal" and a compensator for the corresponding design model. Furthermore, $F_{o,i}(j\omega)$ symbolizes the frequency response of the open-loop system consisting of the aforementioned compensator for the design model "3DOF physical" or "3DOF modal" and the detailed plant model "8DOF". The absolute value of the respective maximum allowed perturbation is calculated as follows

$$|P_{i,max}(j\omega)| = \frac{|1 + F_{o,i,0}(j\omega)|}{|F_{o,i,0}(j\omega)|} \quad . \tag{4}$$

It is equal to the reciprocal of the absolute complementary sensitivity of the nominal plant. Consequently, for perturbation $P_i(j\omega)$ whose absolute value fulfill the condition

$$|P_i(j\omega)| < |P_{i,max}(j\omega)| \tag{5}$$

for all frequencies, stability is guaranteed. This stability prediction is only valid in one direction, since a closed-loop system can be stable although it does not fulfill the condition (5). Therefore, the condition is conservative. [8]

The absolute maximum perturbation $|P_{i,max}(j\omega)|$ is a measure for the allowed difference between the real plant model and the design model, whereas the absolute multiplicative perturbation $|P_i(j\omega)|$ indicates how well the design model actually matches the real plant model. The better the match, the smaller is the value for $|P_i(j\omega)|$.

By comparing Fig. 8 and the frequency response of the detailed plant model "8DOF" in Fig. 2, it can be seen that the minima of the absolute perturbation $|P_{mod}|$ of the control system with compensator for the design model "3DOF modal" are exactly located at the frequencies of the considered eigenmodes. This means that the plots of the nominal frequency response $F_{o,mod,0}$ and the actual frequency response $F_{o,mod}$ are well matching at these frequencies. In contrast, the control system with compensator for the design model "3DOF physical" does not map those eigenfrequencies of the detailed plant model so accurately. Therefore, there are two maxima for the absolute perturbation $|P_{phys}|$ in the low-frequency range. In the high-frequency range far above the bandwidth of the control, both curves of the absolute perturbations of the different control designs show the same maximum as both designs do not consider the high-frequency eigenmodes. Nevertheless, the curves have a sufficiently large distance to the corresponding absolute value of the maximum allowed perturbation in this frequency range so that a satisfactory robustness is still given. In general, the degree of robustness of the control system with compensator for the design model "3DOF modal" is higher.

Those parameter uncertainties in the plant model – which significantly change the dominant eigenfrequencies and notch frequencies considered in the compensator design – are highly critical regarding the robustness characteristics of the control system. However, if the eigenfrequencies and notch frequencies of the design model correspond well with those of the real plant, the control system is extremely robust. Hence, the good robustness behavior shown above can also be achieved by scheduling of the compensator parameters. Nonetheless, the subsequent paragraphs point out to what extent sufficient robustness is given by a variation of plant parameters without such an adaptive control. Fig. 9 to 11 depicts the Nyquist plots of the open-loop systems with variation of the moments of inertia J_W of the wheels, all stiffnesses and the gear ratios between rack and wheels in the plant model. The respective compensator is always designed with the same constant design model parameters. For an increased robustness against parameter uncertainties in the plant model, the compensators were designed with minimum or maximum values instead of mean values of the plant parameters. In [6] it is presented that the robustness characteristics of the control systems are particularly favorable for this choice of design model parameters.

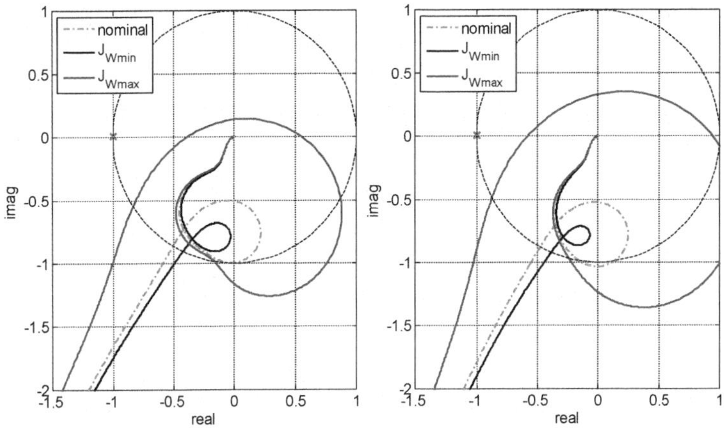

Fig. 9: Nyquist Plots of the Open-Loop Systems with Variation of the Moments of Inertia of the Wheels in the Plant Model ($J_{Wmin} = 0.8$ kg·m², $J_{Wmax} = 1.7$ kg·m²) and Compensator for the Design Model "3DOF physical" (left) and "3DOF modal" (right)

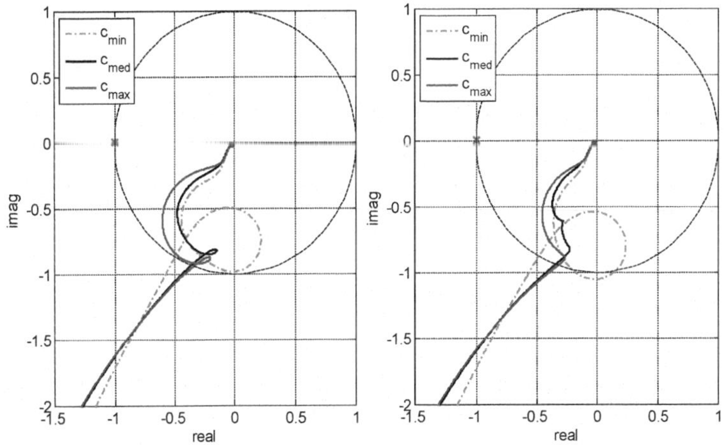

Fig. 10: Nyquist Plots of the Open-Loop Systems with combined load-dependent Variation of the Stiffnesses in the Plant Model and Compensator for the Design Model "3DOF physical" (left) and "3DOF modal" (right)

The Nyquist plots illustrate that the control systems are stable for all tested conditions and that the robustness is always satisfactory. Even for a combined load-dependent variation of the stiffnesses in the plant – for example, as a result of an instantaneous operating point characterized by a specific steering torque as well as external forces and torques acting on the tires – the robustness characteristics of the control systems are good. In the case of a higher steering torque and nonlinear spring characteristics in the real plant, this leads to increased stiffnesses for the springs in the linearized model, which is displayed in Fig. 10 for a small, medium and large steering torque with the corresponding stiffnesses c_{min}, c_{med} and c_{max}. In this way, the nonlinear system characteristics of the plant are accounted for the analysis with linear models.

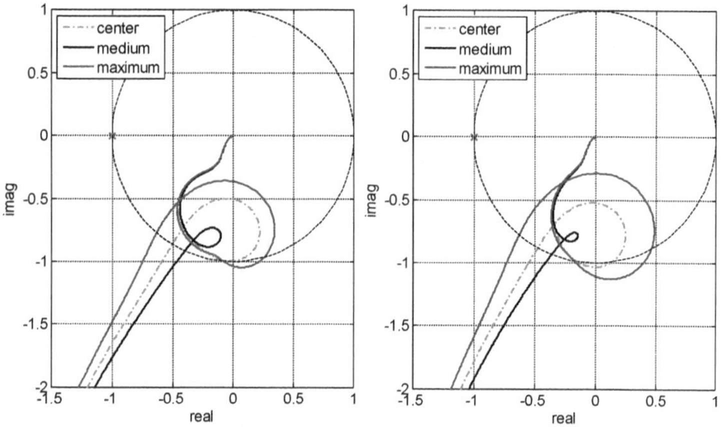

Fig. 11: Nyquist Plots of the Open-Loop Systems with Variation of the Gear Ratios between Rack and Wheels in the Plant Model (not, medium and maximum deflected rack) and Compensator for the Design Model "3DOF physical" (left) and "3DOF modal" (right)

In general, the robustness characteristics of the control system with compensator for the design model "3DOF modal" against parameter uncertainties in the plant model are slightly better than those of the control system with compensator for the design model "3DOF physical". This has already been concluded from Fig. 7 and 8. The parameter changes described here are a selection of the performed parameter variations for the detailed plant model "8DOF". All corresponding control systems with unchanged parameters for the compensators show similar robustness characteristics. Therefore, these results are not presented in this paragraph.

5 Conclusion and Outlook

In this article, two steering torque LQG-controls for EPS systems were presented and analyzed. The control systems remain stable for all tested operating points. The high robustness against parameter variations in the plant which is required for the steering control was achieved in the observer design by means of the method of Loop Transfer Recovery (LTR). The feedback of the state variables of the plant and the compensation of the effect of the disturbance variables acting on the plant lead to a good control and disturbance behavior. Due to the active damping of elastic vibrations in the plant, the controlled variable T_{tb} is settled quickly and precisely to the specified reference variable T_{req}, also during dynamic steering operations. Even if other measured variables are used (e.g. motor angle φ_M instead of motor angular velocity Ω_M), a similar behavior as described above is achieved. Thus, the presented control approaches fulfill the requirements for a modern steering system in terms of comfort and safety. In the next step, these approaches will be transferred to steer-by-wire systems.

References

[1] H. Henrichfreise, J. Jusseit, H. Niessen: „Optimale Regelung einer elektromechanischen Servolenkung“, VDI Mechatroniktagung, Fulda, 2003

[2] O. Graßmann, H. Henrichfreise, H. Niessen, K. v. Hammel: „Variable Lenkunterstützung für eine elektromechanische Servolenkung“, Elektronik im Kfz, Haus der Technik, Stuttgart, 2003

[3] M. Haßenberg, G. Wittler, H. Briese, T. Schubert, H. Henrichfreise: „Systematische modellbasierte Schwingungsanalyse einer elektrischen Servolenkung“, AUTOREG, VDI Verlag, Berlin, 2017

[4] G. Wittler, M. Haßenberg, H. Briese, T. Schubert, H. Henrichfreise: „Systematic model-based vibration analysis of a controlled electric power steering system“, 8th International Munich Chassis Symposium, Springer Vieweg, München, 2017

[5] M. J. Balas: „Active Control of Flexible Systems”, Journal of Optimization Theory and Applications, Plenum Publishing Corporation, New York, 1978

[6] M. Haßenberg: „Auswirkungen parasitärer Eigenbewegungen auf die Regelung einer elektrischen Servolenkung”, Köln, 2018

[7] H. Henrichfreise: „Prototyping of a LQG Compensator for a Compliant Positioning System with Friction”, TransMechatronik, HNI-Verlagsschriftenreihe, Paderborn, 1997

[8] T. Hrycej: „Robuste Regelung – Ein Leitfaden für sicherheitskritische Anwendungen”, Springer Vieweg, Lorch, 2018

Further understanding of steering feedback and driver behaviour through the application of an instrumented steering wheel

Xabier Carrera Akutain, Kimiaki Ono,

TOYOTA MOTOR EUROPE, Brussels, Belgium

Francesco Comolli, Massimiliano Gobbi, Giampiero Mastinu

POLITECNICO DI MILANO, Milan, Italy

Introduction

Vehicle manufacturers try to enhance driver to steering dynamics as a key product power element for their customers. In parallel, on the long path to fully autonomous driving, these dynamics will become more complex in the interaction with driving aid systems and as driver's engagement gets more intermittent.

Steering dynamics are traditionally analysed with torque, angle and their differentiated signals in open loop manoeuvers, relying on the accuracy and repeatability of steering robots [1]. Non-linear nature of human driver is studied conducting psychophysical experiments in order to determine the relationship between physical and perceived values in [2]. Previous work shows how those established metrics combined with simulation tools and optimization techniques can be combined in a streamlined virtual process for an important front loading of the EPS calibration tasks [3].

In reality, drivers act on the steering wheel to control the vehicle by applying forces and moments in all the directions and not only tangentially to rotate the steering wheel and direct the vehicle. The drivers' actions are disturbed by the reaction forces from the steering wheel and the inertial forces due to the vehicle dynamics, forcing the driver to grab and push/pull the steering wheel in order to counteract these disturbing forces [4].

Understanding driver behaviour on closed loop condition becomes however complicated and usually there's limited amount of recorded information of the driver's behaviour. The driver's body motion to support the steering wheel control has been studied in terms of displacements, velocities, and accelerations of defined body segments [5], or by visualizing the lines of action of the forces applied by the drivers [6]. In [7], the authors introduced a biomechanical driver model which takes into account the muscle co-activation to consider the driver response to the steering torque feedback. In the presented model the driver acts on the steering wheel rotation to control the vehicle. By using the information on the forces exerted by the driver, innovative biomechanical driver models allow to consider more realistically the actual driver behaviour. A different approach has been described in [8] in which authors used an inverse dynamic analysis, by adopting a multi-body model of the driver, calculating the body motion once the reaction forces were measured by several sensors.

In the literature there are few instrumented steering wheels which allow obtaining complete information on the loads applied. In [9], the authors fit a steering wheel on a six axis load cell on a static driving simulator and calculate the three force components applied by each hand. In [10], the authors split a commercial steering wheel in three sectors and connect each one at the steering hub by means of instrumented supports to measure the forces and moments applied on each sector. This steering wheel has been used on a static driving simulator to study the driving behaviour of disabled people.

In this work, we introduce an Instrumented Steering Wheel (ISW) which measures the grip force and the six components of force and moment for each hand, providing additional insight on the driver control loop during steering task actions. The ISW has been used in [11], [12] and [13] to identify some basic force patterns among different drivers in different steering situations. The ISW, in addition, allows measuring the grip strength with which the driver holds the steering wheel. The fictitious forces due to the steering wheel rotation and the inertia to the vehicle dynamics are compensated in the data acquisition software, making possible to use such a instrumented steering wheel both on static driving simulators and on real vehicles running on test tracks. The ISW allows a much deeper exploration and understanding of the driver to steering loop and allows a more human centric approach to the improvements of steering control & feel for new vehicles.

In the paper, the information on the loads exerted by the driver are used to rank the steering behaviour of three different vehicles, evaluating the differences while changing the vehicle setup in terms of chassis and steering configurations. By analysing the trend of the forces during specific handling tasks, we identify force-based metrics closer to the human haptic feedback and action, hence enabling a richer assessment of the handling qualities.

Experimental Setup

Three different vehicles (A, B, C) in different configurations have been tested in order to have a wide range of steering behaviours. Each of the three vehicles was tested in at least two different setups by changing tyres, electric power steering (EPS) maps and chassis settings. Vehicle A was tested switching the vehicle chassis and EPS settings from "normal" (A1) to "sport" (A2), vehicle B was tested by modifying EPS parameters from the original one (B1) to a modified map (B2). The last vehicle was tested with three different tyres sets, the original summer tyres set (C1) and two different sets of all-season tyres (C2,C3).

Two different type of manoeuvres were analysed: the first is an on-centre sinusoidal manoeuvre varying the steering wheel angle up to ±5°, with a frequency of 0.5 [Hz], running at constant speed (100 kph).

A professional driver tested each vehicle and performed the subjective evaluation too. The on-centre sine inputs were performed in several runs for each vehicle configuration. The same type of manoeuvre was performed by holding the steering wheel with one hand. The second type of manoeuvre analysed was a selection of corners on a handling track, considering both right and left corners with curvature between 0.01 and 0.025 1/m, performed at an approximate steady state lateral acceleration value of 5 m/s^2, aiming for dynamic yet consistent manoeuvres. Two laps were performed for each vehicle setup.

The Instrumented Steering Wheel (ISW) has been installed on the three vehicles to measure the forces exerted by each driver's hand. The ISW is shown in Figure 1. The two six-axis load cells, the two handles and the electronic box are highlighted. The two handles are the only sensitive sectors of the steering wheel.

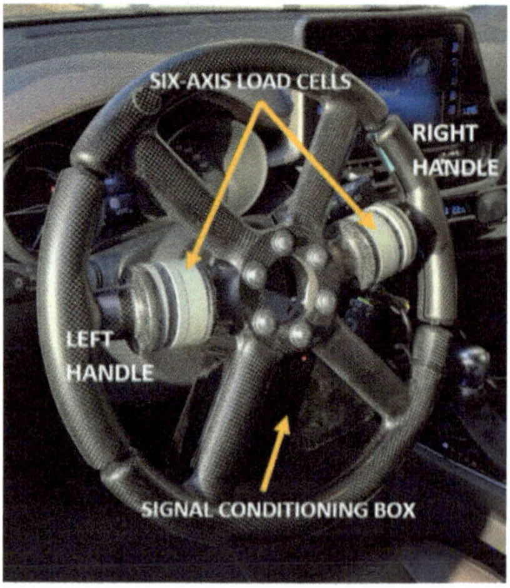

Figure 1. Instrumented Steering Wheel.

The two six-axis load cells are instrumented with twelve strain gauges each, connected to form half Wheatstone bridges in order to compensate the thermal deformation and double the sensitivity. The output voltages of the strain gauges bridges $(\underline{\Delta V})$ are linked to the three force components and the three moments $(\underline{F})$ applied at the load cell centre by an experimental calibration matrix $[C_b]$ as in equation (1):

$$\underline{F} = [C_b]\underline{\Delta V} \tag{1}$$

The output signals are sent to an electronic board on the steering wheel, which digitally processes the signals by calculating the load components applied at each load cell and sends the data via a CAN bus to a data logger.

Each handle accommodates three mono-axial load cells (Figure 2) to measure the grip strength F_{grip} according to Eq. (2), where $F_{x_{six-axis\ load\ cell}}$ is the force measured by the six-axis load cell in the ISW axial direction and F_i are the forces measured by each of the mono-axial load cells in the ISW axial direction.

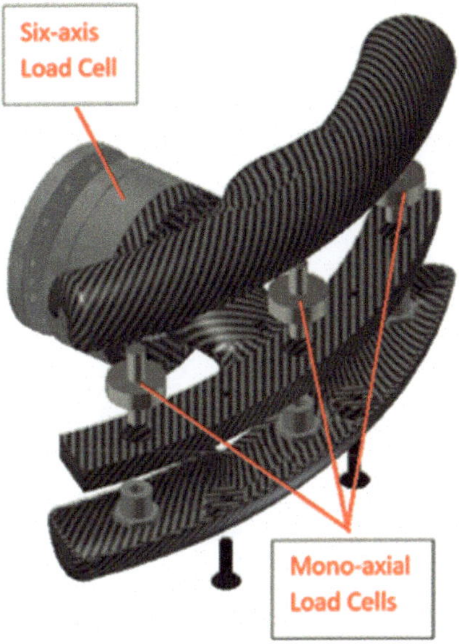

Figure 2. Grip strength meas.

$$F_{grip} = \min\left(\left|F_{x_{six-axis\,load\,cell}} - \sum_{i=1}^{3} F_i\right|, \left|\sum_{i=1}^{3} F_i\right|\right) \tag{2}$$

The fictitious force components due to the rotation of the ISW and the inertia of the handles subjected to the vehicle accelerations while driving have been compensated. The forces applied to the ISW by the driver's hands $(\vec{F}_{applied})$ are calculated as in equation (3), where $\vec{F}_{measured}$ are the forces measured by the load cells, $\vec{F}_{weight}$ are the forces due to the mass of the handles (m_H) and function of the ISW rotation angle (δ), $\vec{F}_{rotation}$ are the fictitious forces which rise from the ISW rotation, and $\vec{F}_{inertial}$ are the inertial forces because of the vehicle accelerations$(\vec{a}_v)$.

$$\vec{F}_{applied} = \vec{F}_{measured} - \vec{F}_{weight} - \vec{F}_{inertial} - \vec{F}_{rotation}$$

$$\begin{cases} \vec{F}_{weight} = \Lambda_{weight}(\delta, \beta) \cdot m_H \cdot g \\ \vec{F}_{rotation} = -\, m_H \underline{\ddot{\delta}} \wedge \underline{r_h} - m_H \underline{\dot{\delta}} \wedge \left(\underline{\dot{\delta}} \wedge \underline{r_h}\right) \\ \vec{F}_{intertial} = -m_H \vec{a}_v \end{cases} \tag{3}$$

where $\underline{r_h}$ is the distance from the load cell centre and the ISW centre.

In Table 1 the technical specification of the Instrumented Steering Wheel is reported.

Table 1 – Technical Specifications of the ISW.

Parameters	Value
Mass	2.7 [kg]
Diameter	375 [mm]
Moment of inertia around rotation axis	<0.030 [kg m^2]
First natural frequency	<240 [Hz]

Table 2 shows the measuring performance.

Table 2 – ISW Measuring Performance.

Load	Rated Capacity [N] – [Nm]	Resolution [N] – [Nm]	Accuracy [%FS]	Linearity Error [%FS]
F_x	250	0.01	±0.29	0.15
F_y	250	0.01	±0.51	0.24
F_z	250	0.01	±0.31	0.16
M_x	25	0.001	±0.26	0.12
M_y	25	0.001	±0.18	0.09
M_z	25	0.001	±0.26	0.12

Each vehicle has been equipped with an Inertial Measuring Unit to collect all the kinematic quantities i.e. spatial and angular positions, velocities and accelerations (longitudinal, lateral, vertical, yaw, pitch and roll). All of these kinematic quantities are used to compensate the inertial forces on the steering wheel due to the vehicle dynamics. In addition, the data from the vehicle embedded sensors as the steering wheel angle and steering torque have been collected via the vehicle CAN bus.

On-Centre regime

During a typical drive on motorway, the driver operates the steering wheel on the on-centre regime most of the time, with relatively low amplitude and frequency values of steering torque, angle and vehicle motion. An ISO standard exists to define the required tests [14], and extensive research has been conducted to study the objective evaluation of on-centre steering [15-16].

However, partially because of inherent signal to noise ratio limitations and mainly because of sensitivity to environment conditions, the study of this area becomes especially challenging. The driver's action is marginally above the required effort to counteract the steering rack forces that are coming from road camber and irregularities, lateral wind fluctuations etc.

In this work, we perform several runs of weave tests for all the vehicle configurations described in the section above in order to better understand the driver behaviour. Several runs are performed for each setting. Every semi-period is cut and processed to fit within certain thresholds of length, absolute and rate values. The entire cut data set is shown in Figure 3.

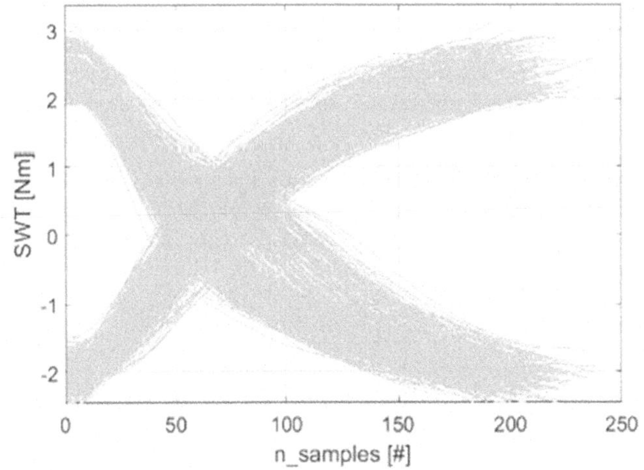

Figure 3. Sine weave test. Cut data

Figure 4 shows the averaged Lissajous curves for a frequency of 0.5Hz and approximate steering angle amplitude of ±5deg. Positive values are result of turning the steering wheel clock-wise, while negative values correspond to anti-clockwise rotation of the wheel. The steering wheel torque (SWT) is on the x-axis, while the steering wheel angle (SWA) and yaw velocity (YV) are represented on the y-axis.

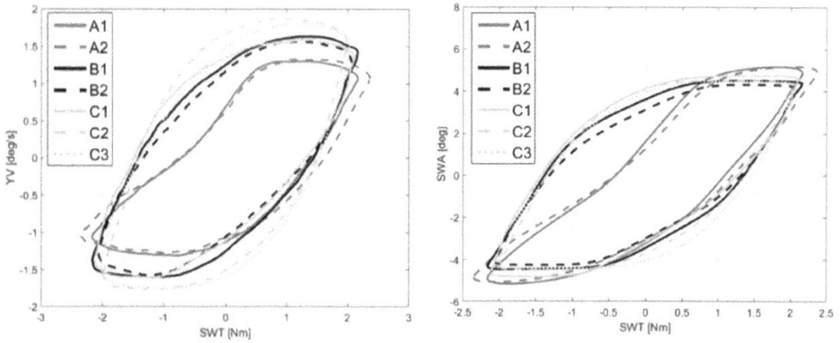

Figure 4. Averaged Lissajous curves for 0.5Hz ±5deg input

Vehicle A shows less hysteresis than B and C. However, A2 presents a clear asymmetry in terms of convexity for left and right turning. Vehicle A presents slightly smaller response, difference that the driver is not able to perceive. The EPS modification of vehicle B2 shows a hysteresis reduction with respect to B1. Vehicle C shows roughly similar curves to those of B, showing the C2-C3 configurations with all-season tires slightly higher hysteresis at the steering wheel and especially a more de-phased steering response, as seen on the SWT-YV curves.

The asymmetry of vehicle A can be explained by Figure 5. In the case of dataset A1b, the vehicle presents evident asymmetry. While both runs 'a' and 'b' are performed on a dynamic platform with a 1 deg side slope, the driver tends to give more input to the right hand side for case 'b', which has as a consequence that the vehicles heads slightly more against the slope.

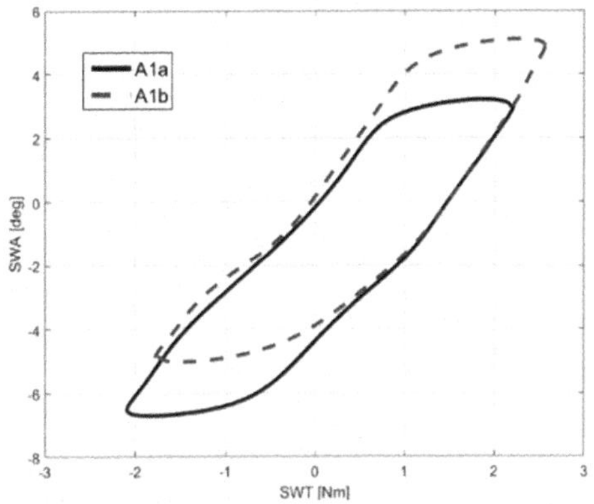

Figure 5. Lissajous curves variability

The sequence of a left to right manoeuvre is displayed taking vehicle A2 as example in Figure 6. Forces are represented for both hands in the shape of red arrow (right hand) and blue arrow (left hand) at each handle. Projections of the forces on the three orthogonal planes are added for a better visualization. The translational offset of the vectors is proportional to the wrist moments in the three directions. In addition, the handle grip on each hand is represented with a dot on the XZ plane, being the dot diameter proportional to the grip level.

a) Start of manoeuvre Leftmost angle. FRZ and FLZ are holding the torque feedback rising from the rack forces. Driver is in addition pushing the wheel inwards longitudinally towards the dashboard with his right hand. Right hand X,Z and grip force values are dominant with respect to left hand.	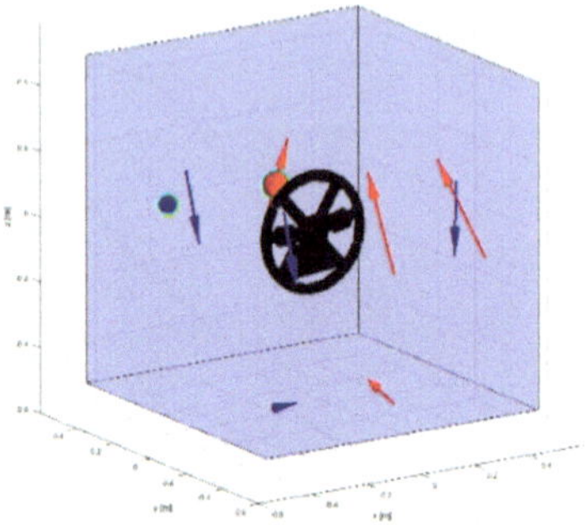
b) Instant of Torque = 0 With all the forces presenting very small values near zero.	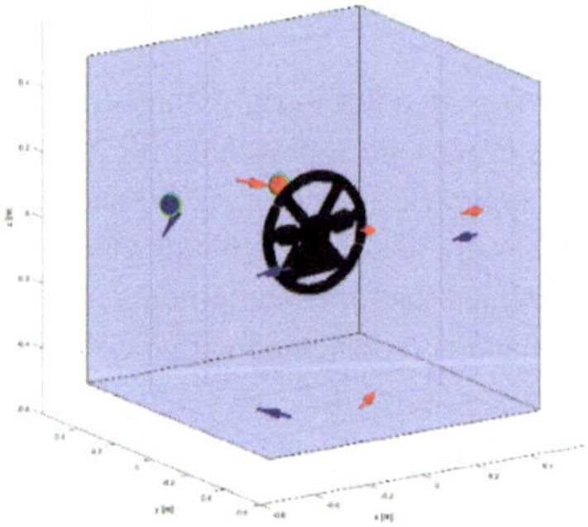
c) End of left to right action Rightmost angle, with both hands exerting similar tangential forces but rather opposite axial forces, right hand pulling the wheel towards the driver, left hand pushing towards the dashboard. Left hand in addition presents a higher final grip and a lateral force leftwards.	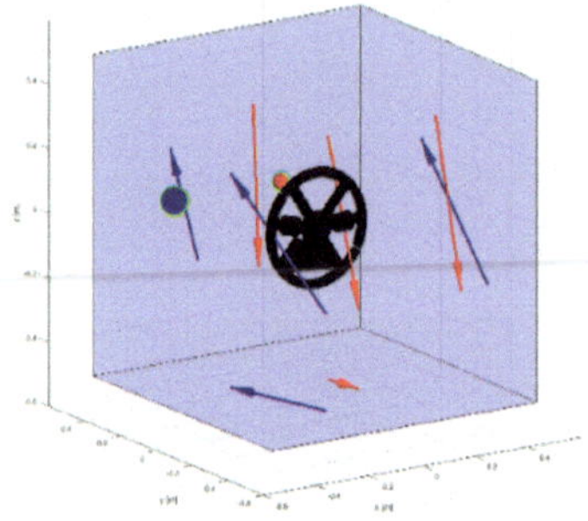

Figure 6. Weave input forces sequence

The test driver is requested to perform the task and to assess the different vehicles on a separate driving exercise, since the simultaneous cognitive task of assessing the subjective qualities prevents the driver from performing the objective tests consistently.

The subjective ranking (from best to worst) of the vehicles for on-centre rigidity and connection feel is A1/A2 >> C1/C3 > C2/B2 > B1.

One entire datarun is plotted for vehicle A2 (shadowed in grey) in Figure 7. The forces are represented in the ISW reference frame, being X axis positive towards the driver for both handles. The solid thick line represents the average for the left to right steering manoeuvres, while the dashed one corresponds to the right to left sense of rotation. Except for few signal spikes, dispersion is small and indicates that the driver keeps the consistency during the entire run (length ~30s) provided that sporadic path correction manoeuvres are washed out in the post-processing.

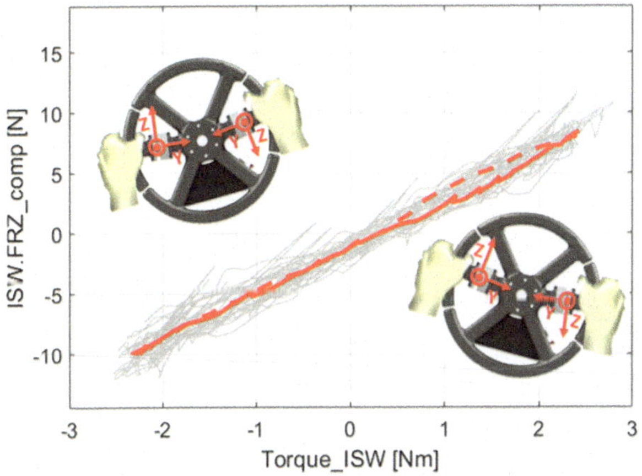

Figure 7. Right hand tangential forces Vs/ measured torque.

When observing the averaged tangential forces for each run of vehicle A2, it's obvious that left hand presents more symmetric behaviour in the tangential direction than the right hand. It can be observed as well that the driver, who is right handed, applies systematically around two times more tangential force with the right hand (see Figure 8). In addition the inclination to the right on the dataset 'b' of Figure 5 is explained by further shift of the right hand to promote pulling force (downwards) than pushing one (upwards), on another example of bias/trait which comes imperceptible even to expert drivers and can strongly influence the data analysis.

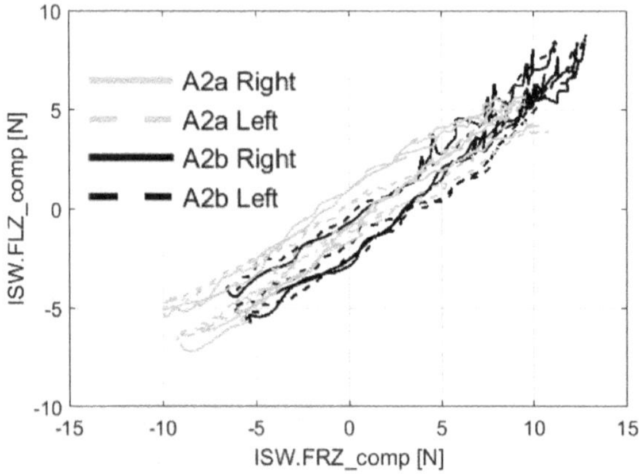

Figure 8. Tangential forces ratio, vehicle A2

When plotting all the vehicles data (Figure 9), it can be noticed that some vehicles present higher offset tangential forces at zero torque than others, indicating that the driver doesn't achieve a complete relaxation state.

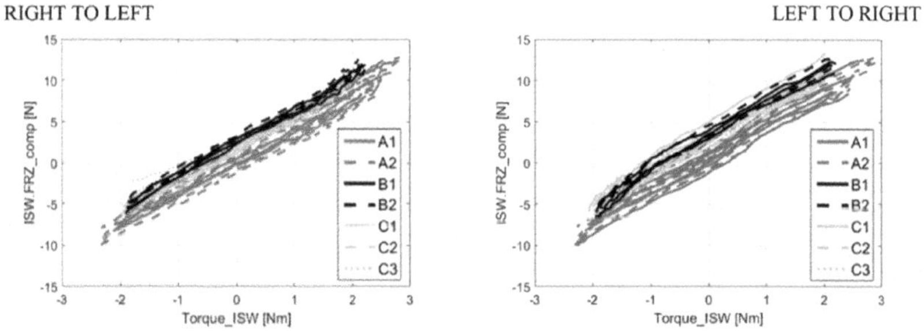

Figure 9. Right hand tangential force as function of the steering torque.

Our hypothesis for the on-centre steering quality is that at zero torque crossing, small amount of tangential forces indicate to the driver a clear separation between passive and active phases, or in other words, a clear end of steering return and start of the "connection" phase feedback. This connection phase, highlighted in Figure 10, is defined as the period between the values of zero torque and zero lateral acceleration. The connection can be further split into two smaller sections if we consider the zero angle crossing. During a normal drive, on the small corrections around centre, the connection or in-between phase by definition is a period of de-phase between action (torque) and vehicle response which needs to be transitioned, therefore its quality and subtle feedback characteristics will highly determine the overall perception of steering feel by the driver.

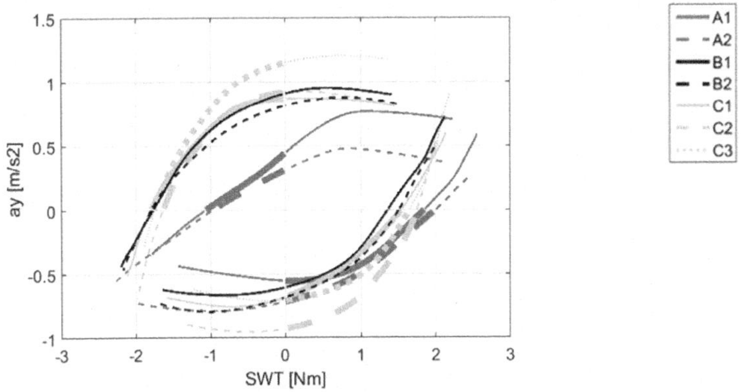

Figure 10. Connection phase on the Lissajous curve

A metric is proposed after plotting all the tangential forces (see Figure 11), consisting of the offset of the measured forces with respect to the origin. For that purpose, let (x,y)∈C where C is the curve defined by both tangential forces ISW.FLZ_comp, ISW.FRZ_comp. The shortest distance from the curve to the origin [0,0] will be given by the minimum of the Euclidean norm $||(x, y)||_2$

Table 3 – Weave test metrics

	A1	A2	B1	B2	C1	C2	C3
mean	1.0	1.1	3.0	3.9	2.2	2.1	1.6
std	0.7	0.7	0.5	0.5	1.1	0.9	0.4

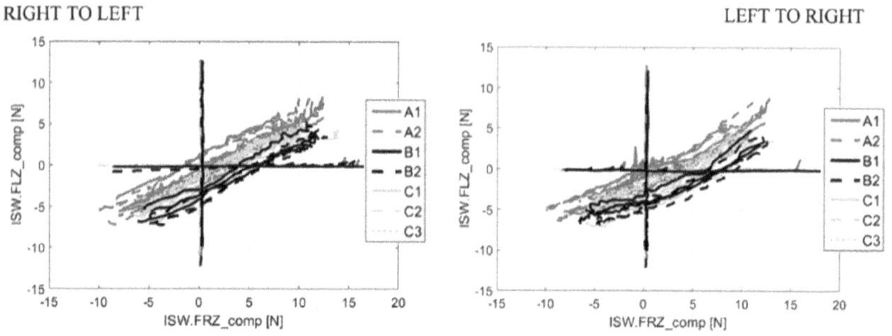

Figure 11. Steering wheel tangential forces.

Quasi-vertical and horizontal lines represent data performed single handed, with the right hand presenting asymmetric behaviour, by pulling (steer to the right) more than pushing (when steering leftwards).

Trying to find a correlation between subjective ranking and this proposed metric can be considered simplistic as it ignores other qualities such as total effort and possibly steering wheel return velocity but it's deemed to be an interesting indicator of the on-centre feel on this exploration of driver-vehicle interaction.

The moments around the steering wheel X axis which create effective steering torque in addition to the moment created by the tangential forces Fz are plotted in Figure 12. In this case, the subjectively better rated vehicles present higher offset on the momentum that the wrists exerts on the handle, roughly cancelling each other. This may indicate that the preferred behaviour from the driver is a transition with a relaxation of forces without fight between them and slight "control" momentum to provide continuity of the steering action. As the neuromuscular mechanisms of arm and hand are different, this seems to open an interesting field of research.

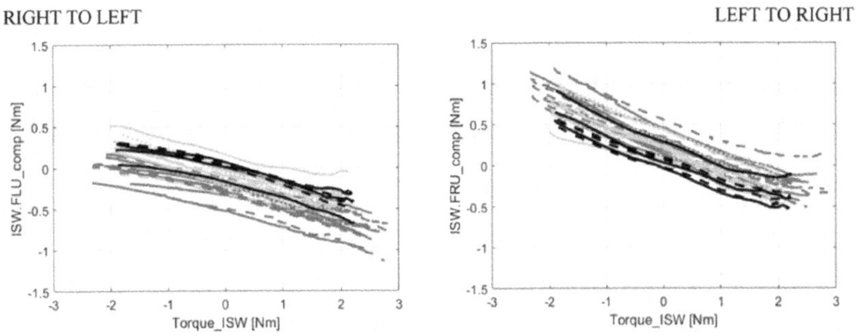

Figure 12. Steering wheel wrist moments.

Concerning the forces on the longitudinal axis (see Figure 13), partial trend is observed between vehicles A and C, having A more pushing force than C. However, the variability is bigger if we include vehicle B, therefore there's no evidence if "pushing" strategy obeys to a natural reaction mechanism or to an arbitrary decision from the driver. While the third quadrant is the most populated, hence both hands pushing towards the dashboard, eventually second and forth quadrants are used too, which can be explained by a pivoting effort around the vertical axis.

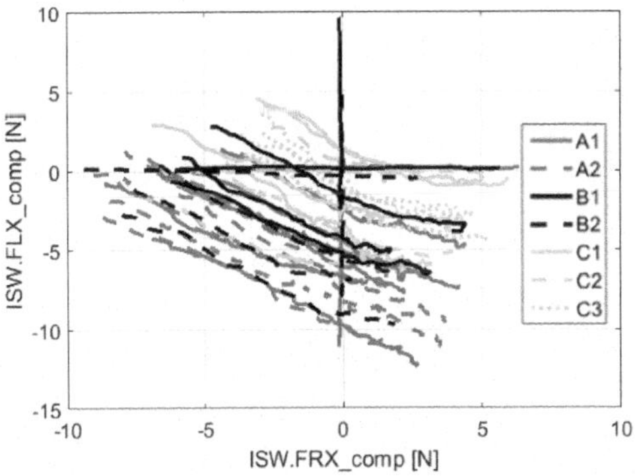

Figure 13. Longitudinal force ratios.

The grip levels in Figure 14 show clear patterns of grabbing and relaxing through the sine manoeuvre but differ significantly between runs. The effect of the intrinsic stiffness and damping of the limbs on the stretch reflex activation and overall steering control performance has been investigated by Cole [17] with an impact on 'reactive' steering control tasks. The intra-subject variability for the small sinusoidal inputs presented and the assumption of no significant reflex loop activation may indicate a low importance of the hand grip level in this regime of inputs.

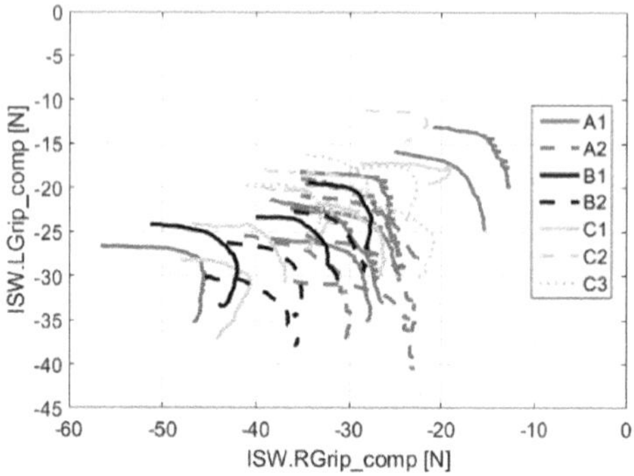

Figure 14. Grip forces.

Handling Track manoeuvres

The selection of corners highlights a very important subjective quality for the expert evaluators and presents little correlation with the standard metrics. The driver feels that with a good vehicle he can "lock" the steering wheel with confidence after the main input into the corner, followed by the required corrections to follow the intended trajectory. The professional driver is familiar with the test track and he tries to consistently follow the same trajectory on all the laps.

A typical example is shown on Figure 15 for corner '2', which is a right turn. The torque values have been magnified ten times for visualization purposes. Right (inner) hand initiates the tangential force and longitudinal moment build-up, followed by the left (outer) hand contribution, which is roughly weaker on all the corners for the right-handed subject. There's a notorious correspondence between certain forces and moments, e.g. FRX & FRW due to handle and load cell disposition. We focused our attention at the end of the steer-in phase towards holding. During this phase, the steering wheel torque stabilizes and the steering angle is almost constant, until the first torque relaxation happens. The thicker width highlights a 0.25s window starting at the end of the steer-in phase, computed as the noise-resistant local torque maximum, aiming to capture the steering stabilization phase suspected to be key on driver's perception of steering settlement.

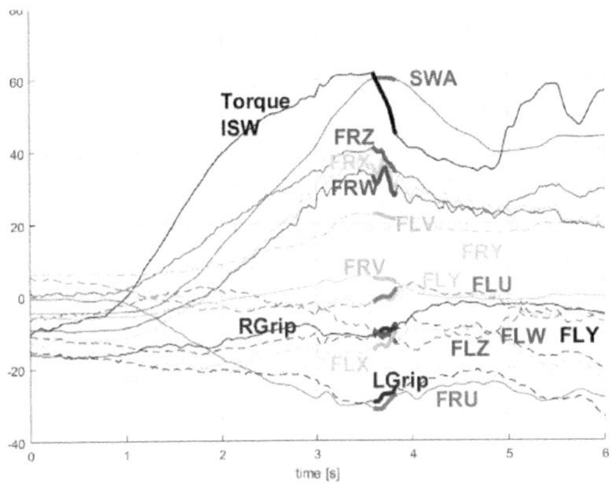

Figure 15. Vehicle response on handling track

The following results correspond to corner 2 and corner3, respectively right and left turns with a final steady lateral acceleration in the range of 5 m/s². The measured vehicles show similar response at corner entry, see Figure 16, eliminating eventual effects originated from very different vehicle dynamics. The highlighted portion is concentrated on quasi-steady steering wheel angle.

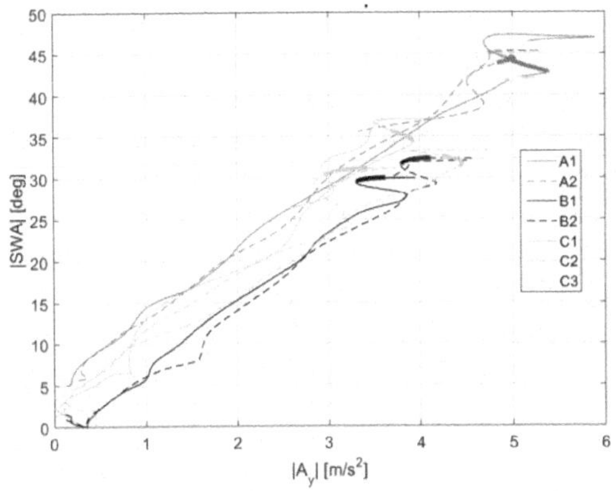

Figure 16. Vehicle response on handling track

The professional driver perceives vehicle A as "easy to settle" and vehicle B as the worst between the tested ones. From best to worst the ranking is: A2>A1>>C1>C3>C2/B2>B1

The plots in Figure 17 show the steering force at the right and left hand (in red) and the resultant of the forces applied by the driver (in blue) measured by the instrumented steering wheel for the best (A2) and worst (B1) vehicles.

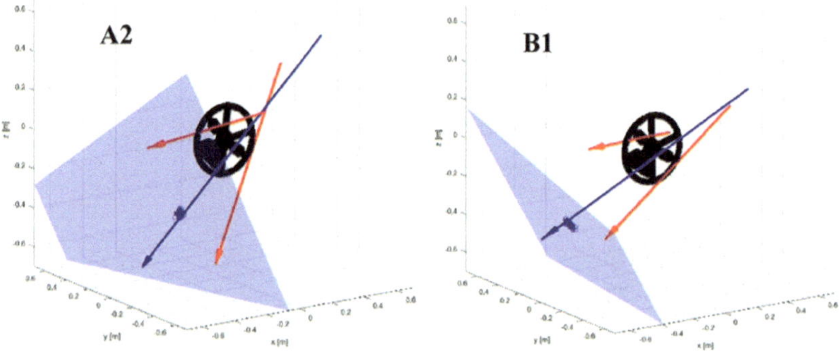

Figure 17. Corner 2. Steering forces resultant at stabilization phase

The blue circles show the migration of the intersection points of the resultant of the forces applied by the driver to the steering wheel with a plane orthogonal to the direction of the average resultant force (distance predefined to 0.5m). Vehicle A1 shows a smaller portion of the plane intersected by the resultant of the forces, while vehicle B2 shows a large intersection area on the orthogonal plane. The performance index σ_I (lower the better) is defined by evaluating the standard deviation of the distances of the intersection points with respect to the average intersection point; see Figure 18 for further details.

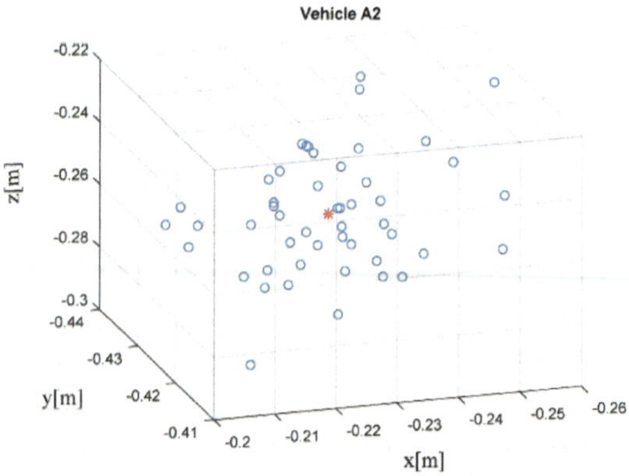

Figure 18. Intersection points and average intersection point. Curve 2, Vehicle A2.

The "steering quality" index σ_I can be used to rank the different vehicles and setups. The results for corner 2, are reported in Table 4. Only the data of the vehicles and setups showing consistent results between first and second lap are reported in the table (i.e. the difference between σ_I computed on two laps has to be smaller than 0.005), as differences may occur when e.g. driver miscalculates the required input and incurs in overshoot leading to early corrections.

The proposed metric is able to correctly reproduce the subjective ranking of the professional driver (A2, A1, C1, C2, B1, from best to worst).

Table 4. Handling Track, Corner 2. Comparison on the basis of the dimension of the set of intersection points of the resultants of the forces applied by the driver to the steering wheel on a reference plane.

Vehicle	Lap 1 – σ_I	Lap 2 – σ_I	**σ_I (averaged)**
A1	0.0095	0.0127	**0.0111**
A2	0.0093	0.0099	**0.0096**
B1	0.0149	0.0124	**0.01365**
C1	0.0128	0.0117	**0.01225**
C2	0.013	0.0121	**0.01255**

The results remain consistent when considering another section of the test track. The left turn 3 is shown on the following table:

Table 5. Handling Track, Corner 3. Comparison on the basis of the dimension of the set of intersection points of the resultants of the forces applied by the driver to the steering wheel on a reference plane.

Vehicle	Lap 1 – σ_I	Lap 2 – σ_I	**σ_I (averaged)**
A1	0.012	0.0147	**0.0133**
A2	0.0109	0.0127	**0.0118**
B1	0.017	0.016	**0.0165**
B2	0.014	0.019	**0.0165**
C1	0.0149	0.0155	**0.0152**

Discussion and Conclusions

The proposed Instrumented Steering Wheel (ISW), which is able to measure the grip force and the six components of force and moment at each hand, has allowed to provide more insight into the driver behaviour while performing steering tasks. Some force patterns of steering operation have been identified and new performance metrics have been developed.

The steering behaviour of three different vehicles and different setups has been analysed. The newly developed force/moment based metrics are closer to the human haptic feedback and action, potentially enabling a richer assessment of the handling qualities.

There are some limitations in this study that should be mentioned; a single driver evaluated the vehicles dynamic behaviour and just a few track laps were available for handling track analysis. Although the results are not statistically significant due to the limited number of observations, they may still provide some insights for further study and can be used to define a first quantization of the steer-in stabilization phase by an enhanced monitoring of driver's action made possible by properly processing the forces/moments measured by the ISW.

The driver feels that while driving a good vehicle he can "lock" the steering wheel with confidence after entering the corner. The hypothesis underneath is that the driver-vehicle interaction is reflected on the force balance applied by the driver and certain vehicle characteristics lead to a different degree of 'neuromuscular hesitation', which cannot be learnt by the driver. The high sensitivity of the handling dynamics context and the different adaptation strategies and cognitive states that the drivers manage makes the subjective assessment a complex and often confrontational task that we are trying to understand by better capturing the driver action. For this reason, focus was set on the settlement of the end of the steer-in phase towards holding. The migration of the intersection point of the resultant of the forces applied by the driver to the steering wheel has proved to be an interesting metric for the steering quality, being able to correctly reproduce the subjective ranking of the professional driver.

Regarding the on-centre steering quality evaluated during steering weave tests, small amount of tangential forces at zero torque crossing at the driver's hands may indicate a clear end of the steering return and the beginning of the "connection" phase. The quality of this phase determines to big extent the driver's overall perception of the steering feel. A metric based on the offset of the measured tangential forces with respect to the origin has shown good correlation with the subjective ranking. Moreover, the subjectively better rated vehicles present higher offset on the moment that the wrists exerts on the handle. As the neuromuscular mechanisms of arm and hand are different, this seems to open an interesting new field of research.

Future studies will focus on the evaluation of the level of reliability of the proposed metrics that can be considered a starting point for enhanced monitoring of the driver's action based on the forces/moments measured by the ISW, considering inter-subject variability and different body motion strategies applied.

References

[1] Kushiro I., Yamamoto M.: Vehicle Behavior Under the Influence of Steering Dy-namics by Means of Low Frequency Torque Input, SAE Technical Paper 2006-01-0557, (2006)

[2] Ono K., Kushiro I., Yamada D., Tomita T.: Steering feel design based on the driver's perceived value of steering reaction force. Proceedings of the 7th International Munich Chassis Symposium, 595-616 (2016)

[3] Carrera Akutain X., Ono K., Ocariz A., Tsitouridis G., Iwamatsu S.: Development of a steering characteristics optimization process and its implementation on vehicle projects. Proceedings of the 9th International Munich Chassis Symposium, 573-596, ATZ chassis.tech plus (2018).

[4] Abe, M. Vehicle Handling Dynamics, Butterworth-Heinemann (2015).

[5] Makita, M.; Kageyama, I. Experimental analysis of driver's head motion on vehicle maneuver from the point of driving pleasure. Vehicle System Dynamics, 44.sup1: 632-640 (2006). https://doi.org/10.1080/00423110600882712

[6] Hada, M.; Yamaguchi, T.; Hattori, Y. Visualization of Driver's Body Motion by Instantaneous Screw Axis and Line of Action. Transactions of Society of Automotive Engineers of Japan, 46.1 (2015). http://dx.doi.org/10.11351/jsaeronbun.46.161

[7] Hoult, W.; Cole, D. J., A neuromuscular model featuring co-activation for use in driver simulation. Vehicle System Dynamics, 46.S1: 175-189 (2008). https://doi.org/10.1080/00423110801935798

[8] Hada, M.; Yasuda, M.; Kobayashi, T.; Yamaguchi, T.; Hattori, Y. Inverse Dynamics Analysis of Driver-Seat-Steering System using Multibody Human Model. In: Proceedings on ECCOMAS Thematic Conference on Multibody Dynamics (2015).

[9] Pick, A.; Cole, D. J. Neuromuscular dynamics and the vehicle steering task. University of Cambridge, Department of Engineering (2003).

[10] Gabrielli, F.; Pudlo, P.; DJemai, M. Instrumented steering wheel for biomechanical measurements. Mechatronics, 22.5: 639-650 (2012). https://doi.org/10.1016/j.mechatronics.2012.02.006

[11] Comolli, F.; Ballo, F. M.; Gobbi, M.; Mastinu, G. Instrumented Steering Wheel: Accurate Experimental Characterisation of the Forces Exerted by the Driver Hands for Future Advanced Driver Assistance Systems. In: ASME 2018 International Design Engineering Technical Conferences and Computers and Information in Engineering Conference. American Society of Mechanical Engineers, 2018. p. V003T01A006-V003T01A006 (2018). http://dx.doi.org/10.1115/DETC2018-85931

[12] Mastinu, G.; Gobbi, M.; Comolli, F.; Hada, M. Instrumented Steering Wheel for Accurate ADAS Development. SAE Technical Paper (2019). https://doi.org/10.4271/2019-01-1241

[13] Gobbi, M., Comolli, F., Hada M., Mastinu, G. An Instrumented Steering Wheel for Driver Model Development, submitted to Mechatronics (2019)

[14] Road vehicle –Test method for the quantification of on-center handling ISO13674-1

[15] Norman, K.D. Objective Evaluation of On-Center Handling Performance.SAE International Congress and Exposition10.4271/840069 (1984).

[16] Kushiro I., Koumura S., Kawai H.: Steering System for the On-center Handling. Transactions of Society of Automotive Engineers of Japan, Volume 40, Issue 2, 233-238, (2009).

[17] Cole, D.: A path-following driver–vehicle model with neuromuscular dynamics, including measured and simulated responses to a step in steering angle overlay. Vehicle System Dynamics – VEH SYST DYN. 50. 573-596 (2012).

NEW STEERING TECHNOLOGIES

Auxiliary power supply system for EPS with high heat-resistant Li-cap

Dirk Ferge

JTEKT Europe

Co-authors: Takumi Mio, Toyoki Sugiyama,
Satou Fumihiko & Satoshi Shinoda

*BR Electrical Power Storage Device Business Department,
JTEKT Corporation, Japan*

List of abbreviations

Abbreviations

AD	Automated Driving
ADAS	Advanced Driver Assistance System
BBW	Brake-by-Wire
C-EPS	Column Electric Power Steering
DP-EPS	Dual Pinion Electric Power Steering
EPS	Electric Power Steering
HPS	Hydraulic Power Steering
Li-Cap	Lithium-ion Capacitor
RP-EPS	Rack Parallel Electric Power Steering
SBW	Steer-by-Wire

Introduction

ADAS and AD are the two major topics for the current and next generations of vehicles. In this context, it is necessary to develop technical solutions to enable larger-size vehicles to follow the trend with an existing 12 V vehicle power supply. EPS can only replace HPS (Fig. 1) as long as this does not degrade performance during any steering maneuvers. On existing EPS, it is also necessary to protect SAE Level 3 and higher against any loss of power during driving. AD functions take over the longitudinal and lateral guidance of the vehicle and must work fail-operational in case of any failure. The Japanese EPS supplier JTEKT has developed a high heat-resistant capacitor, which can be used as an auxiliary power supply system for the electric power steering to support these technical challenges.

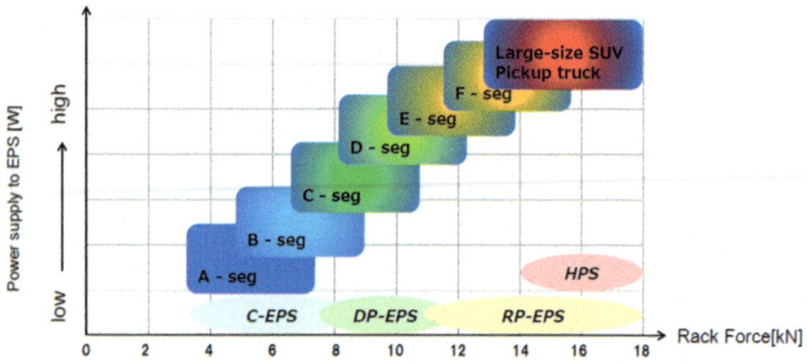

Figure 1: Larger-size vehicles, such as SUVs or pickup trucks, required increased power, so these vehicles currently still use HPS depending on steering speed and stroke.

1 High heat-resistant Li-Cap

Figure 2 shows an overview of the existing technologies for power storage with their typical operating temperature range. Existing storage devices have a narrow temperature range and do not match automotive requirements. Thermal issues always lead to undesirable effects and can only be compensated for by additional measures. To avoid additional costs, weight or packaging problems, it was necessary to investigate the different temperature limitations of existing chemical substances. In conventional Li-Caps, irreversible deterioration occurs because the electrolyte salt thermally decomposes at high temperatures, and reversible deterioration occurs on the low temperature side. Chemical processes are slower, and the viscosity of the electrolytes used in Li cells increases strongly. Internal resistance also increases in lithium-ion batteries in cold temperatures, which reduces power output. In addition, the electrolytes used can freeze at temperatures of around -25°C.

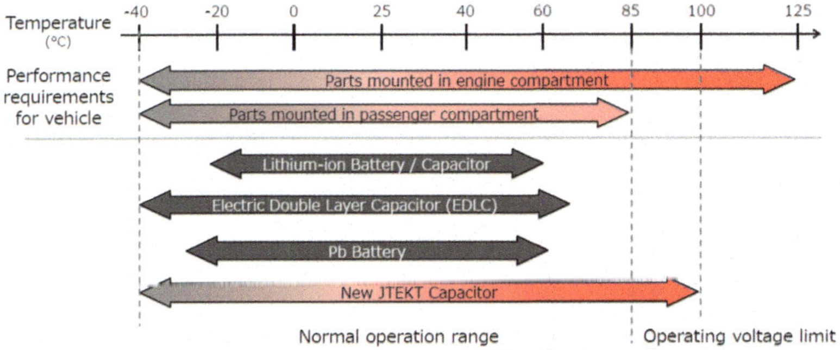

Figure 2: Power storages with their typical operating temperature range

JTEKT has developed a high heat-resistant Li-Cap that can operate from -40°C up to 100°C. This capacitor is based on an active-carbon-based material for the positive electrode and a graphite-based material for the negative electrode with a specially designed electrolyte. The capacitor offers outstanding temperature performance and can fulfill automotive temperature requirements virtually without any additional measures, such as external cooling. This allows installation of the complete power device close to the steering system to avoid long wiring distances between both units, which create additional losses. Li-Cap is particularly suitable for vehicle applications due to its power density. Additional benefits of using a Li-Cap are the rapid charge and discharge times for safety-critical applications, as well as the capacitor's maintenance-free operation and long service life (Fig. 3).

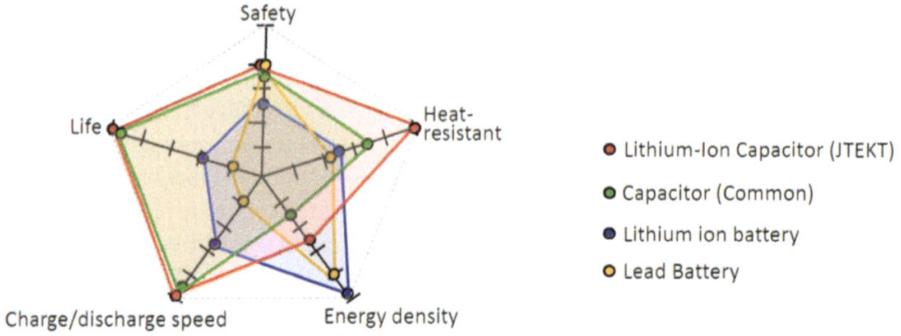

Figure 3: Key objectives for the Li-Cap

2 Electrical power consumption of EPS

Existing vehicles have a limited 12 V power supply, which supplies electric power to various electrical components (Fig. 4). EPS is one of the components that illustrates the challenges faced by vehicle electrical systems to satisfy the variable power demand during stationary steering and straight-line driving.

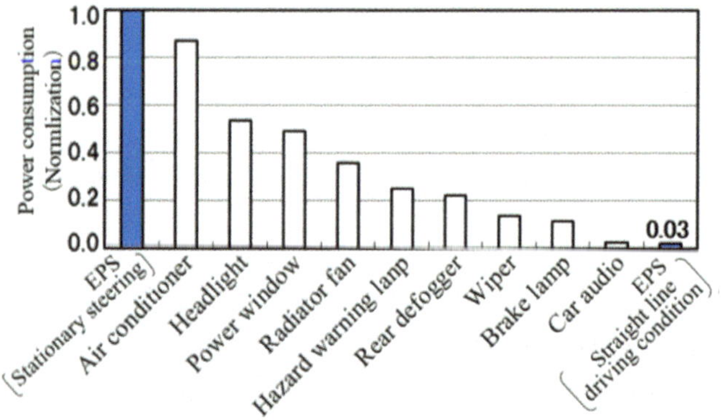

Figure 4: Illustrates the spread of the two extreme conditions of an EPS in stationary steering, such as parking, on the one hand, and during straight-line driving on highways on the other hand.

Compared with the maximum consumption of air conditioning equipment or headlights, EPS has an extremely low power demand most of the time. The challenge of large-size vehicles, such as SUVs or pickup trucks, is to ensure full performance during stationary steering maneuvers. A temporary voltage boost would be necessary to increase power to provide the driver with the expected steering assistance. This lack of power mostly occurs during a limited period and could be solved by a higher supply voltage. In the field of existing technologies, a 48 V electrical system could be one option to solve the power shortage for the EPS, but this would require modification of all electrical components' hardware. The auxiliary power supply system can compensate for the lack in vehicle power supply temporarily, without any impact on other consumers in the vehicle.

3 Auxiliary power supply

3.1 Auxiliary power supply system for power boost

The auxiliary power supply system can balance the power required by the EPS between the maximum given capacity of the battery and the vehicle generator during stationary steering. A capacitor module that includes a charge and discharge controller is integrated between the 12 V vehicle battery and the EPS controller. The block diagram (Fig. 5) shows a simple structure with a separate auxiliary power supply system connected in series with the existing power supply.

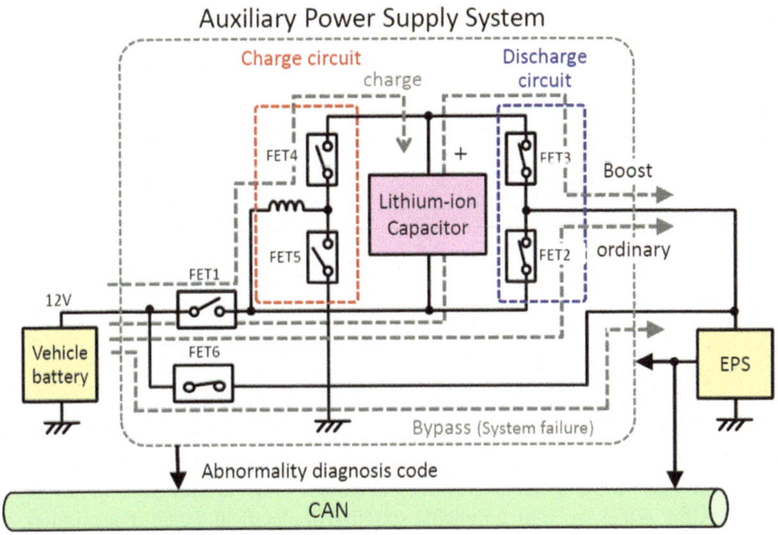

Figure 5: Configuration of auxiliary power supply system and device with lithium-ion capacitor

The electric power storage device consists of two high heat-resistant Li-Caps connected in series to satisfy the required fast charge/discharge characteristics. The operating voltage of the capacitors is about 3 V per cell, 2 series modules together provide 6 V. By limiting the power supply to 18 V, existing MOSFETs can be used. Direct communication between the EPS controller and the charge/discharge controller triggers the additional power supply on demand. Abnormality diagnostics information can be monitored separately on the vehicle network. Fig. 6 illustrates the effect of the auxiliary power supply in a larger-size vehicle. The voltage boost generates the necessary output to keep the steering speed constant during stationary steering. A smooth steering feeling from lock to lock provides perfect steering performance, including for SUVs and pickup trucks. Figure 4 illustrates how the increased output power supports the driver with greater assistance.

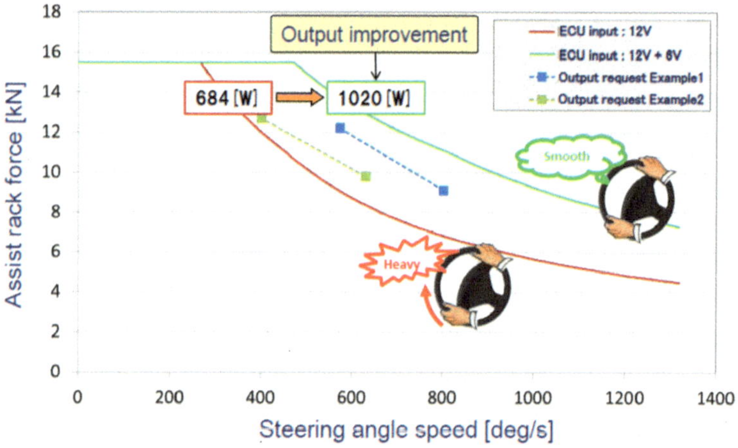

Figure 6: Output improvement on SUVs/ pickup trucks

3.2 Auxiliary power supply system as redundancy system

More and more traditional mechanical systems are being replaced by electronic systems such as SBW and BBW systems. However, electronic systems have a number of safety and reliability issues, such as unreliable stability in electrical signal transmission. In addition, the next generation of vehicles will extend AD functions, in line with SAE Level 3 or higher. In this context, it is mandatory to have redundant hardware and power supply architectures to keep the EPS fail-operational.

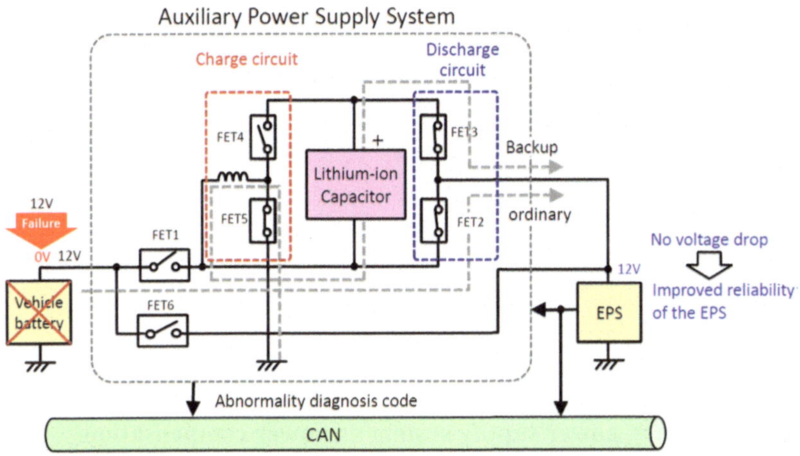

Figure 7: Power cutoff

The high heat-resistant Li-Cap device can be used to implement redundant power supply for the EPS. In the block diagram (Fig. 7), a charge/discharge controller is directly connected with the EPS controller and the vehicle battery. In case of a failure of the vehicle's onboard electrical system, a switch opens the direct connection between the EPS and the vehicle power supply. The stored energy on capacitors is discharged quickly to provide direct power to the EPS and ensure vehicle control for a limited time. A private communication channel ensures smooth transition to avoid any loss of assistance in those safety-critical circumstances. Steady circle turning (Fig. 8) illustrates the performance of the auxiliary power device during a power cutoff. At low vehicle speeds of 20 km/h and a radius of 10 m, the supplied energy supports driver assistance for over 3 minutes. In the case of a power cutoff during highway driving, the driver is supported for approx. 5 minutes, in city driving for 1 minute, and during a parking maneuver for 6 to 10 seconds. These figures are an example and depend on the capacitor size. The capacity of the energy storage must be designed to meet the safety requirements for a safe stop in all situations.

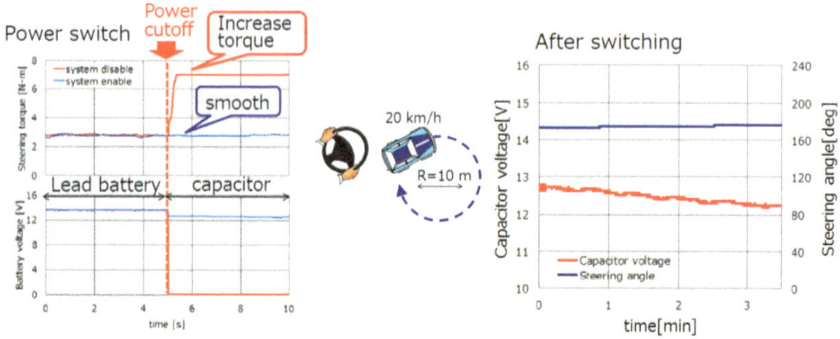

Figure 8: Circle turning with auxiliary power supply

3.3 Auxiliary power supply system as power compensation

Every new generation of vehicles is equipped with more and more electronic assistance and comfort functions. The common features of all these systems are supplied via the onboard electrical system of the vehicle and thus by the vehicle battery. Stable vehicle electrical system voltage is vital to ensure reliable EPS. With the Li-Cap and its fast charge and discharge characteristics, voltage drops (Fig. 9) can be balanced quickly.

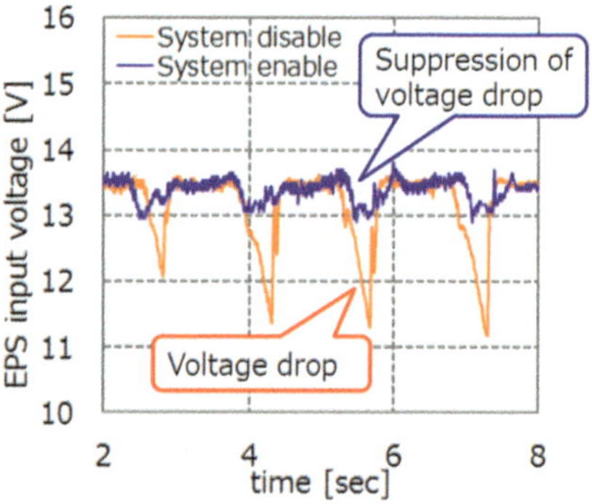

Figure 9: Continuous stationary steering (peak 600 deg/sec)

4 Conclusion

A power storage device for a temperature range from -40°C up to 100°C without additional costs for cooling opens up new possibilities for integration into existing vehicle structures. The maintenance-free, high heat-resistant Li-Cap with its fast charge and discharge times provides a wide range of options for automotive power device applications.

The auxiliary power supply system balances electric power shortage at 12 V and enables equipment of larger-size vehicles with EPS instead of HPS. An increased rate of passenger cars equipped with driver assistance systems and automated driving functions contributes noticeably to making traffic safer. Fuel efficiency also improves by around 3% when hydraulic power steering is replaced with electric power steering. Furthermore, this provides redundant power supply for highly automated vehicles and SBW applications to retain full control over safety-critical functions in certain situations.

References

(1) F. Sato, M. Higashi, T. Sugiyama: Development of Auxiliary Power Supply System for Electric Power Supply Steering, JSAE Annual Congress (Spring) Proceedings, no. S307, 20155307, p1625-1628 (2015)

(2) T. Mio, Y. Komatsubara, N. Ohmi et al.: Development of high heat-resistant Lithium-ion Capacitor, JSAE Annual Congress (Spring) Proceedings, no.194, 20185194 (2018)

(3) T. Mio et al,: Auxiliary power supply system for Electric Power Steering (EPS) and high heat-resistant Lithium-ion Capacitor, EVS31&EVTeC2018, no. E2-1, 20189259 (2018)

New steer-by-wire concept to simplify steering system integration

Bertram Möller, Nexteer Automotive Germany GmbH;

M. Nakielski, Nexteer Automotive Poland Sp. z o.o.

nexteer AUTOMOTIVE
a leader in intuitive motion control
New Steer-by-Wire Concept to simplify Steering System Integration
Chassis Tech 2019
ELECTRIC POWER STEERING
COLUMNS & INTERMEDIATE SHAFTS
DRIVELINE SYSTEMS
HYDRAULIC POWER STEERING
ADAS & AUTOMATED DRIVING
Copyright 2018, Nexteer Automotive Corporation. All rights reserved.

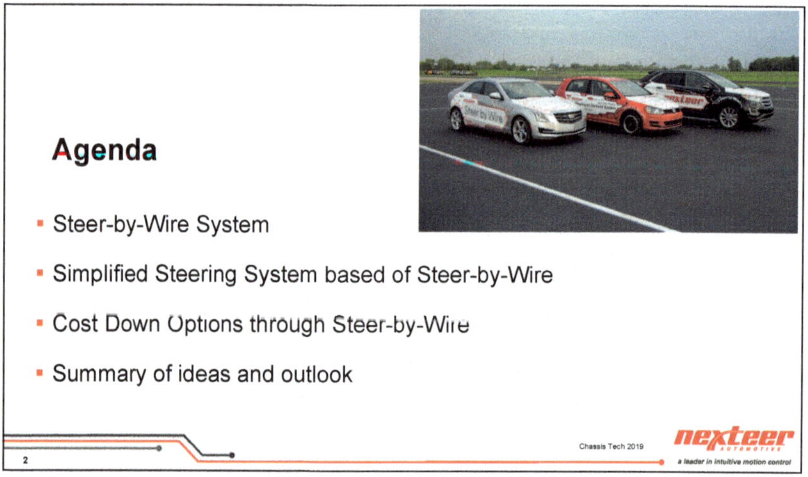
Agenda
Steer-by-Wire System
Simplified Steering System based of Steer-by-Wire
Cost Down Options through Steer-by-Wire
Summary of ideas and outlook
2
Chassis Tech 2019
nexteer AUTOMOTIVE
a leader in intuitive motion control

- Steer-by-Wire Steering System

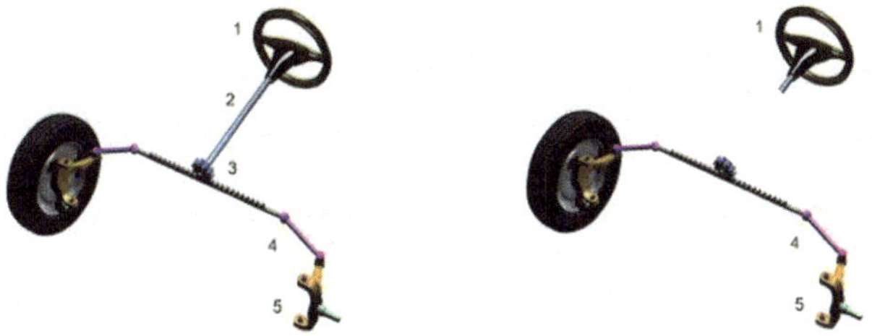

Chassis Tech 2019

nexteer AUTOMOTIVE

a leader in intuitive motion control

3

Evolution of EPS

Electric Power Steering for all vehicle segments and loads from Column to Rack EPS, from A-Segment to HD-SUV,

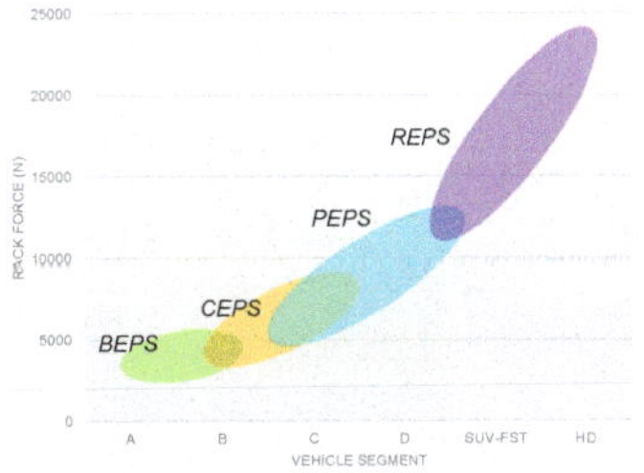

Chassis Tech 2019

nexteer AUTOMOTIVE

a leader in intuitive motion control

4

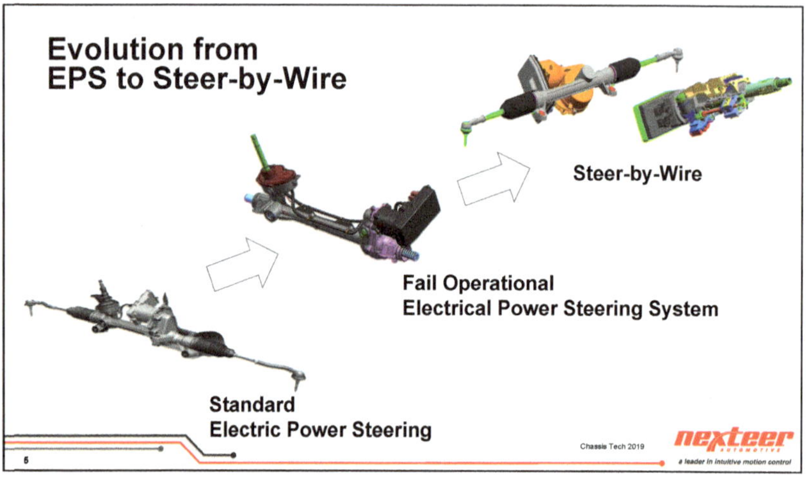
Evolution from
EPS to Steer-by-Wire
Steer-by-Wire
Fail Operational
Electrical Power Steering System
Standard
Electric Power Steering
5
Chassis Tech 2019
nexteer
AUTOMOTIVE
a leader in intuitive motion control

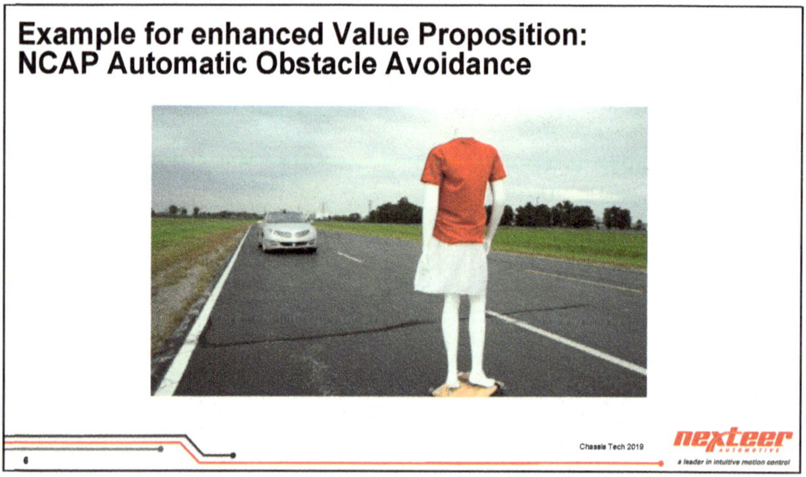
Example for enhanced Value Proposition:
NCAP Automatic Obstacle Avoidance
6
Chassis Tech 2019
nexteer
AUTOMOTIVE
a leader in intuitive motion control

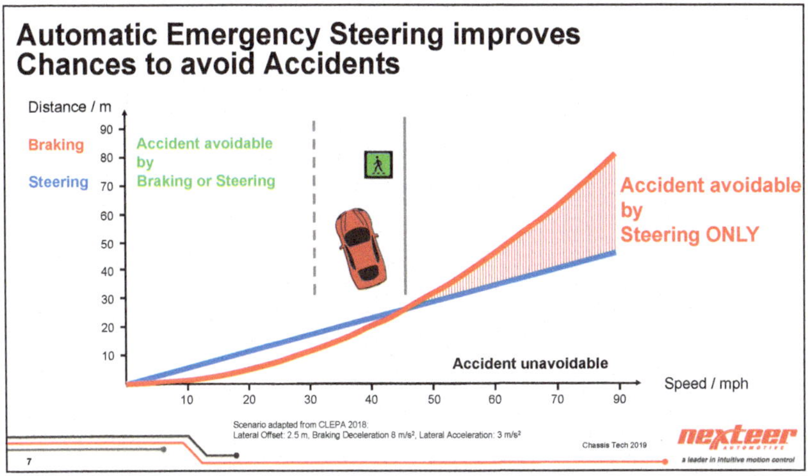
Automatic Emergency Steering improves Chances to avoid Accidents
Distance / m
Braking
Steering
90
80
70
60
50
40
30
20
10
Accident avoidable by Braking or Steering
Accident avoidable by Steering ONLY
Accident unavoidable
Speed / mph
10
20
30
40
50
60
70
90
Scenario adapted from CLEPA 2018:
Lateral Offset: 2.5 m, Braking Deceleration 8 m/s², Lateral Acceleration: 3 m/s²
Chassis Tech 2019
nexteer AUTOMOTIVE
a leader in intuitive motion control
7

Quiet Wheel

Roadmap to Automated Driving

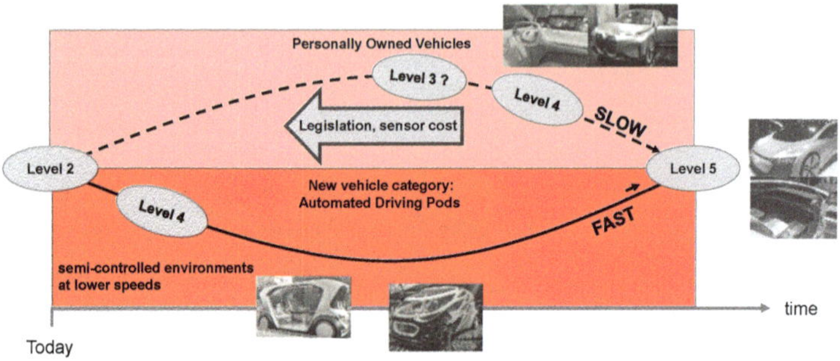

Adapted from: Toyota, R. Eustice, VP of Autonomous Driving

Chassis Tech 2019

a leader in intuitive motion control

9

Simplified Steering System based of Steer-by-Wire

- What generates complexity today ?
  - geometry, packaging, diversity, loads, powertrain versions, assembly, functions, communication network, specifications, legislation
- How could we change this by Steer-by-Wire change
  - Geometry e.g. simplified geometry, easier to adapt, no link between column and steering
  - Packaging e.g. less volume, improved flexibility, where are heat sensitive parts
  - Diversity e.g. reduce number of systems/developments ; e.g LHD/RHD
  - Loads e.g. reduce loads, manage loads smarter
  - Assembly e.g. less assemble steps and considerations

Chassis Tech 2019

a leader in intuitive motion control

10

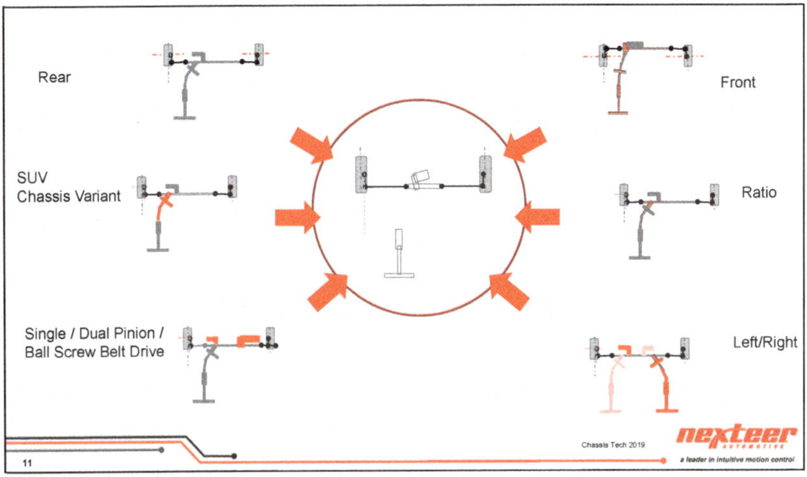

How to enable ONE Steer-by-Wire system

Need to find a cost effective solution to cover all loads

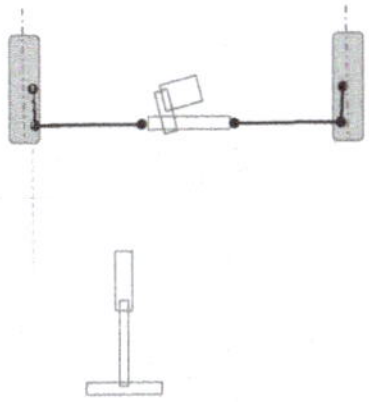

- Lower Loads,
  look at steering loads, reduce loads overall, mainly peak loads
- Easy to package
  packaging flexible cost effective design

12

Chassis Tech 2019

nexteer AUTOMOTIVE

a leader in intuitive motion control

Cost Down Options through Steer-by-Wire

- Steering system and vehicle layout for steering loads

System Sizing is done to meet
A.) maximum load at parking conditions,
(e.g. high force, mid speeds)

B.) evasive maneuvers in rolling conditions
(e.g. mid forces, high speeds)

C.) mechanically withstand vehicle misuse cases
(e.g. rack bar size)

What can be changed using Steer-by-Wire flexibility ?

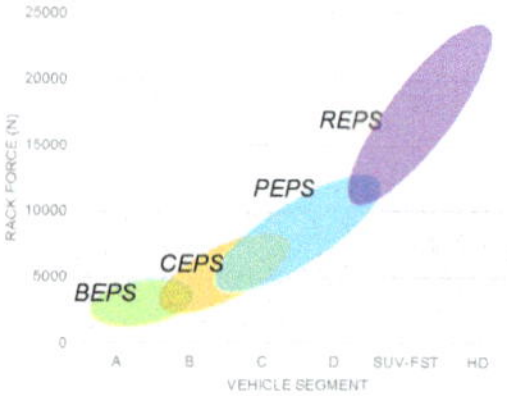

13 Chassis Tech 2019 nexteer AUTOMOTIVE a leader in intuitive motion control

Steering System and vehicle layout for steering loads

Design is defined for 160 % load of what is needed to steer the first 360°, typically on center to peak load is a ratio of 190 – 150 %

Inner wheel is the main contributor to this, in this example inner is 12 kN of 15 kN total, changing from 4 to 12 kN

- Can we change geometry like castor to reduce loads ?
  - ➢ Effects steering feel and vehicle dynamic,
    - maybe not, vehicle dynamics can not be compromised
- Can we change geometry just beyond the 360 deg hand wheel angel area ?
  - ➢ Such area is not really critical for vehicle dynamics

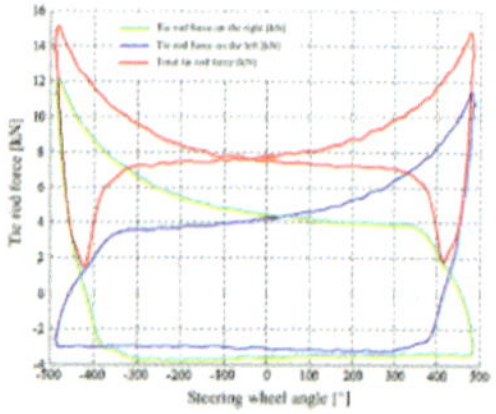

14 Chassis Tech 2019 nexteer AUTOMOTIVE a leader in intuitive motion control

Reduced parking loads

- Longer steering arm = reduced force
  - Increased length reduces torque, Steer-by-Wire ratio compensates effect
  - Trend to larger rims sizes and reduced tier width due to energy saving may allow this for higher loaded vehicles
  - For midsize vehicles we don't need this, as we try to get to one common system
  - **In addition,** longer steering arm = at end of travel the reduction better effective arm length
- Shorter inner ball to ball distance = reduced peak forces
  - Shorter Ball to ball distance, longer tie rods = lower peak load, less effective arm reduction at larger end of travel conditions
  - Connection point to driver pinion no longer needed
  - Need to be discussed in terms of bump steer effects
  - **In addition,** shorter housing /rack with longer tie rods = lighter and lower component costs

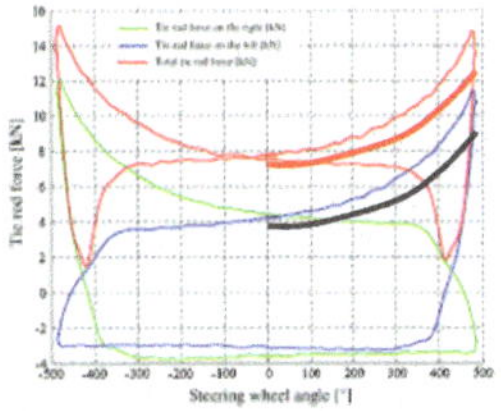

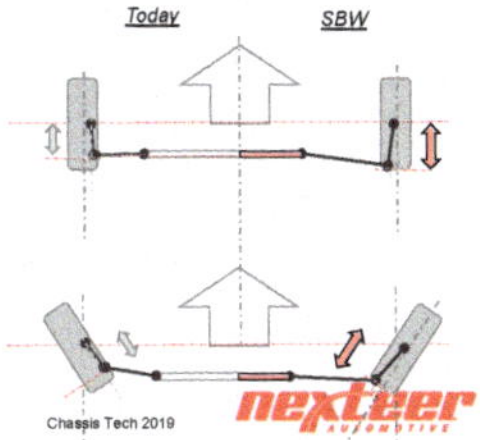

15 Chassis Tech 2019 nexteer AUTOMOTIVE a leader in intuitive motion control

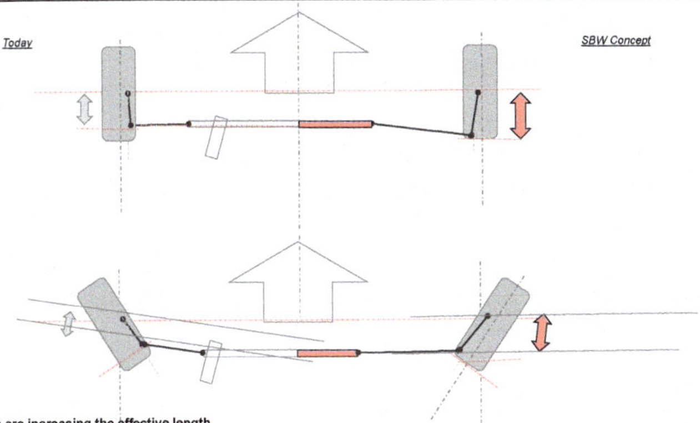

Two effects are increasing the effective length

I. Arm length it self at all positions
II. Effective arm length reduction at large angles is lower with longer arm and longer inner tie rod

16 Chassis Tech 2019 nexteer AUTOMOTIVE a leader in intuitive motion control

Smart reduced parking loads

- Autonomous driving and parking
  - Situation based function, system can sense and predict loads early and adapt maneuver accordingly
  - E.g. assume 12kN capability versus 15 kN max load
    - Reduction of 90 deg Hand wheel angel, or reduced travel by 90deg or by 18%
    - Smart function adapts by rolling wheels or adjusts the maneuver

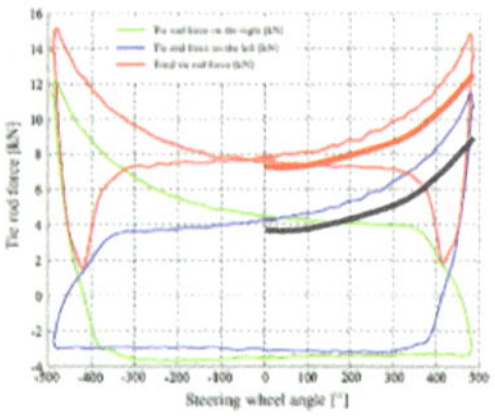

Continental Automotive, Automatisiertes Einparken

- 'Manual' parking
  - Situation based function, system supports the driver how to park
  - Simulates soft stops according to current situation

17 Chassis Tech 2019 nexteer AUTOMOTIVE a leader in intuitive motion control

Smarter Misuse cases

Misuse cases for steering, are cases, in which the mechanical system sees a peak load which could lead to a damage which is safety critical,

I. Drive through a put hole, large forces at worst angles

II. Car on the hoist, 'steer' by the wheel and kick against stop

III. Slide against a curb

Positive effects through Steer-by-Wire

- No inertia effect by the hand wheel
- Better geometry possible (as described in previous chapter)
- Reduces cost and weight

➢ Can we apply smart functions to reduces the cases, e.g. steer to center position when parked

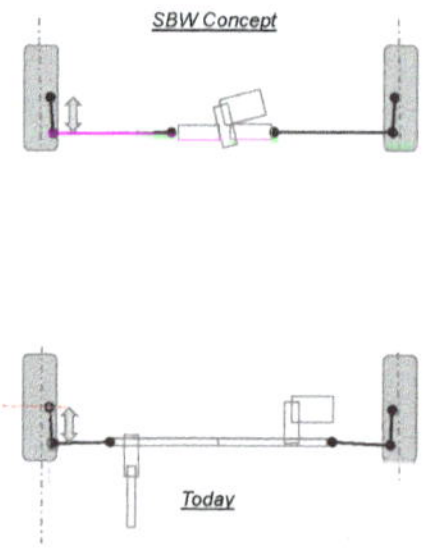

18 Chassis Tech 2019 nexteer AUTOMOTIVE a leader in intuitive motion control

Roadwheel Actuator architecture cost impact

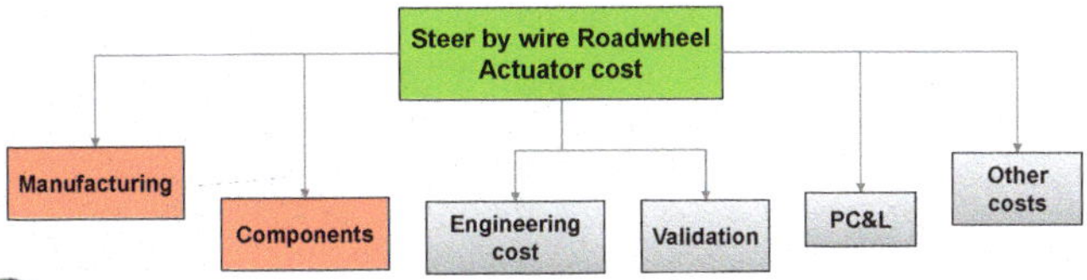

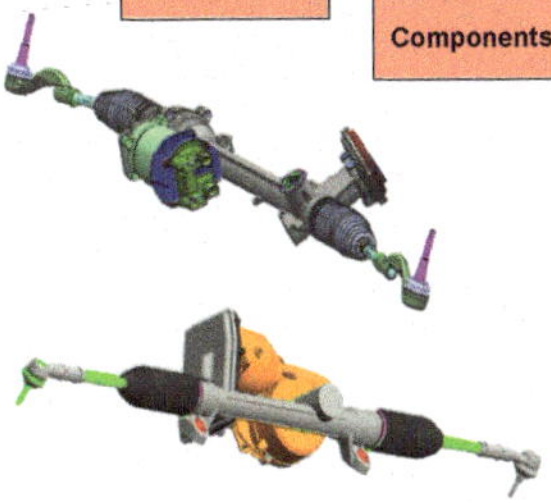

Steer-by-Wire roadwheel actuator architecture can be based on currently known technology of Pinion EPS or Rack EPS.

There are several criteria's to select an architecture of roadwheel actuator. Cost of the system is one of the most important. Comparing the cost of Pinion and Rack based EPS system there is significant benefit of choosing Pinion EPS.

Cost difference between both is driven mostly by cost of components and more complex manufacturing process.

Chassis Tech 2019

nexteer AUTOMOTIVE
a leader in intuitive motion control

19

Current and future steering system architecture

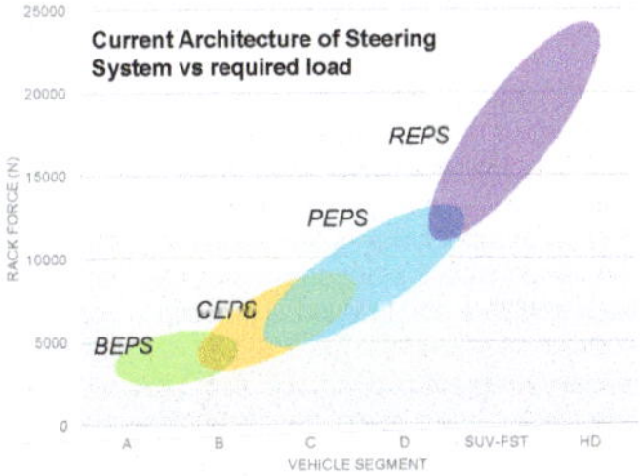

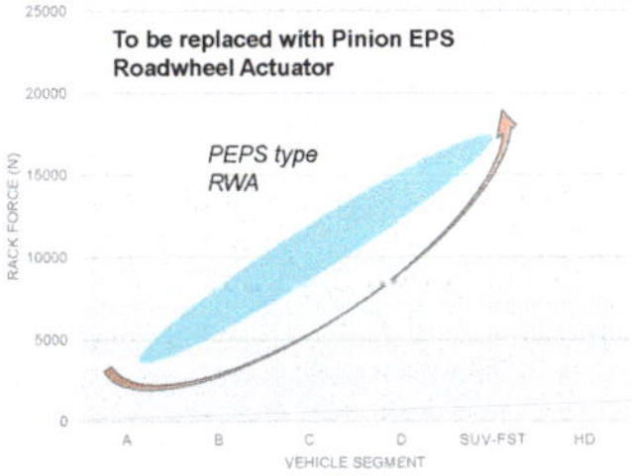

Pinion EPS system have some design restrictions, related to load capabilities, that prevent using it in all vehicles segments. Some of the constrains at steering system level are discussed in greater details in this presentation.

Bringing load requirements down from vehicles side and increasing load capabilities of Pinion EPS unit will enable to use this architecture as steer by wire roadwheel actuator through all vehicles segments.

Chassis Tech 2019

nexteer AUTOMOTIVE
a leader in intuitive motion control

20

Options to enable Pinion EPS type road wheel actuator

- Steering system validation requirements

Traditional Pinion EPS system is design to meet strict validation requirements, to ensure robustness to mechanical and electrical components.

Steer-by-Wire application, due to lack of mechanical connection between steering wheel and wheels in some cases could allow for less severe validation. This could enable steering systems to reach higher loads capabilities.

21 Chassis Tech 2019 nexteer AUTOMOTIVE a leader in intuitive motion control

Options to enable Pinion EPS type road wheel actuator

- Reduction of end of travel bottoming impact

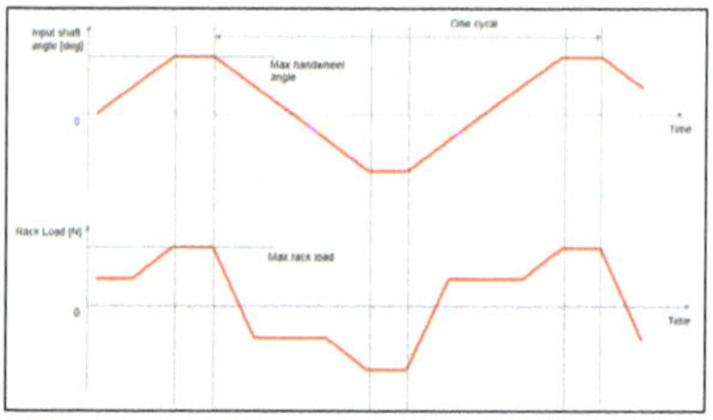

Pic. Example of the bottoming cycle

Software stop function could limit rack travel and velocity. This controls the level of assist from the motor as the steering system approaches the learned (or mechanical) end of travel

For mechanical components, like worm and worm gear or rack and pinion, end of travel bottoming load is most aggressive for endurance stress.

Using advantage of lack of mechanical connection to the steering wheel the bottoming load can be damped through software and learned end of travel positions.

Potentially programmed variable stop position might reduce the localized stress on mechanical gear components.

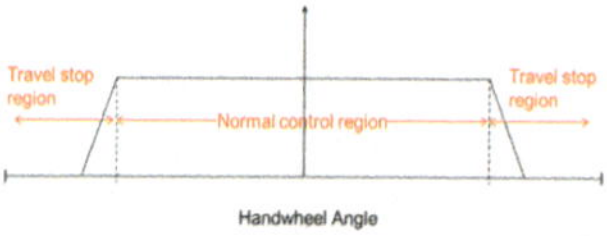

Pic. Controlled bottoming cycle example

22 Chassis Tech 2019 nexteer AUTOMOTIVE a leader in intuitive motion control

Options to enable Pinion EPS type road wheel actuator

- Reduction of impact and misuse loading during validation

Steering system validation includes frequently special cases to simulate events like: high speed breaking on poor road, obstacles or bump crossing, curbstone impact, etc. This test usually simulates high tie rod loading, frequently above maximum calculated tie rod loads within short time.

No mechanical connection allows to reduce the impact of those load on steering system through software logic and other dampening features.

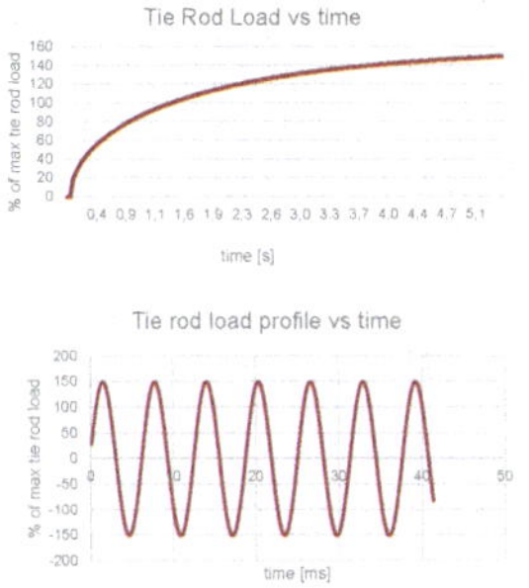

Pic. Example of impact loads profile to simulate special cases

Chassis Tech 2019

nexteer AUTOMOTIVE
a leader in intuitive motion control

23

Options to enable Pinion EPS type road wheel actuator

- Reduction of the temperature impact on durability

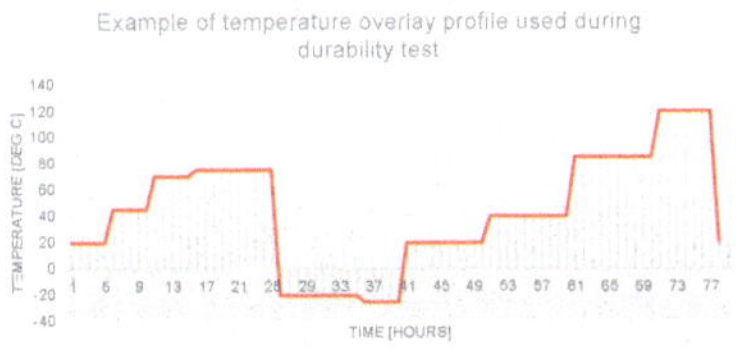

Pic. Example picture of temperature overlay profile used during durability testing

Both electrical and mechanical components are the exposed to the temperature while being tested for durability. Temperature overlay profile assumes exposure to cold (-20 deg C) and hot (up to 120 deg C) environment and checks if the steering systems design works with this condition.

Temperature exposure is a part of every validation plan but it is not always a case in vehicles environment (for example BEV).

Also Pinion EPS roadwheel actuator, being more compact can be placed in the vehicle away from the heat sources.

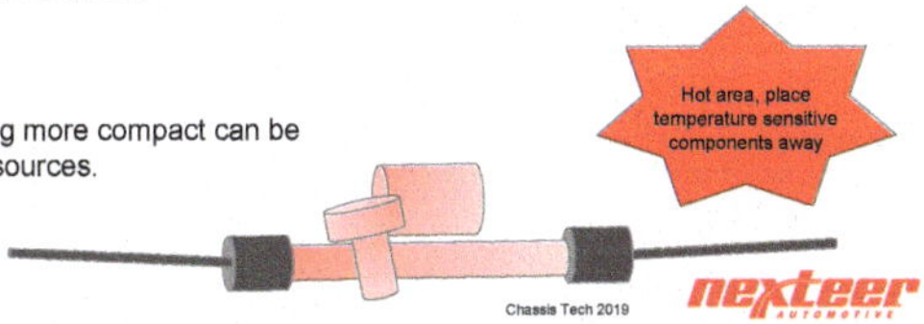

Chassis Tech 2019

nexteer AUTOMOTIVE
a leader in intuitive motion control

24

Options to enable Pinion EPS type road wheel actuator

- Use variable ratio rack with Pinion EPS

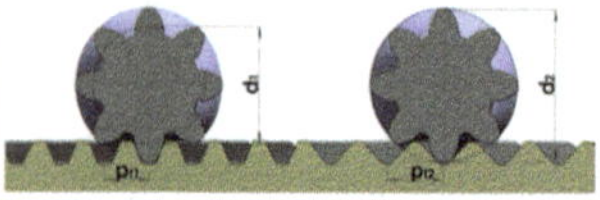

Pic. The different pressure angles of the pinion for different pitch circles diameter

Application of the variable gear ratio allows mechanical advantage for the steering system.
The whole ratio of steering system drops when the front wheel angle rises towards end of rack travel.

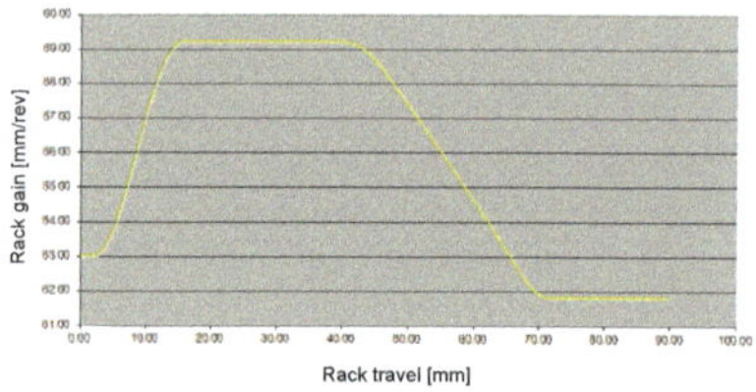

Pic. Example of the rack gain versus rack travel curve

This solution is applied in traditional Pinion EPS. In such case assist torque provided to the rack and pinion significantly increases ability to overcome end of travel loads. Same benefit can be achieved with Pinion EPS roadwheel actuator for steer by wire applications.

25 Chassis Tech 2019 nexteer AUTOMOTIVE a leader in intuitive motion control

Options to enable Pinion EPS type road wheel actuator

- Use variable ratio rack with Pinion EPS

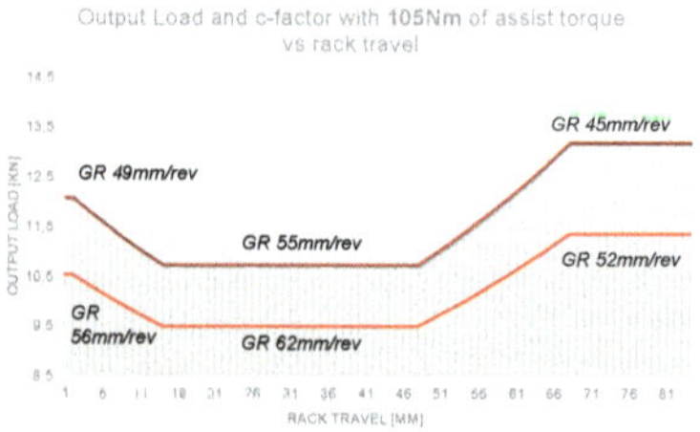

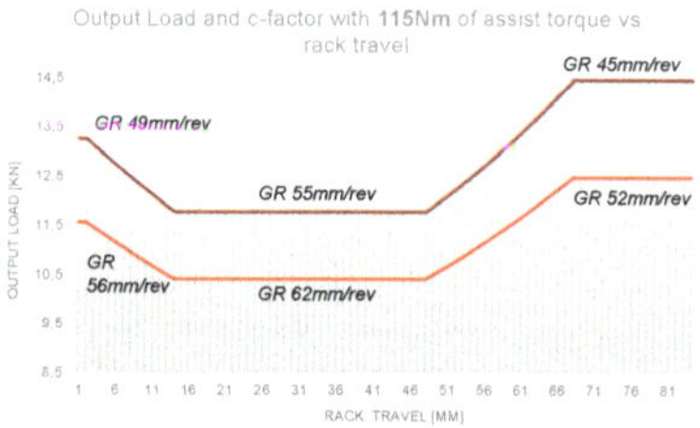

Cases presented above show relation between output load and gear ratio with the assumption of 105 Nm and 115 Nm assist torque provided. Proper selection of the rack gain towards end of travel allows to achieve up to 2.5 kN difference.

26 Chassis Tech 2019 nexteer AUTOMOTIVE a leader in intuitive motion control

Summary of ideas and outlook / enabler for new values

- **Packaging** ,
  more space, no column to I shaft connection needed, no second pinion area needed, hot and cold areas, where to place ONE steering ECU
  - how much value has space ?
- **Time To Market**
  one system developed, keep system, add function
  e.g. can adapt ratio by software, no upfront simulation and design, can adapt last minute with tier development,
  e.g. functions added, with or with AD mode , can adapt feedback , can allow customized feedback solutions
  - software as a product, how much value has time to market, new marketing models possible
- **NVH**
  no connection between wheel and interior, closed firewall
  - Value of NVH improvements, with EV NVH from wheels gets more important, SBW can deliver a significant step in the right direction

27 Chassis Tech 2019 nexteer AUTOMOTIVE a leader in intuitive motion control

Summary of ideas and outlook / enabler for new values

- **Assembly**
  no need to connect, other assembly process possible, module based vehicle assembly
  - Reduced assembly cost, new assembly models, preassembled chassis or EV modules,
- **Vehicle dynamics**
  adapt ratio, sport versus comfort, closed loop control of ratio (on center response with soft mounts)
  - Have optimum branding, reduced development time, can adapt to tier and actual chassis easily
- **Marketing**
  - Have AD and NON AD on same platform, sell AD service modules on demand like time based features
- **ONE Crash design**
  - Do not need to consider intermediate shaft in crash concept
  - One column angle, one design, not dependent on the architecture with different positions and connections
  - More space between pedals and engine can be used for energy absorption
  - Less effort to develop, shorter to market

➢ *Can develop more models based of same architecture in reduced time and effort*

28 Chassis Tech 2019 nexteer AUTOMOTIVE a leader in intuitive motion control

Encourage Early Adoption Through Expanded Value Proposition

Multifaceted Benefits

DRIVERS: SAFETY & PERFORMANCE

Preferred enabler of…

Vehicle Performance

- **Automatic Emergency Steering** (AES)
  Euro NCAP 2025 tech roadmap
- **Enhanced stability control**
- **Weather & road condition compensation**
- **Improved braking distance**
- **A "branded" steering feel,**
  - Luxury to sporty range – A Nexteer difference

AV Complement

- **Safe, transitions** b/w driver input & automated driving - Nexteer's **Steering on Demand™ System**
- **Stowable column technology & Nexteer's Quiet Wheel™ Steering** – Steering wheel remains still during automated driving

OEMs: FLEXIBILITY, BRAND DIFFERENTIATOR

Value Enhancement

- **Enables re-use of components** across platforms (LHD/RHD, EVs, etc.)
- **Enhances flexibility** of under-hood packaging & front of dash design
- **Enables intelligent, cost-saving** steering strategies – such as reduced motor size requirements, etc.
- **Potentially saves assembly time** with automatic alignment of the roadwheel & handwheel
- **Applies to manual, automated & mixed mode driving**
- **Brand differentiator** by delivering advanced safety & performance w/ uncompromising steering feel

29 Chassis Tech 2019 nexteer AUTOMOTIVE a leader in intuitive motion control

nexteer AUTOMOTIVE

a leader in intuitive motion control

30 Chassis Tech 2019 nexteer AUTOMOTIVE a leader in intuitive motion control

OmniSteer – multidirectional chassis system based on wheel-individual steering

Dominik Nees, Schaeffler Technologies AG & Co. KG

Jannick Altherr, Schaeffler Technologies AG & Co. KG

Dr. Marcel Ph. Mayer, Schaeffler Technologies AG & Co. KG

Dr. Michael Frey, Institute for Automotive Engineering (FAST), KIT

Sebastian Buchwald, Schaeffler Technologies AG & Co. KG

Philipp Kautzmann, Schaeffler Technologies AG & Co. KG

1 Introduction

By 2050, 70 percent of the world's population will live in cities as a result of ongoing urbanization. This leads to various challenges such as increasing traffic within cities and less available space. Therefore, multidirectional vehicles with high maneuverability are an advantageous vehicle type for urban areas. Outside the urban traffic of the future, however, existing multidirectional chassis concepts still have significant disadvantages with regard to driving dynamics. The consortium in the "OmniSteer" project, funded by the BMBF, researched a novel chassis and steering system. It combines the advantages of existing multidirectional chassis concepts in terms of maneuverability and those of conventional chassis concepts, characterized by driving safety and pronounced comfort. This could extend the application area of future vehicles to interurban mobility. (**Figure 1**).

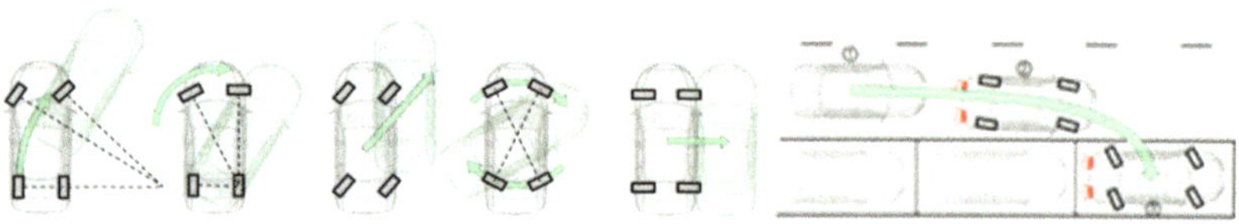

Figure 1 – Possible driving maneuvers using the "OmniSteer" concept

The objective of the joint OmniSteer project was to research a functionally safe, mechatronic longitudinal and lateral guidance system. This is only made possible by electrifying the drive train and integrating the drives into the wheels. The consideration was not limited to a single component of the vehicle, but rather approached the vehicle as an overall system, consisting of the planning and the sensory level, the associated mechanical or mechatronic subsystem as well as the interaction with humans. In order to fully control and utilize the higher number of degrees of freedom of the chassis system, the vehicle must be operated fully automatically.

Within the scope of the collaborative project, efficient driving functions as well as the necessary intelligent electrical/electronic systems, environmental sensors and mechatronic subsystems in the area of chassis and powertrain were developed, and functional verification was performed on the demonstrator. Efficiency here does not only include pure energy consumption, but also, for example, the duration of the process and the maneuvering space. Thereby the potential exists to use the vehicle for maneuvers such as parking or driving in spatially cramped environments, especially in urban environments.

2 State of the Art

This chapter provides an overview of vehicle concepts with increased steering angle, according to the current state of the art. Not only concepts are presented which can perform omnidirectional driving maneuvers, but also any concepts which represent increased value compared to conventional vehicles through increased steering angles. This overview is not to be considered complete.

SpeedE, RWTH Aachen

In 2011, the Institute for Automotive Engineering at RWTH Aachen University developed the SpeedE vehicle [1]. This vehicle features two near-wheel electric motors on the rear axle and a front axle suspension with extended steering angle. (**Figure 2**). It is based on a double wishbone suspension, with the lower wishbone spreading only shortly before the body connection. This creates space for the rotating wheel. To allow movement in the z direction, the wheel carrier is attached to the upper wishbone with a universal joint. This allows the wheel carrier to rotate in relation to the upper wishbone. The steering actuator is mounted on the wheel carrier. This suspension enables the wheel to reach steering angles between -60° and +90°. Omnidirectional driving maneuvers are therefore not feasible, but the increased steering angle offers an extension of the possible driving maneuvers with regard to conventional vehicles, such as "turning around a rear wheel".

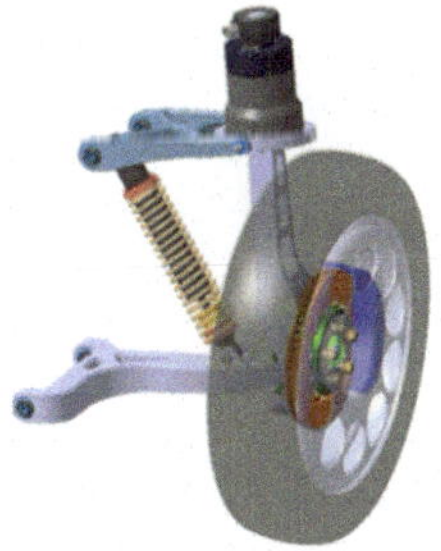

Figure 2 – SpeedE (RWTH Aachen) Suspension [1]

Figure 3 – RoboMobil (DLR) Suspension [2]

RoboMobil, German Aerospace Center (DLR)

The RoboMobil, developed by the German Aerospace Center [2] features four individually steerable wheels with four wheel hub motors that can be driven independently.

All actuators are placed in the wheel. This includes the wheel hub motor and the steering actuator, which controls the wheel via a gear stage (**Figure 3**). Furthermore, the brake system and the connection of the two wishbones are located in the wheel. These are triangular links and are primarily pulled to one side. This provides free space for the wheel on one side, which allows large steering angles in one direction. On the other side the opposite is the case. Therefore, the RoboMobil is able to provide steering angles between -25° and 95°. Omnidirectional driving maneuvers cannot be performed or cannot be performed continuously due to this suspension system.

Advanced Urban Vehicle, ZF Friedrichshafen AG

In addition to the patent specifications of ZF Friedrichshafen AG (DE102016211385A1 & DE102016208790A1), the Advanced Urban Vehicle [3] exists, which, similar to the SpeedE concept, is driven by two near-wheel electric motors on the rear axle and has an increased steering angle on the steered front axle (**Figure 4**). One customized feature of the vehicle is the driver assistance function, which enables autonomous parking. The increased steering angle of a maximum of 75° is achieved by the fact that the lower wishbone spreads only shortly before the body connection and the upper wishbone is mounted above the wheel. The steering system is also mounted above the wheel and is guided to the wheel hub via a swan neck. As a result, no omnidirectional driving maneuvers are possible with this vehicle either, even though driving possibilities are extended compared to conventional vehicles.

Figure 4 – Advanced Urban Vehicle – Suspension [3]

Figure 5 – Modular Robotic Vehicle (NASA) – Suspension [4]

Modular Robotic Vehicle, NASA

The Modular Robotic Vehicle [4] is equipped with four independently steerable and drivable wheel modules. Thanks to the L-profile suspension, the vehicle achieves a steering angle of ±90° and as a result is able to perform omnidirectional driving maneuvers (**Figure 5**). The vehicle's field of application is both the execution of various driving maneuvers in urban space, as well as in the terrain of space. Driving, steering and braking are performed via actuators directly on the wheel. The wheel is rotatably mounted on the body via an L-shaped mounting with integrated rail guide for wheel-stroke movements. It does not provide adequate lateral and longitudinal support at high speeds.

Schaeffler Mover, Schaeffler Technologies AG & Co. KG

Highly integrated corner modules such as the "Schaeffler Intelligent Corner Module" [5] are also suitable for urban mobility applications. This concept comprises the wheel hub motor, the wheel suspension including spring and damper, and the actuator for the electromechanical steering. The "Schaeffler Mover" utilizes this module for all four of its wheels, which greatly increases the number of achievable driving maneuvers and thus the maneuverability of the vehicle.

3 Chassis Concepts

Within the scope of "OmniSteer", different concepts for the realization of an omnidirectional chassis have been developed and evaluated. The following is an excerpt of these concepts.

Sequential Steering System (Concept 1)

Figure 6 shows a double wishbone wheel suspension concept. The pivot point of the upper triangle wishbone is located above the wheel by the swan-neck-shaped wheel carrier. On the chassis side, it is mounted parallel to the vehicle's longitudinal axis and can be rotated. An extension forms a rocker mechanism for the spring and damper unit.

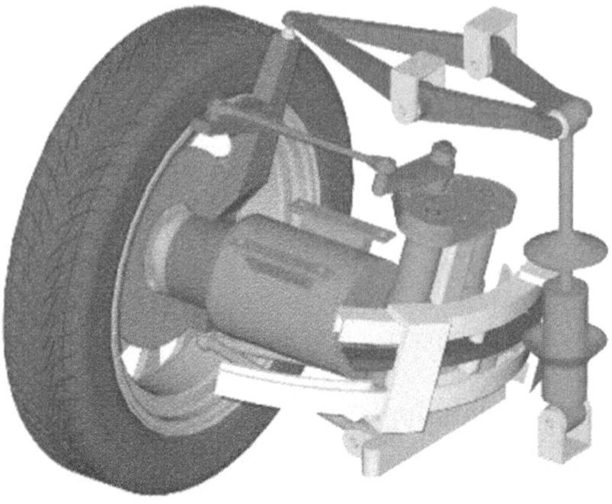

Figure 6 – Suspension concept with sequential steering system

This arrangement prevents a collision between the upper wishbone and the spring damper unit and the wheel in the event of an increased wheel angle. The lower wishbone is also designed as a wishbone and connected to the wheel carrier by a spherical-head joint. In order to achieve a small scrub radius, the lower pivot point is located inside the rim. On the chassis side, the wishbone is attached to a circular segment bearing.

Table 1 – Advantages and disadvantages of the sequential steering system

Advantages	Disadvantages
○ Small scrub radius ○ Small disturbing force lever arm ○ Good longitudinal and lateral force support	○ Constant steering axis position ○ Extreme camber at a steering angle of ± 90° (small wheel contact surface) ○ Complex mounting of the lower wishbone ○ Connection of spring-damper unit difficult

Strut Wheel Suspension (Concept 2)

An essential part of this idea is to reduce the unsprung mass. The drive unit and the steering system are not attached to the wheel carrier, but to the shock absorber bracket. This greatly reduces the unwanted translational mass forces caused by the wheel stroke.

However, a complex drive system is required to transmit the forces to the wheel hub. The steering actuator is attached to the upper end of the strut. The steering movement is realized via a spur gear stage, which is connected to the bodywork via a torsion-resistant joint. An additional wishbone above the wheel restricts the required degrees of freedom. The extreme positive camber due to the steering axis inclination and the steering angle prevents a collision between the wishbone and wheel at ± 90°. All other wheel-guiding and driving components move with the wheel, so there is no risk of collision.

Table 2 – Advantages and disadvantages of the strut wheel suspension

Advantages	Disadvantages
o Simple controllability	o Constant steering axis position
o Only one steering actuator and one drive unit required per wheel	o Extreme camber at a steering angle of ± 90° (small wheel contact surface)
o Low unsprung mass	o Complex construction of the powertrain
	o Low longitudinal force support
	o Large space requirement around vertical axis

Steering Wishbones (Concept 3)

In this concept, four rod steering arms are each hinged to the body and to the wheel carrier. If the wheel is turned via a steering mechanism, the wishbones swivel to the body side (**Figure** 7). The wishbones are specially shaped to prevent a collision between wishbone and wheel during steering. In addition, both lower and upper wishbones are offset in the z-axis to prevent a collision between the wishbones during steering. Due to the movement of the wishbones, the usual connection of the spring-damper unit is not possible. This can be solved, for example, by means of a rocker mechanism similar to the concepts described above. The mounting points on the wheel carrier are arranged in such a way that sufficient space is available for a wheel hub motor.

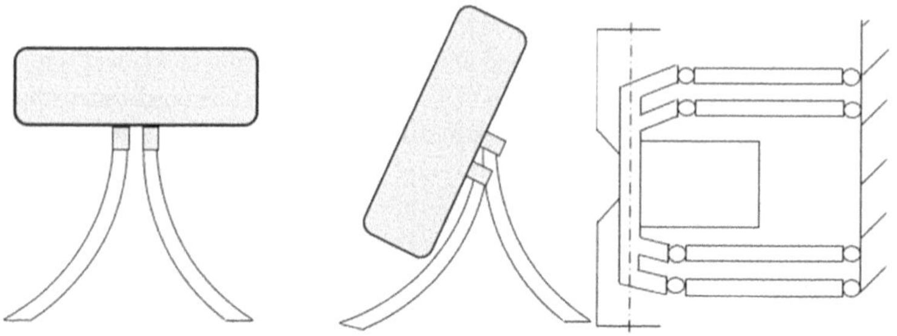

Figure 7 – Steering wishbones concept in plan view in design position (left) and in steered position (middle), and the concept in front view (right)

Table 3 – Advantages and disadvantages of the steering wishbones

Advantages	Disadvantages
o Good longitudinal and lateral force support o Simple controllability o Only one steering actuator and one drive unit required per wheel	o All wheel-guiding components move during the steering process o Complex connection of the spring-damper unit

Conclusion Chassis Concept

The evaluation of the various concepts using the evaluation criteria defined in an expert workshop [6] is presented using an evaluation table (see **Table 4**). The first columns list the evaluation criteria and their associated weighting factors.

Table 4 – Concept evaluation incl. criteria, weighting factor g_k and score p

			Concept 1		Concept 2		Concept 3	
No.	**Evaluation Criteria**	g_k	p	pg_k	p	pg_k	p	pg_k
1	Kinematic changes and moving parts during the steering process	0.04	2	0.09	2	0.09	1	0.04
2	Complexity	0.13	1	0.13	1	0.13	3	0.39
3	Minimum design space	0.04	2	0.09	2	0.09	1	0.04
4	Number of construction parts	0.04	1	0.04	2	0.09	2	0.09
5	Longitudinal force support	0.26	3	0.78	1	0.26	4	1.04
6	Lateral force support	0.26	4	1.04	1	0.26	4	1.04
7	Number of actuators	0.09	2	0.17	3	0.26	3	0.26
8	Control engineering effort	0.13	2	0.26	3	0.39	4	0.52
	Score			2.61		1.57		3.43
	Technical value			0.65		0.39		0.86

The concept of "steering wishbones" achieves a technical value of 0.86 and is therefore superior to the other concepts. The multi-link design allows the replacement of double wishbone or five-link suspensions in a conventional vehicle with sufficient space in the wheel arch without major effort. In addition, this concept is characterized by very good longitudinal and lateral force support.

In order to further develop the concept of "steering wishbones", the connection points on the wheel carrier and body side must be positioned in such a way that a collision between the steering wheel and wishbone is prevented. In addition, a steering actuator system must be developed which enables a wheel steering angle of ±90°. The connection of the spring-damper unit is also another point of development.

4 Suspension Development

The designated concept of "steering wishbones" is basically designed like a multi-link independent wheel suspension and should enable an increased wheel angle of ±90° due to its geometry. During a steering process, the respective wishbones avoid the turning wheel. To avoid collisions between moving parts, they must have a special shape. In addition to the geometry, the positioning of the connection points is a decisive factor. The aims of the kinematic dimensioning are:

- Reaching ±90° wheel steering angle
- Favorable camber over wheel stroke
- Minor change of scrub radius, caster and steering axis inclination
- Sufficient distance between wishbone and wheel
- Minimized movement of the wheel contact point
- Precise measurement of the steering angle on the wheel for reliable control

Different analyses for the determination of the respective coordinates of the individual wishbone mounting points were necessary in order to enable a wheel steering angle of ±90°. The connection points were systematically varied in order to qualitatively describe and understand the relationships between the variable geometric influencing factors and the associated movement of the wheel. Investigations are divided into two main categories – symmetrical configuration and asymmetrical configuration.

Symmetrical Design

Two rods are located above and two below the wheel center. All four wishbones have the same length, i.e. the distances in vehicle longitudinal direction (x-axis) and in vehicle lateral direction (y-axis) are identical. The virtual steering axis intersects the wheel contact point in the design attitude (wheel angle 0°) and is perpendicular to the x-y plane (travel path) over a steering angle of ±90°.

Considering these boundary conditions, there are three possible variations (see **Figure 8**).

1. Displacement of the mounting points on the wheel carrier and on the chassis in the longitudinal direction of the vehicle
2. Displacement of the mounting points on the wheel carrier in the lateral direction of the vehicle
3. Displacement of the mounting points on the chassis in vehicle transverse direction

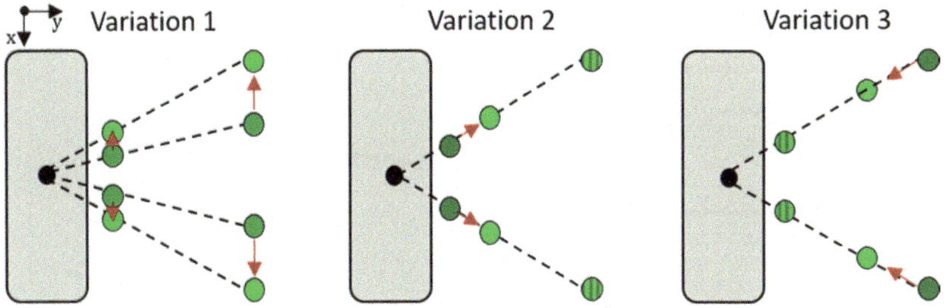

Figure 8 – Evaluated variations of a symmetrical position of the kinematic points

Determination of kinematic Points and Wishbone Shape

The conclusions derived from variations 1-3 of the kinematic point positions can be summarized as follows: the closer the points are to the chassis, the lower the wheel movement in the lateral direction of the vehicle. This, however, considerably increases the further the wheel-sided points lie apart in the longitudinal direction of the vehicle. Increasing the distance between the wheel and the chassis-side points results in greater wheel movement in the longitudinal direction of the vehicle.

Using these variations, the laws for the wheel motion during the steering process were identified. In addition, the effects mentioned above overlap independently of one another. Subsequently, a basic position of the wheel and chassis-sided kinematic points was determined, which served as the starting point for the iteration process for the suspension design.

Main objective of the iterative process is to achieve a steering angle of ±90° taking into account the possible wishbone shapes and minimal wheel movement in longitudinal and lateral vehicle direction. A further important factor is the fulfilment of the requirements of the kinematic chassis parameters over wheel stroke and steering angle.

Through several optimization loops, an arrangement was found which delivers satisfactory results for the intended use.

Wheel Lift Cams

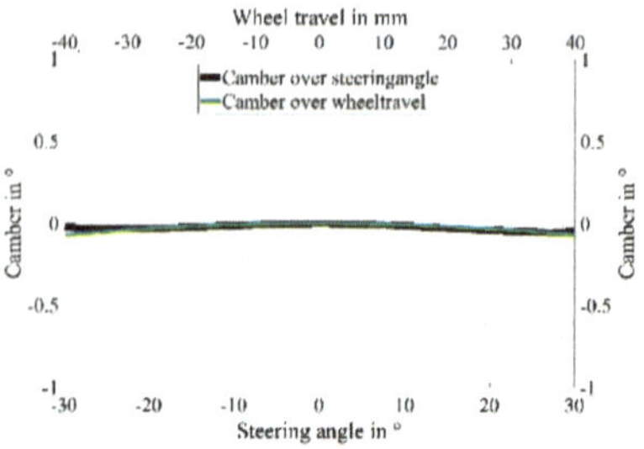

Figure 9 – Camber over steering angle and wheel travel

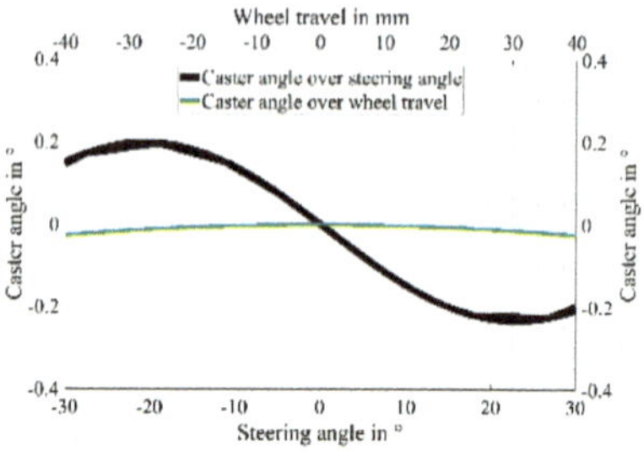

Figure 10 – Caster angle over steering angle and wheel travel

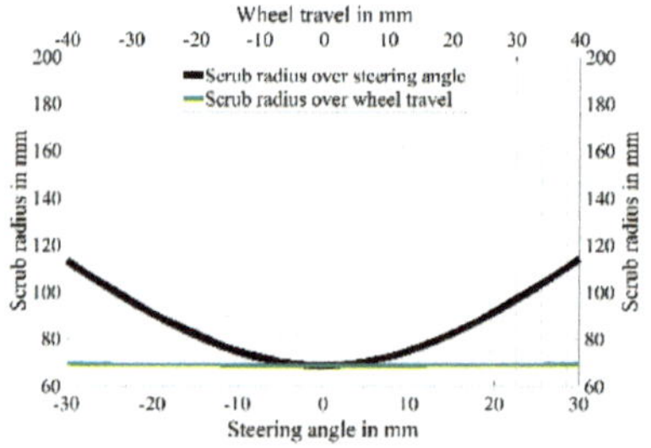

Figure 11 – Scrub radius over steering angle and wheel travel

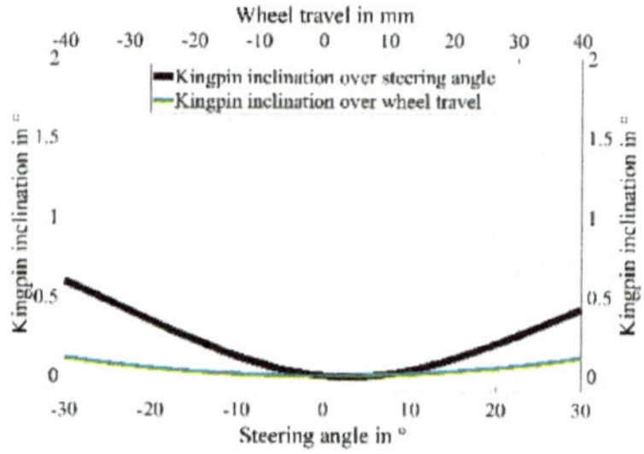

Figure 12 – Steering axis inclination angle over steering angle and wheel travel

Figure 9 through **Figure 12** show the wheel travel and steering cams for a conventional steering angle range of ±30°. The curves of the chassis parameters are similar to those of conventional chassis according to the current state of the art, which indicates favorable effects on the vehicle dynamic behavior in higher speed ranges. When driving at low speed and higher steering angles up to 90°, such as during a parking maneuver, the curves of the chassis parameters play a subordinate role.

The optimization of the suspension link points was realized under the boundary conditions of the steering angle range to be achieved and the available installation space. As the above graphs confirm, the suspension parameters behave almost neutrally over wheel travel. In addition, the camber over steering angle is kept constant.

During optimization of the resolved wishbones, a compromise was found for the resulting scrub radius and wheel movement during the steering process in the x-y-direction.

Load Profile and Dimensioning

The geometric shape of the wishbones is limited by the high steering angle. The requirement for sufficient longitudinal and lateral stability, which in conventional vehicles is also achieved by high stiffness of the suspension components, and the available design space influence the dimensioning to a large extent. The design scenario is derived from triple wheel load assumed as a result of a pothole overrun. From this overload and the coefficient of friction, the maximum transmittable lateral force is determined via Kamm's circle. In addition, a calculated braking force is superimposed on the assumed load. In addition, the individual wishbone loads are evaluated over the entire steering angle and wheel stroke range and included in the dimensioning. In the following, the force application directions in **Figure 13** and **Figure 14**, the achieved safety factors and maximum deformations of the suspension arms are shown in **Table 5**. **Figure 15** shows the weight-optimized wishbones.

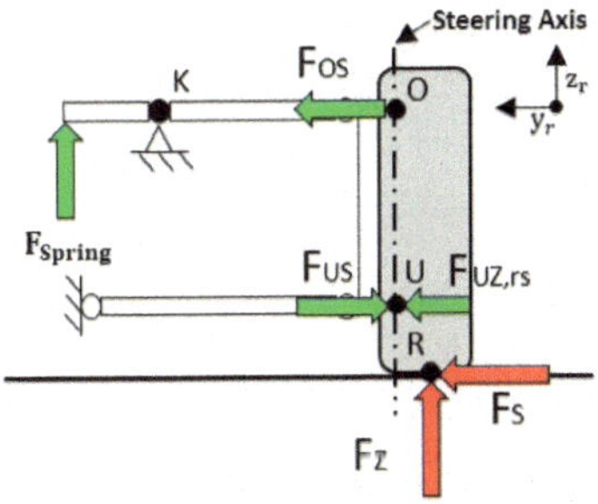

Figure 13 – Front view of the acting joint forces on the upper and lower wishbones

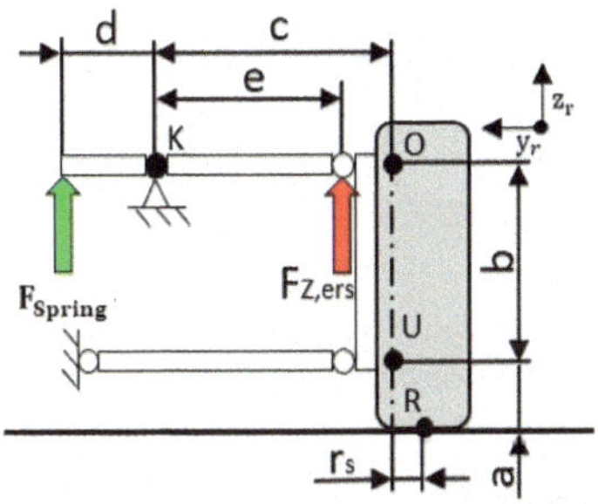

Figure 14 – Lever arms for force calculation in front view

Table 5 – Maximum stresses and deformations, safety factors and wishbone weight

Parameters	Top Front	Top Rear	Bottom Front	Bottom Rear
Max. stress [MPa]	340	130	280	230
Safety factor [-]	1.5	4.1	1.9	2.3
Max. Deformation [mm]	2.9	2.9	2.8	3.2
Weight [kg]	1.5	0.7	1.7	1.7

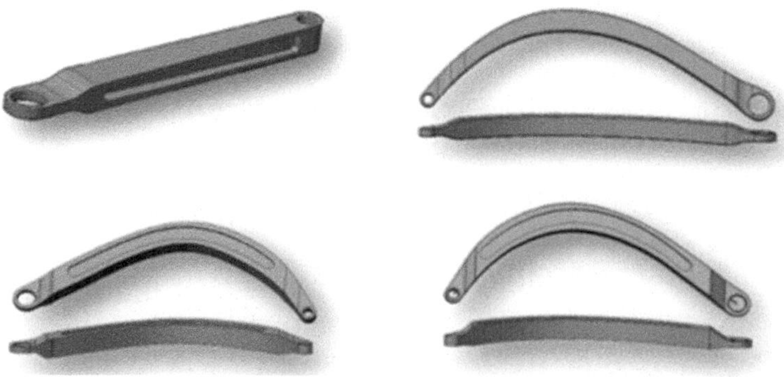

Figure 15 – Weight optimized wishbones (1st row: Top Front, Top Rear; 2nd row: Bottom Front, Bottom Rear)

Steering System

The resolution of the wheel-sided pivot points causes a displacement of the wheel-sided pivot points of all wishbones during the steering movement. This results in a steering axis that moves perpendicular to the x-y plane in the vehicle coordinate system. A constructive solution to transmit a steering torque to the wheel carrier consists of a steering actuator and a lever system (**Figure 16**). This lever system is represented by an actuator control arm and a track control arm, which introduce the torque to the wheel carrier. Investigations have shown that the effective steering torque on the wheel carrier side depends strongly on the steering angle and the selected lever ratios. To achieve a favorable control behavior, however, it is advantageous to have as linear a correlation as possible between the steering angle and the actuator angle of rotation. In order to achieve a constant effective steering torque (**Figure 17**) and simultaneously keep the unsprung mass low, an alternative concept was developed.

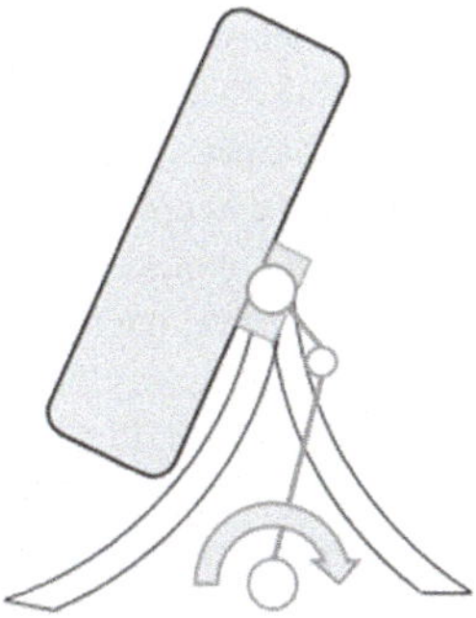

Figure 16 – Top view of the lever handle system in design position

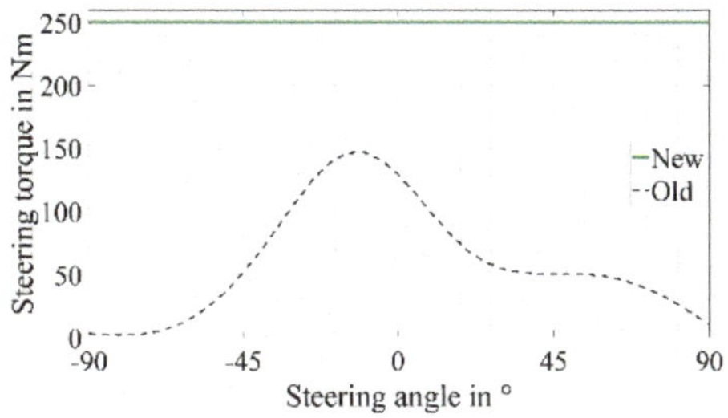

Figure 17 – Comparison of the steering torque curve of the lever steering system with the bevel gear steering system

As **Figure 18** shows, the wheel carrier for this new concept is actuated by a bevel gear stage and cardan drive. The steering actuator is attached to the frame and transmits the steering torque via the universal joint. A parallel guidance system, which can represent both angular and translational displacement, supports the steering torque on the chassis. An angle sensor fixed to the wheel carrier detects the steering angle.

Figure 18 – Wheel suspension with new steering system in -90°, 0° and +90° pose

This enables a constantly high effective steering torque with constant transmission over the entire wheel steering angle range and simple controllability.

5 Results and Outlook

During the suspension development process, as described in chapter 4, a high steering angle range of ±90° was realized. The suspension is capable of absorbing high longitudinal and lateral forces and at the same time ensuring high stiffness. The intelligent development of the steering system has kept the unsprung mass as low as possible. In addition, camber-neutral behavior over the entire steering angle and wheel stroke range was achieved. With its conventional connection, this spring and damper system allows the use of an anti-roll bar, a shock absorber with level control or, for example, an active suspension strut to increase driving safety and comfort. Thus, the OmniSteer chassis has the potential to increase driving comfort, which is one of the central development topics in today's chassis research. Important fundamental research work was therefore carried out in order to develop a future-proof as well as marketable product. In addition, it is possible to easily adapt the driving behavior by changing the spring rate, damping in rebound and compression stages, and anti-roll bar stiffness at the front and rear axles. Thus, the potential is also given to adapt the chassis to today's conventional chassis in terms of driving dynamics.

Figure 19 – Final Chassis Corner Module of the OmniSteer Demonstrator

The high functional integration during parking and the extraordinary maneuverability enable a high level of comfort and significant time savings during drives in urban areas.

In contrast to chassis concepts in the field of robotics, the OmniSteer concept with its roots in automotive chassis underlines that urban flexibility, comfort and driving safety do not have to limit each other in future vehicle concepts. This suspension concept enables safe driving at high and low speeds in urban and interurban use. For future concepts, it is planned to integrate an appropriate steering actuator into the carrier plate and thus provide a space-optimized corner module for urban movers. These movers could be used in inter-urban traffic and offer a new level of achievable driving safety and dynamics.

6 References

[1] B. S. M. K. L. E. Lars Hesse, „The Wheel-Individually Steerable Front Axle of the Research Vehicle SpeedE,“ Aachen, 2013.

[2] L. M. H. A. S. C. S. J. T. J. B. G. H. Jonathan Brembeck, „ROMO – THE ROBOTIC ELECTRIC VEHICLE,“ Wessling, 2016.

[3] J. C. A. N. V. V. D. V. Gerhard Gumpoltsberger, „Advanced Urban Vehicle: The ZF Innovation Car 2015,“ in *24th Aachen Colloquium Automobile and Engine Technology 2015*, Aachen, 2015.

[4] NASA, „MRV – Modular Robotic Vehicle,“ Hampton, 2016.

[5] C. Harkort, D. Kesselgruber, M. Kraus und et. al, „Mobil in der Stadt von morgen – Die Verschmelzung von Antrieb und Chassis,“ 2018. [Online].

[6] J. Altherr, M. D. Mayer, P. Kautzmann und M. D. Frey, *Expertenworkshop zur Findung neuartiger omnidirektionaler Fahrwerkkonzepte,* 2015.

HUMAN-MACHINE INTERFACE

Straight running – stability analysis with a driving simulator

dott-ing Danilo BIGGIO
Politecnico di Milano, now with Ferrari GES

dr-ing Fabio DELLA ROSSA
Politecnico di Milano

dott-ing Marco FAINELLO
Danisi Engineering

prof dr-ing Giampiero MASTINU
Politecnico di Milano

Abstract

The straight running of the system composed by a car plus driver is studied. Straight running is an important case study for analysing stability. Despite the lateral slip angles of the tyres are small, the system is highly non linear, due essentially to the driver action. Following the simple model of McRuer, later developed by Mistschke and revised by many other authors, we have developed a mathematical model of a car plus driver. The dynamic behaviour of the mathematical model has shown the presence of limit cycles generated by so called Hopf-bifurcations. The mathematical model predicts that, despite the understeering vehicle is globally stable, the driver can make the whole system (car plus driver) unstable. This occurs in case an external disturbance is sufficiently strong. If the external disturbance is small, the understeering vehicle plus driver remains stable. There is a speed above which the understeering car plus driver is unstable, usually such a speed is much greater than the maximum speed of the car on high grip surface. The statements introduced above have been validated by employing the driving simulator of Danisi Engineering, Nichelino, Italy. We experimentally saw that limit cycles do exist and that the driver can make the understeering vehicle model of the simulator quite unstable. We were able to validate the mathematical model by including two humans in the driving loop. One driver was a professional driver, the other one was a novice. The same non linear behaviours were highlighted for the two drivers, however, the amplitudes of the limit cycles and the ability of controlling the car were higher for the professional driver. A question arises whether an electronic power steering (EPS) may reduce or cancel instability. The answer is that there are a number of possible solutions for ESP to counteract the effect of unstable limit cycle.

Introduction

The paper focuses on straigh running of cars on an even or slightly uneven road. In particular stability is studied.

Common drivers feels that a car, running straight ahead, may become unstable if a disturbance of sucient level occurs. To describe accurately such an intuitive fact, an in-depth mathematical reasoning is needed, which addressed in the paper.

Academic books deal with the stability of cars referring to linearized models [10, 25]. We introduce a linear model of a car providing analytical formuale giving the front and rear lateral slips, the yaw angle and the steering angle [38] which is needed to run straight ahead. Road vehicles run straight with a side-slip angle, which is due to some inherent tyre characteristics, namely ply-steer and conicity. Additionally, road banking acts to deviate sideways the straight motion of the vehicle. Normally drivers, to run straight, have to slightly steer and to apply a torque at the steering wheel. The influence of tire ply-steer and conicity on straight motion have been dealt with by Lindenmuth [39] who indicated the existence of a relationship between steering wheel rotation during straight running and tyre conicity and ply-steer . Topping [40] stated that vehicle pull or steering wheel rotation is originated by tyres. A sound influence on Ply-steer Residual Aligning Torque (PRAT) by tire tread was highlighted experimentally by Matyja [41]. The mentioned tyre effects are just the very basic issues referring to vehicle stability since they influence the linearization of the equations of motion at steady state running.

The linearization of the equations of motion is the very preliminary step for stability analysis. Actually, studying stability by considering a limited disturbance is misleading. In other words, classical academic hypotheses on straight running stability of cars cannot provide any prediction if the disturbance is strong. We resorted to Bifurcation theory to allow a proper understanding of the phenomenon, as shown in [8, 6, 5, 7, 18, 26, 27, 34]. In the paper we provide an early experimental substantiation of bifurcation theory referring to straight running stability of cars.

In the literature, a number of papers have been written on bifurcations of road vehicle models [8, 3, 13, 15, 17, 22, 30, 31, 35, 36]. We focus here on the influence of an actual driver in the loop.

The paper is structured as follows. At first the linearized expressions of relevant kinematics characteristics of straight running of cars are given. Then the nonlinear model of a simple car and driver is introduced to provide an easy dealing with bifurcation theory. Then actual experiments at Danisi driving simulator are given and interpreted accoring with bifurcation theory. The behaviours of a novice driver and that of a professional driver are compared. The possibility of solving stability problems by an Electronic Power Steering system (EPS) are discussed.

1 Steady-state straight ahead running

In Fig.1 a simple model of a car is depicted. Geometric parameters and the forces acting at the tyres are reported.

Let us consider tyre caracteristics, which read

$$F_{yF} = C^{*}_{eff,F}\alpha_F + F_{y0,F} \quad (1.1)$$

$$F_{yR} = C^{*}_{eff,R}\alpha_R + F_{y0,R} \quad (1.2)$$

$$M_{zF} = -C_{M\alpha,F}\alpha_F + M_{z0,F} \tag{1.3}$$

$$M_{zR} = -C_{M\alpha,R}\alpha_R + M_{z0,R} \tag{1.4}$$

where $C^*_{eff,F}$ and $C^*_{eff,F}$ are the effective axle cornering stiffnesses [24, 27], front or rear, respectively. The offsets due to ply-steer and conicity are $F_{y0,F}$, $F_{y0,R}$, $M_{z0,F}$, $M_{z0,R}$.

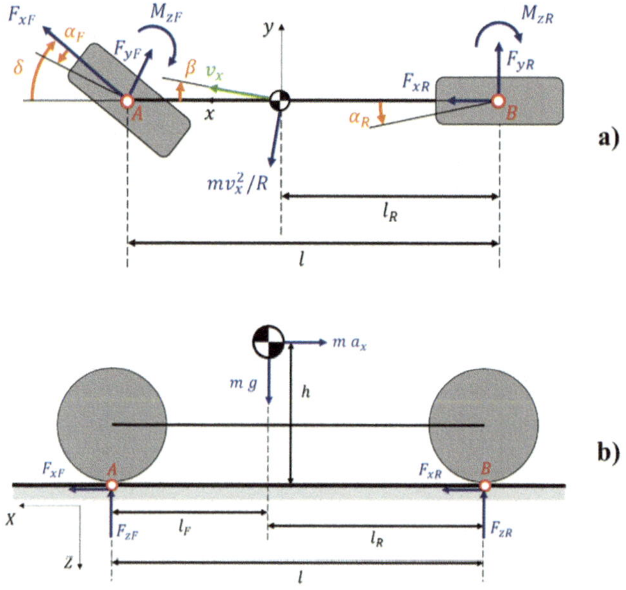

Fig.1. Simple model of a car, a): top view, b): lateral view

After [39], it is possible to derive analytically the lateral slips at the front and rear axle, α_F and α_R, respectively. Such lateral slips are given as funcion of lateral acceleration (divided by gravity) $v_x^2/R\,g = a_y/g$, yielding to Eqs.1.5-1.10

$$\alpha_F = K_{\alpha F} \cdot \left(\frac{a_y}{g} - \varphi_r\right) + \alpha_{F,st} \tag{1.5}$$

$$K_{\alpha F} = \frac{lF_{zF}C^*_{eff,R} + C_{M\alpha,R}mg}{lC^*_{eff,F}C^*_{eff,R} - C_{M\alpha,F}C^*_{eff,R} + C_{M\alpha,R}C^*_{eff,F}} \approx \frac{F_{zF}}{C^*_{eff,F}} \tag{1.6}$$

$$\alpha_{F,st} = \frac{-C^*_{eff,R}(M_{zo,F}+M_{zo,R}+lF_{yo,F})-C_{M\alpha,R}(F_{yo,F}+F_{yo,R})}{lC^*_{eff,F}C^*_{eff,R}-C_{M\alpha,F}C^*_{eff,R}+C_{M\alpha,R}C^*_{eff,F}} \approx -\frac{F_{yo,F}}{C^*_{eff,F}} \tag{1.7}$$

$$\alpha_R = K_{\alpha R} \cdot \left(\frac{a_y}{g} - \varphi_r\right) + \alpha_{R,st} \tag{1.8}$$

$$K_{\alpha R} = \frac{lF_{zR}C^*_{eff,F} - C_{M\alpha,F}mg}{lC^*_{eff,F}C^*_{eff,R} - C_{M\alpha,F}C^*_{eff,R} + C_{M\alpha,R}C^*_{eff,F}} \approx \frac{F_{zR}}{C^*_{eff,R}} \tag{1.9}$$

$$\alpha_{R,st} = \frac{C^*_{eff,F}(M_{zo,F}+M_{zo,R}-lF_{yo,R})+C_{M\alpha,F}(F_{yo,F}+F_{yo,R})}{lC^*_{eff,F}C^*_{eff,R}-C_{M\alpha,F}C^*_{eff,R}+C_{M\alpha,R}C^*_{eff,F}} \approx -\frac{F_{yo,R}}{C^*_{eff,R}} \tag{1.10}$$

Given the analytical expressions of α_F and α_R, the vehicle side-slip angle β

$$\beta = -\alpha_R + l_R/R \tag{1.11}$$

and the steering angle δ

$$\delta = l/R - \alpha_F + \alpha_R \tag{1.12}$$

can be derived [24, 25, 27].

In [39] the validation of the above formulae is discussed with satisfactory results.

2 Models of vehicles and driver model

Let us now introduce the models that have been used to assess the nonlinear behaviour of car and driver. We introduce a simple car model and a simple driver model to deal with Bifurcation theory easily. Then we use a full nonlinear model to be implementented into the driving simulator and use it with a humand driver in the loop.

2.1 Simple car and simple driver model

The equations of motions of the system in Fig. 2 read

$$\begin{cases} \dot{v} &= \frac{1}{m}(F_{y_f} + F_{y_r}) - ur \\ \dot{r} &= \frac{1}{J}(aF_{y_f} - bF_{y_r}) \\ \dot{\delta} &= \frac{1}{\tau}(-\delta + ke + k_d\dot{e}) \\ \dot{y}_G &= u\sin(\psi) + v\cos(\psi) \\ \dot{\psi} &= r \end{cases} \tag{2.1}$$

where the symbols are reported in Fig.2.

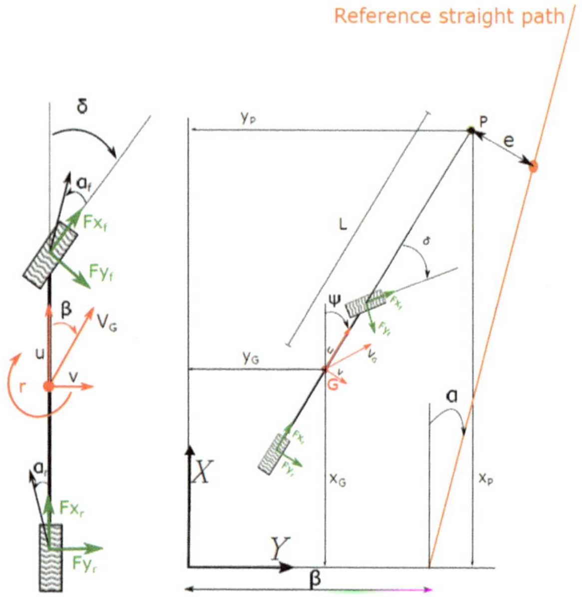

Fig.2. Simple vehicle and simple driver model.

The tyre characteristic is nonlinear, the vehicle is understeering. The tyre data are reported in Tab.1.

Tab.1 Data of Magic Formulae of tyre characteristics

	B	C	E	D
Front	6.92	2.34	0.83	9493.94
Rear	10.31	2.30	1.02	9805.56

2.2 Complex car model

In order to let the human driver at the driving simulator interact with a close-to-reality car model, the VI-Grade model of a car has been introduced. The same tyre characteristics and chassis parameters of the simple model introduced in the previous section were adopted. Basically, the complex car model is a 14 degrees of freedom multi-body model, in which the bodies are the vehicle chassis and the four wheels. The car body has six degrees of freedom and each wheel has two degrees of freedom, one corresponding to the rolling rotation and one corresponding to the relative displacement between the wheel and the car body. Two reference systems are used, one global, fixed with respect to the ground and one local, fixed with the vehicle, with the origin fixed at the mid point of the front axle, the longitudinal axis is parallel to the vehicle centreline, the normal axis is orthogonal to the ground and directed upwards. The lateral axis constitutes a right-hand reference system with the other mentioned axes. The reference systems are represented in Figure 3. The directions of both the lateral and vertical axes of the simple car with simple driver model are reversed with respect to the corresponding axes of the complex car model. The results will be reported using the reference systems of the simple car with simple driver model.

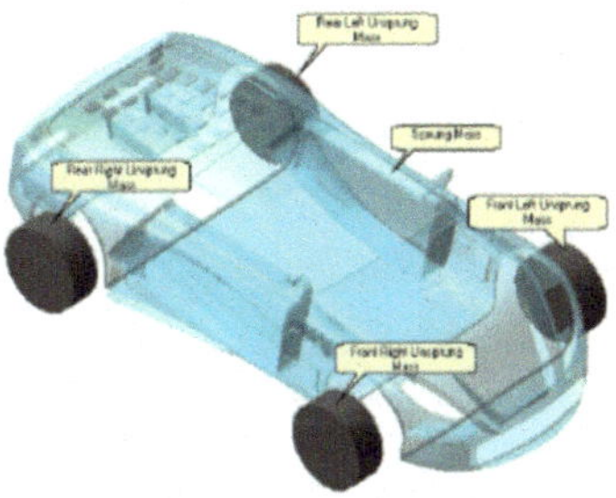

(a) 14 dof vehicle model

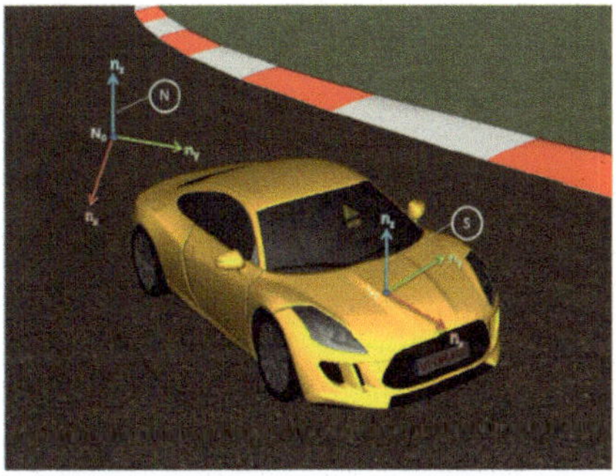

(b) Reference systems

Fig.3. VI-Grade Complex model and reference systems.

3 Nonlinear behaviour of simple car model with simple driver model – Bifurcations

Let us focus on the simple car model with simple driver model. We consider Eq. (2.1). We do not integrate the equations of motions since this would be tedious. We are interested in the limit value that a disturbance may reach to induce a spin of a car initially running straight ahead. We derive the bifircation diagram by a proper continuation alorithm [42,43]. By applying Matcont [42,43], the bifurcation diagram in Fig.4 is found. In Fig. 4 (left figure) we report a three-dimensional projection of the bifurcation diagram of the simple car model with simple driver model. The mathematical model is described by Eq. (2.1) and has 5 state variables. In Fig.4, solid blue lines indicate equilibria which are stable. They could be either steady-state or limit cycles. Dashed red lines indicate unstable equilibria or unstable limit cycles. We see that, increasing the forward speed, a bifurcation speed can be encountered. At longitudinal speeds lower than the bifurcation one, an unstable limit cycle exists in the state-space. The unstable limit cycle repulses the state space trajectories. The motion of the vehicle can be repulsed towards the straight running equilibrium or can diverge, depending on the initial state-space condition. In Fig. 4 (right figure) the maximum lateral error with respect to the reference path of the unstable limit cycle has been reported. The amplitude increases by decreasing the speed, so at lower speeds the perturbation able to make the response of the vehicle unstable has to be very high, as common sense suggests. The lateral position disturbance or the yaw angle disturbance are initial states for the model.

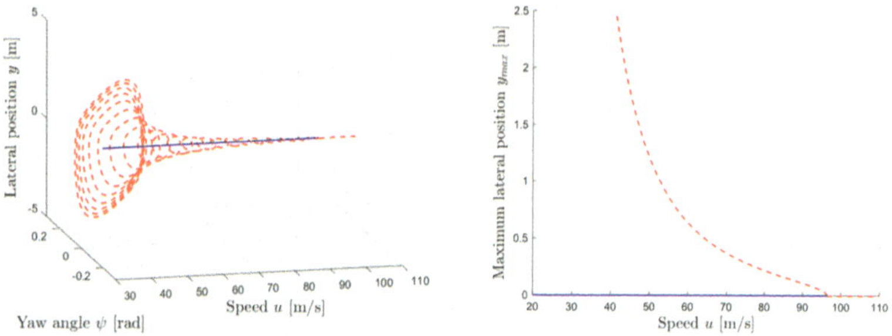

Fig.4. Bifurcation diagram.

4 Straight running experiments at the driving simulator

We checked whether the theoretical results reported in Fig.4 are close to reality. Actually the combination of the simple vehicle model with the simple driver model may provide false information. The driving simulator at Danisi Engineering (located in Nichelino, Italy) has been used (Fig.5). The complex model described in Sect. 2.2 has been employed.

The experimental tests at the driving simulator were perfomed by simulating the straight running of a car. The car was subject to a lateral force acting at a certain distance from the centre of gravity. Such disturbance corresponds to initial states of the model in Sect.2.1. Namely lateral position disturbance and/or yaw angle disturbance are a consequence of the lateral force applied at the car body.

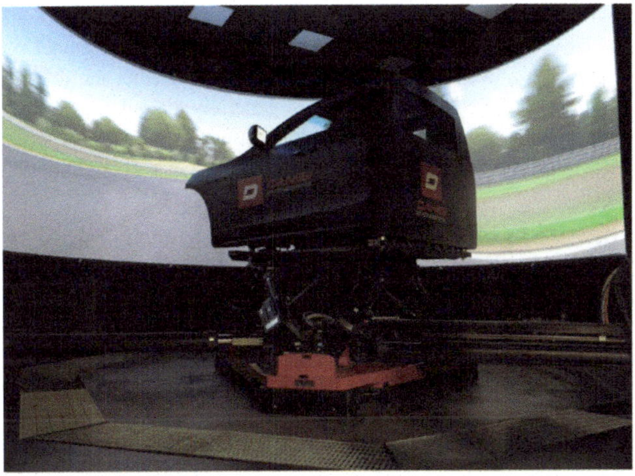

Fig. 5. Danisi Driving Simulator

Similar results with respect to the ones shown in Fig.4 have been found at the Danisi Driving Simulator, the results are reported in Fig.6. Fig.6 is to be compared with Fig.4. The stable or unstable responses of the vehicle and driver are reported, as function of the disturbance parameter, namely the moment arm of the lateral force applied at the vehicle body as described above. The force is 10000N.

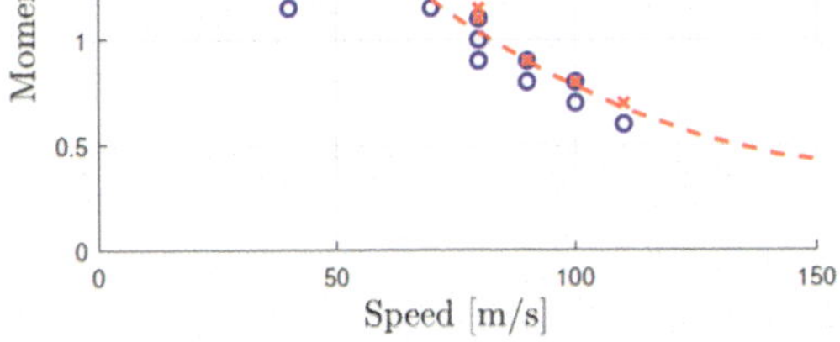

Fig. 6. Stable and unstable vehicle and driver motion as recored at the Danisi Driving Simulator, corresponding to the bifurcation diagram in Fig.4.

We see that the level of disturbance is crucial to define a stable or an unstable behaviour. We can state that qualitativeli the simple car model with the simple drive model do predict what is experienced in a driving simulator. The question whether the driving simulator is close to reality is a topic for further studies, generally the outomes form a driving simulator are trusted as reasonable. We notice that testing stability at high speed may be quite unpractical and even impossible due to the safety issues that are implied.

5 Comparison of experienced driver with novice driver

While performing experiments at the driving simulator we did compare an experienced driver with a novice driver. This was made to avoid a bias. We did want to obtain results as general as possible. To make the control more tricky, we made the vehicle oversteering by switching the front with the rear tyres (see Tab.1).

In Fig.7 the driving activity of the novice driver is reported. The speed is increased up to more than 100m/s. At t=45s and speed=47m/s the bifurcation speed is reached. The system composed by car and driver becomes unsteady and undergoes into a chaotic motion. We see that the lateral displacement is very big. The frequency content of the steering angle is concentrated in the neighbourhood of 0.4Hz. The driver is not able to suppress the chaotic motion.

In Fig.8 the driving activity of the experienced driver is reported. Again, the speed is increased up to more than 100m/s. At t=120s and speed=47m/s the bifurcation speed is reached. Also the experienced driver is not able to suppress the chaotic motion that arises.

Let us compare the driving activity of the experienced driver with the one of the novice driver. We notice that the experienced driver may control the steering wheel angle not only in the range nearby 0.4 Hw but also in the range from 1Hz to 1.5Hz. This changes the time history but the amplitudes of both lateral position and yaw rate are still quite dangerous.

6 Role of EPS on stability of straight running car

We address the problem of avoiding instability by employing an EPS in the driving loop. We only address the problem of exixtence of a solutions. We cannot provide here a solution that would take a proper study to be developed. We just suggest a future topic for research.

The theoretical hypothesis, still to be confirmed, is that the EPS, in case of an hard disturbance, recognizes the occurrence and takes over the driver control. In [44] such control transfer was studied. With this basic assumpion, in Eq.(2.1) two alternative or combined controls could be added.

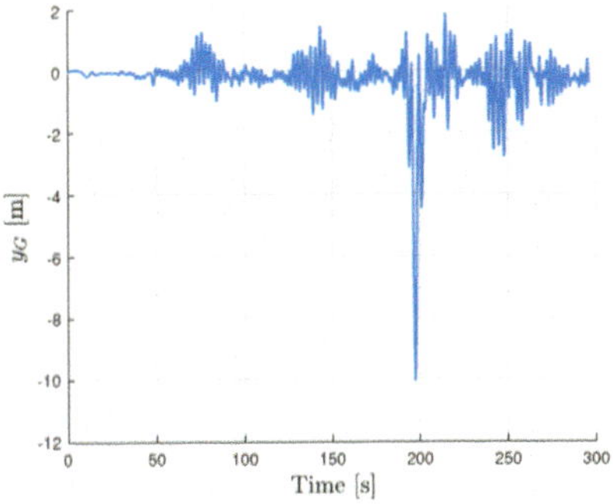

(a) Lateral position

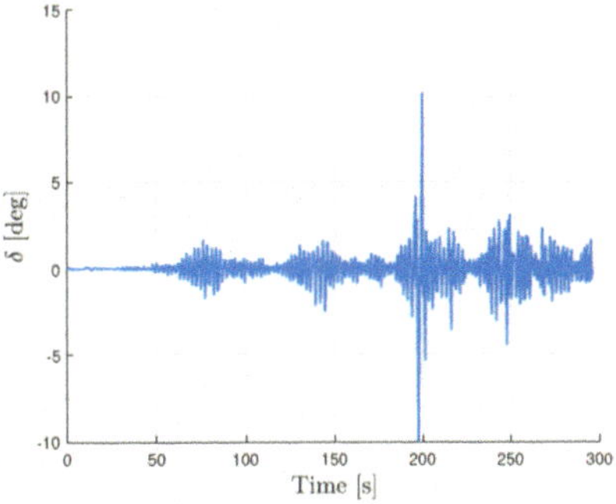

(b) Steering angle

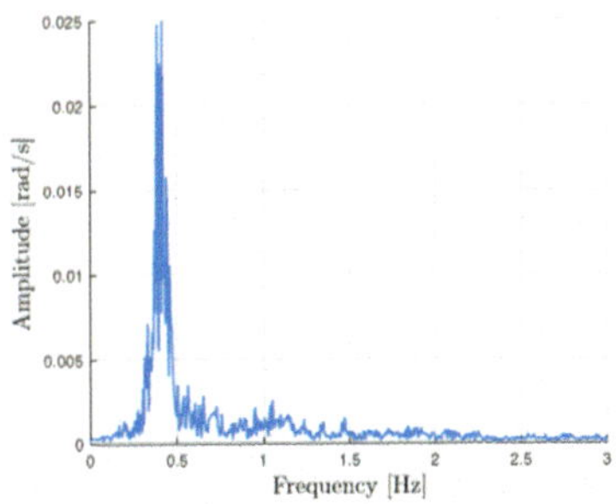

(c) Frequency spectrum: Yaw rate (0-170 s)

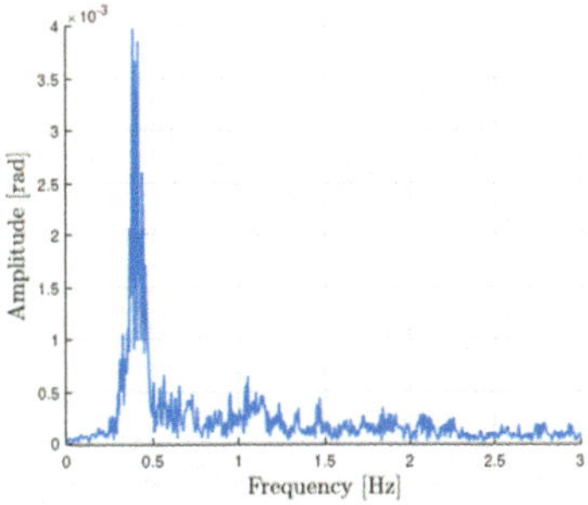

(d) Frequency spectrum: Steering angle (0-170 s)

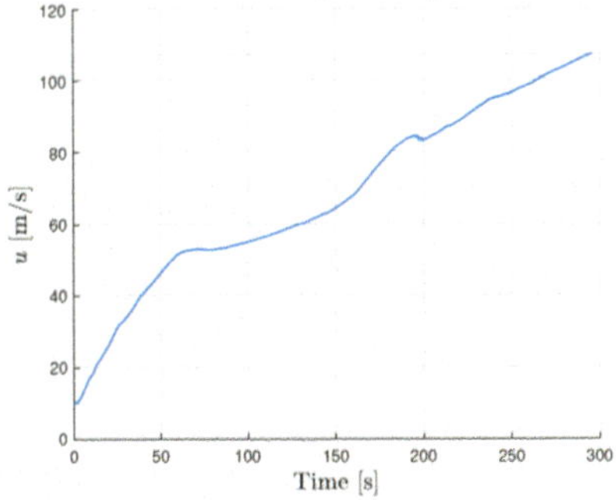

(e) Longitudinal speed

Fig.7. Novice driver with an oversteering vehicle running straight, driving simulator results. The speed is increased up to more than 100m/s. At t=45s and speed=47m/s the bifurcation speed is reached. The system car and driver become unsteady undergoing into a chaotic motion.

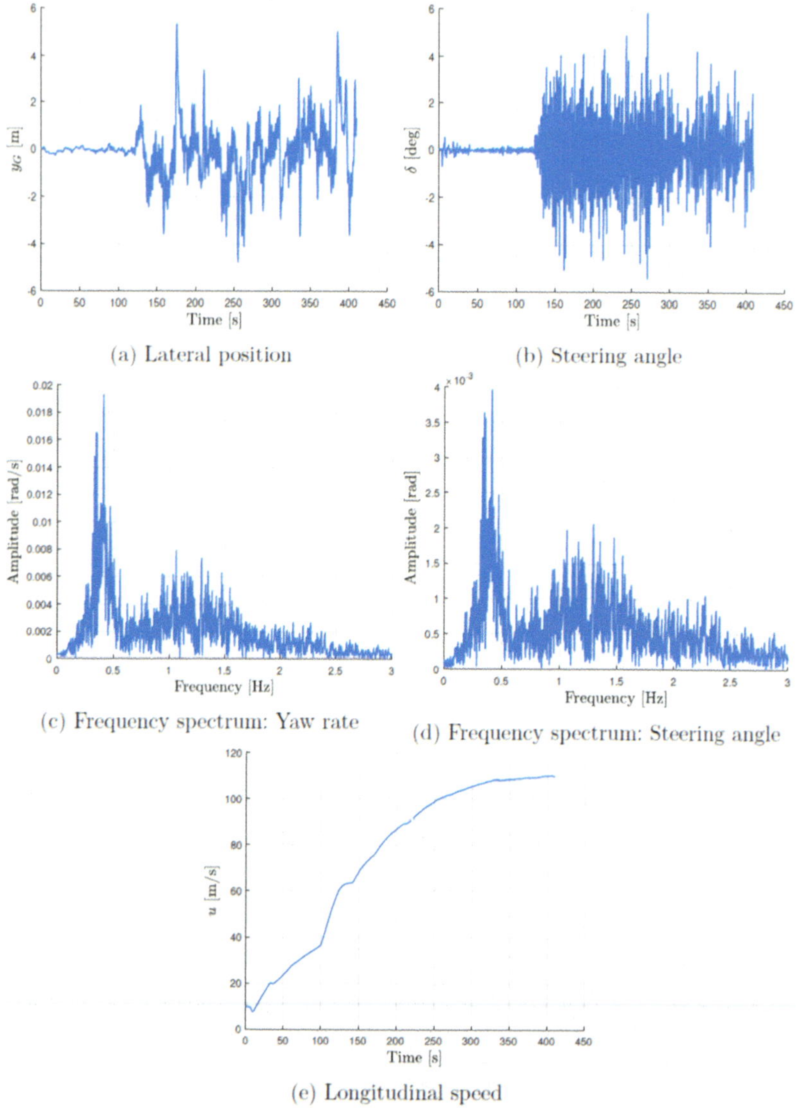

(a) Lateral position

(b) Steering angle

(c) Frequency spectrum: Yaw rate

(d) Frequency spectrum: Steering angle

(e) Longitudinal speed

Fig.8. Experienced driver with an oversteering vehicle running straight, driving simulator results. The speed is increased up to more than 100m/s. At t=120s and speed=47m/s the bifurcation speed is reached. The system car and driver becomes unsteady undergoing into a chaotic motion.

The first control aims at destroying the unstable limit cycle nearby the critical or bifurcation speed. The normal form of the bifurcation, in case of a subcritical Hopf, reads

$$\dot{x}_1 = px_1 - x_2 + x1(x_1^2 + x_2^2)$$
$$\dot{x}_2 = x_1 + px_2 + x_2(x_1^2 + x_2^2)$$

in which x_i (i=1,2) and p are proper combinations of system's states and system's parameters, respectively.

By introducing non linear control terms in Eq.7.1 that make vanish the non linear part, namely

$$x1(x_1^2 + x_2^2)$$
$$x_2(x_1^2 + x_2^2)$$

we could inherently destroy the limit cycle. This is obviuosly a very preliminary hint for a future research.

The seconnd control may be more easy to be implemented and refers to adding a more strong derivative (damping) coefficient to the steering control in Eq. 2.1, in other words we increase the value of the parameter k_d in Eq.2.1.

In Fig. 9 the bifurcation or critical speed is plotted as function of both the driver's delay time and the parameter k_d. For normal driver's lag time (0,14÷0,20, [25]) we see that increasing k_d make the the bifurcation or critical speed increase. This implies also a considrable increase of the amplitude of the unstable limit cycle, which corresponds to improved stability.

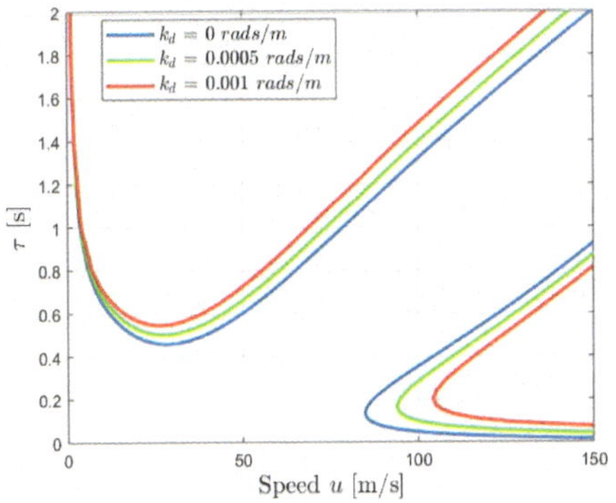

Fig. 9. Bifurcation or critical speed as function of the driver's delay time τ and of the steering control derivative coefficient k_d. Such a coefficient may be associated not only to the driver action but also to ESP.

7 Conclusion

Bifurcation theory is able to describe the the stabiliy of the system vehicle plus driver. Actually, by said theory, we can explain why a car looses stability due to a disturbance of sufficient extent. A mathematical modelling of the driver's behaviour is still missing, so the matching between bifurcation theory (i.e. simulation by nonlinear models) and actual driving simulator results cannot be given in a quantitative sense. Despite this inherent limitation, we highlighted that the source of instability is related to a disturbance the car is subjected. This sheds some light on a number of issues related to active safety and on the design of future autonomus vehicles. We have investigated the influence of driver's experience on the ability to drive after the bifurcation speed has been reached. We discovered that, despite the experienced driver is more effective than the novice one in controlling the steering angle, still the chaotic motion, and thus instability, could not be supressed.

It seems that the crucial problem of straight running stability could be studied at the driving simulator. Actual experiments seem too dangerous to be perfomed on a track. We also tried to answer the question whether or not EPS could solve the problem of stability. We just addressed the exixtence of a solution. Providing an actual solution is out of the scope of this paper. There are two possible control schemes that could be used, either single or combined, to fight against instability. One could be effective nearby the critical or bifurcation speed, the other could be effective at any speed. The development of such controls is here proposed for further studies.

Literature

[1] Danisi engineering, via ippolito nievo 62, nichelino (to) { italy. www.danisieng.com.

[2] Vi-grade: Driving simulation | dynamic driving simulator. www.vi-grade.com/en/products/vi-carrealtime/.

[3] C. Bobier-Tiu, C. Beal, J. Kegelman, R. Hindiyeh, and J. Gerdes. Vehicle control synthesis using phase portraits of planar dynamics. Vehicle System Dynamics, 2018.

[4] B. Catino, S. Santini, and M. Di Bernardo. Mcs adaptive control of vehicle dynamics: an application of bifurcation techniques to control system design. In 42nd IEEE International Conference on Decision and Control, volume 3, pages 2252-2257. IEEE, 2003.

[5] F. Della Rossa, M. Gobbi, G. Mastinu, C. Piccardi, and G. Previati. Bifurcation analysis of a car and driver model. Vehicle System Dynamics, 52(sup1):142-156, 2014.

[6] F. Della Rossa and G. Mastinu. Analysis of the lateral dynamics of a vehicle and driver model running straight ahead. Nonlinear Dynamics, 92(1):97-106, 2018.

[7] F. Della Rossa and G. Mastinu. Straight ahead running of a nonlinear car and driver model-new nonlinear behaviours highlighted. Vehicle system dynamics, 56(5):753-768, 2018.

[8] F. Della Rossa, G. Mastinu, and C. Piccardi. Bifurcation analysis of an automobile model negotiating a curve. Vehicle System Dynamics, 50(10):1539-1562, 2012.

[9] A. Dhooge, W. Govaerts, and Y. A. Kuznetsov. Matcont: a matlab package for numerical bifurcation analysis of odes. ACM Transactions on Mathematical Software (TOMS), 29(2):141-164, 2003.

[10] T. D. Gillespie. Fundamentals of vehicle dynamics. Technical report, SAE, 1992.

[11] J. Guckenheimer and P. Holmes. Nonlinear Oscillations, Dynamical Systems, and Bifurcations of Vector Fields. Springer, 1983.

[12] K. Guo and H. Guan. Modelling of driver/vehicle directional control system. Vehicle System Dynamics, 22(3-4):141-184, 1993.

[13] Z. Hao, L. Xian-sheng, S. Shu-ming, L. Hong-fei, G. Rachel, and L. Li. Phase plane analysis for vehicle handling and stability. International journal of computational intelligence Systems, 4(6):1179-1186, 2011.

[14] H. Hu, Z. Gao, Z. Yu, and Y. Sun. An experimental driving simulator study of unintentional lane departure. Advances in Mechanical Engineering, 9(10):1687814017726290, 2017.

[15] H. Hu and Z. Wu. Stability and hopf bifurcation of four-wheel-steering vehicles involving driver's delay. Nonlinear Dynamics, 22(4):361-374, 2000.

[16] Y. A. Kuznetsov. Elements of applied bifurcation theory, volume 112. Springer Science & Business Media, 2013.

[17] D.-C. Liaw, H.-H. Chiang, and T.-T. Lee. Elucidating vehicle lateral dynamics using a bifurcation analysis. IEEE Transactions on Intelligent Transportation Systems, 8(2):195-207, 2007.

[18] Z. Liu and G. Payre. Global bifurcation analysis of a nonlinear road vehicle system. Journal of Computational and Nonlinear Dynamics, 2(4):308{315, 2007.

[19] Z. Liu, G. Payre, and P. Bourassa. Nonlinear oscillations and chaotic motions in a road vehicle system with driver steering control. Nonlinear Dynamics, 9(3):281-304, 1996.

[20] Z. Liu, G. Payre, and P. Bourassa. Stability and oscillations in a time-delayed vehicle system with driver control. Nonlinear Dynamics, 35(2):159-173, 2004.

[21] C. C. Macadam. Understanding and modeling the human driver. Vehicle System Dynamics, 40(1-3):101-134, 2003.

[22] G. Mastinu, F. Della Rossa, and C. Piccardi. Nonlinear dynamics of a road vehicle running into a curve. In Applications of Chaos and Nonlinear Dynamics in Science and Engineering-Vol. 2, pages 125-153. Springer, 2012.

[23] G. Mastinu, A. Lattuada, and G. Matrascia. Straight-ahead running of road vehicles analytical formulae including full tyre characteristics. Vehicle System Dynamics, in press, 2018.

[24] G. Mastinu and M. Ploechl. Road and off-road vehicle system dynamics handbook. CRC press, 2014.

[25] M. Mitschke and H. Wallentowitz. Dynamik der kraftfahrzeuge, volume 4. Springer, 1972.

[26] E. Ono, S. Hosoe, H. D. Tuan, and S. Doi. Bifurcation in vehicle dynamics and robust front wheel steering control. IEEE transactions on control systems technology, 6(3):412-420, 1998.

[27] H. Pacejka. Tire and vehicle dynamics. Elsevier, 2005.

[28] M. Ploechl and J. Edelmann. Driver models in automobile dynamics application. Vehicle System Dynamics, 45(7-8):699-741, 2007.

[29] G. Reymond, A. Kemeny, J. Droulez, and A. Berthoz. Role of lateral acceleration in curve driving: Driver model and experiments on a real vehicle and a driving simulator. Human factors, 43(3):483-495, 2001.

[30] S. Shen, J. Wang, P. Shi, and G. Premier. Nonlinear dynamics and stability analysis of vehicle plane motions. Vehicle System Dynamics, 45(1):15-35, 2007.

[31] S. Shi, L. Li, X. Wang, H. Liu, and Y. Wang. Analysis of the vehicle driving stability region based on the bifurcation of the driving torque and the steering angle. Proceedings of the Institution of Mechanical Engineers, Part D: Journal of Automobile Engineering, 231(7):984-998, 2017.

[32] S. H. Strogatz. Nonlinear Dynamics and Chaos with Student Solutions Manual: With Applications to Physics, Biology, Chemistry, and Engineering. CRC Press, 2018.

[33] S. Tousi, A. Bajaj, and W. Soedel. Finite disturbance directional stability of vehicles with human pilot considering nonlinear cornering behavior. Vehicle System Dynamics, 20(1):2155, 1991.

[34] H. True. On the theory of nonlinear dynamics and its applications in vehicle systems dynamics. Vehicle System Dynamics, 31(5-6):393-421, 1999.

[35] X. Wang and S. Shi. Analysis of vehicle steering and driving bifurcation characteristics. Mathematical Problems in Engineering, 2015, 2015.

[36] X. Wang, S. Shi, L. Liu, and L. Jin. Analysis of driving mode e ect on vehicle stability. International journal of automotive technology, 14(3):363-373, 2013.

[37] D. H. Weir. Models for steering control of motor vehicles. In Proceedings, Fourth Annual NASA-University Conference on Manual Control, 1968, pages 135-169. US Government Printing Office 1968

[38] A Lattuada, G Mastinu, G Matrascia, Tire Ply-Steer, Conicity and Rolling Resistance – Analytical Formulae for Accurate Assessment of Vehicle Performance during Straight Running, Technical Paper 2019-01-1237 ISSN 0148-7191, DOI: https://doi.org/10.4271/2019-01-1237, Published April 2, 2019 by SAE International in United States

[39] Lindenmuth BE. Tire conicity and ply steer effects on vehicle performance. SAE Technical Paper 740074. 1974

[40] Topping RW. Tire induced steering pull. SAE Technical Paper 750406. 1975

[41] Matyja FE. Steering pull and residual aligning torque. Tire Science and Technology. 1987 Jul-Sep;15(3):207–240

[42] Y. A. Kuznetsov.Elements of applied bifurcation theory, volume 112. Springer Science &Business Media, 2013

[43] A. Dhooge, W. Govaerts, and Y. A. Kuznetsov. Matcont: a matlab package for numerical bifurcation analysis of odes. ACM Transactions on Mathematical Software (TOMS),29(2):141–164, 2003.

[44] K Yoshimoto, X Tang, Research of Driver Assistance System for Recovering Vehicle Stability from Unstable States, SAE Technical Paper2001-01-1276, DOI: https://doi.org/10.4271/2001-01-1276

Future steering wheels for automated vehicles

Dr.-Ing. Christian Strümpler, Team Leader HMI / Human Factors

Daniel Alt, Head of Core Engineering Steering Wheel Innovations

Heiko Ruck, Head of PBU Steering Wheel Engineering

Joyson Safety Systems

Abstract

In the context of automated driving a variety of different technologies will hit the market. Increasing levels of automated driving will have an influence on functions and styling of future steering wheels. Based on a general view on possible automation levels and boundary conditions, three different automated vehicle types have been identified. Those needs and requirements for steering wheel systems will be discussed in this paper. These range from increased sensing functions for driver state, additional input and communication technologies to new steering wheel shapes and foldable steering wheels.

Levels of Automation and potential vehicle types

According to SAE there have been 6 levels of automation defined. These range from Level 0 with no driver assistance to Level 5, which is a fully autonomous system able to operate in any condition as good as an experienced driver. The lower levels of Automation support the driver in operating the vehicle either in longitudinal or lateral control (Level 1) or in both (Level 2). While using these systems the driver needs to stay in the loop to monitor the system and correct any system malfunctions. The driver's role will change with Automation Level 3 and 4, when the system is able to operate safely under certain circumstances. The two levels differentiate by the fact that a Level 3 system is not capable of handling any situation that can occur while using the system and therefore requests the driver to take over imminent control within a short amount of time. This limited time to take over control requires the driver to monitor the system carefully and limits the execution of non-driving related side tasks. A Level 4 automation system is also able to handle critical situations under certain conditions (e.g. weather or road type) and enables the driver to divert attention from the system and the driving situation. This means that the handover time increases significantly and is limited to the operating conditions of the system. Current production vehicles are equipped with driver assistance systems of Level 1-2. Several vehicle manufacturers announced the introduction of the first Level 3 functions within the next 2 years especially for highways, while the introduction of Level 4 systems is announced for 2021 to 2024 timeframe.

Based on the levels of automation and the use of the vehicle, there are three likely groups of automated vehicles being introduced within the next 5 years:

1. Mid-size and economy vehicles, which are privately owned and equipped with Level 1 to 3 systems, that mainly assist on highways and rural roads.
2. Luxury vehicles, which are privately owned or used with Level 4 functions limited to highways.
3. Driverless shared vehicles, which are remotely operated and used for short distance urban transport with Level 4 functions limited to a certain area (e.g. inner city).

Within these three vehicles types, the requirements for steering wheels will vary significantly. Ranging from a traditional steering wheel for manual driving to being totally obsolete. In addition the steering system will face a couple of new requirements:

- Electric Power Steering (EPS) with failure prediction for Level 3 to ensure a sufficient hand over time
- Redundant power steering to ensure that the automation is able to steer the vehicle after steering system errors and to bring the vehicle to a safe stop or to ensure sufficient handover time
- Introduction of Steer-by-Wire (SbW) systems to enable advanced user interfaces for Level 4 vehicles.
- Reduced maximum steering wheel stroke due to higher levels of automation and the introduction of advanced steering systems (Active Front Steering (AFS) and SbW).

Impact on steering wheels

The introduction of these technologies will change the way how the user is interacting with the vehicle in general and the steering wheel in particular. While a shared driverless vehicle for urban short distance transport doesn't need a steering wheel at all, a steering wheel is still used for manual driving in privately owned or operated vehicles with Level 2-4 systems. In those vehicles the steering wheel gets an additional new role, as it is a very present and well known control device within the car. People feel in control of their vehicle when touching the steering wheel. For example, with higher perceived risk of the traffic situation most people tend to touch their wheel with two hands on a higher position on the rim. This means that the steering wheel is an important interface between automation and the driver in case of handing over control or communicating the automation status. Not only does the driver need to be informed, but also the automation system needs to have a certain knowledge about the driver and therefore needs an according monitoring function. Finally, when it comes to Level 4 highway automation for luxury vehicles, the end user probably will expect to use the newly owned free time for non-driving related tasks. No matter whether the driver is willing to work, enjoy the free time or eat behind the steering wheel, situations will occur where the space to the steering wheel is not sufficient. This will lead to the demand of a changing interior to make the automated journey more convenient. For driverless shared vehicles with Level 5 automation a steering wheel will probably be obsolete during normal operation. Nevertheless in certain situations, e.g. moving the vehicle in car service shops, an auxiliary control device may be necessary to move the vehicle. This is probably not a steering wheel but a control device relaying on the autonomous driving system, which can also be a remote controller. Table 1 shows a couple of steering wheel and steering system technologies, which are relevant for different automation levels. Those are explained and discussed in the following sections in detail.

Table 1: Steering wheel and steering system technologies for different levels of automation

SAE Level	**0** Manual	**1** Assisted	**2** Partly automated	**3** Highly automated	**4** Fully automated	**5** Autonomous
Steering System and Steering Wheel Stroke	approx. +/- 600	EPS (AFS) > +/-360		Reliable EPS	Redundant EPS or SbW > +/-180°	-
Steering Wheel Functions		Hands on/off detection (HoD)		HoD, Light element on rim, Driver monitoring camera, Steering wheel as input device		Auxillary control device
					Vital sign sensing, Foldable steering wheel	

Driver State Sensing

As described above for the different levels of automation, the driver is required to stay in the control loop (L 1,2) or to monitor the system to intervene in a short time frame. In case of a Level 1 system the driver is still responsible to control the vehicle either in lateral or longitudinal direction. To perform this task attention needs to be paid to the road environment and actively controlling the vehicle is required. Thus, an additional monitoring of the driver by the system is not required. Nevertheless, with an analysis of the driver's input on steering wheel or pedals it is possible to estimate the driver's anticipation in the driving tasks. This is nowadays used to detect driver's fatigue and to assist by suggesting a break.

With Level 2 systems the driver is no longer required to give any input to the control devices but needs to be ready for any immediate intervention in case of a system failure. To ensure the imminent takeover by the driver especially for lane keeping ECE-r79, a hand off detection system for the steering wheel is required. A system based on the torque sensor of the power steering system measuring the torque input of the driver on the steering wheel, is widely used in series applications. With the Addendum 78 of ECE-r79 in November 2017, new testing provisions have been created requiring the detection of a hands off situation within the full speed range of the systems operation, within 5 seconds on smooth road surfaces. Under these conditions current systems struggle in detecting the hands off condition in a timely manner. To fulfill the requirement, a direct measurement of hands on the steering wheel is already in use by several OEMs and will be adopted by further OEMs within the next two years. Among the three currently known detection methods, the capacitive measurement is superior to camera based visual detection and skin contact detection, as it is independent of lightning scene, coverage of field of view and the use of gloves.

Going one-step further with the introduction of partly automated vehicles, the driver is only required to pay attention to the road environment and to intervene within a short time frame. The appropriateness of handover time is not finally defined yet. Current research is discussing a time period of 3 to 10 seconds. Some user studies even indicate longer time frames to ensure that the driver fully anticipates the current traffic scenario and is aware of the situation. Nevertheless, it has been proven that the take over time among different drivers varies significantly depending on individual factors. As these systems take over the driving task, the driver is not needed to touch any control device while using the system. In parallel, the driver increasingly trusts the system as these are getting more advanced in handling many situations. This will result in an extended allowance of side tasks and a lower involvement in monitoring tasks. To overcome this, a monitoring of driver's visual attention becomes necessary. First series applications are using infrared driver monitoring cameras located close to the steering wheel (see Fig. 1), for example on the steering column shroud, to monitor drivers' head orientation and eye gaze. This position of the system enables a maximum detection range because the camera is looking from a lower straight frontal direction in the drivers face and thus reduces eye occlusion due to drivers' head pitch and yaw rotation. In the future, visual attention monitoring systems will be introduced on a wider range, as NTSB[1] investigations on an accident with a Level 2 automation system and planned amendments for the ECE-r79 recommend the detection of a driver's visual attention. Nevertheless, a camera alone is not able to detect any lack in driver's perception of the road environment, e.g. due to high workload. A driving simulator study showed that the take over time can be significantly reduced when the technology of hands off detection is used to detect the driver putting both hands on the steering wheel to disable the automated driving system. For systems, allowing hands off driving in Level 3 or 4 the hands off detection is changing to hands on detection. The requirements will also change: to ensure that an unintended deactivation of the system is omitted when touching the steering wheel accidently i.e., the system should differentiate between touching the steering wheel and a full grip with both hands. This requires at least three different sensing zones on the steering wheel: one in the rear and two on the front (left/right side).

1 National Transportation Safety Board: NTSB/HAR-17/02, PB2017-102600, Notation 56955, September 2017

Especially when Level 4 systems will be introduced in the market and the driver is allowed to extend the side tasks, visual attention detection systems can come to their limit as they can be covered by objects related to side tasks (e.g. newspapers). Formally, it is not necessary to monitor the driver's attention during automated driving, but it can be of interest to check the driver's attention before handing back the driving task to the driver, or distinguishing between intended an unintended driver initiated inputs to the control devices. Electrocardiogram measurement on the steering wheel (see Fig. 1) can be used to monitor drowsiness or workload of a driver before handing over the driving task. EuroNCAP is currently planning to integrate advanced driver monitoring systems in their future ratings, which could require vital sign sensing to classify driver states. It is also very likely that manufacturers will bring non –driving related monitoring systems to the market. For example, a permanent non-intrusive heart rate detection to detect medical emergencies of drivers, who are travelling on their own with hands off the wheel on long journeys. This measurement would enable an automated emergency call and a safe stop on the roadside.

As driverless shared urban transport vehicles are likely to come without a steering wheel, there are no related additional functions. From a driver monitoring perspective, functions like identification of occupants and detection of persons and objects left behind can become relevant to ensure passenger safety and to maximize user experience when using the service.

Figure 1: Steering Wheel with sensing systems for Level 3/4 automated driving and light element for status communication.

Communication of Warnings and Automation Status

The steering wheel is already widely used in Level 1 and 2 systems to indicate lane departure warnings or blind spot warnings by a vibration. A rotating unequal mass or a linear mass exciter usually realizes this function. In the future, exciters, which are programmable in vibration amplitude and frequency, can enable different warning levels. Furthermore, an active haptic feedback on the steering wheel for secondary vehicle functions can be realized.

Additionally the steering wheel rotation angle and torque are providing information on the current lateral control of the vehicle. In environments where a moderate steering wheel angle is necessary such as highways, slight movements of the steering wheel can indicate the current operational status of the system. On windy roads and urban environments, the steering wheel movement increases and could be perceived as disturbing. Furthermore, fast rotations of the steering wheel can cause an additional injury risk. Concepts based on AFS or SbW have been presented to either reduce the rotation of the steering wheel or to eliminate it during autonomous driving. In this case, further adoptions of the ECE-r79 legislation and further research is necessary to overcome the steering wheel and steered wheel angle difference when a takeover is initiated either by the driver or the system.

To communicate automation status and warnings, a visual light element (Lightbar, see Fig. 1) in the steering wheel rim can be used. A first application is using a light element in the 12 o'clock position to indicate forward collision warnings, automation status and handover processes. The 12 o'clock position of the steering wheel offers the possibility to have a salient warning function in the near- to mid-peripheral view of the driver while driving or while engaged in side tasks. Additionally, if the driver is operating with secondary devices in their lap, the position of the light element will guide the drivers' attention back to the road environment. This can lead to significantly faster reaction times in handover scenarios as well as in frontal collision warnings. In a driving simulator study, frontal collision warnings supported by a Lightbar element reduced the average distance to stop the vehicle from 40 kph cruise speed by more than 10 meters (equal to 39%). Besides the first application in series production in 2018, further applications are expected within the next years. Latest concepts show applications of a light element on the complete rim. These are mainly used for decorative purposes, hand position or grip status indication and activation of secondary vehicle functions. Further use cases in the context of automation need to be investigated. Based on cost and integration complexity, a moderate amount of applications is expected in the future.

Increasing driver's space for side tasks

Expert and user interviews show the desire to execute a broad range of side tasks, when the driver is no longer required to stay in the control loop or to monitor the system. For example, these tasks can vary from working, entertainment and eating to sleeping. All these situations have in common that the driver will need more space to execute the task. At present, the steering wheel is in close proximity to the driver, even if the seat is moved backwards and the backrest angle is inclined. Ideally, the steering wheel can be retracted to provide more space. But this is only possible for Level 4 systems where no imminent system initiated take over requests occur. As there will be collisions with surrounding interior parts while retracting the steering wheel, a simple retraction based on steering column adjustment is very limited. Several concepts have been presented to fold the steering wheel in order to retract it. This will require AFS or SbW to ensure that the steering wheel is not moving when folded, as it would cause collisions in its park position. While the introduction of any joint in the steering wheel rim will decrease the steering wheel rim stiffness significantly, it is recommended to install the rotating mechanism in the spoke area. For example, the complete rim can be rotated along an axis close to the steering wheel hub so that the steering wheel can be stowed under the steering column shroud and instrument panel (see Fig. 2). The design of an appropriate user interaction during transitions needs to be investigated as well as impacts on trust in the automation system or mode awareness issues in alignment with regulatory requirements.

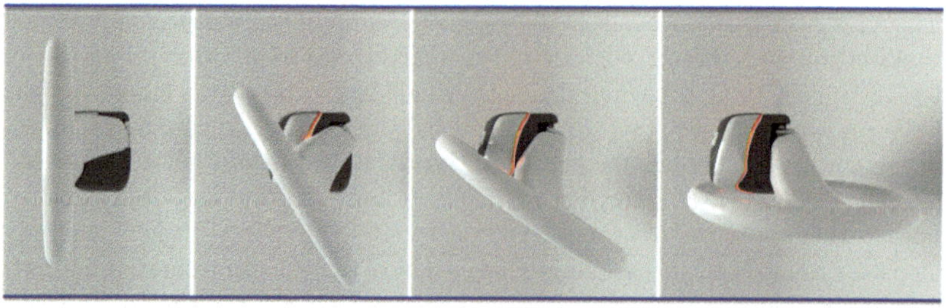

Figure 2: Option for stowing the steering wheel under the steering column shroud and instrument panel.

As a simple but very limited alternative, several OEMs already introduced smaller steering wheel geometries for Level 1 and 2 vehicles (see Fig. 3). Driving simulator studies on maximum required steering wheel stroke as well as possible shape modifications show that a closed geometry is necessary to avoid driver's grabbing onto nothing while steering. Furthermore, flattened steering wheels in 6 o'clock and 12 o'clock rim area, as used for sportive designs, have already been introduced in Europe.

Figure 3: Smaller and flattened steering wheel for assisted and highly automated vehicles.

Using the steering wheel as additional input device for side tasks

Some concept vehicles indicate that instead of folding the steering wheel away it can also serve as an additional input device for side tasks while driving autonomously. In any case, this requires the input device to be non-rotating. This can be realized by a technology used for non-rotating driver airbags or with a SbW system, which enables the decoupling of steering wheel and steered front wheels. In both cases, it is possible to integrate a touch interface or a touchscreen in the center of the steering wheel. For a current common steering wheel diameter of 360 mm the display size is limited to approximately 10 inches. The steering column angle of 19 to 21° is a non-favorable position for the display in terms of view angle and wrist comfort while using the touch functionality. Portable devices like tablets and smartphones are normally held and used at a 45 to 60° angle, thus an additional angular adjustment of the display or input device can increase comfort and user acceptance. Additionally, these systems need to be carefully designed concerning driver distraction in manual drive and lower levels of automation.

Summary

In this paper several technologies which can be integrated in steering wheels have been discussed in the context of automated driving. It is very likely that functions for hands on/off detection will be widely used for future assisted and automated vehicles to ensure a safe operation of the system as well as safe handovers between automated and manual driving. In addition, several concepts for status and warning communication like steering wheel vibration and light elements on the steering wheel rim will be used in future vehicles. Looking at higher levels of automation where the driver is not required to pay attention to the driving task at all but the vehicle still can be driven manually there are options to stow the steering wheel or to reuse them for non-driving related tasks. The introduction of such functions is expected by 2024 or later. To be able to introduce systems like these, a variety of further investigations and changes in regulations are necessary.

The battle of the senses: application of future control devices as a fallback solution for highly automated driving

Dr. Marcus Perner, Christian Dreher, Ronny Lepke,
Viktor Reich, Malte Stock, Angelo Niendorf, Moritz Lampe,
Martin Gebhardt, Simon Heine

IAV GmbH, Vehicle Dynamics, Chassis Development

Introduction

There are several categories defined by SAE automation levels on the way to autonomous driving [1]. Nowadays, automated functions associated with the 2nd and 3rd level exist in cars, e.g. the lane keeping assist and the traffic jam pilot system. This paper looks forward and focusses on the degree of high automation, respectively the 4th level. Imagine you are travelling by highly automated car or shuttle and you are sitting on a comfortable seat. Somehow, a certain traffic constellation or hazard appears and human has to take over the control. If there are no windows to front or the view is obstructed, smart strategies have to be developed to cope with this issue. Since human needs to see and to react based on the sudden situation, the use of additional video googles is seen as part of the solution. In the event, there is no steering wheel or other conventional control device an alternative concept is necessary to achieve safe driving. At its best, the device is part of or integrated into the vehicle interior and immediately ready to hand. Therefore, this paper illustrates a feasibility study of an innovative drive controller. The control device is proposed as a fallback solution to guarantee controllability and safety in particular situations.

On the way to autonomous driving

The use of steering wheels in future vehicles is highly discussed in research and application. Since the level of information that is provided by assistance functions is high, the driver needs to filter. In 2015, NHTSA investigated human factors in case of level 2 and level 3 automated driving [2]. The paper concludes that there is a huge improvement using nonvisual components in case of a take-over request. Moreover, a faster take-over reaction is achieved in specific situations if multiple alert types are used simultaneously, likewise visual and haptic signs. The paper of NHTSA [3] provides a guidance to develop a valid kind of interface that copes with driver limitations, capabilities and expectations. The contribution of Meier Arendt [4] confirms the need for more investigations towards optimized driver-machine interfaces. Different issues have to be considered likewise the take-over request for autonomous driving, the interface during autonomous driving and the take-over request for human driving. The paper of Harblik et al. concludes that there should be some kind of training prior using assistance systems to prevent harm in case of a sudden loss of functionality [5].

In addition, the discussion of suitable control devices and balance of information level is ongoing. Ruck and Stottan [6] propose a steering wheel design that consists of a touch display, several ergonomic switches and buttons including hands-on detection. Kienle [7] point out that measures are necessary to guarantee that the driver is still part of the control loop. A complete human passivity during highly automated driving should be avoided. Nevertheless, Kienle pronounces that automated actions should be communicated to the driver.

Despite of this, different geometry of steering wheels are investigated. Strümpler et al. [8] established, that the so-called 6 p.m. area could be flattened. Nevertheless, the steering wheel rim still should be closed for crash safety.

In summary, many disciplines are working hard to reach the necessary technology step for highly automated and autonomous driving. Mainly, these approaches are located in the chassis area with focus on steering system and steering wheel. Nevertheless, acceleration and deceleration management are needed, as well.

Approach for future control concepts

Control concepts as an interface between the driver or customer and the vehicle will be still needed on the way to autonomous driving. Despite of all challenges, it is expected that the driver or customer will be the fallback solution in case of functionality shortcomings due to road works, unclear road marking and extreme traffic.

In contrast to that, we expect control devices that are different from the current steering wheel will have their impact for highly automated driving. Therefore, IAV GmbH proposed an innovative control device in 2018 [9]. Due to the innovative design, primary functionality can be control of infotainment system or light concept. If automated driving mode reaches functional limits, the customer needs to take over driving control. Therefore, the control device can be handled as a drive stick.

Drive stick is a hand-operated device to control vehicle motion in case of no steering wheel or foot pedal. Moreover, control is realized via cable or wireless connection. The device provides two independent motion directions. A translatory movement of the drive stick initiates either acceleration or deceleration to control longitudinal motion of the vehicle. Then, drive stick in middle position applies steady vehicle velocity. A rotatory movement of the drive stick realizes steering of the vehicle. Maximum angle range is about +/- 20 degrees around the central position. Using a progressive steering characteristic improves controllability. Besides, maximum steering angle depends on the current vehicle velocity. Haptic feedback is possible for response of driving and vehicle state. Therefore, the realization concept can be mechanical or electrical based.

Feedback through mechanical mechanism applies force or moment that is reverse to the initial motion. Contrary force or moment increases proportionally. It is an easy and economical approach. An electrical-based feedback actuator gives more clearance since individual characteristic can be realized, e.g. different operational profiles or velocity-dependent counterforces. During standstill, short translational movement of the drive stick enables gear selection. Pull to central position reactivates acceleration commands.

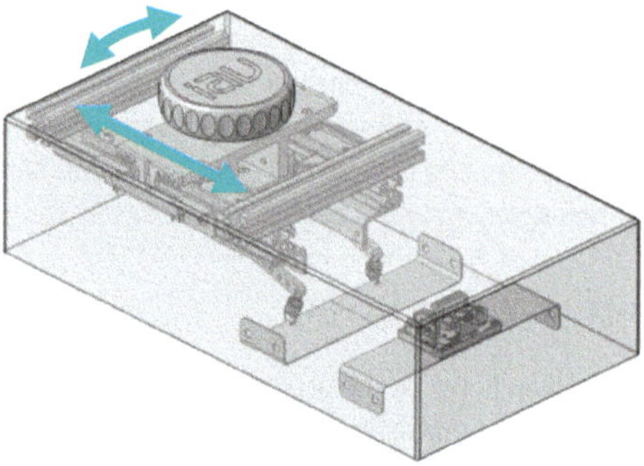

Figure 1: Functional design of the drive stick

Experimental set-up

Functional Design

The functional design of the approach is classified into two sections: driver and vehicle (cf. Figure 2). Within the study, the test person uses the drive stick as the control device. The control transmitter on driver side sends information to the control receiver on vehicle side. The control unit separates information to steering and power train system. In addition, a camera is attached to the RC car and the video is transmitted to the first person view goggles (FPV) that the test person is using.

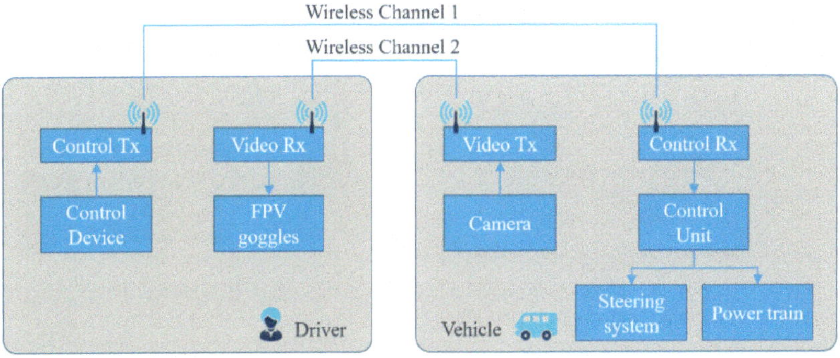

Figure 2: Functional design of the study approach using two wireless channels to support control of the vehicle and current camera view to the test person

Test scenarios

To evaluate usability of the proposed drive stick five scenarios are defined:

- drive through a gate (A),
- drive through a gate with stop after passing (B),
- avoiding an obstacle by lane change and return (C),
- drive a curve (D) and
- forward parking into 90 degrees parking space (E).

Test case A requires driving through a 450 mm wide gate. Test case B adds the task to stop right after the gate. For test case C the test persons have to avoid an obstacle. Therefore, the RC car needs to perform a lane change and a return to the prior lane. Pylons are used to realize the gates and the obstacle. The challenge in test case D is to avoid lane crossing while driving through a curve. Front parking is the task of test case E. Therefore, the test persons need to follow a lane and to perform 90 degree parking.

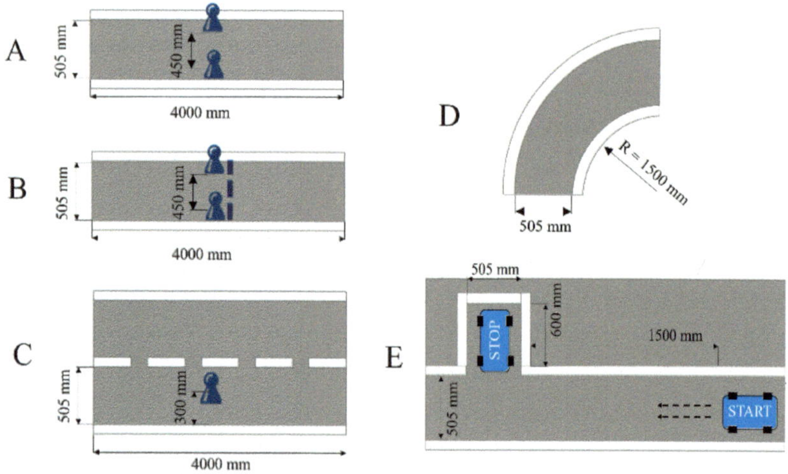

Figure 3: Overview of the test cases to evaluate drive stick: A - gate, B - gate and stop, C - avoid obstacle, D - curve driving, E - front parking

Evaluation criteria

Two different evaluation approaches are combined. First of all, lane crossing is observed and counted. Secondly, a questionnaire is designed for the test person to answer. The test person classifies impression after each test case using different answer categories. Therefore, a common rating scale is used (cf. Figure 4). By finishing all test scenarios the test persons are asked for further judgement. Therefore, the points are stress level, controllability, handling and steering feel using the drive stick. In addition, applicability for actual vehicles is questioned.

Fulfillment degree of the driving task				
☐ very low	☐ low	☐ medium	☐ high	☐ very high

Figure 4: Example of the subjective evaluation of the driving tasks by using several questions and grading (very low = 1 ... very high = 5)

Study remarks

First of all, the camera view of the RC car needs to be discussed. That is why Figure 5 shows the comparison to an actual test vehicle. Despite of the different camera angle, the view seems equal. Moreover, if the obstacles around the RC car are true to scale, the view becomes even more comparable. If the driver is not located in the car, there will be no effect of acceleration, as well. That is why we expect a certain transferability of the results over to an actual vehicle.

For testing, two groups are defined with different test case order. In summary, 70 people performed the tests. Each of them has a certain knowledge of using controllers, e.g. for video games. That is why we addressed an equal level of experience. We achieved this selection due to a prior questionnaire. Distribution in age and gender is related to the statistic of driver licenses in Germany [10].

In fact, there is no training for the test persons before test-driving with the drive stick.

Figure 5: Camera view from RC car (left) and vehicle (right)

Results

All subjective ratings and lane crossings are assessed for each group and each test case. At first, it is necessary to know how many test persons had the impression that they managed the task. Therefore, the ratings that are higher than two are counted. The results are listed in Table 1. It can be seen that the success of the test result depends on the test case. To give an example, 78 % of the test person think that they managed the task of parking. In contrast, 50 % of the test person think that they managed it to avoid an obstacle and return to the original lane.

Using the objective approach of counting lane crossing during the test-driving gives an enhanced result. Therefore, all ratings lower than three are analyzed. That means, the test person got the feeling, that the success of the driving task was low or very low. In addition to that, these ratings are reviewed whether there was a lane crossing. If there was no lane crossing at all, the test person managed the task from objective point of view. Finally, these ratings are summed up to get an overall result. Then the mean value of the group result is calculated with respect to the test case. The average shows that the rating of the success differs. Overall, test persons of both groups managed the driving tasks, in figures with at least 62 % (test case C – obstacle) and at its best 85 % (test case B – gate and stop). Nevertheless, the result is promising because the test persons had no prior training. Investigation of several important criteria show a further result. Therefore, stress level, controllability, handling and steering feel are assessed after all test cases. In addition, applicability for actual vehicles is asked. The results are listed in Table 2. Although the drive stick already provides intuitive handling, the test persons see the need to improve controllability and steering feel. Ongoing tuning and application will improve these two characteristics. The opinion of test persons is diverging due to the question whether the approach is applicable to actual vehicles. Today, further integration tests are already going on to show proof of concept on proving ground driving actual vehicles.

Test case	A	B	C	D	E
Description	Gate	Gate and stop	obstacle	curve	parking
Group I					
Rate > 2	58 %	67 %	50 %	69 %	78 %
Rate < 3, no pylon and lane crossing	6 %	11 %	6 %	0 %	0 %
Sum group I	64 %	78 %	56 %	69 %	78 %
Group II					
Rate > 2	85 %	91 %	68 %	65 %	76 %
Rate < 3, no pylon and lane crossing	6 %	0 %	0 %	3 %	3 %
Sum group II	91 %	91 %	68 %	68 %	79 %
Mean	78 %	85 %	62 %	69 %	79 %

Table 1: Rating the success of the driving task by two separated groups of test persons, each group with different test case order

Criteria	Stress level	Controllability	Handling	Steering feel	Applicability
Group I					
Rate > 2	72 %	42 %	86 %	44 %	58 %
Group II					
Rate > 2	76 %	53 %	88 %	47 %	41 %
Mean	74 %	47 %	87 %	46 %	50 %

Table 2: Rating of several criteria by two separated groups of test persons, each group with different test case order to investigate drive stick approach

Conclusion

On the way to autonomous driving takeover scenarios still need to be taking into account. Due to road works, differing road marking or extreme traffic, unclear situation can arise. This paper proposed a drive stick as a fallback solution in case of functionality shortcomings. Mechanical and functional aspects were explained. The concept is adaptable for use via cable or wireless connection. In addition, a test study was carried out to analyze usability of the drive stick. The majority of the test persons succeeded their drive tasks. Nevertheless, potential for improvements come up. Steering feel and controllability are two directly linked criteria. Based on improvements in steering feel by application, we expect the controllability rating to increase, as well. Beside of this, the integration into an actual test vehicle is going on.

References

[1] Society of Automotive Engineers (SAE), *J3016: Taxonomy and Definitions for Terms Related to Driving Automation,* 2018th ed.: SAE International.

[2] Myra Blanco et al., *Human Factors Evaluation of Level 2 and Level 3 Automated Driving Concepts*: NHTSA.

[3] NHTSA, "Human Factors Design Guidance for Level 2 And Level 3 Automated Driving Concepts,"

[4] G. Meier-Arendt, "Interaktionskonzepte für das automatisierte Fahren," *ATZ - Automobiltechnische Zeitschrift*, vol. 120, no. 4, pp. 18–23, 2018.

[5] Joanne L. Harbluk, Peter C. Burns, Darryl Malone & Jeremy Hamilton, *Power Steering Assist Failures: Driver Behavior, Safety Impacts, and Implications for Automated Vehicles*.

[6] H. Ruck and T. Stottan, "Interaktives Lenkrad für eine bessere Bedienbarkeit," in *ATZ / MTZ--Fachbuch, Fahrerassistenzsysteme und Effiziente Antriebe*, W. Siebenpfeiffer, Ed., Wiesbaden: Springer Vieweg, 2015, pp. 18–26.

[7] Martin Michael Kienle, "Kooperative Interaktionskonzepte zum Fahren eines teilautomatisierten Fahrzeugs," Dissertation, Fakultät für Maschinenwesen, Technische Universität München, München, 2014.

[8] C. Strümpler, D. Alt, and S. Sohr, "Reduzierte Lenkräder für die Automatisierung," *ATZ - Automobiltechnische Zeitschrift*, vol. 120, no. 4, pp. 28–33, 2018.

[9] Ronny Lepke, "Verfahren und Vorrichtung zur Steuerung der Längs- und Querbewegung eines Fahrzeuges mit einem von einem Bediener betätigbaren Stellelement," DE 10 2018 104 522 B3, May 2, 2019.

[10] Kraftfahrt-Bundesamt, *Bestand an allgemeinen Fahrerlaubnissen im ZFER am 1. Januar 2018 nach Geschlecht, Lebensalter und Fahrerlaubnisklassen.* [Online] Available: https://www.kba.de/DE/Statistik/Kraftfahrer/Fahrerlaubnisse/Fahrerlaubnisbestand/2018/2018_fe_b_geschlecht_alter_fahrerlaubniskl.html?nn=2218648.

BRAKE.TECH SECTION

SYSTEM REQUIREMENTS AND BRAKE FEEL

NCAP, legislation, and real world requirements require high-pressure dynamics in brake systems

Dr. Alexander Brandt, Robert Bosch GmbH

This manuscript is not available according to publishing restriction. Thank you for your understanding.

End of oversizing: smart designing of brake systems for BEVs

Dr. Jens Hoffmann, K. F. Wörsdörfer,
Continental Teves AG & Co. oHG

This manuscript is not available according to publishing restriction. Thank you for your understanding.

Virtual brake design and influence of brake pedal feel in a front-loading process

Dirk FRERICHS, GROUPE PSA, Rüsselsheim

Samuel LAGO PLACES, GROUPE PSA, Rüsselsheim

Karl Michael HAHN, GROUPE PSA, Rüsselsheim

Introduction

The automotive industry is in a radical change because of business fields. Topics like emission free mobility as well as highly automated driving are the new key topics of the industry. To be robust for the future the industry already optimizes its budgets and adapts its development processes to reduce the time to market. To reduce the costs for a vehicle development project, the number of prototypes is limited as well as the test trips for the development e.g. for an application engineer for Electronic Stability Control (ESC). Furthermore the variant and functional complexity is increasing which results in an increasing validation effort. Both, the reduction of costs and the increasing functional complexity are contradictory to reducing the time to market.

To find ways out of this dilemma the importance and responsibility of the system simulation increases. So the motivation is to apply system simulation to create results for all the available variants which are not available as a prototype. This moved validation tests from the *Road* to the *Lab*. For example Hardware- in- the- Loop (HiL) systems are used for the homologation of ESC [1], which happens very late in the development process. In a next step system simulation is used to support the calibration by simulation. Based on *Math* models same tests are executed on the computer only in Model- or Software- in – the – Loop (M/SiL) simulations [2]. This overall strategy can be called *Road- Lab- Math* [3].

In this contribution an approach is presented to extend this strategy. System simulation can be used to frontload the development process by virtualizing the real world processes for component design. So system simulation can support the design decisions already in a very early development phase where potential critical paths in the design can be identified, such that late design changes can be prevented. This supports the cost goal as well as the launch excellent. Also it is possible to support the goal of reducing the time to market due the fact that system simulation can start with an uncertain parameter space directly at the first milestone of the project. In this application field a cultural mindset like *agile* will support a resource optimal development. Focusing for example on the hydraulic brake system, based on the experience of [4], the brake sizing process as well as the analyses of the pedal feel can be virtualized to close the gap between virtual testing and virtual component design.

In the following the virtual bench concept will be introduced which can be derivate from the real world processes. Subsequently the virtual bench concept will be applied in the first step for the brake system sizing and afterwards in a demonstration how it effects the pedal feel.

Virtual Test Bench Concept for Hydraulic Brake System

In this chapter we want to give an overview about the real test rig setup which will be the typically use for wheel brake investigations. From this state of the art we want to derive the concept of virtual test benches in general. The usability and scalability of this approach will be demonstrated for a hydraulic brake system.

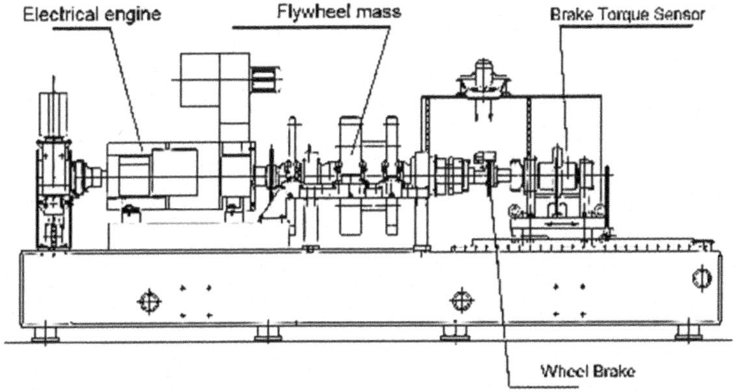

Figure 1: Flywheel Brake Bench in the real world illustrated like in [5]

A typical flywheel brake bench is shown in Figure 1. It is the classical experimental approach in the physical world. The typical use case for such a test bench is to check the functionality and performance of the wheel brake. Other possible applications fields are durability tests, wear analysis as well as NVH investigations e.g. brake rubbing [5].

The bench consists of an electrical engine which increases the load torque M_{Load} to accelerate the flywheel mass J_{Wheel} to a specific target velocity $v_{Wheel,d}$. The flywheel mass J_{Wheel} can be decoupled from the electrical engine, e.g. if a specific target velocity $v_{Wheel,d}$ is reached, so that the wheel brake and the engine do not influence each other. If the electrical engine is decoupled, which means that M_{Load} equals zero Nm, the pressure in the wheel brake $p_{Caliper}$ is increased. This leads to a torque M_{Brake} which affects the velocity of the flywheel v_{Wheel}. As result deceleration a_{Wheel} of the flywheel increases. The torque of the wheel brake M_{Brake} is measured by a torque sensor.

Not shown in the Figure 1 is the control of the test bench. It can be assumed that there are two main control parts in the bench [5]. The first is the velocity of flywheel mass v_{Wheel} and the other one is pressure control of the wheel brake $p_{Caliper}$ based on the target declaration of $a_{Wheel,d}$. Furthermore the states of the bench have to be observed which is done by the bench control. The different states are (1) accelerating the flywheel mass J_{Wheel} to the specific target velocity $v_{Wheel,d}$, (2) decoupling the electrical engine from the flywheel mass J_{Wheel} and (3) increasing the pressure in the wheel brake $p_{Caliper}$ with a target deceleration of the wheel $a_{Wheel,d}$.

This flywheel brake bench can be abstracted into 4 different partitions which is illustrated in Figure 2.

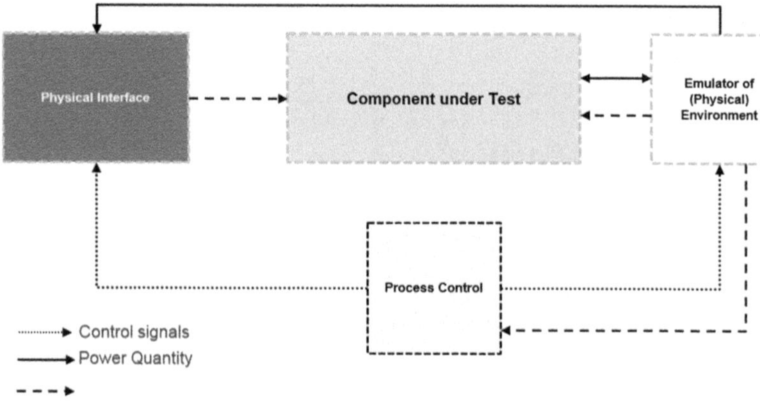

Figure 2: Abstraction level of benches

The main focus is the *Component under Test* which is the wheel brake in our described use case. This is the device to prove the properties and/or attitudes. With the *Physical Interface* the connection between the test bench and the *Component under Test* is characterized. The *Physical Interface* has to stimulate the *Component under Test* to follow the wanted loadcase which describes the actuation of the bench. In the case of the wheel brake it is the pressure in the caliper $p_{Caliper}$. According to this there are also inputs that are reactions of the *Component under Test*. Due to the fact that the *Component under Test* has influence on the connected environment an *Emulator of the Physical Environment* is needed. This emulator is just a simplification of the real environment. In the illustrated flywheel mass bench in Figure 1 the flywheel mass is the *Emulator of the Environment*. In this case the flywheel itself is the simplification which represents the mass of the quarter vehicle [6]. The *Physical Interface* as well as the *Emulator of the Physical Environment* will be controlled by a process control unit. This *Process Control* describes the interface between the real test rig and the test rig operator. The interface enables the control of the emulator of the environment according to the load profile. Furthermore it collects all required information which is needed to analyze and assess the *Component under Test*. To describe the different classes of information flow we define three types: control signals, states and power quantities. Thereby all control signals are used to start, send or stop requests to the controller. A power quantity $P(t)$ is defined as the product of the general potential $e(t)$ and a general flow $f(t)$ [7]. Note that the exchange of the quantities is expressed as a vector by

$$\underline{P}(t) = \begin{bmatrix} f(t) \\ e(t) \end{bmatrix}. \tag{1}$$

State quantities $\underline{x}(t)$ describe all non- potential and non- flow variables which are needed as stimulus to solve the mathematical descriptor system.

The formalization in Figure 2 can be used to derive a virtual flywheel mass test bench for a virtual wheel brake system. The partitioning in different subsystems is shown in Figure 3.

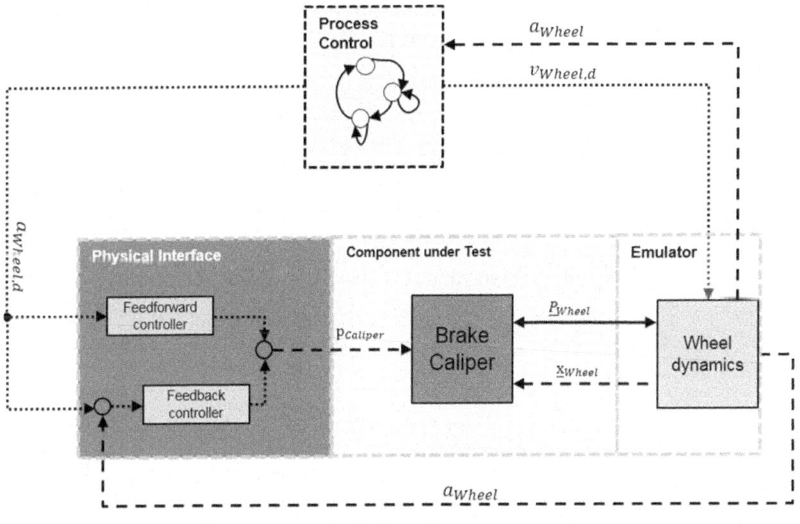

Figure 3: Virtual Flywheel Mass Test Bench for a Wheel Brake

The *Component under Test* is a virtual component model, which is managed in GitHub, a decentral version control system. A virtual component model consists of the model structure self and the parameter set [8]. Based on the Automotive X-in-the-loop object-oriented model framework (AXIOM) the *Model Configurator* supports the exchange of model structure or set of parameters [9]. The component models themselves can have their sources in different modeling tools and can be simulated in co- simulation or model exchange mode e.g. [10]. The emulator type can be also configured via the *Model*

Configurator. During the simulation the emulator and the *Component under Test* exchange the power quantities $\underline{P}_{Wheel} = \begin{bmatrix} \omega(t) \\ M(t) \end{bmatrix}$ referencing to Equation (1) and the state quantity $\underline{x}_{Wheel} = \omega_{Wheel}$. Here the state quantity is needed for the numeric solution.

To stimulate the input a *Physical Interface* is needed to build pressure $p_{Caliper}$ causing the required deceleration $a_{Wheel,d}$. A feedforward with a feedback controller is designed [11]. The feedforward controller is based on the formula of the total brake torque

$$F_{in} = \frac{a_{x,Wheel}\, J_{Wheel}\, r_{Wheel}\, A_{MC}}{2\, \mu_{eff}\, A_{Pist}\, r_{eff}\, i_{ped}\, i_{Booster}} \tag{2}$$

with the radius of the wheel r_{wheel},area of the Master Cylinder A_{MC}, effective friction between brake disc and brake pad μ_{eff}, area of the piston A_{Pist}, effective radius of the disc r_{eff}, transmission of the pedal box i_{ped} and the transmission of the booster.

Because of math based assumption in Equation (2) a continuous error between $a_{Wheel,d}$ and a_{Wheel} will remain so that a feedback controller is required. For a feedback controller a simple PI- controller or state controller (e.g. LQR) can be used [11].

In the *Bench Contol* a state machine is observing the different states of the bench which can activate the different controller. If a specific state is reached, the bench control can turn on or off the different controllers and start various cycles.

Each part of the virtual bench – *Component under Test*, the controllers, the bench control with the load case and emulator of the physical environment – can be parameterzied independly via the *Model Configurator*. Also it is possible to save a complete application configuration which enables the possibilty to start the simulation remotely.

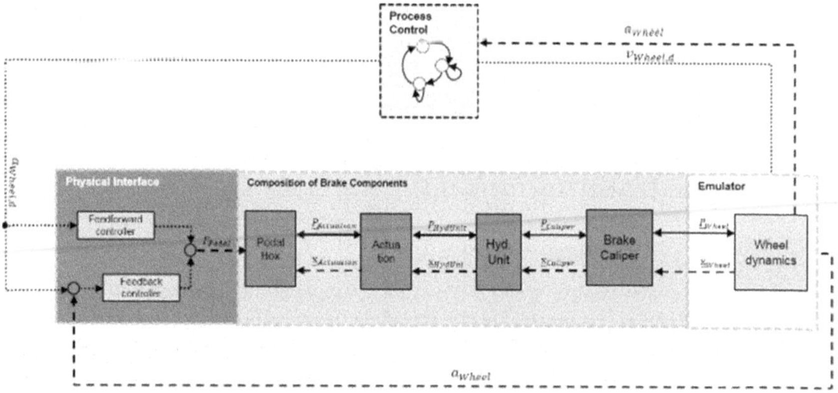

Figure 4: Virtual Bench for a composition of brake components

The applicability of this approach is of course not limited to the use case of wheel brakes as *Component under Test.* To stay at the brakes domain, it is e.g. possible to exchange the wheel brakes by a master cylinder or a composition of a set of components like it is illustrated in Figure 4.

The scaleability supports the identification of the composition behaviour already in early stages of vehicle development because hardware does not have to be available to make a composition rating. This approach enables to check the functional behaviour of the composition and also allows to perform a first assessment of the *Component under Test.*

For this it is possible to use different load cases with different metrics.

In the following chapter the concrete application field of Figure *3* and Figure 4 for the typical brake system sizig and pedal feel will demonstrated.

Brake System Sizing

Today a conventional brake system still consists of a brake pedal, booster, master cylinder, controller, hydraulic lines and wheel brakes. But beside of this more and more variations of the common brake system are developed. New technical solutions like electronic boosters and recuperative wheel brakes are investigated and find their way into serial vehicles. This increases additionally the complexity of brake system sizing. The main criteria considered are legal requirements of brake pedal forces and the resulting deceleration, thermal aspects including wear especially at wheel brakes and comfort criteria such as noise and vibration or brake pedal feel [5] .

With the increasing complexity in vehicle programs and due to cost and weight constraints demand for simulations of typical brake test increased significantly. For this reason the most OEMs and also suppliers developed in-house tools and methodologies to predict the behavior of brakes for different vehicle and brake combinations. The developed solutions vary from simple 0-D equation based calculations up to fully coupled CFD approaches [12]. The question, if a brake test can be simulated is no longer the main issue. It is more about which level of data is available, the amount of time to respond to the request and the required validity of results. The application of this virtual test bench is based on 1-D models, which represent the functional dynamics in a physical way. The approach shows a compromise between the simple ideal equation based description of the components and the high end CFD methodology.

In many cases the bottle neck is not only the parameterization of the Component under test. In early phases of vehicle projects, it is often possible to deduce parameters and models from predecessor projects and engineering judgement. The same is more complicated for parameters coming from other departments which might not have the needed parameters available at that phase. Parameters like vehicle masses are not very accurate and based on rough estimations.

In some cases it is possible to scale the environment to a certain level which still provides the required validity of results in the beginning of a vehicle project. Detailed vehicle and tire dynamics are candidates which have a limited influence on some of the brake tests used for brake system sizing. A typical brake bench for example represents the vehicle mass by a rotating inertia. The tire is deduced to the one geometrical parameter rolling radius, which helps to create a link from the inertia rotation speed ω_{Wheel} to the theoretic longitudinal velocity of a vehicle. When investigating thermal aspects at the wheel brakes, dynamics of pressure generation are not of major interest. With that an evaluation of fulfilling the thermal needs at the wheel brakes requires only a small set of parameters.

Virtualizing the typical brake bench decreases the time to answer questions of thermal requirements and allows to run DoE's of mass, tires and wheel brake parameters. Figure 5 shows a typical heating over a series of brake snubs on the left and the pressure change on the right side, which is caused by friction value change over temperature. An additional advantage of the virtual brake bench over the real brake bench is the consideration of air stream cooling the brakes. On virtual side it is easy to simulate an in-vehicle representative air stream. The same is hardly possible at the real brake bench.

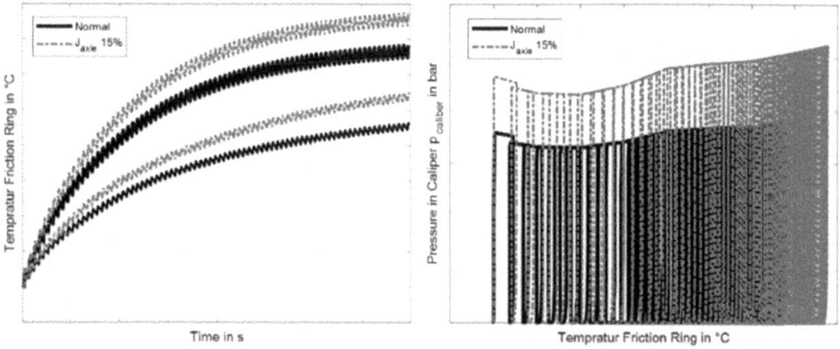

Figure 5: Left: Temperature behavior for 80 cycles Right: Pressure in caliper while increasing temperature

The concept of using virtual benches creates the possibility to test the performance of each component separately. Usually a simulation engineer would accept a system like the brake system when input to output of the system shows a realistic behavior. But due to the several components in a brake system it could for example happen that a higher friction at the wheel brakes compensates a lower performance of the booster. Exchanging the wheel brakes without checking the overall system performance again could lead to an error in the overall output of the brake systems. By testing each component to its supposed performance like it is done at real test benches, mistakes can be avoided and trust in simulation results increased. This is a key especially in early stages of vehicle developments to establish simulation driven development. Functional decomposition of systems to component level and testing of the proper performance provides the needed quality to enable reliable front loading of design decisions.

Brake Pedal Feel

With regards to brake pedal feel all components of the brake system are of major interest. Therefore it is necessary to ensure a realistic behaviour of each component of the brake system. Recalling the example of a higher torque output from wheel brakes which compensates for the lower performance of the booster, it is possible to use the virtual bench to test both components isolated for their performance. A typical test scenario of the booster, including the master cylinder, would be to test a pressure rise caused by a certain force gradient at the booster input-rod. In most cases a quasi-static measurement is performed showing the common design factors like jump-in pressure, different slopes of assistance and run-out pressure, shown in Figure 6. For brake pedal feel investigation also dynamic measurements at different apply speeds are in scope.

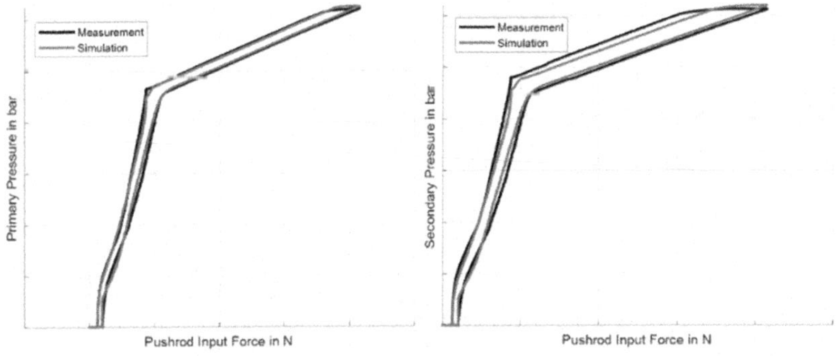

Figure 6: Quasi-static booster test comparison of measurement and simulation

In such cases the characteristic values of a booster are distorted and design values like the jump-in pressure deviates from the quasi-static application. Comparing the output of the two measurements Figure 6 and 7 shows that a simple model will hardly satisfy the needs of simulating brake pedal feeling in all conditions. With regards to brake pedal feel the validity of the booster model performance is shown. To assure a confident prediction overall performance of the brake pedal feel, it is needed to focus on wheel brake performance in a next step.

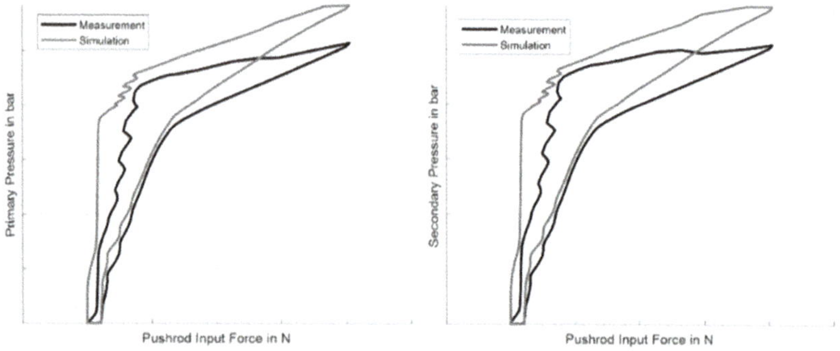

Figure 7: Dynamic booster test comparison of measurement and simulation

One of the main criteria of interest at the wheel brakes is the volume consumptions by all brake corners. Similarly, the torque generation which is very dependent on the friction value between rotor and pads is crucial. Both depend strictly depend on the applied pressure. Once the performance of the wheel brake as well as the performance of all other brake components is validated a reliable prediction of brake pedal feel is possible.

This approach is just mirroring the test bench scenarios which are performed for the components in the real component development.

But besides this model quality assurance the virtual bench concept can be abstracted to a scenario not existing in the real test bench world. It is easily possible to simulate a complete brake system on a test bench. And so brake pedal feel can be studied on a test bench without any environmental influences. Meaning our *Component under Test* is a complete brake system still connected to a flywheel brake bench. This allows to study the brake pedal feel without aerodynamic drag of vehicle, rolling resistance from tires or residual torques from the powertrain. Typically designing the pedal feel is always an act of balancing. Pedal force and pedal travel, which are both key factors of the pedal feel, are in conflict to each other. In general no matter which parameter is tuned in the brake system, one will find the one criteria impacted positively and the other negatively. In a moderate way for example the brake pedal ratio could be tuned. Decreasing the pedal ratio would cause a shorter pedal travel at same deceleration. On the other hand a higher pedal force is needed to establish the same rate of deceleration as shown in Figure 8. As the pedal ratio is just one parameter influencing the brake pedal feel it makes absolute sense to run a DOE including all relevant parameters and their theoretical tuning bandwidth.

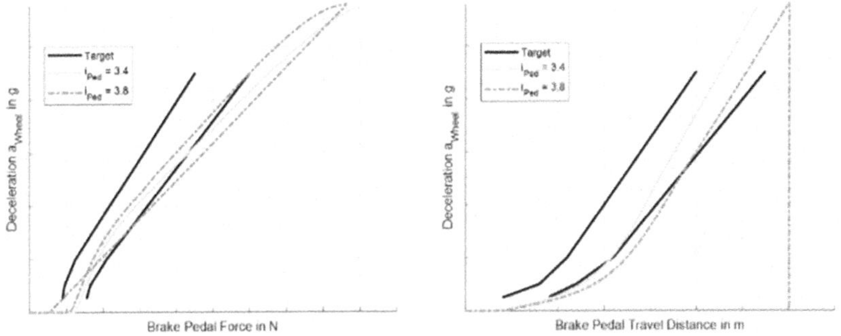

Figure 8: Simulation of brake pedal feel with different pedal ratios

Both the virtual bench concept as the complete vehicle simulations are based on AXIOM [9]. So running a simulation of the same brake system in a vehicle as studied on the virtual bench is an easy task.

Conclusion and Outlook

An approach to front load the process of designing the components based on virtual benches has been presented and demonstrated exemplary based on brake system sizing and pedal feel. Starting point was the real world flywheel bench which was introduced. Derived from this a generic virtual concept was proposed which was applied to the use case of designing the brake system. It was shown that this approach enables to scale the complexity during design. This was demonstrated for the brake system sizing as well as for the pedal feel.

Due to the fact that the virtual bench concept is based on a simulation framework several simulation departments can benefit from the increase in component model quality. All participants of the framework are able to integrate the designed brake system into their simulations by knowing that the correct behaviour of the system is in the interest of each team. So the gain in simulation results goes way beyond the aim of a robust simulation for the own discipline. As all teams in AXIOM develop and simulate all the different vehicle systems, like steering or tire, an application of this approach to all other systems is mandatory – and easily possible.

References

[1] Henning Holzmann, Karl- Michael Hahn, et al., "Simulation-Based ESC Homologation for Passenger Cars," *ATZ worldwide,* vol. 9, pp. 40-43, 2012.

[2] Markus Bauer, Roman Mannale, M. Kochem, A. Ravi, "CaliSim – Calibration by Simulation," in *VDI- Bereichte 2292*, Berlin, 2017.

[3] Michael Kochem, Henning Holzmann, "Einsatz der Road-Lab-Math Strategie bei der simulationsbasierten Entwicklung von Fahrdynamikregelsystemen," *at – Automatisierungstechnik,* vol. 55, no. 7, p. 368–374, 2007.

[4] Samuel Lago Places, Karl Michael Hahn, M. Alirand, "Braking circuit simulation – on a hardware-in-the-loop test bench – a first step to complete virtual testing," in *7th International Munich Chassis Symposium 2016*, Munich, 2016.

[5] Bert Breuer, Karlheinz H. Bill, Bremsenhandbuch, Wiesbaden: Vieweg+Teubner Verlag, 2012.

[6] Manfred Mitschke, Henning Wallentowitz, Dynamik der Kraftfahrzeuge, Wiesbaden: Springer Berlin Heidelberg , 2004.

[7] H. Czichos, Mechatronik, Wiesbaden: Vieweg+ Teubner Verlag, 2006.

[8] Dirk Frerichs, Matthias Borsdorf, "Quality for Vehicle System Simuation," in *SimVec*, Baden- Baden, November 2018.

[9] Andreas Erbes, Dirk Frerichs, Stefan Sinsel, Jochen Zäpf, "Collaboartive Model Development for System Simulation," in *Matlab Automotive Conference 2019*, Stuttgart, 2019.

[10] Dirk Frerichs, Matthias Borsdorf, Karl Michael Hahn, "FMI for Simulation of Vehicle Dynamics Control Systems," in *VDI- Berichte 2292*, Berlin, 2017.

[11] O. Föllinger, Reglungstechnik, Berlin: VDE VERLAG GMBH, 2016.

[12] S. Yigit, "An enhanced CFD approach to simulate brake fluid temperature rise during mountain descent and soaking phase," in *SimVec 2016*, 2016.

INNOVATIVE BRAKE COMPONENTS

Investigation of the run-in and corrosion behavior of a PEO-coated aluminum brake disc

F. Gulden[1], B. Reinhold[3], S. Gramstat[3], A. Stich[3], U. Tetzlaff2, H. W. Höppel[1]

[1] Materials Science & Engineering, Institute I, Friedrich-Alexander-Universität Erlangen-Nürnberg (FAU), Germany;

[2] Technische Hochschule Ingolstadt (THI), Germany;

[3] AUDI AG, Germany

Introduction

Plasma electrolytic oxidation (PEO) is a relatively new electrochemical surface treatment process for the generation of oxide coatings on metals such as Aluminum or Magnesium. Most of the papers, referring to the PEO process, were published in the eighties of the previous century [2, 3]. Nowadays, the process is used for a wide range of applications in various industrial fields [4]. In plasma electrolytic oxidation higher voltages are applied than in the anodizing process. Due to the high voltage and current, intense plasma is created, which oxidizes the surface of the metal mainly by subsurface oxidation. This high local input of energy transforms the oxide into crystalline Alumina, whereas conventional anodizing can only produce amorphous Aluminum oxide. The resulting layers are hard, wear-resistant, dense and well adhered oxide coatings [5, 6].

The electrical conditions are selected in such a way that millions of microscopic discharges arise through the oxide layer, which result in plasma temperatures of up to 10.000 K. These high temperatures cause the Al_2O_3 layer to melt, fill up the porous and amorphous oxide structure before re-solidifying and leading to a quite compact and dense coating. The coating itself consists of a thin porous layer near the metal/oxide-interface with trapped gas and solidification shrinkage, a rather dense and hard main body containing micro-cracks, and a porous top layer [6].

The characteristics of the coating can be influenced by adjusting the electrical process parameters, by modifying the chemical composition and concentration of the electrolyte and by varying the treatment time. The hardness of PEO coatings exceed the values of anodized layers. The Young's modulus is microscopically higher compared to anodized coatings, but the fine crack network reduces the macroscopic stiffness to roughly 40 GPa compared to about 370 GPa in dense Alumina [7, 8].

Referred to brake discs, the fine network of cracks may minimize the risk of delamination, because it reduces tensile stresses in the coating system, which occur during brake applications due to thermal and mechanical loads. Past investigations regarding wear performance showed that the hard, thick and well adhered Alumina coating is fundamentally predestined for friction systems [5]. Nie et al. [9] for example tested the wear performance of unpolished PEO-coatings on Al 6082 in contact with austenitic steel 316L in a dry rubber wheel abrasive wear test. They showed that after a run-in period, after which the porous top layer was worn off, the wear of 316L exceeded the wear of the coating significantly. Further research groups investigated the wear behavior of Alumina ceramic coatings and confirmed the statement of a high wear resistance [7, 10-11]. However, investigations in the context of a brake disc application were rarely carried out in the past.

The research group of Barton et al. [12] tested PEO coatings against an aligned automotive friction material on a 2 kW and a 45 kW small-scale brake dynamometer. They investigated a 30 to 50 μm thick PEO-coating on an Al 6082 alloy to evaluate the withstanding of the material system to real braking.

The measured coefficient of friction was $\mu = 0.25\text{-}0.30$ and therefore slightly lower compared to the reference disc made of grey cast iron ($\mu = 0.32$). The roughness of the disc surface, expressed by the R_a value, decreased during the brake applications. The wear of the coating was only 1 µm. The low wear can be traced back to the high hardness of the coating of about 1400 Vickers and the formation of a thin tribolayer of 2 to 4 µm thickness, which mainly consists of wear debris of the lining material. The maximum operating temperature and the failure mechanisms of the coated Al disc were investigated in drag braking tests. After reaching a temperature of 550 °C the disc failed catastrophically.

The PEO process mainly works for light metals, but cannot be applied on grey cast iron brake discs. However, the ongoing change towards an electrified powertrain accompanied by regenerative braking offers the opportunity to change the requirements for mechanical brake systems. In recent and future battery-electric vehicles (BEV) the disc temperature will decrease significantly and lower melting substrate materials like Aluminum alloys may be pushed into the field of view as alternative brake disc material.

The low density of 2.7 g/cm³ can improve driving comfort and dynamics by reducing rotating and unsprung masses. Furthermore, the corrosion resistance of the PEO-coating and the Aluminum itself are advantageous. Corrosion might become a more important factor for the lifetime of brake discs in BEVs, due to the reduced use of the brake disc and the lower disc temperature, which prevents the disc from drying and increases the exposure time to wetness and to highly corrosive electrolytes especially in the winter term. The low wear of the coating could also reduce the brake particle emissions, which is very important today and in the future [13].

This work, is a follow up to an earlier study [1] and takes a closer look at the formation and degradation of the tribolayer on PEO-coated brake discs. Based on the previous results, the corrosion behavior was investigated in an application relevant environment.

Methodology

Brake System

Brake disc concept

The inside-ventilated disc concept consists of an Al 6082 base alloy coated with a PEO layer of 40 to 50 µm. The cross-section of the initial state is shown in Figure 1. The coating can be divided into three areas as described in the literature [6]. It shows a high porosity in the region close to the substrate alloy and a main body with several microcracks. The appearance of the coating in the cross-section is the result of the applied preparation process with an ion mill ("curtaining effect").

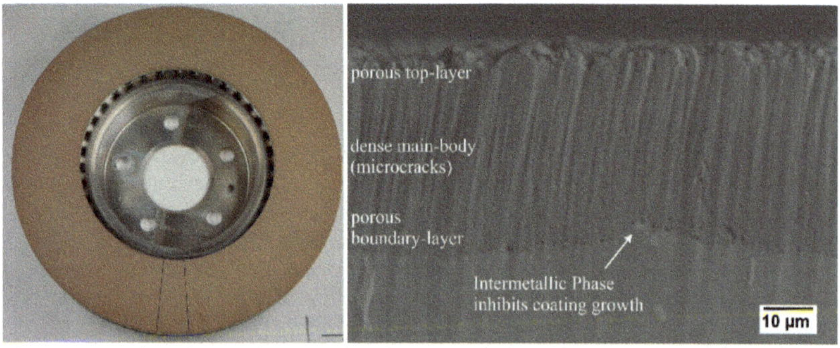

Figure 1: Overview photo and BSE-image of the cross section of the PEO-coating.

The BSE-image (Backscattered Electron) reveals the influence of intermetallic phases on the growing behavior of the coating. The coating grows mainly into the substrate and due to the greater volume of the formed oxide also a bit out of the material. The intermetallic phases seem to hinder the oxide formation and result in a locally reduced coating thickness. The coating quality does not seem to suffer from the intermetallic phases, but they might be a starting point for long-term damage and delamination. Previous investigations suggest that a chain reaction can be expected, if a significant flake of the coating is delaminated: Judder spots form on the surface, which result in increased shear stresses and ongoing delamination [1].

Brake lining material

The selection of a serial low steel quality as brake pad material, was based on preliminary tests. The low steel showed the best performance in friction and tribolayer formation in pin-on-disc tribometer tests [1]. Microscopical images of the pad material are given in Fig. 2a and b. The most represented elements in the EDX meas-urement (Fig. 2c) are Carbon (organic components and graphite, in total 52 %), Oxygen (16 %), Iron (15 %), Magnesium (4,8 %), Aluminum (3,1 %) and Tin (2,1 %).

The brake dynamometer tests were performed with a Continental 4MN 16" 42 brake caliper.

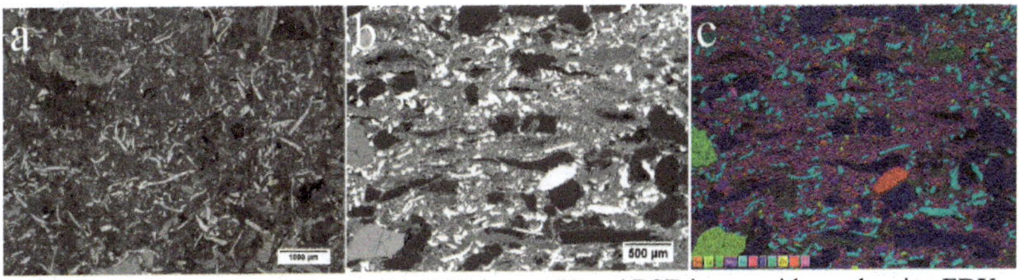

Figure 2: Optical microscope image (a), BSE-image (b) and BSE-image with overlapping EDX-measurement (c) of the Brake lining material.

Brake dynamometer testing and Interrupted Monitoring studies

The braking tests were conducted on a Horiba GIANT 5200 brake dynamometer located at the AUDI AG in Ingolstadt, Germany. The underlying boundary conditions of the vehicle are listed in Table 1. A mass moment of inertia (MMOI) of 50 kgm^2 and an assumed brake force distribution between front and rear axle of 50 % were applied.

Table 1: Specifications of the units under test.

Parameter	Value
Vehicle weight m_v	2000 kg
Brake force distribution ϕ	50 %
Dynamic tyre radius r_{dyn}	0,316 m
MMOI J	50 kgm^2

The experimental test program on the dynamometer consisted of four blocks, each being a deceleration controlled synthetic loading spectra (Table 2). The thermomechanical load was increased consecutively between the four individual load sections. This enables the examination of the run-in behavior and of the damage progress of the friction surface during every day brake applications at rising load. The final velocity in Block 3 is marked as variable, because it depends on an adjusted temperature maximum in the brake rotor. The brake is released when the measured disc temperature exceeds a given temperature of 350 °C.

Table 2: Interrupted Monitoring test program for the coated Al-brake disc on the brake dynamometer.

Experiment	Simulated braking situation	Number of stops	v_{start} / km/h	v_{end} / km/h	T_{start} / °C	a / m/s²
Block 1	City / Run-In	150	50	5	100	-2,5
Block 2	Road	100	100	50	100	-3,5
Block 3	Highway	50	200	var.	200	-4,5
Block 4	Emergency braking	10	100	5	100	-10

A special feature of the test program is the interruption of the test after every load block. The brake disc is removed from the dynamometer, cleaned in an ultrasonic bath and subsequently examined in a Large Chamber SEM (LC-SEM). Afterwards, the disc is reinstalled on the test bench with the previously used brake pads. As a result of this procedure, the test program is specified, in the following, as Interrupted Monitoring test (IM).

Corrosion studies

Following the Interrupted Monitoring studies, the brake discs was placed in a climatic chamber and exposed to the VW PV 1210 test for 120 h (five runs over 24 h). Each test run consists of 4 h salt spraying at 35 °C, 4 h ventilation at room temperature and 16 h condensation water environment at 40 °C. The brake disc chamber and the friction ring on the side of the brake disc chamber were protected by tape in order to prevent these surfaces from corrosion. The comparison of both friction surfaces allows an evaluation of the corrosive attack on the pre-stressed coating system (Fig. 3).

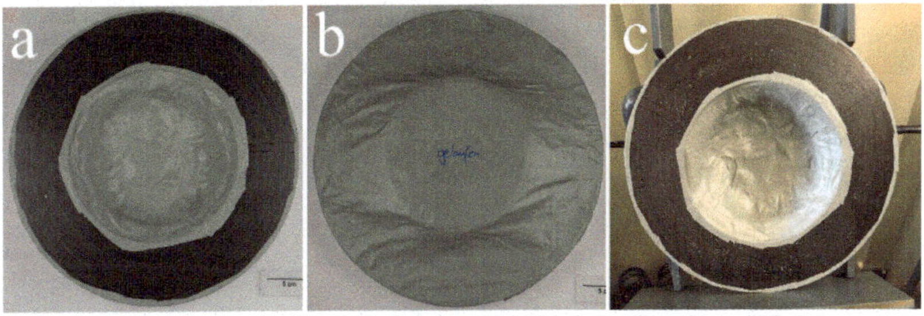

Figure 3: Taped brake disc (a, b) and friction ring after 48 h in the climatic chamber (c).

After the corrosive exposure, the disc was tested on the brake dynamometer again, using the load sections 3 and 4 of the IM test followed by SAE J2522 brake test. This sequence of corrosive and thermomechanical exposure was repeated a second time with the aim to investigate a possible effect of increased tribolayer formation, advanced crack propagation and further crack formation in the coating. In the second loop, the corrosion test was conducted for 240 h (ten runs) to further increase the corrosive attack. Fig. 4 illustrates the complete test sequence of braking and corrosion tests.

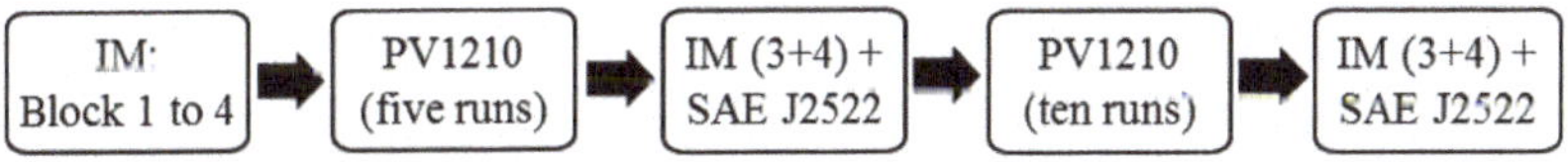

Figure 4: Sequence of the complete test program.

Electrochemical measurements

To gain a better understanding of the corrosion characteristics of the material concept, the electrochemical potential was measured in a measuring cell positioned at the quality management of the AUDI AG in Ingolstadt, Germany (Fig. 5). With respect to the use of road salt in the winter term, the tests were performed at 20 °C in a one molar

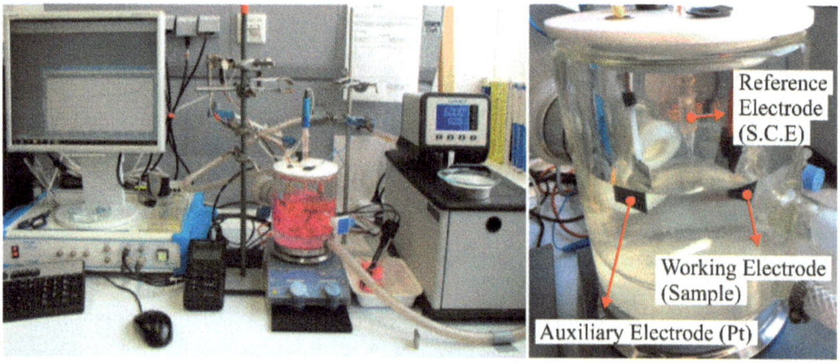

Figure 5: Measurement setup for the electrochemical measurements.

aqueous NaCl solution (pH level of 6.9). The solution is continuously stirred with 600 rpm and the test is conducted with a potential feed rate of 1 mV/s. For sample conditioning, the specimens were inserted into the electrolyte 60 min before the test starts. The reference electrode used was a saturated calomel electrode.

Analysis Methods

The surfaces of the discs were investigated in the LC-SEM at the ZMP (Joint Institute of Advanced Materials and Processes) of the University Erlangen-Nürnberg. The further microstructural investigations were carried out with a Zeiss EVO MA25 SEM in variable pressure mode, which was required due to the low conductivity of the ceramic coating. The investigations were supplemented by roughness measurements conducted by a Leica DCM8 confocal microscope and by wear measurements with a Mettler PM16 balance and a Mitutoyo QuantuMike 0-25 mm measuring screw.

Results

Brake dynamometer and interrupted monitoring studies

To get an idea of the run-in behavior of the disc concept and the formation of the expected tribolayer, the interrupted monitoring test program was conducted. Figure 6 shows the disc after each block of the IM test in an overview and in detailed views of the friction surface.
In the 'initial state' the unmachined surface is visible, where craters and discharge channels are shaping the friction surface. The dark discoloration of the friction surface after the city breakings (block 1) seems to indicate that a consistent tribolayer respectively third body formed on the entire friction surface already after the first brake applications, but the investigations in the LC-SEM showed that this was not the case. The dark appearance of the brake disc can be associated to slight brake pad material transfer on the primary contact plateaus.

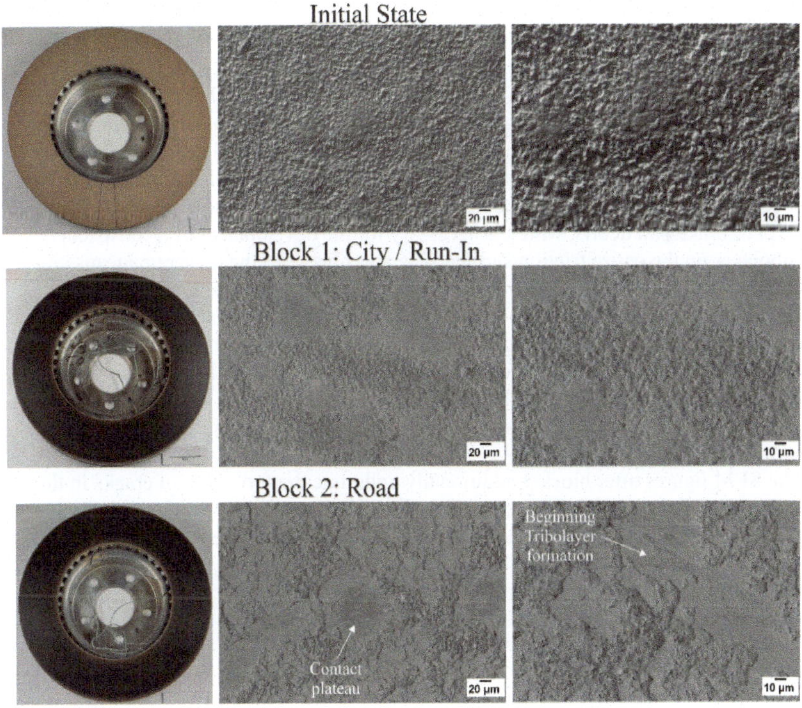

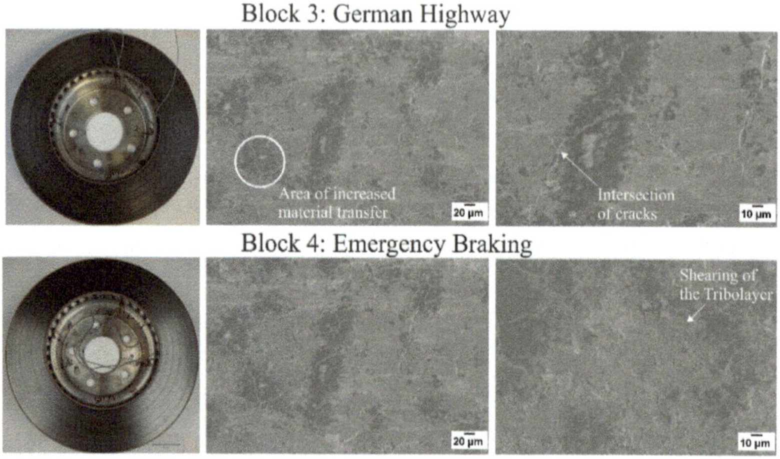

Figure 6: Overview photos and BSE-images of characteristic surface appearances after the different blocks of the Interrupted Monitoring test program.

The porous top layer of the coating has been worn off more and more, and the contact area between pad and disc has grown steadily. The higher initial velocity during road braking (block 2) caused more abrasion and thus a more intense material transfer with beginning tribolayer formation. The dark spots on the contact plateaus are the result of this increased pad material transfer. During highway brake applications, the rotational speed and temperature are for the first time high enough to cause a sufficient high level of brake pad debris, which is necessary for the development of a consistent tribolayer on the entire friction surface. The areas between the contact plateaus are largely filled with brake dust, which has been compressed by the brake pressure. Nevertheless, contactless areas with mainly original surface structure remain over the entire surface of the friction ring after load block 3 (see Fig. 7a). During emergency braking of block 4, the brake pressure further increases. Basically, the appearance of the friction surface is comparable to block 3, but the high brake pressure leads to high shear stresses and smears the tribolayer, expressed by the formation of a multilayer-layout (see Fig. 7b).

The SEM figures after block 3 additionally underline the formation of cracks in the ceramic oxide layer, which are caused by the high disc temperatures in both load sections and the mismatch between the thermal expansion coefficients of substrate and oxide material of about $15 \cdot 10^{-6}\ K^{-1}$ [8]. The cracks form at high temperatures, get filled with pad material and therefore seem to protrude compared to the rest of the coating. This leads to increased pad material transfer close to the cracks.

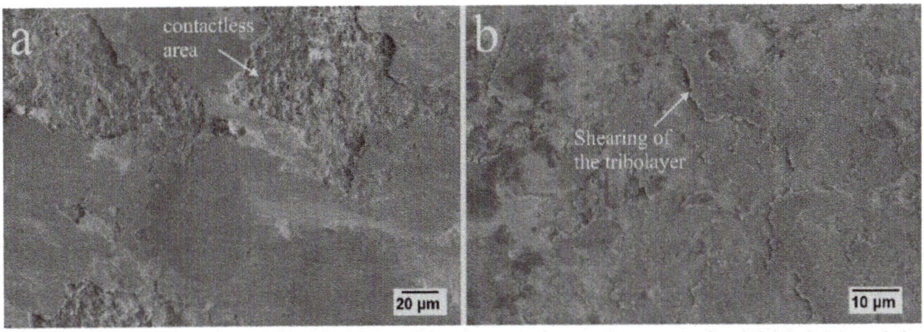

Figure 7: The SE-images of the friction surface shows a contactless area after Block 3 (a) and the multilayer layout after Block 4 (b), which indicates the strong shearing of the tribolayer.

The wear of the disc is neglectable, due to the pad material transfer, thickness and weight even increase. The good wear resistance is underpinned by an earlier study with a more demanding "suitability test", which contains the IM test program and shows a coating wear of below 10 µm [1].

Fig. 8 shows the evolution of the coefficient of friction (COF) within load section 1, 2 and 3. The red square symbols represent the investigated friction combination of PEO coating with low steel pad material. Additionally, the COF values of the run-in behavior of a standard grey cast iron disc against the same low steel brake lining (grey

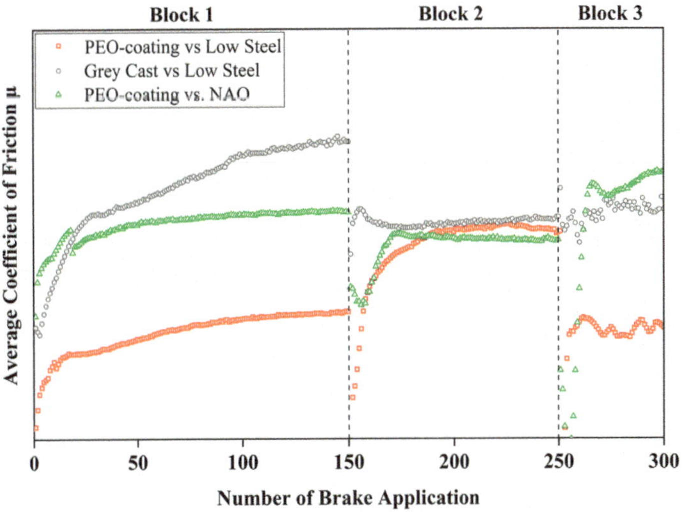

Figure 8: Evolution of the coefficient of friction in the first three blocks of the interrupted monitoring test program.

round symbols) and the PEO-coating against a NAO material (green triangles) are recorded. Section 1 shows the same tendency with increasing number of braking operations, but on very different friction levels. During the road braking (block 2) the friction levels of the two different disc materials approach each other. During the highway braking, the COF of the PEO coating falls back to the level of block 1, while grey cast iron keeps its high level. The picture changes, if the PEO coating is paired with a NAO brake lining material. The friction level is higher and develops faster during run-in compared to the low steel lining, and the variation is smaller. The COF is similar to grey cast iron and after fading effects during the first brake applications in Block 3, the COF even increases slightly. It should be noted that both PEO coated discs were removed from the test bench after each test block to enable the SEM examinations and then reinstalled on the brake dynamometer. The removal of wear dust from the pads during disassembly and the cleaning of the discs may partly explain the low starting value for the COF at the beginning of a new test section as well as its quick recovery within a few braking operations. The grey cast iron disc was tested without interruption.

Corrosion tests

Combination of brake dynamometer studies and corrosion tests

After the interrupted monitoring test on the brake dynamometer, the brake disc passed a 120 h corrosion test in the climate chamber (five runs over 24 h). Figure 9 shows that the friction surface is covered with crystallized salt and corrosion products derived from the tribolayer and presumably from the disc. Unexpected is the abstinence of red rust, despite the high amount of iron in the brake pad material.

Figure 9: Brake disc (a) and close-up view of the friction surface (b) after 5 cycles of the PV1210 corrosion test.

After the corrosive stress, the disc was tested on the brake dynamometer again, in order to evaluate whether corrosion had weakened the bond strength between substrate alloy and PEO-coating. Figure 10 shows the evolution of the friction surface during the dynamometer testing. Within five bedding brake applications, the corrosion products and the salt were worn off. The measurement data prove that there is no significant influence of the corrosion products on the coefficient of friction or any other relevant parameter. In addition, it is important to mention that dimples became visible in the friction surface after a few brake applications.
Figure 11 shows the confocal measurement of the friction surface after 5 cycles PV1210 and the subsequent dynamometer tests. The profile section shows a depth of up to 50 µm, which exceeds the nominal thickness of the PEO-coating.

Figure 10: Mounted brake disc after 5 cycles of PV1210 corrosion test (a), after 5 brake applications of the bedding block (b) and the friction surface after the whole dynamometer test (c).

According to Figure 11 the dimples can extend into the substrate material and might thus increase the risk for a corrosive attack of the Aluminum disc at the interface to the oxide layer. To achieve a better understanding of the formation of dimples and the associated damage potential, the disc was exposed to a second PV1210 test for 10 cycles respectively 240 h. After this corrosive stress and the final brake dynamometer testing the number of dimples increased, but delamination or even catastrophic failure of the coating was not observed. This result confirms the good adhesion between substrate material and PEO coating.

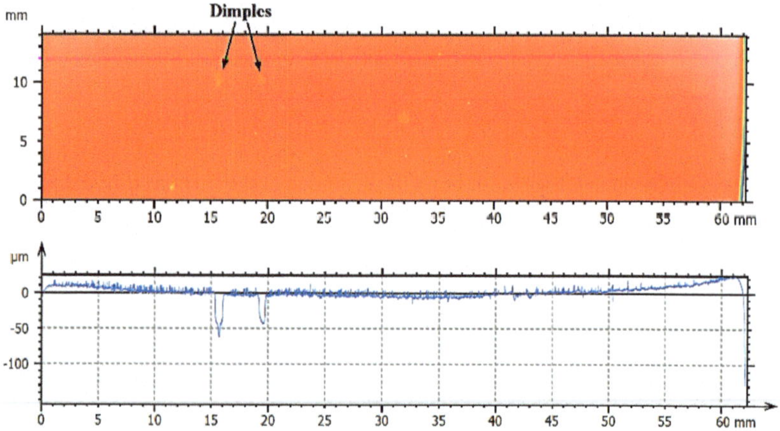

Figure 11: Confocal measurement and profile section of the friction surface after 5 cycles of PV1210 and the dynamometer tests.

Microstructural investigations

Fig. 12a shows a typical surface area of the friction ring after the first interrupted monitoring test. Fig. 12b provides a statement of the change in the surface structure after further corrosive and mechanical load consisting of five cycles PV1210 test and subsequent IM and SAE J2522 braking operations. A comparison of both pictures reveals that the corrosive stress and the subsequent thermomechanical load resulted in a more pronounced network of cracks and in material outbreaks in the peripheral crack area. Findings, however, giving a clear indication of a corrosive damage could not be detected in the scanning electron microscope.

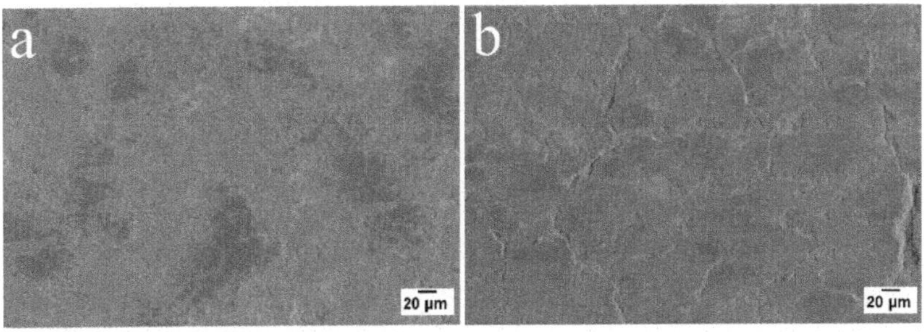

Figure 12: BSE-images of the friction surface after Block 4 of the Interrupted Monitoring (a) and after the combined brake dynamometer and corrosion test program (b).

The corroded and the corrosion-protected side of the friction ring do not differ significantly in their basic appearance (Fig. 13 a, b), and the higher loaded suitability test described in [1], which was performed without corrosion, caused comparable failure features. It is noticeable, that the amount of dimples is significantly higher in the corroded friction surface and that these dimples are surrounded by a fine network of cracks (Fig. 13d). In areas with clearly visible abrasive wear and deeper wear tracks, the network of cracks is also very pronounced and the primarily rough surface of the coating as well as the tribolayer located above are worn off (Fig. 13c). However, a delamination of coating segments did not appear.

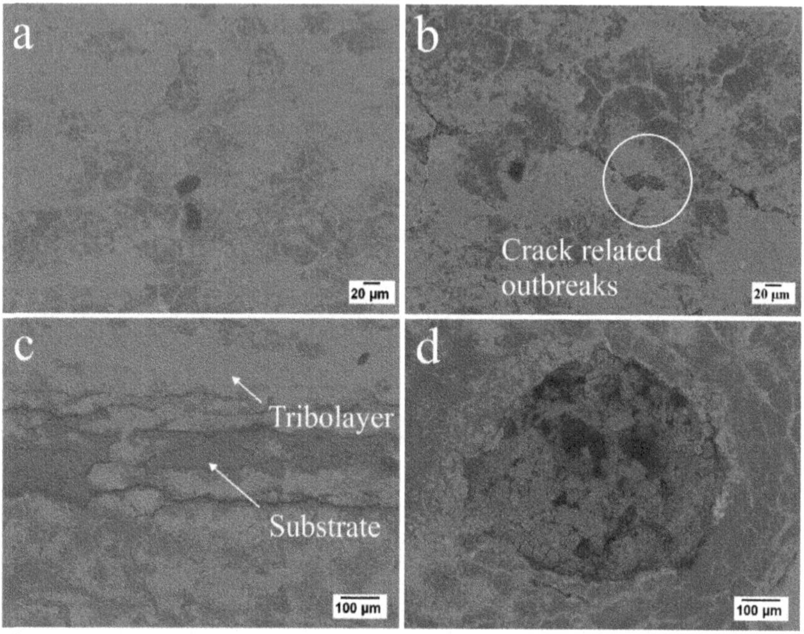

Figure 13: BSE-images of the taped (a) and corroded (b) friction surfaces after the test program and an overview of typical appearances on the tested brake disc surfaces (c, d).

Figure 14a shows a metallographic section through the disc, which was prepared transverse to the corrosive stressed friction surface after ending the test program. Surprisingly, the porous top layer of the coating still exists. This porous top layer is covered continuously with a tribolayer, which has a two-layer structure in this example. The bright zone consists mainly of wear debris from the brake pad and the darker upper zone of heavily oxidized abrasion products from the pad material.

Figure 14b illustrates a cross section through a dimple. The damage extends deep into the Aluminum substrate. The bright appearance of the dimple surface results from insoluble remains of silver conductive paste, which was applied in the ion milling process within the sample preparation works. The optical impression that Aluminum oxides have developed on the substrate alloy, was confirmed by EDS measurements. The PEO coating in the surrounding of the dimples seems to be subject to more intensive wear.

There are dimples on both sides of the friction ring, but the number of dimples on the corroded friction surface is many times higher compared to the taped one. The fact that dimples mainly form on the corrosive stressed friction surface supports the assumption that corrosion of the substrate weakens the interface and increases the risk for layer delamination. The undoubted cause for the formation of dimples and material damages, however, is difficult to establish. A junction within the crack network, a progression of the damage by crack related outbreaks in combination with a supernormal amount of local iron transfer from the inhomogeneous brake pad, and the influence of corrosion, could be an explanation for the extent of the damage. Lastly, it is necessary to examine, whether and under which conditions dimples can lead to a widespread delamination of the coating. Cracks close to the dimple (see Fig. 13) give rise to the assumption that the delamination process is not finished yet. The assumption is supported by Figure 14b, which shows ablation of the coating at the edge of the dimple.

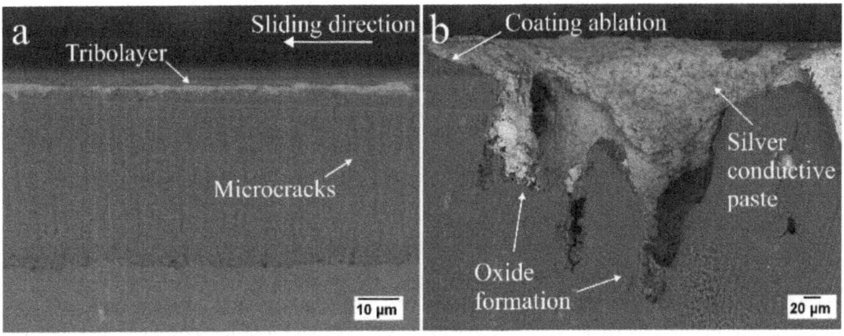

Figure 14: Typical cross section after the test program (a) and the cross section of a dimple (b).

Electrochemical measurements

The Aluminum substrate, the PEO coated Aluminum brake disc in the condition "new" and "braked", and two brake pad materials (low steel and NAO) were tested in a one molar aqueous NaCl salt water solution. The "braked" condition was subject to a suitability test without wet test sections [1]. The Tafel-plot in Figure 15 indicates that the PEO oxide coating is dense and impermeable for the electrolyte (red line), and thus has an excellent resistance to corrosion, which is noticeably better than that of the Aluminum substrate (black line). In comparison to this new state of the coating, the pre-stressed coating shows a higher corrosion tendency in contact with the electrolyte (blue line). The associated exchange current density, however, is still very similar to that of the substrate Aluminum alloy. For this reason, the braked condition can be considered as very well corrosion-resistant. In contrast, the low steel pad material (green line) exhibits a more than ten times higher current density attributed to its high content auf base metals.

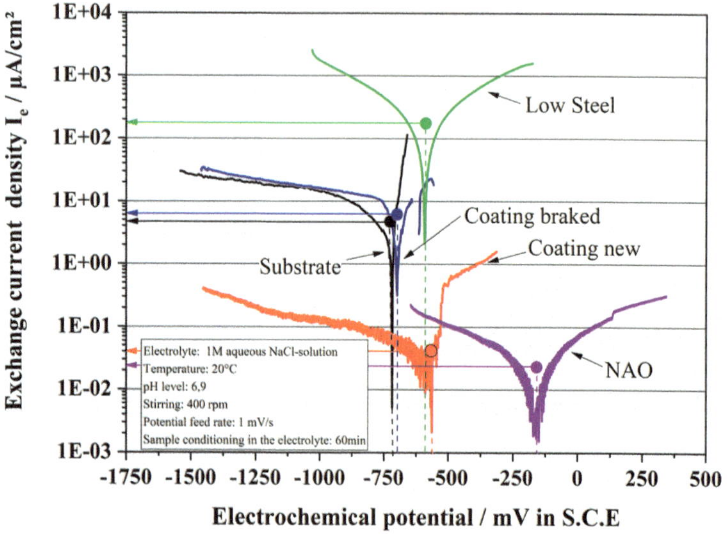

Figure 15: Tafel-Plot of the relevant components of the tribological system.

The decline of the corrosion resistance of the pre-stressed coating can be explained with the formation of cracks, which penetrate the coating. Cracks occur at high disc temperatures due to the different heat expansion of the ceramic coating and the Aluminum base material. Repeated braking at high temperatures can cause crack opening and filling of the opened cracks with wear debris from the pad material. Additionally, substrate material seems to flow into the opened crack promoted by the brake pressure and the very low strength of the 6082 Aluminum alloy at high temperatures. If salt water penetrates into such a crack, the base metals of the lining can built an electrochemical element with the underlying Aluminum alloy and result in Aluminum corrosion. Figure 16 shows such an electrochemically critical state of the friction surface, where the coating is cracked and brake pad material reaches the substrate alloy. It should be noted however that the large number of cracks in the coating remain closed and do not penetrate the coating, which makes them less critical in terms of corrosion. This might be one reason for the continued good corrosion resistance of the pre-stressed coating. Furthermore, an extensive third body formed on the friction surface, which may hinder the entry of corrosive media and thus protects the substrate from extensively corrosive attacks. According to Figure 16, this dark appearing tribolayer traverses the crack and seems to adhere quite certainly to the underground.

The use of a NAO lining material instead of low steel seems to open the opportunity to reduce the susceptibility of the pre-stressed PEO coating to corrosion. Due to the lack of metallic components in the pad material composition, the NAO lining exhibits an excellent corrosion resistance in the electrochemical measurements (Fig. 15). Therefore, cracks in the coating, which open under tensile stresses, will not contain corrodible pad elements. The same applies to the oxide rich tribolayer.

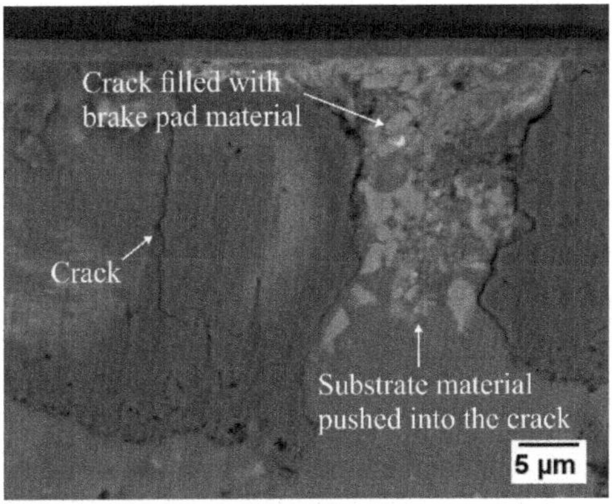

Figure 16: Cracked friction surface covered with a fully extended tribolayer.

Conclusion

The tested PEO coating on a 6082 brake disc exhibits a good wear resistance in everyday braking situations as well as in emergency stops. However, a series of braking movements at high temperatures results in a growth of the production-related cracks, the development of new cracks and in the formation of a continuous crack network. Depending on the disc temperature and the brake pressure, some of these cracks penetrate the PEO coating, open partially under repeated braking and are filled with wear debris of the pad material. Moreover, the Aluminum base alloy seems to be too soft at high temperatures and undergoes plastic deformation in the region around a crack, arising the need for a more temperature resistant substrate material. On the other hand, the present and the previous studies do not show extensive delamination or severe material damage, and thus prove the fundamental suitability of the material system for braking applications.

If the PEO coated disc is paired with a low steel lining material, a continuous third body only forms at high disc temperatures, which occurs for the first time during

highway braking operations of the applied load test. This tribolayer consists of oxidized wear debris from the base elements of the brake pad material and mainly appears in a single-layered form. It is supposed, that the composition of the tribolayer is responsible for the low friction values within the highway braking movements. Overall, it appears that the hard ceramic PEO coating does not support the abrasive wear character of the low steel pad material, and thus results in a low friction level. In contrast, an organic lining material promises benefits with respect to the height and stability of the COF, an early formation of the wear reducing tribolayer and a higher corrosion resistance. Besides, the low-strength Aluminum substrate does not allow high disc temperatures, which would require a more heat-resistant low steel lining.
The investigated Aluminum brake disc concept is very resistant to corrosion due to the ceramic character and the good adhesion of the PEO layer. This is especially the case, when the brake disc runs in a low temperature range, as expected in regenerative braking, and thus forms no cracks. Even if cracks penetrate the layer and are filled with oxidized base elements from the pad material, the corrosion resistance in electrochemical measurements is similar to that of the underlying Aluminum alloy. However, SEM investigations of a pre-stressed brake disc, which was subsequently exposed to a sequence of corrosion and braking tests, showed the formation of dimples in the friction area. The mechanism of dimple formation seems to base on a strong local expression of the crack network, a local breakthrough of the coating followed by an opening of the cracks during repeated braking at high temperatures, and a transfer of corrodible elements of the brake pad into these holes. Crack opening seems to be promoted by the low strength of the Aluminum substrate or by material outbreaks at the crack flanks and junctions. If salt-water infiltrates these areas, local corrosion of the Aluminum substrate may arise, causing pitting corrosion and a weakening of the interface between substrate and coating due to the higher volume of the formed Aluminum oxides. The result is a delamination of the coating in areas of a width of about several hundred microns. With reasonable certainty, the development of these dimples can be attributed to the impact of corrosion, because comparable damages did not occur in such a large number under dry conditions. This damage mechanism calls for a deeper analysis in future studies with the aim of investigating the progress of the damage under further corrosive and thermomechanical load, and defining possible improvement measures, such as the use of a NAO lining.

References

[1] F. Gulden, S. Gramstat, A. Stich, H.W. Höppel, U. Tetzlaff, "Properties and Limitation of an Oxide Coated Aluminum Brake Rotor," SAE Technical Paper 2018-01-1877, 2018, doi: 10.4271/2018-01-1877.

[2] P. Kurze, H.G. Schneider, T. Schwarz, W. Krysman, Metalloberfläche 40 (12) (1986) p. 539.

[3] A. L. Yerokhin, X. Nie, A. Leyland A. Matthews, S. J. Dowey, Surface and Coatings Technology 122 (1999) 73-93.

[4] S. Shrestha and B. D. Dunn, "Advanced plasma electrolytic oxidation treatment for protection of lightweight materials and structures in a space environment", Surface World, 2017.

[5] C. Liu, D. He, Q. Yin, Z. Huang, P. Liu, D. Li, G. Jiang, H. Ma, P. Nash, D. Shen, Surface and Coatings Technology 280 (2015) pp. 86-91.

[6] R. O. Hussein, X. Nie, D. O. Northwood, A. Yerokhin, A. Mattthews, Journal of Vacuum Science & Technology A 28 (4) (2010) pp. 766-773.

[7] U. Malayoglu, K. C. Tekin, U. Malayoglu, S. Shrestha - Materials Science and Engineering A 528 (2011) pp. 7451– 7460.

[8] J.A. Curran and T.W. Clyne, Surface & Coatings Technology 199 (2005) pp. 168 – 176.

[9] X. Nie, E.I. Meletis, J. C. Jiang, A. Leyland, A. L. Yerokhin, A. Matthews, Surface and Coatings Technology 149 (2002) pp. 245-251.

[10] Y. Jiang, Y. Zhang, Y. Boa, K. Yang, Wear 271 (2011) pp. 1667-1670.

[11] T. Wei, F. Yan, J. Tian – Journal of Alloys and Compounds 389 (2005) pp. 169-176.

[12] A. Alnaqi, S. Kosarieh, D. C. Barton, P. C. Brooks, S. Shrestha, Journal of Materials Design and Applications, 232 (7). (2016) pp. 555-565.

[13] S. Gramstat, R. Waninger, D. Lugovyy, M. Schröder, T. Grigoratos, "Brake particle emissions – A global challenge", 9th International Munich Chassis Symposium 2018, pp.649-661.

At source brake dust collection system

Matthieu Hascoet, Tallano Technologie, project manager

Loïc Adamczak, Tallano Technologie, CTO

1 Introduction

With the improvements made on exhaust particle emissions, non-exhaust and especially brake particles are pulled out from the dark. Tallano Technologie has developed a brake particles collection system, named TAMIC. TAMIC was designed to trap at least 80% of brake particles directly at the pad-disc interface without altering braking efficiency.

This paper summarizes the studies implemented on automotive applications. Vehicle behaviour and performances have been verified on a F32 BMW 4 Series.

2 Outdoor air quality – brake dust emissions

Air quality is a growing concern in the world. The WHO (World Health Organization) Air Quality Guidelines (2005) provides an assessment of health effects of air pollution and thresholds for health-harmful pollution levels. These thresholds are given for PM2.5 and PM10 in table 1. In 2016, 91% of the world population was living in places where the WHO air quality guidelines levels were not met.

Table 1 – WHO Particulate matter guideline values

Guidelines	
$PM_{2.5}$:	**10 μg/m³ annual mean** **25 μg/m³ 24-hour mean**
PM_{10}:	**20 μg/m³ annual mean** **50 μg/m³ 24-hour mean**

Ambient (outdoor air pollution) in both cities and rural areas was estimated by WHO to cause 4.2 million premature deaths worldwide in 2016.

Moreover, airborne PM have been classified as carcinogenic level 1 to humans by IARC in 2013. Meaning there is enough evidence of carcinogenicity in humans.

Brake wear particulate matter is one of the sources of particle emissions [1]. Today, in countries like Germany or UK, brake wear contribution to PM10 pollution is higher than exhaust particles contribution (Figure 1).

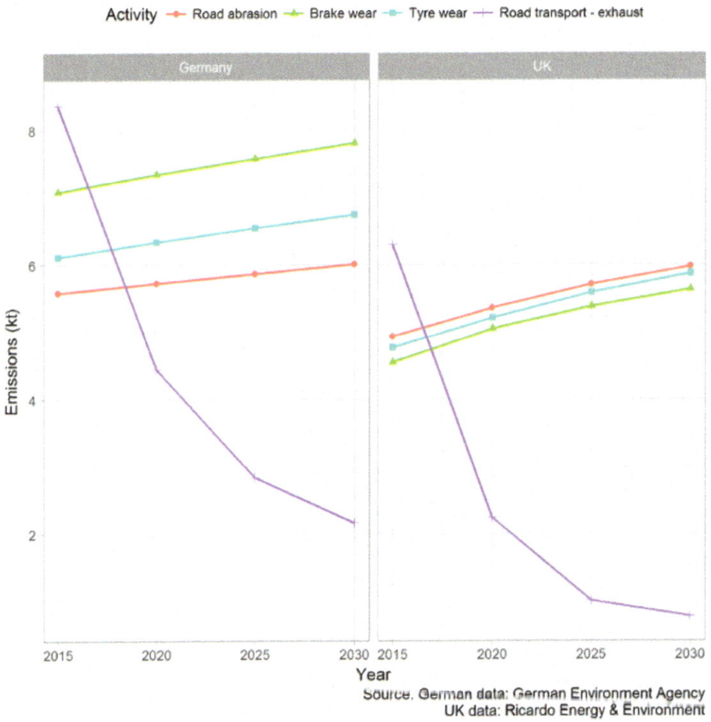

Figure 1: PM2.5 emissions from road transport sources according to the UK and German Inventories. [2]

Braking is a quite complex process. Depending on speed, weight and deceleration levels, and so on pressure and temperature induced, different amounts of gases are emitted. Metals oxidation may occur or not. And different amounts of ultrafine, fine and coarse particles are emitted. A fraction of about 40-50% of brake wear becomes airborne [3] while the remaining fraction deposits either on the vehicle or on the road. The former fraction contributes to the pollution and health concern. The latter fraction contributes to an environmental one. As these pollutants are carried off through a wash-out process till rivers and sea.

Concerning airborne particles emitted during braking, more and more studies are now available. They all show that mainly PM10 particles are emitted during braking and the size distribution varies with the braking conditions from nano to micro particles [4 – 6]. So, there is a need to collect and retain such a broad distribution.

3 At source collection system: Operating principle

The brake pads and disc wear out during contact and release particles (Figure 2). To remedy this, it has been developed an at source collection system to meet the challenges of automotive of today and tomorrow.

Figure 2: Visualization of particles released during braking

To collect these particles, we did study the prior art and different solutions were developed. The best results were obtained with the solution presented on Figure 3 and Figure 4. A groove is machined at the trailing edge of the pads. Particles are collected before they leave the pad-disc contact.

Figure 3: Modified brake pads

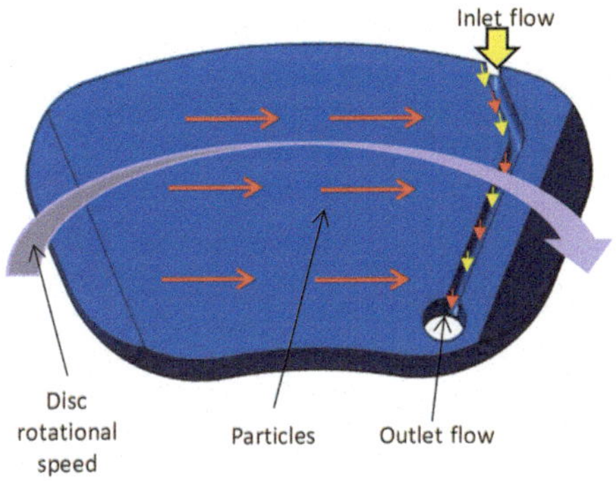

Figure 4: Automotive groove design

Based on CFD simulations and designs of experiments, we optimized the shape and location of this groove. And we defined the negative pressure and the flow rate levels needed to collect most of the big and small particles.

This allowed us to design our aspiration system. It relies on pipes connected to an electrical suction turbine equipped with a specifically developed filter to trap wear particles (figure 5).

Figure 5: Turbine and filter

The filter developed with our filtration partner allows to cope with the large distribution of brake dust particle size. Figure 12 shows the efficiency levels for different particle sizes obtained with normalized ISO/TS 11155-1 testing procedure at 10 cm/s, which is the actual air speed in our filters.

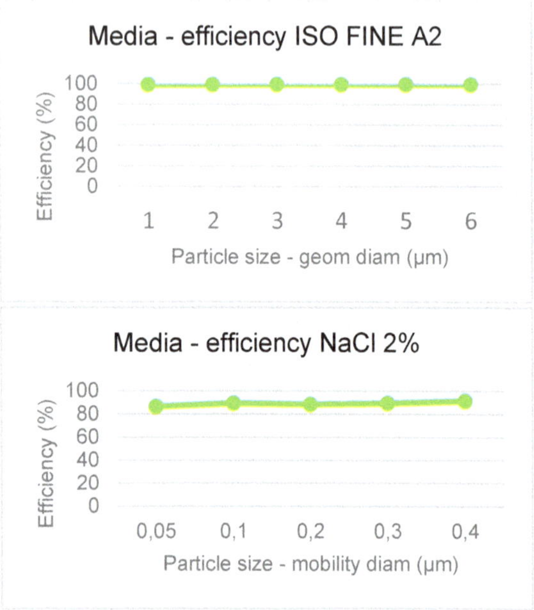

Figure 6: Media filter efficiency at 10 cm/s

The whole system is driven by an embedded electric brushless motor powered by the vehicle battery (Figure 13)

Figure *1:* System mounted on the F32 BMW 4 series

4 Performance evaluation of the brake dust collection system

All presented results here are coming from tests done on a F32 BMW caliper. The system has been tested on brake bench and on a BMW 4 series car.

As our system was added on existing callipers, and to avoid any influence of residual drag torque over the results, the entire tests presented here were performed with a continuous aspiration of the system. It has been shown that the residual drag torque of a caliper can represent up to 34% of the total brake dust emission [3]. The serial version of our system will only aspirate during braking events.

4.1 System test on brake rig

4.1.1 Mass collection efficiency

Figure 8: System mounted on brake test bench

Collection rate in mass protocol consists in weighing friction components mass before and after the test, and to compare it with the mass collected in the system. This procedure is quite simple to implement and gives rapidly a reliable and reproducible testing evaluation.

Different testing cycle were used to confront the system to various braking conditions.

The first test cycle used for the evaluation was the well-known Los Angeles City Traffic (3h-LACT). The cycle consisted of 217 brake stops with initial braking speeds from 16.9 to 154.3 km/h and decelerations from 0.2 to 2.88 m/s².

This 3h cycle is repeated 6 times to make a 18h long test comprising 1302 braking events that allow to generate enough brake dust mass to avoid weighing bias.

The second test cycle conducted was the new WLTP. This test of 30 minutes long comprises 48 stops. The cycle is repeated 36 times to reach 18h of testing and so 1728 braking events, to have here again enough brake dust mass generated for the measurement.

The last test cycle used, called urban driving test, is the one applied also for the car testing. As the cycle on the car has to be repeatable in terms of speed and deceleration, we conducted a specific test cycle composed of braking from 80 to 0 km/h and 50 to 0 km/h at deceleration of 1.2m/s² and 2.2m/s² with a specified cooling distance between each braking events. These different braking combinations are repeated to reach around 600 braking events.

The results regarding mass reduction for the 4 series caliper with ECE pads are presented on figure 9:

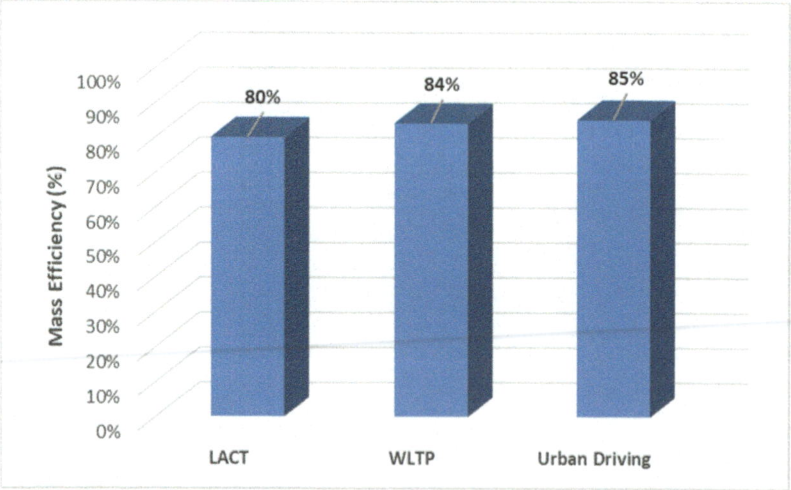

Figure 9: Average mass collection efficiency measured on bench

The slightly lower efficiency obtained during LACT test can be explained by a higher thermal exposure during the test compared to the two other ones.

The max temperature measured inside the pads during a LACT cycle reaches 250°C when it reaches 170°C during the 2 other tests.

Brake pads are made of organic materials. When heated at high temperature these organic materials degrade and gazes are emitted. Figure 10 presents a thermogravimetric analysis of a F32 pad sample. The weight loss observed is accompanied by gas emissions, mainly CO_2. As our filters are efficient for particles but not gases, this could explain the difference between LACT and other cycles.

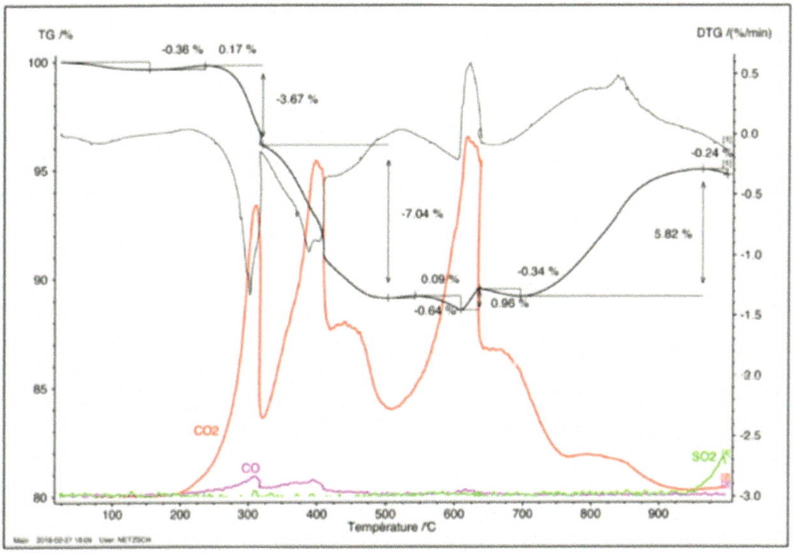

Figure 10: F32 pad TGA done under air

In order to confirm this hypothesis, 2 new tests were implemented:

- Urban driving 80°C: Braking from 50 to 0 km/h; deceleration level: 0.12g; initial temperature before each brake: 80°C (measured in the pad at 10mm from the contact surface)
- Urban driving 200°C: Braking from 80 to 0 km/h; deceleration level: 0.22g; initial temperature before each brake: 200°C

The results of these 2 tests are presented on figure 11.

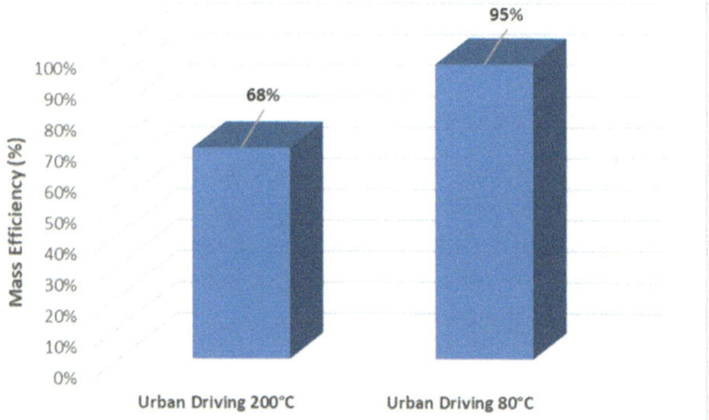

Figure 11: Influence of average temperature on mass collection efficiency

As the 2 tests were performed in quite similar conditions (speed and deceleration differences were not so important) we can conclude temperature has a high impact on mass collection efficiency.

So, a special attention must be paid to the testing conditions.

All test results show a quite constant and repeatable collection efficiency of the system among different braking conditions on bench. They also pointed out that a particular attention must be paid off when it comes to compare cycles in terms of pads temperature when a mass measurement is made.

4.1.2 Number collection efficiency

We did implement particle number efficiency tests at Ilmenau University. The protocol is described in [7]. A constant volume sampling system has been used with an inlet air filtered by an HEPA filter. The measurement devices were the following:

- Horiba Mexa-2100SPCS: Condensation Particle Counter – size range: 10 – 2500nm / Modified sub 23nm version
- Dekati ELPI+: Electrical low-pressure impactor – size range: 6 – 10000nm

A standard serial brake calliper with non-modified ECE pads has been used for the test. We did compare the serial pads and the pads and caliper modified to integrate our system.

For each test, a first run-in procedure of 150 brakes AK-Master section 3 (80->30km/h; 30 bar) was performed (SAE International, 2003). Then section 4.1 (40->5km/h; 10-80 bar), 4.2 (80->40km/h; 10-80bar), 4.3 (120->80km/h; 10-80bar) and 4.4 (160->130km/h; 10-80bar) of AK master were performed. System efficiency compared to serial pads is displayed on Figure 12. A reduction potential of 50-90% with low mechanical stress and 50-80% with high mechanical stress could be proved. The highest reduction potentials were obtained in urban driving conditions. Which is very positive as brake frequency and related pollution concerns are the highest in cities.

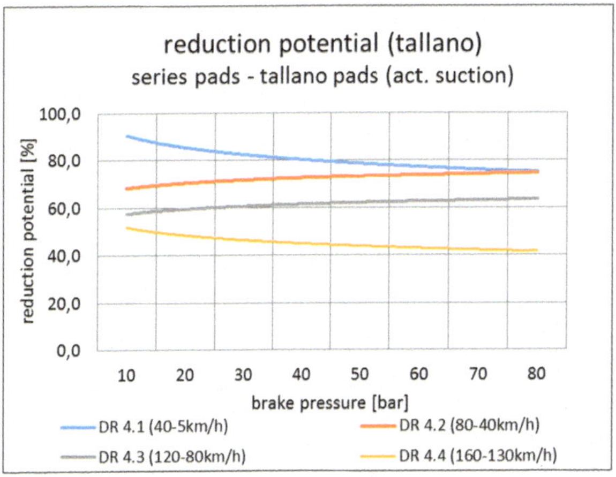

Figure 2: Number reduction potential in different driving conditions

We also did perform Laser Images in Ilmenau University in order to visualize the efficiency of our system. We could visualize the particles emitted during a braking event with and without our system (Figure 13). The high efficiency measured with the particle counters is confirmed.

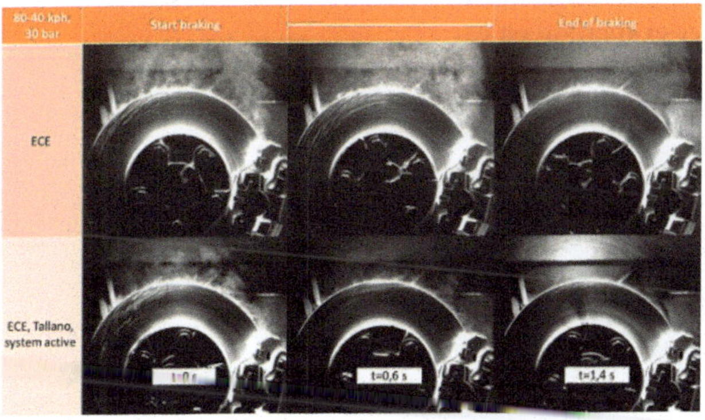

Figure 3: Laser Images with and without our system

4.1.3 Performance homologation of modified pads

An overriding point to check also with the modified brake pads was their impact on the braking efficiency compare to standard pads. As concessions can not be made on the safety of vehicle passengers, we ran a R90 homologation procedure between standard and modified ECE pads to measure the modifications impact.

The results show that there is no significant variation in braking performance between the two pads (Figure 14).

Modifying the brake pads to integrate the grooves does not affect their performances and safety.

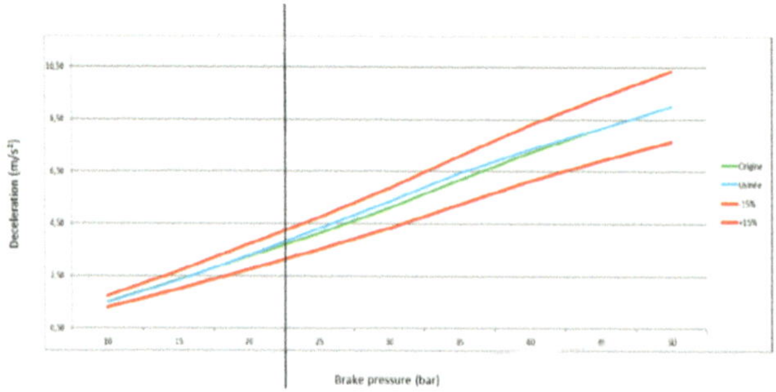

Figure 14: R90 efficiency regarding pressure. The green curve is the original pad and the blue one the pad modified to integrate the groove.

4.2 System tests on vehicle

4.2.1 Performance evaluation on car

Figure 15: UTAC CERAM test track

Once characterised on the bench, the second part of the collection evaluation was done on the BMW 4 series car. It has been conducted in two phases, one done in France by Tallano, and a second one done by BMW in Munich.

First car tests done by Tallano have been carried out at the French UTAC CERAM testing centre in Monthlery. The procedure used here is the car testing protocol composed of around 600 braking events at two different speed and deceleration explained in point 5.1.

To avoid any influence of the water over the mass weighing, all the tests here were performed under dry road conditions.

The efficiency measured during these first phase of testing is in average 84% (figure16)

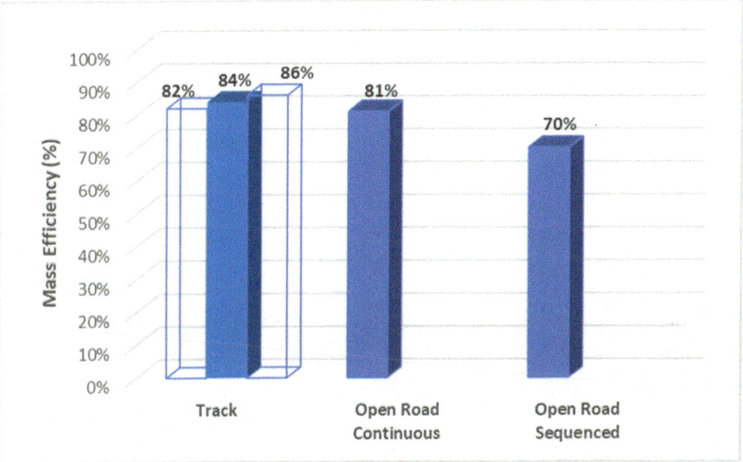

Figure16: Vehicle test results

We can notice here a good correlation between the collection efficiency measured on the bench and on the car among the same testing cycle.

The second phase of the collection evaluation on the car was conducted by BMW (Munich Figure 16). It has been performed on open road on a defined circuit around Munich to assess the collection efficiency in real world driving conditions. This testing circuit was composed of around 1/3 of city road condition, and 2/3 of country road conditions.

For this evaluation, one of the runs was made with a system working only when braking (in sequenced conditions). This was made to assess the potential difference between continuous and sequenced aspiration regarding real driving conditions on the car.

As we can see there is a collection efficiency difference of 11 points between the continuous and sequenced aspiration conditions. This difference can be explained by the residual drag torque on the car, it means that the drag torque would be responsible for around 14% of the total particle mass emission. This is much less than what the 34% shown in the paper [3], but it has been measured on brake bench that the F32 caliper has a low residual drag torque.

What can be seen with these results is that a good correlation is obtained between the tests done on the bench and on the car. Thanks to its design, the TAMIC system is not impacted by the aerology of the car in the wheel arch. Moreover, efficiency measured on open road is close to the one measured on track.

4.2.2 Environmental testing

To verify the functioning of the brake dust collection system under different weather conditions, test in the climatic wind tunnel of the CSTB in Nantes under heavy rain conditions and cold conditions have been performed.

The goal of the water test was to evaluate the amount of water aspirated within the system via the grooves in the pad under heavy rain conditions. To simulate these conditions, a rain intensity of about 36mm/h was applied on the car and a water film of approximately 3mm thickness was also simulated by injecting water directly at the wheel / road interface.

Figure 17: BMW 4 series installed in the climatic wind tunnel

Figure 18: Injection of water at the wheel / road interface

The system ran four hours under these rain conditions with a braking actuation every 2 minutes with a deceleration of $2m/s^2$.

After this test no trace of water were found in the filter body or in the filter, everything was dry.

The goal of the cold conditions test was to evaluate the functioning of the system at very low temperatures.

To do so a first test was performed on a sub system alone composed of the turbine and the filter together, that were placed for a night at -30°C after having been water sprayed in functioning and stored for 4h in a 90% ambient humidity chamber. The system started up at the first starting request with no significant impact on the filter pressure drop.

The second cold test was conducted right after the water test in the climatic wind tunnel. After the run under rain conditions, the temperature of the wind tunnel chamber has been lowered down to -20°C. After a maceration at -20°C the system restarted up again immediately at the first braking actuation with no significant impact on the filter pressure drop.

4.2.3 System noise level measurement

Another environmental integration work done on the system is its noise emission. The system has been specifically designed to minimize its global noise emissions.

The solution tested and implemented on the car consists of a volute disposed around the rotor and of specific absorbent foam, as well as a quarter wave resonator added to absorb the noise.

To evaluate the system noise, measurements were performed inside the car with microphones disposed in the area of the passenger head. These measurements show that the acoustic level measured in the car with the system operating and the engine running at idle is of about 46.7 dB(A) when the acoustic level with the engine running at idle and the turbine shut down is 46.0 dB(A).

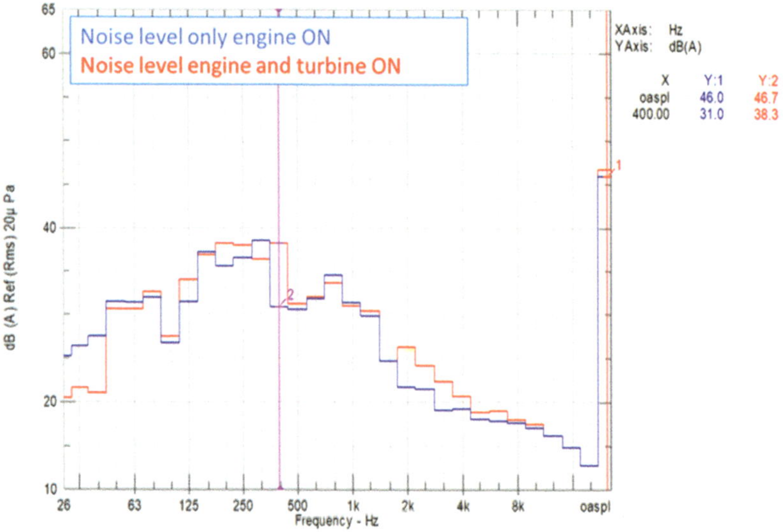

Figure 19: Noise level comparison between engine alone and engine + turbine inside the car

5 Conclusions

The TAMIC groove collection design developed by Tallano to collect brake particles at source for automotive applications have been tested in different conditions on a BMW F32 (4 series) calliper. A mass efficiency above 80% has been proved on bench and on vehicle.

Meaning that TAMIC efficiency does not depend on the aerology and in particular on the relative wind speed prevailing in the immediate vicinity of the brake pads.

It is known that a fraction of about 40-50% of brake wear becomes airborne [3] while the remaining fraction deposits either on the vehicle or on the road. With its high efficiency level, TAMIC does then not only contribute to a reduction of the health concern linked to airborne particles but also of the environmental one linked to on road deposited particles.

Concerning the health problem, the highest reduction potentials of airborne particles were obtained in urban driving conditions. So, in condition where a solution is most needed. As brake frequency and related pollution concerns are the highest in cities.

Five first filling patents covering this technology have already been granted. Twelve additional patent applications are pending.

TAMIC efficiency has been proved in different conditions. The wheel arch positioning is not the final one. It was chosen to faster the proof of concept. Tallano is now working on an integration under the hood targeting a storage capacity of 30000 km. The first vehicle will be presented at the end of this summer.

[1] Grigoratos, T., Martini, G., Brake wear particle emissions: a review. Environ. (2015) Sci. Pollut. Res. Int.

[2] Wakeling D., Murrells T., Carslaw D., Norris J., Jones L., The Contribution of Brake Wear Emissions to Particulate Matter in Ambient Air. FTA (2017).

[3] Ferdinand H. Farwick zum Hagen, Marcel Mathissen, Tomasz Grabiec, Tim Hennicke, Marc Rettig, Jaroslaw Grochowicz, Rainer Vogt, and Thorsten Benter (2019) Study of Brake Wear Particle Emissions: Impact of Braking and Cruising Conditions; Environ. Sci. Technol.

[4] Olofsson, U. (2010) A study of airborne wear particles generated from the train traffic – block braking simulation in a pin-on-disc machine. Wear, 2010, 271(1–2)

[5] Kukutschova, J.; Moravec, P.; Tomasek, V.; Matejka, V.; Smolik, J.; Schwarz, J.; Seidlerova, J.; Safarova, K.; Filip, P., (2011) On airborne nano/micro-sized wear particles released from low-metallic automotive brakes. Environ. Pollut.

[6] Namgung H. G., Kim J. B., Kim M. S., Kim M., Park S., Woo S. H., Bae G. N., Park D., Kwon S. B., (2017), Size distribution analysis of airborne wear particles released by subway brake system, Wear 2017 372-373, 169-176

[7] Augsburg K., Hesse D., Kuhn U., Leicht R., (2018) Comparison of the run-in and emission behavior of different disc concepts, PMP-47-20 rev1

Authors

Matthieu Hascoet (mh@tallano.eu), Tallano Technologie, project manager

Loïc Adamczak (la@tallano.eu), Tallano Technologie, CTO

98, route de la Reine
92100 Boulogne Billancourt
Phone number: +33 6 49 28 71 97

Advanced rayon brake hose reinforcement

Rudolf Einsiedel

Cordenka GmbH & Co. KG
Germany

1 Introduction

Cordenka rayon is a high tenacity man-made fiber based on cellulose extracted from wood. Its key property is a high Young's modulus that is performed constantly in a temperature range between -40 and +120 °C. This feature as well as rayon's low thermal shrinkage, its good resistance to chemicals and its sustainable raw material origin have made rayon popular as reinforcing material for high performance automotive tires and brake hoses.

Among the many challenging demands placed by consumers on today's automobiles, safety definitely remains the core topic. When constructing tires and brake systems for performance vehicles, therefore no compromises should be made on quality.

The excellent volume constancy – even at higher temperatures – is one of the major benefits of brake hoses reinforced with Cordenka rayon. It constitutes an advantage of rayon over thermoplastic fibers in this application. Rayon reinforced brake hoses make sure that the braking force directly gets to where it is needed and that a touch of the brake pedal – in all driving conditions – gives the driver the sense of safety that car makers aim to provide. In 2012 Cordenka MA1, 1100 den was commercialized as advanced rayon brake hose reinforcement, representing the state-of-the-art rayon material for this application since then. Cordenka MA1, 1100 den provides excellent volumetric expansion uniformity at elevated temperatures on the one hand and optimized whip test durability on the other. Having proved its worth in premium sports car applications, Cordenka MA1 is expected to capture significant brake hose reinforcement market shares.

This paper compares Cordenka MA1 fiber properties with competing materials and introduces brake hose benchmark test results that demonstrate the state-of-the-art rayon brake hose grade's potentials.

2 Rayon fiber performance

The tenacity / elongation behavior in the operational elongation range of brake hose reinforcements illustrates rayon's constant dimensional stability which is widely independent of ambient temperatures (figure 1). The constantly high Young's modulus at temperatures up to 120 °C is visualized here. The tenacity / elongation behavior of thermoplastic PVA fibers – a material applied in commodity-type brake hoses – shows some remarkable differences in comparison with rayon. Figure 2 shows a very good initial modulus for PVA fibers at room temperature, but a significantly lower performance at temperatures of 60 °C and higher.

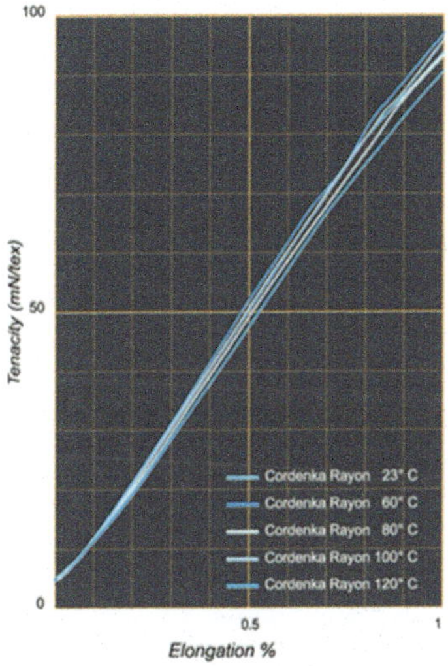

Figure 1: Rayon's tenacity / elongation behav.

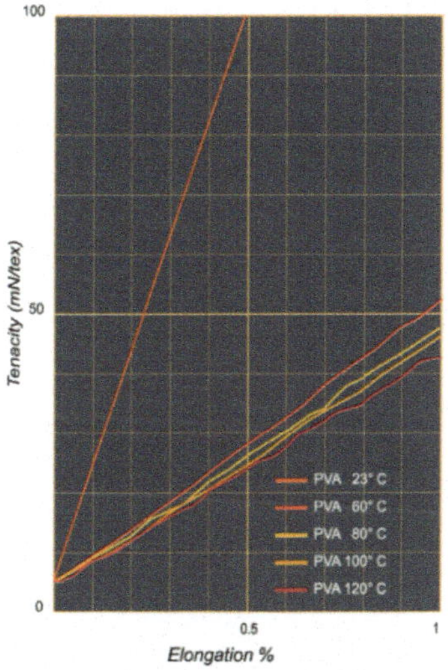

Figure 2: PVA's tenacity / elongation behav.

Its Young's modulus indicates the tensile stiffness of a reinforcement material. This parameter is the crucial factor for the dimensional uniformity of a reinforced hose under pressure load. The maintenance of high Young's modulus values at elevated temperatures separates rayon (topmost curve in fig. 3) from thermoplastic reinforcements.

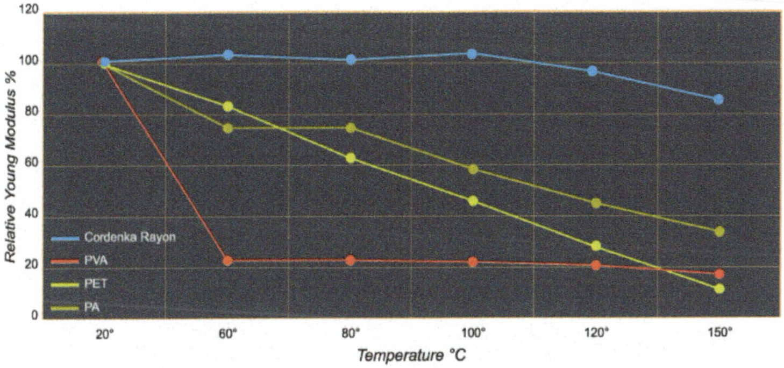

Figure 3: Reinforcement's Young's modulus as function of the testing temperature

3 Brake hose performance

Brake hose volumetric expansion tests at different temperature levels confirm the superior dimensional stability of rayon reinforced brake hoses and correspond with the yarn measurement results discussed before. Volumetric expansion tests were carried out according to the FMVSS 106 method. This test measures the volume of brake fluid present in a brake hose at a predefined pressure. Figure 4 shows that rayon and PVA reinforced brake hoses exhibit similar levels of volume expansion under increasing pressure at room temperature.

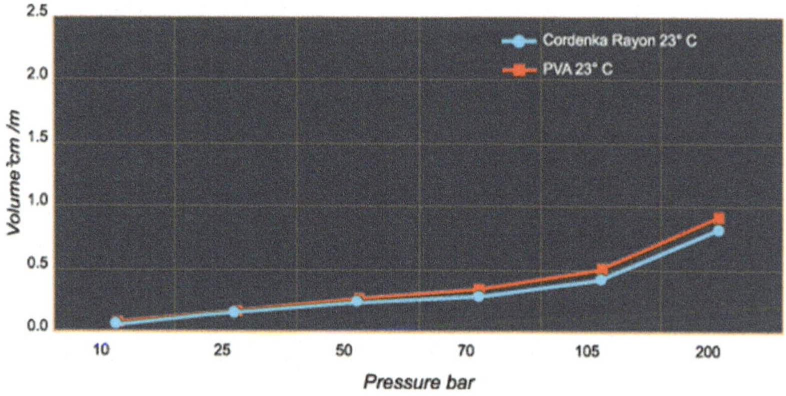

Figure 4: Volumetric brake hose expansion with rayon and PVA reinforcement at 23 °C

The comparison of two brake hoses applied by two different European car manufacturers demonstrates the typical small advantage of rayon reinforced brake hoses at room temperature. The volume expansion of the rayon reinforced hose compared to that of the PVA reinforced hose is 10 – 20% lower.

Figures 5 and 6, however, reveal the superior dimensional stability of rayon reinforced brake hoses at common temperature levels for brake hoses in actual use. At 60° C and 50 bar pressure, the rayon reinforced hose contains a volume of brake fluid 49% greater than that measured with the same hose at room temperature and 50 bar.

The volume increase of 161% exhibited by the PVA reinforced brake hose is more than three times higher. The stroke distance of the test cylinder inducing the pressure reflects this relation in linear fashion.

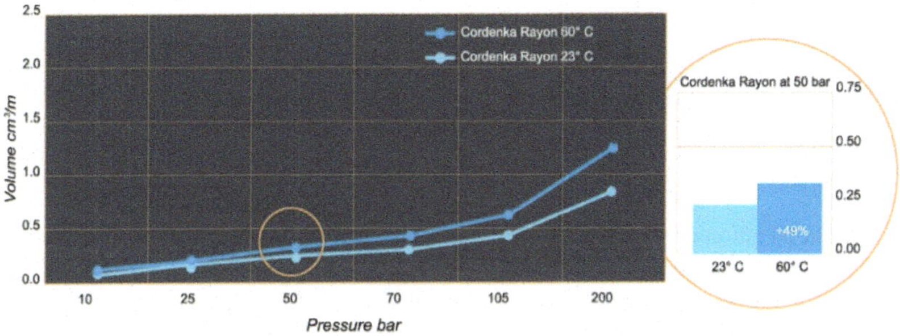

Figure 5: Volumetric brake hose expansion with rayon reinforcement at 23 and 60 °C

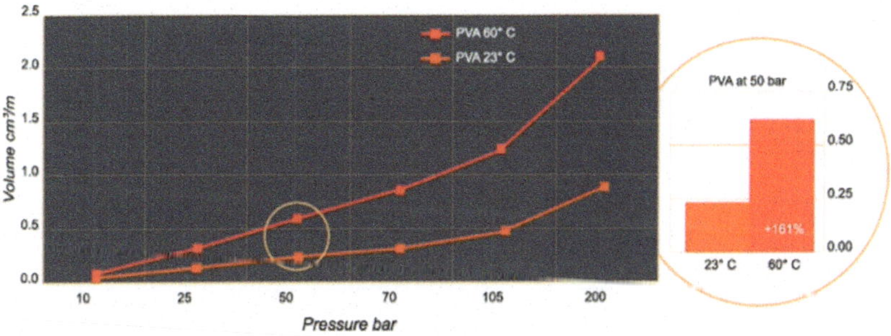

Figure 6: Volumetric brake hose expansion with PVA reinforcement at 23 and 60 °C

4 Application relevance

The necessary elasticity of the system's single parts poses a challenge for brake-system designers. The deployment of braking force requires a steady volume of brake fluid under pressure. The volume consumed for pressure generation should therefore be as small as possible, in order to minimize pedal travel and facilitate short braking distances. That's why a tight brake system provides a sense of safety at a touch of the brake pedal. Another driver for the minimization of pressure dissipation within the brake systems of modern cars are the requirements of short response Electronic Stability Control Programs (ESCs). Within fractions of a second, ESCs steer the braking of individual wheels to stabilize cars in critical situations, like skidding. Rayon-reinforced brake hoses contribute to the maintenance of fastest reaction times, independent of temperature conditions.

Given

- Rayon's superiority with regard to dimensional stability at higher temperatures and
- Common brake-hose service temperatures of greater than 40 °C

it is comprehensible that rayon is the material of choice for the reinforcement of performance brake hoses in the premium segment of Europe's automobile industry.

Yet the application of rayon-reinforced brake hoses is not limited to the premium segment alone. High performing, rayon reinforced brake hoses can also be found in compact cars made by demanding manufacturers around the world. The compliance of rayon-reinforced brake hoses with all relevant standards and specifications has been proved with their application in the cars of leading OEMs.

To further validate the application relevance of the foresaid, additional comparative Young's modulus measurements have been carried out on rayon and PVA brake hose grade yarns. Figure 7 displays absolute Young's modulus values of rayon and PVA brake hose yarns measured also at 40 °C.

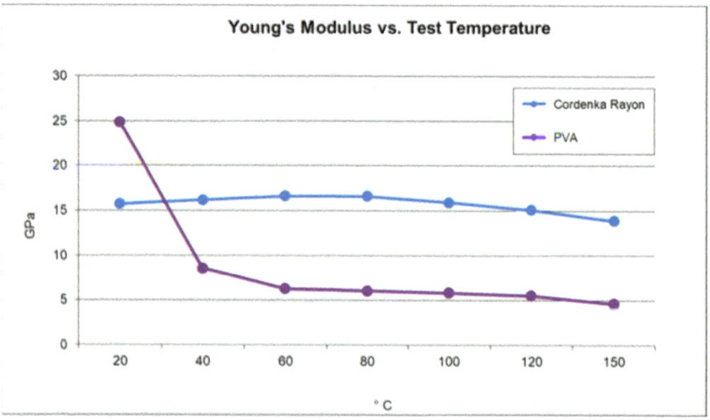

Figure 7: Rayon's & PVA's Young's modulus as function of the testing temperature incl. 40 °C

5 Driving tests

Temperature measurements also were carried out on brake hoses in car driving tests. In brake systems of motor vehicles, brake hoses are exposed to mechanical and thermal loads. Both, direct heat conduction from the caliper and heat transfer by convection occur.

The aim of the experiments was to show how high these thermal loads actually are in a normal car under typical operating conditions. The temperatures were measured directly at the brake hose as well as in its environment.

For the driving test a Ford S-Max 2.0 Duratorq (diesel / 103 kW) was used. The installed alloy wheels allow for good ventilation conditions at the wheel brake. The temperature was measured using Type K thermocouples on the left front axle wheel brake

- directly at the point of attachment of the brake hose to the caliper (Position a)
- at a distance of 5 mm from the brake hose near the saddle (Position b)

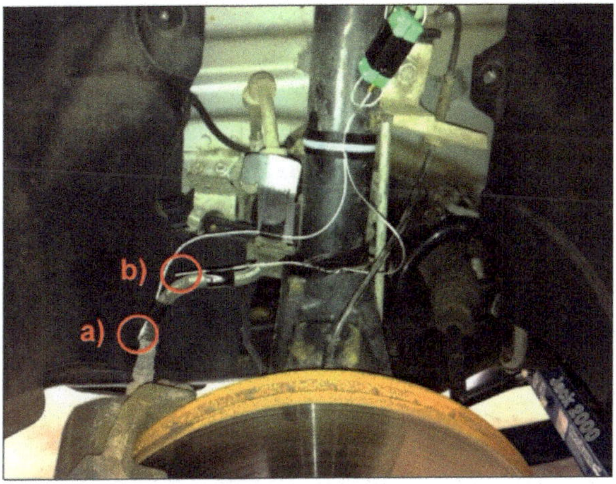

Picture 1: Installation situation of the thermocouples on the brake hose

The temperature measurements were carried out respectively at or after sporty driving on the following test tracks:

City: 15 km, brisk drive / stop-and-go
Country: 31 km, country road in mountain location
Motorway: 48 km, adaptive braking with high delays
Mountain: 3 km, average gradient 9% / maximum gradient 16.5%

All driving test were carried out at ambient air temperatures of 16-20 °C and dry roads. Figures 8 and 9 illustrate the temperatures measured at the brake hoses during the test runs as well as those measured with the engine running at the same places five minutes after the end of the test runs.

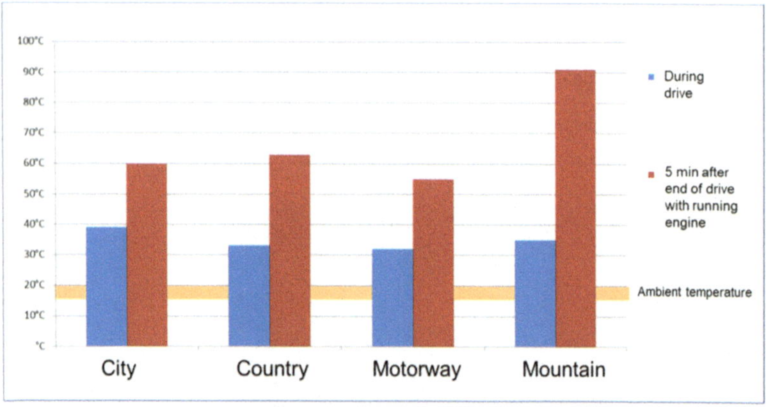

Figure 8: Driving test brake hose temperatures measured at sensor position a)

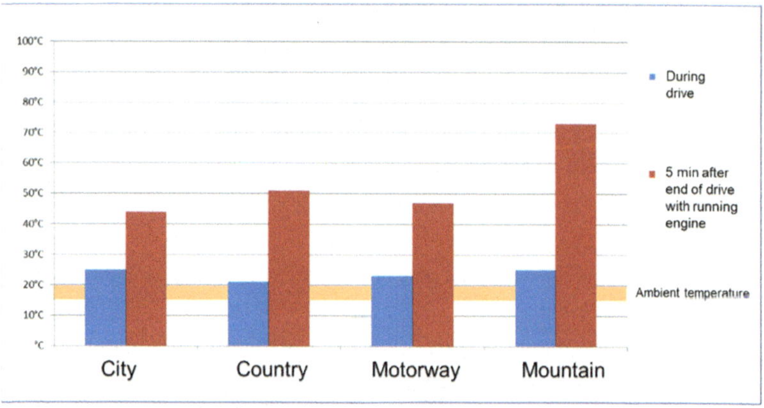

Figure 9: Driving test brake hose temperatures measured at sensor position b)

6 Brake hose benchmark – MA 1 in comparison

Cordenka MA 1 rayon brake hose reinforcement is found in state-of-the-art brake hoses in two braids with linear densities of 1100 den (1220 dtex) each. Competing brake hoses typically are reinforced with two braids of Kuralon 1239 PVA with linear densities of 1200 den (1330 dtex) each. In benchmark tests of according brake hoses the data in tables 1 and 2 was determined.

Table 1: Brake hose performance overview: Rayon MA 1 vs. PVA Kuralon 1239

Braid		Volumetric Expansion, 200 bar, RT, cm³/m hose	Target FMVS S 106	Volumetric Expansion, 200 bar, 60°C, cm³/m hose	Target FMVS S 106	Burst Pressure, bar	Target FMVS S 106	Whip-Test Durability, h	Target FMVS S 106
outer	inner								
PVA 1330	PVA 1330	**0,79**	< 2,00	**1,86**	n.a.	**920**	≥ 517	**> 100**	≥ 35
Rayon 1220	Rayon 1220	**0,66**	< 2,00	**0,96**	n.a.	**870**	≥ 517	**78**	≥ 35

Table 2 data was determined at the same hose types used for table 1 analysis. It proves that 1100 den Cordenka MA 1 can be regarded as efficient brake hose reinforcement also.

Table 2: Brake hose masses and composition

Braid		Mass of 10 cm hose (rubber + yarn), g	Mass of 10 cm hose (rubber + yarn), %	Yarn proportion of hose mass, %	Rubber proportion of hose mass, %
outer	inner				
PVA 1330	PVA 1330	**7,62**	**100**	**39**	**61**
Rayon 1220	Rayon 1220	**6,89**	**90**	**39**	**61**

7 Conclusions

Its lasting high Young's modulus at higher temperatures is the remarkable advantage of rayon as reinforcement for high performance mechanical rubber goods. Rayon reinforced brake hoses feature a superior dimensional stability at every day's driving conditions and contribute to the driver's sense of safety when steering his or her car.

Moreover it can be shown that 1100 den Cordenka MA 1 rayon is an efficient brake hose reinforcement, the use of which leads to advantageous and balanced brake hose properties.

8 Acknowledgements

The author would like to thank Univ.-Prof. Dr.-Ing. K. Augsburg and DI L. Heidrich from the TU Ilmenau / FG Kraftfahrzeugtechnik for carrying out the driving tests presented in this article.

Authors

Rudolf Einsiedel

Cordenka GmbH & Co. KG
Industrie Center Obernburg
63784 Obernburg
Germany

+49 6022 81 3261
rudolf.einsiedel@cordenka.com

SYSTEMS FOR ELECTRIC / CONNECTED VEHICLES

Use of a driving simulator to develop a brake-by-wire system designed for electric vehicles and car stability controls

M. Montani; R. Capitani; M. Fainello; C. Annicchiarico

Abstract

Vehicles with electrical propulsion and energy recovery need specific brake systems able to guarantee the correct integration of electrical energy recovery and high braking capability. The use of advanced controls with brake by wire systems allows to integrate such functionalities leaving the correct feedback to brake pedal and the possibility to enhance vehicle dynamics. We present a study done with simulation and simulators with human in the loop to show the potentiality of specific controls to enhance vehicle dynamics and their integration with electric motor controls. In detail, we will show how a physical brake-by-wire unit can be installed on a simulator environment obtaining an enhancement in the driving realism and enabling the virtual development of advanced vehicle functions. The same testing architecture can be used also for fault-injection activities and precise pedal feeling calibration. The case-study test bench has been installed on a static driving simulator equipped with a concurrent real-time machine, a vi-grade car real time model, the complete vehicle network and the full steering system.

Introduction

In the last fifty years with the increase of the electronical components on vehicles, the automotive industries have invested in development of stability control systems to ensure passenger safety. Consequently, all the commercial vehicles are equipped by now well-known ABS stability systems which can prevent wheels lock during full braking and preserve handling; since 2011, the European Commission decided that all new vehicles have to be equipped with a functional Electronic Stability System, too. The ESP has improved the active safety system passing from a wheel behavior control to a vehicle behavior control; in fact, it can keep the vehicle dynamic close to an expected nominal one, managing the torques on wheels to produce a correction yaw moment on the car [1]. However, the ESP acts only in emergency situations, while driving at limit of adhesion.

The goal of this study is to build an advanced control that is able to increase the dynamics, ensuring stability and better continuous performance during vehicle motion through the adoption of 4 single wheel braking actuators, each one with its own low actuation logic. The use of this architecture allows to increase the responsiveness of the system, its accuracy, due to the reduction of the inertia related to each caliper and to the possibility of continuously tracking four different pressure targets, and its reliability: being inherently redundant, if one actuator goes to fault conditions, the others can ensure the safety of the brake plant. This has allowed to develop a high-level control system ensuring stability and safety at any time and condition. To achieve this aim, it has been built a reference model that provides the values of the yaw rate and side-slip angle that the vehicle has to have to ensure stability. To make the vehicle following the targets and remaining in safety conditions in any driving condition, an optimal controller is implemented to actuate the single brake wheel actuators. Because of their capability to detect the vehicle instability the yaw rate and side-slip angle are chosen as states of control. The yaw rate is the angular velocity with which the vehicle moves around yaw axis. Its value is given by a standard IMU sensor already available in all commercial vehicles. The side slip angle, instead, is the angle between the direction of the vehicle velocity and the longitudinal barycentric axis of the car. For this application a new architecture was developed and published [2]. In literature there are many studies about side slip angle estimation because of its large non-linearity and uncertainty. In fact, it can be measured directly only by expensive sensors and model-based estimators cannot be done good results. In [3-5] there are some examples of machine learning technics to estimate side-slip angle thanks of their good capability of learning and to approximate non-linearities. However, to improve their performance, Neural Networks need to be trained with a great amount of data, not so easy to have. To overcome these problems, a Neural Network was trained with vehicle inertial values, longitudinal, lateral accelerations and yaw rate, given by the low cost IMU sensor. In this way it can be possible to estimate in real time the side-slip angle trend. Then a non-linear state observer, UKF [6-9], was added to improve the network estimation extending the data domain. This type of architecture has allowed to decrease the computational time, having a real-time measurement of side slip angle, and to decrease the economic cost, not using more sophisticated sensors.

In the state of the art different typologies of vehicle stability control can be found. Because of the great non-linearity of dynamic vehicle many authors chose to implement a Non-linear Model Predictive Control (NMPC), that is able to take into account the non-linearities, the mutual interaction of each dynamic states and the possibility of using constraints for both the states and the inputs [10-14]. The main problem of this type of control is the computational cost that precludes real-time usage. For this reason, other papers proposed to use Neural Network, capable of reproducing the vehicle dynamics and acts in real time reducing computational cost [15-17]. However, this type of control based on an end-to-end approach, without any analytic correlation with the dynamics of the system: this is critical for the safety assessment of the control system, not allowing to ensure the robustness of control under any condition. Then, a Linear Quadratic Regulator (LQR), lean and easy to implement, was chosen. It permits an optimal control of the dynamic model and a robust control thanks to precise representation of the vehicle dynamics. Compared to the few papers find in the state of the art [18-21] that acts leading the side-slip angle and the yaw rate under a saturation limit, not having a continuous control on their values, we implemented a refence model that continuously provides the target values and therefore the errors to be nullified, exploiting the possibilities of the brake actuators. In this way, the onset of instability is prevented and a control delay between the upper controller and the pression controllers is not leaded.

To validate the control, it is shown the results achieved on a static driver simulator, equipped by single wheel brake actuator and a car real-time complex multi-body model built up with the Vi-Grade software. This structure can be used not only to validate our model, but also to replicate a real vehicle motion condition and then be able to calibrate pedal feeling and optimize other systems.

The paper is dived as follow: a first part in which the vehicle dynamic model and controller architecture are explain in detail, splitting in a reference model and in a control model; then it's shown the driving simulator structure including the brake actuators; in the last part it's presented the types of simulations and the results achieved.

Conventions

m	vehicle mass
u	vehicle longitudinal velocity
v	vehicle lateral velocity
β	side-slip angle
r	yaw rate
F_{xij}	longitudinal force at the front/rear-left/right wheel
F_{yij}	lateral force at the front/rear-left/right wheel
δ	front steering wheel angle
I	vehicle inertia
$I_{\omega ij}$	front/rear left/right wheel inertia
ω_{ij}	front/rear left/right wheel angular velocities
M_{bij}	front/rear left/right wheel braking torque
p_{bij}	front/rear left/right wheel braking pressure
C_{yl}	front/rear cornering stiffness
α_i	front/rear slip angle
L	wheelbase
a_i	front/rear wheelbase
Δ	understeering gradient
A_c	cylinder area
R	wheel radius
μ	contact grip

Reference Model

The core aim of this study is to ensure that our vehicle follows at any time an ideal behavior and achieves stability and optimal performances.

During a maneuver, lateral and yawing dynamics of the vehicle change over time according to initial condition, steering wheel angle and longitudinal velocity imposed by the driver. To extrapolate the inner characteristics of the vehicle and be able to act the brakes to compensate the understeering, or the oversteering behavior of the vehicle, the steady state values of the yaw rate and side-slip angle are calculated.

The understeering gradient, Δ, is the difference between actual and kinematic steering. Taken difference between front and rear slip angles, it's the gradient to lateral acceleration that represents Δ. In this way, the yaw rate of the vehicle can be expressed as a function of the vehicle parameters (shown in tab 1), of the cornering stiffness and of the driver's inputs, steering and speed (1-3).

VEHICLE PARAMETERS	QUANTITIES
a_1	1.315 [m]
a_2	1.505 [m]
M	1751 [kg]

The steady-sate values of the side-slip angle can be estimated by the understeering gradient too, using the linearized congruence equations (4).

$$\alpha_f = \frac{m}{L}\frac{a_2}{C_{yf}} ru \; ; \; \alpha_r = \frac{m}{L}\frac{a_1}{C_{yr}} ru \tag{1}$$

$$\delta - \frac{L}{R} = \alpha_f - \alpha_r = \frac{m}{L}\left(\frac{a_2}{C_{yf}} - \frac{a_1}{C_{yr}}\right) ru \tag{2}$$

$$\Delta = \frac{m}{L}\left(\frac{a_2}{C_{yf}} - \frac{a_1}{C_{yr}}\right) \tag{3}$$

$$\alpha_2 = -\beta + \frac{r a_2}{u} \tag{4}$$

$$r_s = \frac{u}{L+u^2\Delta}\delta \; ; \; \beta_s = \frac{a_2 - \frac{m a_1 u^2}{L C_{yr}}}{L+u^2\Delta}\delta \tag{5}$$

These values of the yaw rate and of the side-slip angle (5) are the values that the vehicle has to follow in order to ensure steady behavior.

On the other hand, these expressions don't ensure to consider the limit of adhesion of the wheels and must be saturated to upper limits (6). In this way, our reference model is implemented to provide the yaw rate and the side-slip angle knowing the steering angle and the longitudinal velocity, ensuring that the vehicle remains in conditions of grip and handling. The logic used is given in (7).

$$r_{up} = \frac{\mu g}{u} ; \beta_{up} = \tan^{-1}(0.02\mu g) \tag{6}$$

$$|r_d| = min(r_s, r_{up}) ; |\beta_d| = min(\beta_s, \beta_{up}) \tag{7}$$

Since these reference values change while the vehicle is in motion, to be able to implement them in the control, we have to represent them in a state-space form. We can do this defining the gradients of r_s and β_s in the Laplace transform derivative, where $s = 1$ and $\tau_i = 0.1$:

$$\dot{\beta}_s = \frac{s\beta_s}{1+\tau_\beta s} ; \dot{r}_s = \frac{sr_s}{1+\tau_r s} \tag{8}$$

So, the state-space of these transfer functions is composed by side-slip angle and yaw rate as states. This represent the reference that the vehicle must follow.

Vehicle optimal control mode

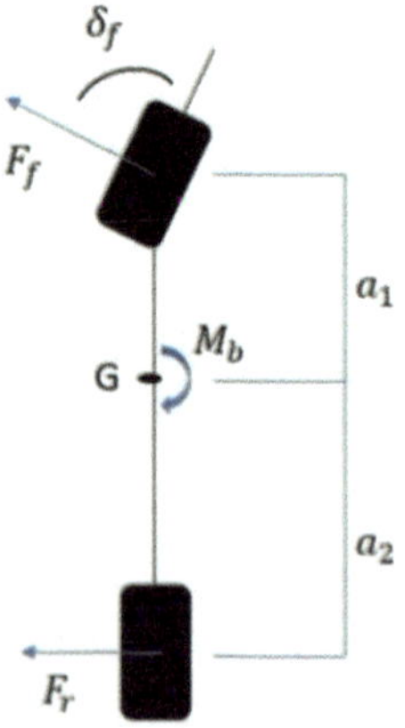

Fig.1

Because we decide to use an optimal controller we must defined a vehicle dynamic model where the correlation between sates and inputs are linearized at each time step (9-10).

$$\dot{X} = AX + Bu \tag{9}$$

$$Y = CX + Du \tag{10}$$

To ensure this aim, a single-track model of vehicle with 2 degree of freedom, lateral and yaw, was implemented (figure 1). The dynamic is expressed by the equations shown in (11-12), where the brake forces, that has to control the vehicle, are defined as a total yaw moment added in the yaw moment balance equation.

$$m(\dot{\beta}u + ru) = F_{yf} + F_{yr} + F_{xf} \tag{11}$$

$$I\dot{r} = (F_{yf} + F_{xf})a_1 - F_{yr}a_2 + M \tag{12}$$

Every time step the following hypothesis are assumed:

- constant longitudinal velocities;
- lateral forces are in linear proportional relation with the lateral slips

so, taken congruence equations, it is possible to define the lateral forces as dependent to side-slip angle and yaw rate. Tanks to this, a State-Space of the vehicle was implemented, where the states (13) are the side-slip angle and the yaw rate, and the input (14) is the total yaw moment given by the brake forces on the wheels that the actuators can ensure.

$$X = [\beta, r] \tag{13}$$

$$u = [M_{tot}] \tag{14}$$

$$A_r = -\begin{bmatrix} \frac{2C_{yf}+2C_{yr}}{mu} & 1+\frac{2C_{yf}a_1-2C_{yr}a_2}{mu^2} \\ \frac{2C_{yf}a_1+2C_{yr}a_2}{I} & \frac{2C_{yf}a_1^2+2C_{yr}a_2^2}{Iu} \end{bmatrix} \tag{15}$$

$$B_r = -\begin{bmatrix} 0 \\ \frac{-1}{I} \end{bmatrix} \tag{16}$$

Using the longitudinal velocity gives by the vehicle and a HSRI tire model that upgrades the cornering stiffness values, every time step the model is optimized to represent the dynamic conditions of the vehicle.

LQR controller design

Our problem is composed by a given system $\dot{x} = f(x, u)$ and a feasible trajectory (x_d), and we wish to design a compensator of the form $u = \alpha(x, x_d)$ such that $\lim_{t\to\infty}(x - x_d) = 0$. This is known as Trajectory Tracking Problem.

To design a controller capable of solving this type of problem, the vehicle model State-Space was concatenated with the reference model State-Space achieving a new dynamic model (17-20). In this way, it was possible to define the outputs as the errors between the actual and the desired values of the states. So, the goal of the control becomes find the right input to minimize these errors.

$$X = [\beta_r, r_r, \beta_d, r_d] \tag{17}$$

$$Y = [\beta_r - \beta_d, r_r - r_d] = [e_\beta, e_r] \tag{18}$$

$$A = \begin{bmatrix} A_r & 0 \\ 0 & A_d \end{bmatrix}; \; B = \begin{bmatrix} B_d \\ B_r \end{bmatrix} \tag{19}$$

$$C = \begin{bmatrix} 1 & 0 & -1 & 0 \\ 0 & 1 & 0 & -1 \end{bmatrix}; \; D = \begin{bmatrix} 0 \\ 0 \end{bmatrix} \tag{20}$$

Once specified the dynamic model, the control is ensured by the Linear Quadratic Regulator that is an optimal linear control. This type of control uses the minimization of a quadratic cost function (21) producing the necessary gains to each state to define the value of the input (22).

$$J = \int_0^\infty (y^T Q y + u^T R u) dt \tag{21}$$

$$u = -Kx - K_d x_d \tag{22}$$

In these equations x and x_d represents respectively the model states and the reference states; u is the input, namely a total brake yaw moment; Q and R are the weigh matrixes respectively for the outputs and the inputs and K is the gain matrix given by the minimization of the cost function, J, solving the Riccati equations:

$$P_k = Q + A'P_{k+1}A - A'P_{k+1}B(R + B'P_{k+1}B)^{-1}B'P_{k+1}A \tag{23}$$

$$K = (R + B'PB)^{-1}B'PA \tag{24}$$

The LQR control is, as already mentioned, an optimal control, i.e. it is optimal compared to an appropriate performance index. This performance index is the cost function, defined in (21), whose minimization allows to solve the problem, in this case of trajectory tracking, finding the law of control in state feedback (22). Having assumed that the dynamic is steady-state, various performance indices have been resolved in relation to different longitudinal speeds and different slip conditions. Once the cornering stiffness and the longitudinal velocity are updated, the Riccati equation optimizes the cost index providing each time step the gains. At this point the control is applied in real time.

So, the controller works in two ways, one of feedback and one in forward, ensuring strength and continuous optimization.

Controller structure

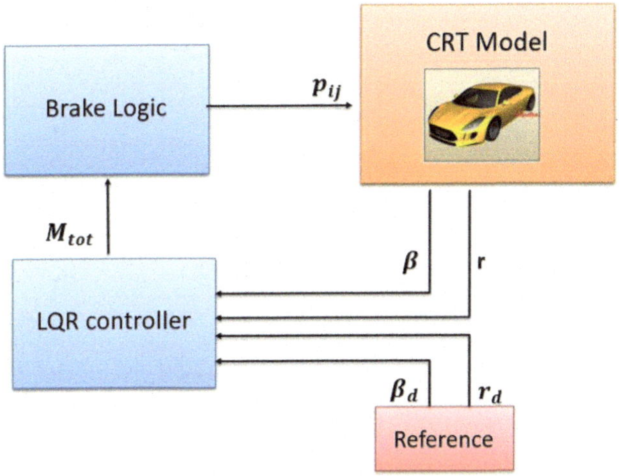

Fig.2

The controller architecture, shown in the figure 3, is composed by a reference model, an LQR controller and a brake split logic. According to the real-time conditions, the control quantities have been discretized with a sample time of 0.001 seconds.

The reference model has the task of providing the desired values of the side-slip angle and yaw rate. It receives, as inputs, the same driver demand steering and the same longitudinal velocity of the car and thanks to the formulations shown in the model explanation section, it reproduces the vehicle behavior and allows side slip angle and yaw rate to remain under low values and ensure stability.

The LQR controller works in real-time, providing, at each time step, the correction gains to compensate the errors of the vehicle compared to the reference model. As has been said before, to ensure that the controller works in any type of driver velocity input and dynamic condition, the cost function of the controller has to be updated each time step. So, the controller reads the longitudinal velocity of the vehicle and lateral engagement of the tires and produces the right law of control to assure the value of the yaw moment that the wheels forces have to achieved, fixing car behavior.

The solution of the control problem is the total barycentric yaw moment that the vehicle must follow to reach the dynamic of the reference model.

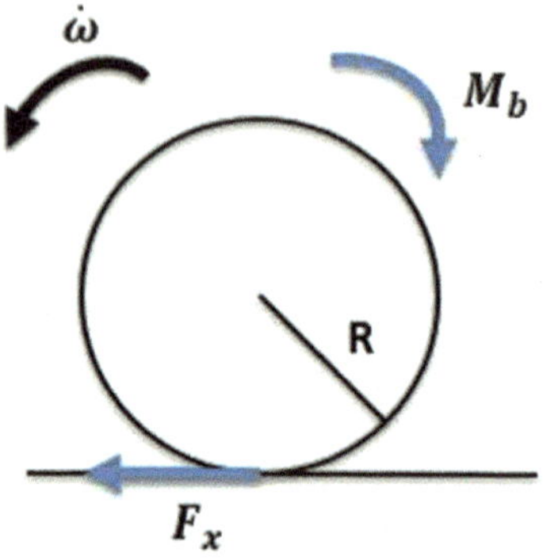

Fig.3

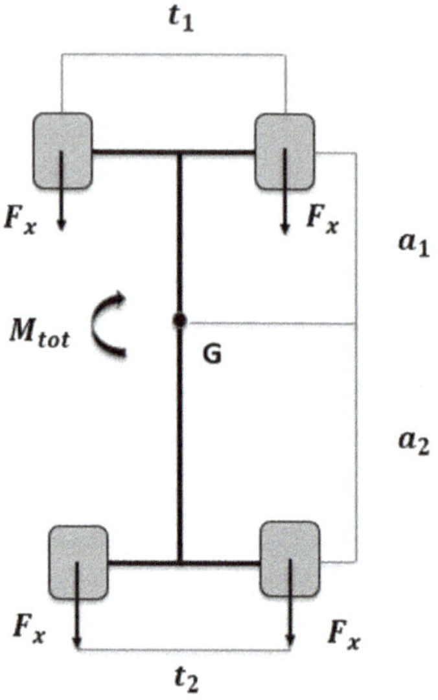

Fig.4

To ensure that the vehicle achieved this moment, it was implemented a logic that splits it in pressures that are given to each wheel actuator. The works of this logic is to determinate which wheel is to be braked and how much is to be braked. To select the wheel, a selector based on the understeering and oversteering behavior of the vehicle was generated. When the vehicle is understeering, it goes to a greater trajectory than that set by the steering. To compensate the error on the trajectory and linearise the vehicle behavior to the driver's input, a braking force must be applied on the inner rear wheel. On the other hand, when the vehicle is oversteering, to compensate the development of the vehicle on a lower trajectory the outer front wheel must be braked (25-28). In this way it is possible to reduce the lateral slip by adding a longitudinal slip component.

$$M_{b11} = \begin{cases} 0, & right\ turn\ and\ understeering \\ \frac{|M_{tot}|R}{\frac{t_1}{2}\cos\delta}, & right\ turn\ and\ oversteering \end{cases} \tag{25}$$

$$M_{b22} = \begin{cases} \frac{|M_{tot}|R}{\frac{t_2}{2}}, & right\ turn\ and\ understeering \\ 0, & rigth\ turn\ and\ oversteering \end{cases} \tag{26}$$

$$M_{b12} = \begin{cases} \frac{|M_{tot}|R}{\frac{t_1}{2}\cos\delta}, & left\ turn\ and\ oversteering \\ 0, & left\ turn\ and\ understeering \end{cases} \tag{27}$$

$$M_{b21} = \begin{cases} 0, & left\ turn\ and\ oversteering \\ \frac{|M_{tot}|R}{\frac{t_2}{2}}, & left\ turn\ and\ aundersteering \end{cases} \tag{28}$$

About the amount of braking, it is linked to the moments balance of the wheel (30) and the track width of the vehicle. Brake actuators can reach a maximum of 100 bar, so, the pressure is saturated at this value. In addition, an ABS logic has been added to ensure that the wheels do not lock up.

$$M_{tot} = \sum_{ij} F_{xij} \tag{29}$$

$$I_{\omega ij}\dot{\omega}_{ij} = -M_{bij} - F_{xij}R \tag{30}$$

$$P_{bij} = \frac{M_{bij}}{A_C \mu R} \tag{31}$$

Driver simulator and brake actuators

The control system has been tested in a static driving simulator, putting the corner braking actuator in the simulation loop. This allowed to consider the real performance of the brake system in tuning the controller.

In detail, the brake actuators are four electro-hydraulic units, capable of building up 50 bar of braking pressure within 50 ms with the considered brake calipers. Being a CAN controlled device, each actuator can track the pressure target imposed by the high-level control system, simplifying the deployment and the integration of the whole control loop.

Fig. 5-Static Driving simulator

Test

To validate the control, a series of test maneuvers was carried out to the simulator. These maneuvers were made to bring the vehicle at limit of handling and in steady-state conditions; they are a Sine with Dwell, given by the ESC test procedure of the ECE-R 13-H European normative [28], and a steering pad.

They are close-loop maneuvers built on car real time Vi-grade software with a sample time of 0,001 second. The characteristics of every maneuver are shown in the table 2:

MANEUVERS	LONGITUDINAL VELOCITIES	STEERING AMPLITUDE	STEERING FRE-QUENCY
SWD	80 [km/h]	270 [deg]	0.7 [Hz]
Steering PAD	100-150 [km/h]	200 [deg]	1 [deg/s]

Results

Steering Pad

This test is intended to assess the maximum lateral acceleration under quasi-stationary conditions that the vehicle can maintain. It involves bringing the vehicle to a certain longitudinal speed and then applying a steering rotation with constant angular velocities.In the graphs below the maneuver done with different velocities, 100 km/h and 150 km/h, are shown

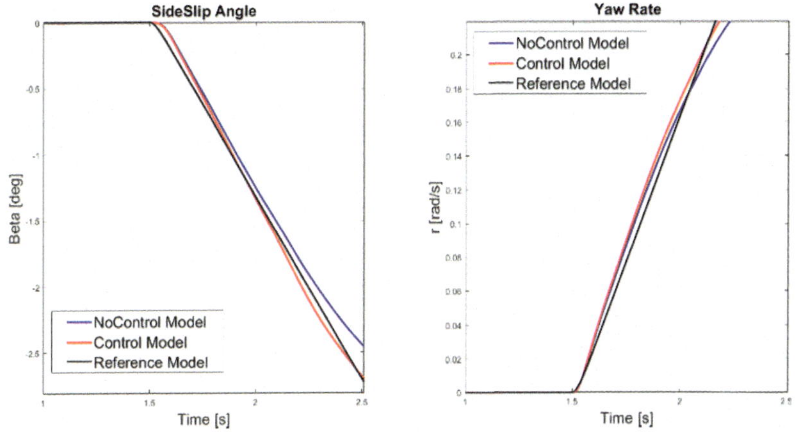

Fig.6.1-side slip angle on left and yaw rate on right at 100 km/h

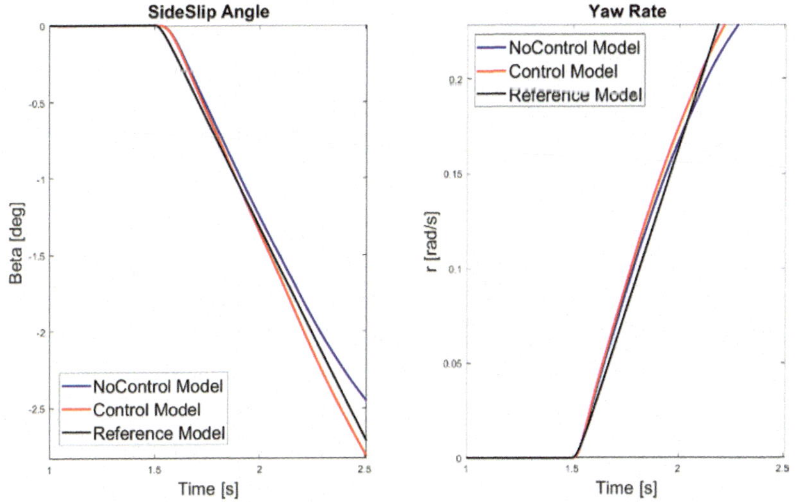

Fig.6.2-side slip angle on left and yaw rate on right at 150km/h

Figure 6.1, 6.2 show the trends of side-slip angle and yaw rate over time at various speed, in case of uncontrolled vehicle, controlled vehicle and reference model. It's possible to see that the reference model, that represent a quasi-stationary state, has a linear trend, in which the vehicle responds to the driver's inputs in a smooth manner, ensuring good handling and stability. The controlled vehicle follows the trend of the refence model ensuring better performance than an uncontrolled vehicle.

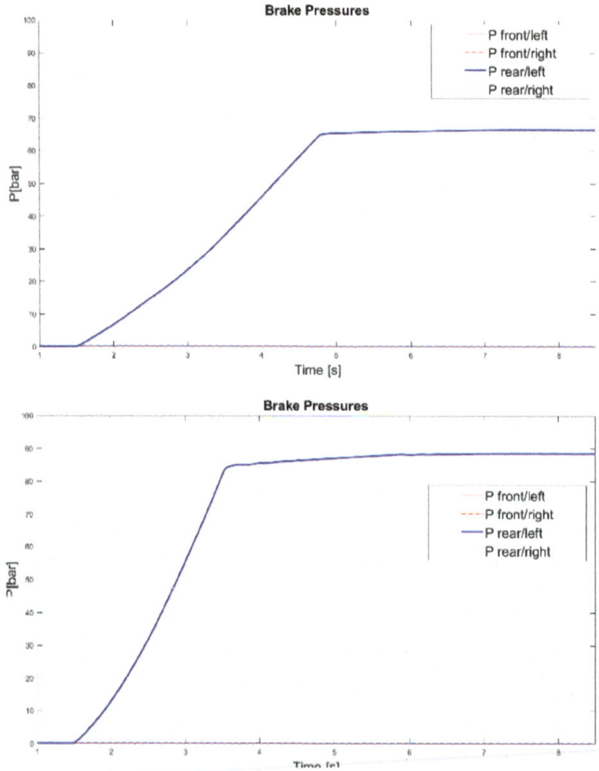

Fig.7-Brake Pressure at 100km/h on left; brake pressure at 150 km/h on right

From figure 7, it is possible to validate the brake logic. The vehicle is in understeering condition thanks of the side slip angle sign. In fact, the controller actuates the inner rear wheel ensuring a closer trajectory than the vehicle without the controller (fig.10). As the speed increases, the brake pressure also increases to counteract the additional lateral force generated.

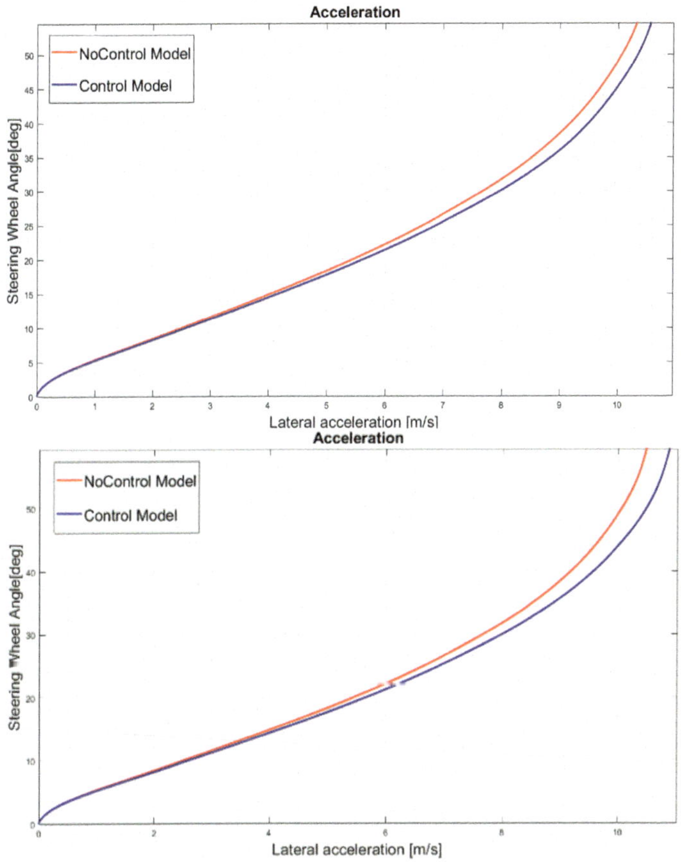

Fig.8-SWA as function of lateral acceleration at 100 km/h on left; SWA as function of lateral acceleration at 150 km/h on right.

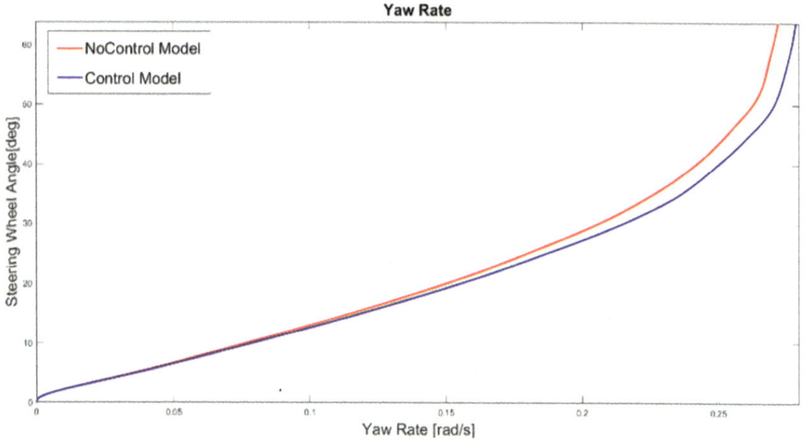

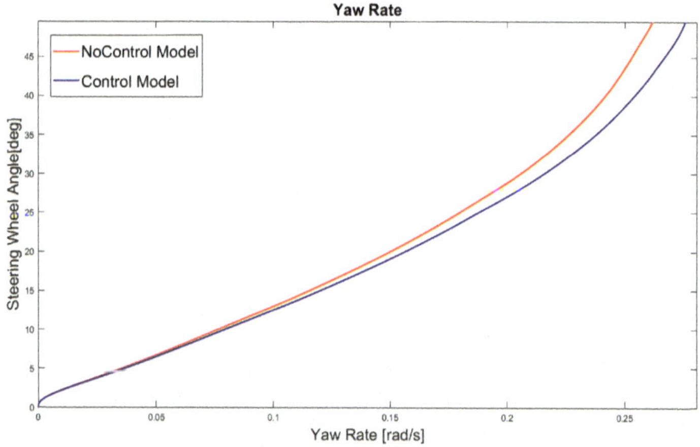

Fig.9-SWA as function of the yaw rate at 100 km/h on left; SWA as function of the yaw rate at 150 km/h on right.

Figure 8 and 9, instead, shows the improvement of the vehicle performances. The trend of the understeering curve (fig.8), where the steering is given as function of lateral acceleration, illustrates that the vehicle equipped with the control in both situations can reach higher lateral acceleration and the slope of the curve is lower. This means an improvement of the performances, because the vehicle is able to maintain stability at higher accelerations and an improvement of vehicle handling, because it replies at the driver steer in more neutral way.

In the end it's shown the trend of the steering as function of the yaw rate. We can see that the steer remains linear for higher values of the yaw rate. This is another index of the better levels of the car and results in the achievement of closer trajectory (fig.10).

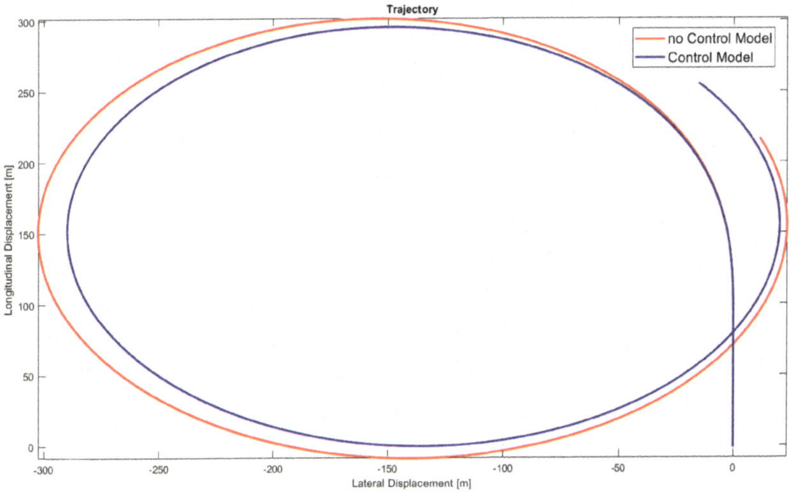

Fig.10-trajectory of vehicle at 150 km/h with and without controller.

Steering With Dwell

This type of maneuver allows to verify the stability of the vehicle as it is very demanding and brings the vehicle to the limit of handling.

The type of test, required by the normative, involves identifying the steering angle, δ_0,at which the car achieves lateral acceleration of 0.3g with a longitudinal speed of 80 Km/h. Than another series of tests involve the SWD steering input with an increasing amplitude of $\delta = k\delta_0$, from k=1.5 to k=6,5. In this study the maximum value of the steering amplitude is 270 deg, and only the results achieved with this steering angle are shown below, as they represent the most critical driving condition.

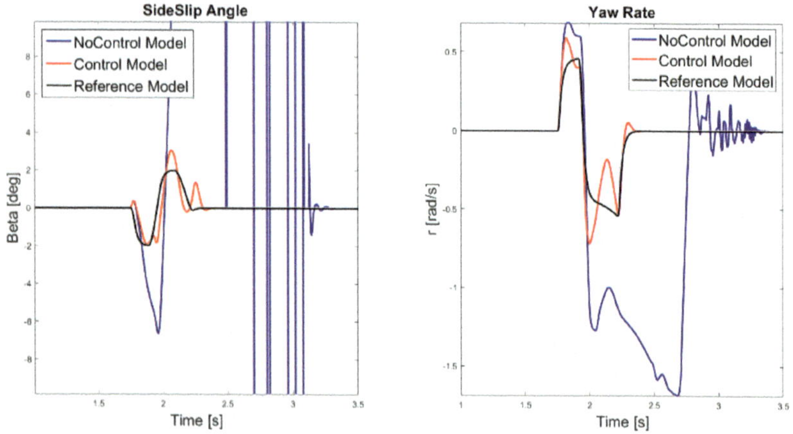

Fig.11- side slip angle on left and yaw rate on right

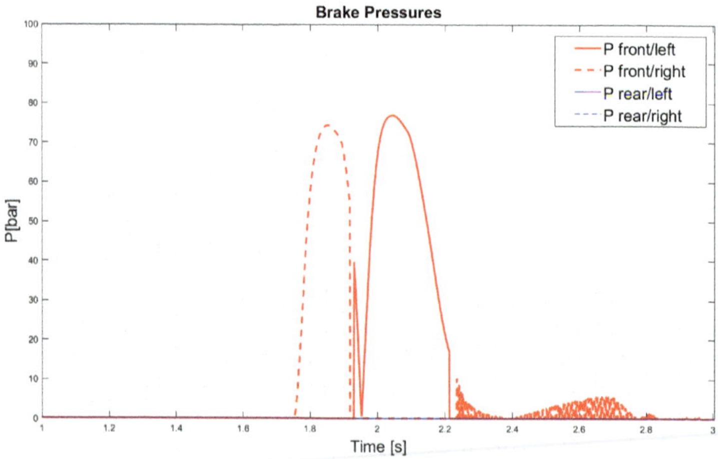

Fig.12-Brake Pressure of the wheels

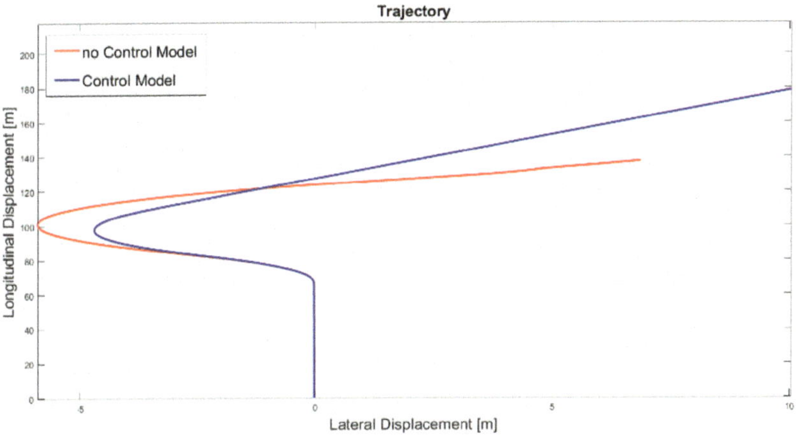

Fig.13-trajectory of vehicle with and without controller.

The plots show how the vehicle without the controller loses its stability and starts to spin.

In the figure 11 the reference, the controlled and the no-controlled model are shown. It is clearly visible that the vehicle checked is able to follow the side-slip angle and yaw rate values given by the reference and is able to maintain stability.

The wheels braked are the outer front ones. In fact, the uncontrolled vehicle shows an oversteering behavior that ends with its spin.

Figure 13 shows the trajectory of the vehicle with the controller and without the controller. It is possible to see how the uncontrolled vehicle can't finish the maneuver and loses its stability. Instead, the controller ensures stability and control also at limit of handling condition.

Conclusions

This paper presents a stability vehicle controller based on a trajectory tracking LQR. The use of this controller involves the schematization of the vehicle as a single-track model and two phases of law controller, a feedback and a feedforward. Every time step the actual and desired values of the side-slip angle and yaw rate are optimize by the controller given the yaw moment input to the vehicle. The vehicle receives the total input as a single pressure that each brake actuator has to give to the wheel. The pressure and reference states values depend on the vehicle behavior during its movement, i.e. the understeering gradient. The values of the reference and pressure are both saturated to an upper limit to take into account the dynamic limits of the tires and the performance limits of the actuator.

This solution ensures the achievement of very good performances thanks to the possibility of use four actuators able to continuously work and optimally manage the individual pressures at the wheels. In fact, it is possible to implement the controller in a real time simulator and ensure a continuous interaction with innovative brake actuators. Tests show that this control results robust in different type of conditions, in steady-state maneuvers and at limit of handling, for different longitudinal velocities and steering angles.

Bibliography

1. van Zanten, A. T. & GmbH, R. B. Evolution of Electronic Control Systems for Improving the Vehicle Dynamic Behavior. 9

2. Novi, T., Capitani, R. & Annicchiarico, C. An integrated ANN-UKF vehicle sideslip angle estimation based on IMU measurements. 13

3. Du, X., Sun, H., Qian, K., Li, Y. & Lu, L. A prediction model for vehicle sideslip angle based on neural network. in *2010 2nd IEEE International Conference on Information and Financial Engineering* 451–455 (IEEE, 2010).

4. Melzi, S. & Sabbioni, E. On the vehicle sideslip angle estimation through neural networks: Numerical and experimental results. *Mechanical Systems and Signal Processing* 25, 2005–2019 (2011).

5. Melzi, S., Sabbioni, E., Concas, A. & Pesce, M. Vehicle Sideslip Angle Estimation Through Neural Networks: Application to Experimental Data. in *Volume 2: Automotive Systems, Bioengineering and Biomedical Technology, Fluids Engineering, Maintenance Engineering and Non-Destructive Evaluation, and Nanotechnology* 2006, 219–224 (ASME, 2006).

6. Hodgson, G. & Best, M. C. A Parameter Identifying a Kalman Filter Observer for Vehicle Handling Dynamics. *Proceedings of the Institution of Mechanical Engineers, Part D: Journal of Automobile Engineering* 220, 1063–1072 (2006).

7. Vargas-Meléndez, L., Boada, B., Boada, M., Gauchía, A. & Díaz, V. A Sensor Fusion Method Based on an Integrated Neural Network and Kalman Filter for Vehicle Roll Angle Estimation. *Sensors* 16, 1400 (2016).

8. Li, L., Jia, G., Ran, X., Song, J. & Wu, K. A variable structure extended Kalman filter for vehicle sideslip angle estimation on a low friction road. *Vehicle System Dynamics* 52, 280–308 (2014).

9. Boada, B. L., Boada, M. J. L. & Diaz, V. Vehicle sideslip angle measurement based on sensor data fusion using an integrated ANFIS and an Unscented Kalman Filter algorithm. *Mechanical Systems and Signal Processing* 72–73, 832–845 (2016).

10. Jalali, M., Khosravani, S., Khajepour, A., Chen, S. & Litkouhi, B. Model predictive control of vehicle stability using coordinated active steering and differential brakes. *Mechatronics* 48, 30–41 (2017).

11. Falcone, P., Eric Tseng, H., Borrelli, F., Asgari, J. & Hrovat, D. MPC-based yaw and lateral stabilisation via active front steering and braking. *Vehicle System Dynamics* 46, 611–628 (2008).

12. Falcone, P., Borrelli, F., Asgari, J., Tseng, H. E. & Hrovat, D. Predictive Active Steering Control for Autonomous Vehicle Systems. *IEEE Transactions on Control Systems Technology* 15, 566–580 (2007).

13. Barbarisi, O., Palmieri, G., Scala, S. & Glielmo, L. LTV-MPC for Yaw Rate Control and Side Slip Control with Dynamically Constrained Differential Braking. *European Journal of Control* 15, 468–479 (2009).

14. Ataei, M., Khajepour, A. & Jeon, S. Model Predictive Control for integrated lateral stability, traction/braking control, and rollover prevention of electric vehicles. *Vehicle System Dynamics* 1–25 (2019).

15. Zhu, B., Piao, Q., Zhao, J. & Guo, L. Integrated chassis control for vehicle rollover prevention with neural network time-to-rollover warning metrics. *Advances in Mechanical Engineering* **8**, 168781401663267 (2016).

16. Ohno, H., Suzuki, T., Aoki, K., Takahasi, A. & Sugimoto, G. Neural network control for automatic braking control system. *Neural Networks* **7**, 1303–1312 (1994).

17. Ćirović, V., Aleksendrić, D. & Smiljanić, D. Longitudinal wheel slip control using dynamic neural networks. *Mechatronics* **23**, 135–146 (2013).

18. Velenis, E., Katzourakis, D., Frazzoli, E., Tsiotras, P. & Happee, R. Steady-state drifting stabilization of RWD vehicles. *Control Engineering Practice* 19, 1363–1376 (2011).

19. Dal Poggetto, V. F. & Serpa, A. L. Vehicle rollover avoidance by application of gain-scheduled LQR controllers using state observers. *Vehicle System Dynamics* 54, 191–209 (2016).

20. Li, L. *et al.* A novel vehicle dynamics stability control algorithm based on the hierarchical strategy with constrain of nonlinear tyre forces. *Vehicle System Dynamics* 53, 1093–1116 (2015).

21. Jagga, D., Lv, M. & Baldi, S. Hybrid Adaptive Chassis Control for Vehicle Lateral Stability in the Presence of Uncertainty. in *2018 26th Mediterranean Conference on Control and Automation (MED)* 1–6 (IEEE, 2018).

22. Alcala, E. *et al.* Comparison of two non-linear model-based control strategies for autonomous vehicles. in *2016 24th Mediterranean Conference on Control and Automation (MED)* 846–851 (IEEE, 2016).

23. Heinzl, P., Lugner, P. & Plöchl, M. Stability Control of a Passenger Car by Combined Additional Steering and Unilateral Braking. *Vehicle System Dynamics* 37, 221–233 (2002).

24. You, S.-S. & Jeong, S.-K. Controller design and analysis for automatic steering of passenger cars. *Mechatronics* 12, 427–446 (2002).

25. Weiss, Y., Allerhand, L. I. & Arogeti, S. Yaw stability control for a rear double-driven electric vehicle using LPV-H∞ methods. *Science China Information Sciences* 61, (2018).

26. Adaptive Vehicle Stability Control by means of Tire Slip-angle.pdf.

24. Zhou, H. & Liu, Z. Vehicle Yaw Stability-Control System Design Based on Sliding Mode and Backstepping Control Approach. *IEEE Transactions on Vehicular Technology* 59, 3674–3678 (2010).

27. Mirzaei, M. A new strategy for minimum usage of external yaw moment in vehicle dynamic control system. *Transportation Research Part C: Emerging Technologies* 18, 213–224 (2010).

28. Regulation No 13-H of the Economic Commission for Europe of the United Nations (UN/ECE) — Uniform provisions concerning the approval of passenger cars with regard to braking [2015/].

Modification of an electronic brake system for the use in an electric bus

Jochen Buck
Udo Rein

Daimler AG

1 Introduction

In 1996, Daimler AG was the first manufacturer who introduced an electronic brake system (EBS) for commercial vehicles above 7.5 t [1]. The brake system includes a function which controls the slip on different axles and therefore replaces the load sensing valve [2], [3]. In comparison to pneumatic brake systems, the EBS has further advantages such as a shorter response time, load independent and inertial referenced deceleration requests and brake wear harmonization. Endurance brakes such as an engine brake and a retarder are seamlessly integrated during braking, which increases the lifetime of the service brake [4], [5].

In response to current trends in the automotive industry as well as future emission and fuel consumption regulations, Daimler has developed multiple e-mobility concepts for the commercial vehicle sector. The electric bus Mercedes-Benz eCitaro is one of those concepts (cf. Fig. 1). The eCitaro has an electric portal axle with two wheel-hub motors, which have a maximum power of 250 kW and a maximum torque of 970 Nm. If the maximum number of batteries is installed, the minimum driving range of the bus is 150 km [6]. The brake system includes all the conventional components of an electronic brake system with the exception of the air-treatment system and the endurance brakes.

Figure 1: eCitaro

Since electric motors can generate a regenerative braking torque, they can be considered as endurance brakes. Therefore, it was decided at the beginning of the concept phase that the recuperation functions are integrated into the EBS with the following primary and secondary objectives:

Primary objective:

- increase of the driving range via recuperation without violating any stability criteria and regulations

Secondary objectives:

- no noticeable transition between service brake and electric motor
- fast response behavior
- high precision

Because of the different characteristics of electric motors and endurance brakes, the EBS has been modified in several ways in order to integrate the recuperation functions and fulfill the secondary objectives. In the following, those modifications are described in detail.

2 Differences between electric motors and conventional endurance brakes

Due to the physical concept electric motors have different characteristics than conventional endurance brakes such as hydro-dynamic retarders or engine brakes. The following four aspects are the most important differences from a braking point of view:

1. The response time of electric motors and especially of electric wheel hub motors is a lot better than the ones from service and endurance brakes.
2. Conventional endurance brakes have their maximum torque at high speeds, whereas electric motors have their maximum torque at low speeds due to the motor characteristic (cf. Fig. 2). As a result, it is possible that electric vehicles can brake fully electrically nearly until standstill. Especially city busses feature a very high recuperation potential as city busses generally drive below 60 km/h and constantly brake until standstill.
3. Electric motors which are used in fully electric vehicles usually have more power and a higher maximum torque than conventional endurance brakes.
4. The torque of electric motors can be controlled more precisely and continuously, especially in comparison to engine brakes.

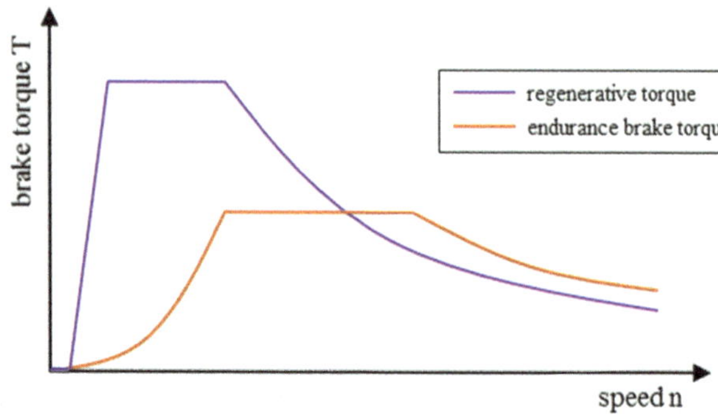

Figure 2: Torque characteristic of an electric motor and an endurance brake

3 Modifications of the EBS

One of the main features of an EBS is the endurance brake integration (EBI) (cf. chapter 1). Since the electric motors of the eCitaro are also part of the powertrain, the basic EBI logic can be used for integrating the recuperation functions into the EBS. However, in order to fully exploit the advantages of electric motors, while still fulfilling the ECE-R13 regulation, the EBS has to be modified in several ways.

3.1 Modifications due to different dynamics

The different dynamics of conventional endurance brakes and electric wheel-hub motors required the biggest modifications in the EBS. On the one hand, electric motors need to be controlled differently than endurance brakes. On the other hand, the different dynamics have to be taken into account when the brake torque is shifted from the electric motors to the service brake.

3.1.1 Control of electric motors

Conventional endurance brakes are controlled by a closed loop system. This is due to the fact that conventional endurance brakes have a response time which can vary between a few hundred milliseconds and more than a second. In addition, the actual torque is not always matching the requested torque, especially in case of engine brakes, since some endurance brakes cannot be controlled that precisely. By using a closed loop control the brake management can always react to the current situation and fill the gap between the actual and the requested brake torque with the faster responding service brake if needed.

On the contrary, electric motors have a very fast response time and can be controlled very precisely. The response time is not only faster than the one of the service brake, but even faster than the signal routing across the existing control chain. Therefore electric motors have to be controlled by an open loop system if the full dynamic shall be exploited. As a consequence, it is necessary to take several measures in order to avoid an undesired braking behavior.

Firstly, clear communication rules have to be set up between the EBS and other electronic control units (ECU). The most important one is that the requested regenerative torque of the EBS is not altered by any other ECU. The only situation where an alteration of the signal is permitted is a sudden decrease of the brake demand, since this can lead to drive train vibrations. Furthermore, the EBS needs to be informed about the availability of the regenerative torque at all times, especially when the torque is limited due to overheating or a failure of one of the electric motors.

However, there are situations where a limitation cannot be detected by the control unit of the electric motors, e.g. mechanical problems in the drive train. Therefore an algorithm is implemented in the EBS which detects situations like these. For this purpose, the actual and requested regenerative torque as well as the actual and expected deceleration are compared in the background during all braking phases. If a mismatch is detected over a certain period of time, the brake management shifts the brake torque from the electric motors to the service brake.

Since the detection of such mismatches takes a few hundred milliseconds, it is necessary that there are also control mechanisms which can react faster to undesired deceleration behavior. One control mechanism is the deceleration controller, which is part of every electronic brake management. The deceleration controller increases or decreases the brake torque depending on the actual deceleration. The other control mechanism is the slip controller, which shifts the brake torque from one axle to the other, if the slip becomes too high on one axle.

3.1.2 Brake torque shifting from electric motors to service brake

At low speeds the efficiency of electric motors, which are operated as generators during recuperation, is limited due to current heat losses [7]. Therefore, the brake torque has to be shifted from the electric motors to the service brake shortly before standstill. Since the response time of electric motors is faster than the one of the service brake, the actual brake force would decrease if the service brake torque demand was simply increased by the amount by which the regenerative torque was decreased [8]. As a consequence, also the deceleration would decrease during the shifting phase, which is uncomfortable for the driver and the passengers (cf. Fig. 3(a)).

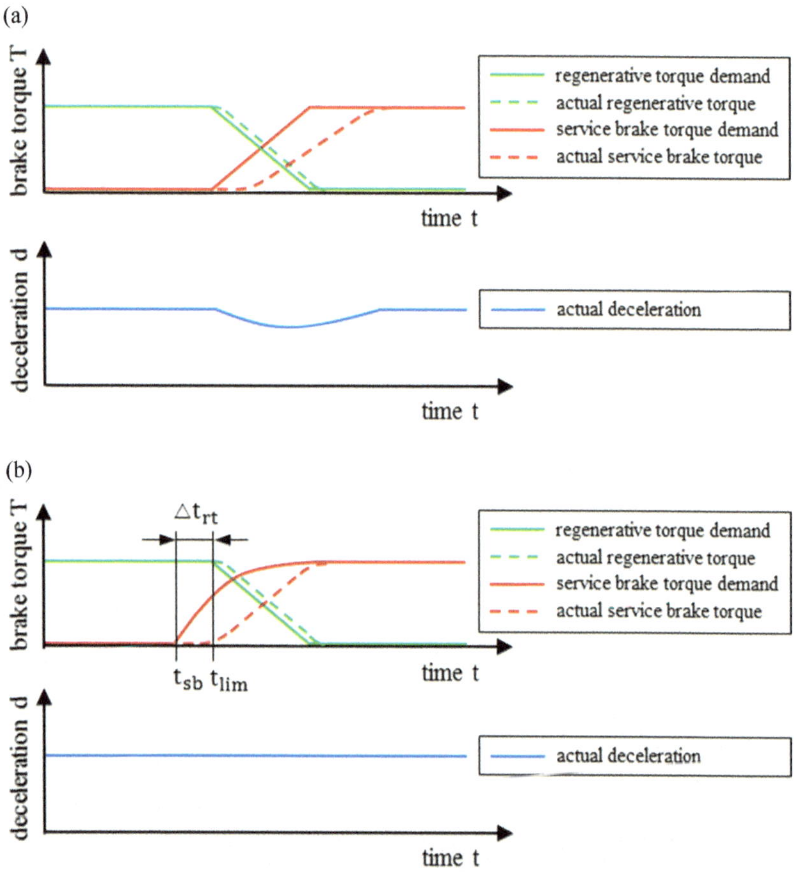

Figure 3: Brake torque shift without (a) and with (b) premature service brake torque demand

In order to avoid deviations in the deceleration, the service brake torque has to be requested prematurely. However, if the service brake is requested too early, the deceleration will increase, which is also uncomfortable for the driver and the passengers. Thus, the point – when the regenerative torque is limited during braking – must be precisely determined. For this purpose, the actual and maximum regenerative torque need to be analyzed depending on the time. In general, the maximum regenerative torque is so high at low speeds that the actual torque is considerably below it. Thereby, the limitation of the maximum regenerative torque starts at a speed at which it does not have any effects on the current torque yet. Since the maximum torque steadily decreases, the constant gradient can already be determined after two time steps. Via extrapolation and by means of the gradient the time t_{lim} when the maximum regenerative torque reaches the same value as the current torque can be precisely predicted (cf. Fig. 4).

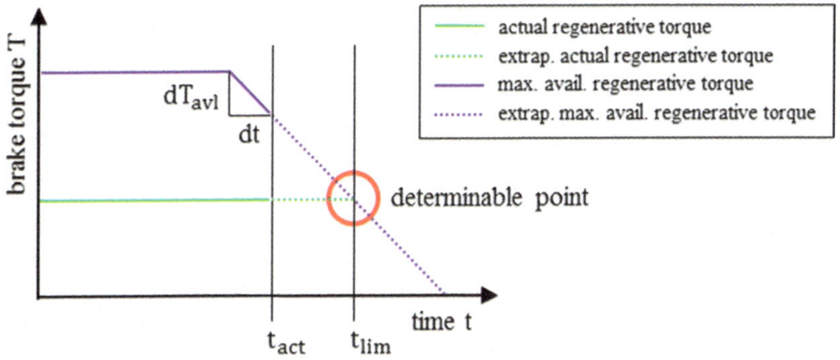

Figure 4: Determination of the point when the regenerative torque is limited

Once the limitation time t_{lim} is known, the time t_{sb}, when the service brake needs to be requested prematurely, can be determined as follows (also cf. Fig. 3(b)).

$$T_{sb} = t_{lim} - \Delta t_{rt} \tag{1}$$

with t_{sb} : time, when the service brake needs to be requested

t_{lim} : time, when the actual regenerative torque is limited

Δt_{rt} : difference between the response times of the electric motors and the service brake

The difference between the response times depends on the drive train and the brake layout. For example, the time difference for vehicles with electric wheel-hub motors is higher than the one for vehicles with a central electric motor and a differential.

3.2 Modifications due to different torque characteristic and controllability

From an EBS point of view, the main objective of the project is the increase of the driving range via maximum recuperation without violating any stability criteria and regulations (cf. chapter 1). Although the stability criteria are the same for electric and conventional vehicles, the better controllability of the electric motors gives the brake management more opportunities to hold back the service brake. Nevertheless, in situations where the service brake cannot be hold back, it has to be ensured that the driving stability is not negatively affected by the use of the service brake. In the following, it is explained which modifications are necessary to maximize the recuperation and how the service brake has to be used in order to have a stabilizing effect on the vehicle behavior.

3.2.1 Increase of the maximum allowed deceleration for endurance brakes

As for all conventional city busses the rear axle of the eCitaro is the driven axle. Consequently, the regenerative torque of the electric motors is only applied to the rear axle of the vehicle (cf. Fig. 5). If only the electric motors are used for the deceleration, brake slip only arises on the rear axle, which is not favourable to the stability of the vehicle. Therefore, the slip controller of the brake management monitors the slip at all times and interferes if the slip becomes critical. However, since busses have a high load on the rear axle, a high amount of regenerative torque can be applied to this axle before the slip limit is reached.

Figure 5: Regenerative torque applied on the rear axle

As described in chapter 2, the maximum torque of electric motors which are used in purely electric vehicles is usually higher than the one of conventional endurance brakes. Thus, energy would be wasted if the maximum allowed deceleration for endurance brakes or electric motors was not raised for electric vehicles. A higher maximum deceleration automatically leads to an increase of the slip difference between front and rear axle. This is feasible as electric motors can be controlled faster and more precisely than endurance brakes (cf. chapter 2). Therefore, the regenerative torque can be quickly adapted in situations, in which the slip on the rear axle endangers the stability of the vehicle.

3.2.2 Overlapping brake requests from brake pedal and retarder lever

Sometimes bus drivers use the retarder lever and the brake pedal at the same time, especially when driving downhill. In those cases the retarder lever is normally pulled to keep the vehicle speed constant. Before curves the driver usually decreases the velocity by additionally pushing the brake pedal.

In conventional vehicles, the EBS normally does not demand any more brake torque from the endurance brake if a different ECU like the Central Powertrain Controller (CPC) has already sent a brake request to the endurance brake. The reason is the slow response time of conventional endurance brakes. If two different ECUs request a brake torque from the endurance brake, both ECUs have to react correctly and coordinated when the vehicle stability decreases. Due to the slow response time, an uncoordinated or wrong reaction would affect the other ECU with a time delay which makes the control of the endurance brake even more difficult. To simplify things, the EBS only sends a brake demand to the service brake if the endurance brake has already been requested by another ECU.

If the same logic is transferred to an electric vehicle, energy is wasted every time when the retarder lever and the brake pedal are used at the same time. For this reason, the interface between the EBS and CPC has been extended by a signal which constantly informs the EBS about the brake request from the CPC to the electric motors. With this knowledge the EBS can request an additional torque from the electric motors if the maximum available regenerative torque has not been reached yet (cf. Fig. 6). Because of the information exchange between the EBS and CPC, both ECUs can see how the other ECU reacts in case of stability issues. Additionally, as electric vehicles have fast response times, wrong or uncoordinated reactions show up right away and thus can be instantly corrected.

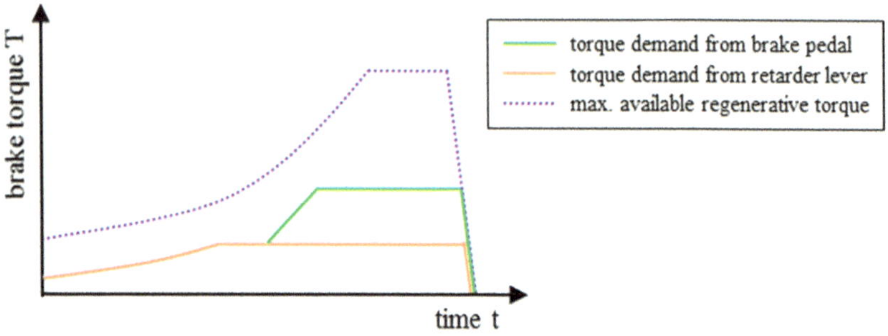

Figure 6: Overlapping of multiple regenerative brake requests

3.2.3 Optimization of the slip and brake pressure distribution controller

Depending on the driving conditions, the torque which is requested from the endurance brakes or electric motors may be limited due to availability or stability reasons. A limitation is mostly caused by stability reasons on low friction and availability reasons on high friction. If the deceleration demand of the driver cannot be fulfilled with the available or maximum allowable torque of the endurance brake or electric motors, additional brake torque has to be requested from the service brake. The brake management has to ensure that the additional torque does not have a destabilizing effect on the driving behavior. For this purpose, the slip and the brake pressure distribution controller have been modified for electric vehicles.

As described above, the regenerative torque is only applied on the rear axle. As a consequence, the front axle is not braked as long as the vehicle is solely decelerated via recuperation. In case that an additional brake from the service brake is needed, this brake torque should be mostly applied on the front axle in order to decrease the slip difference between the front and rear axle. Therefore, the more regenerative torque is applied on the rear axle, the more brake pressure is generally shifted to the front axle.

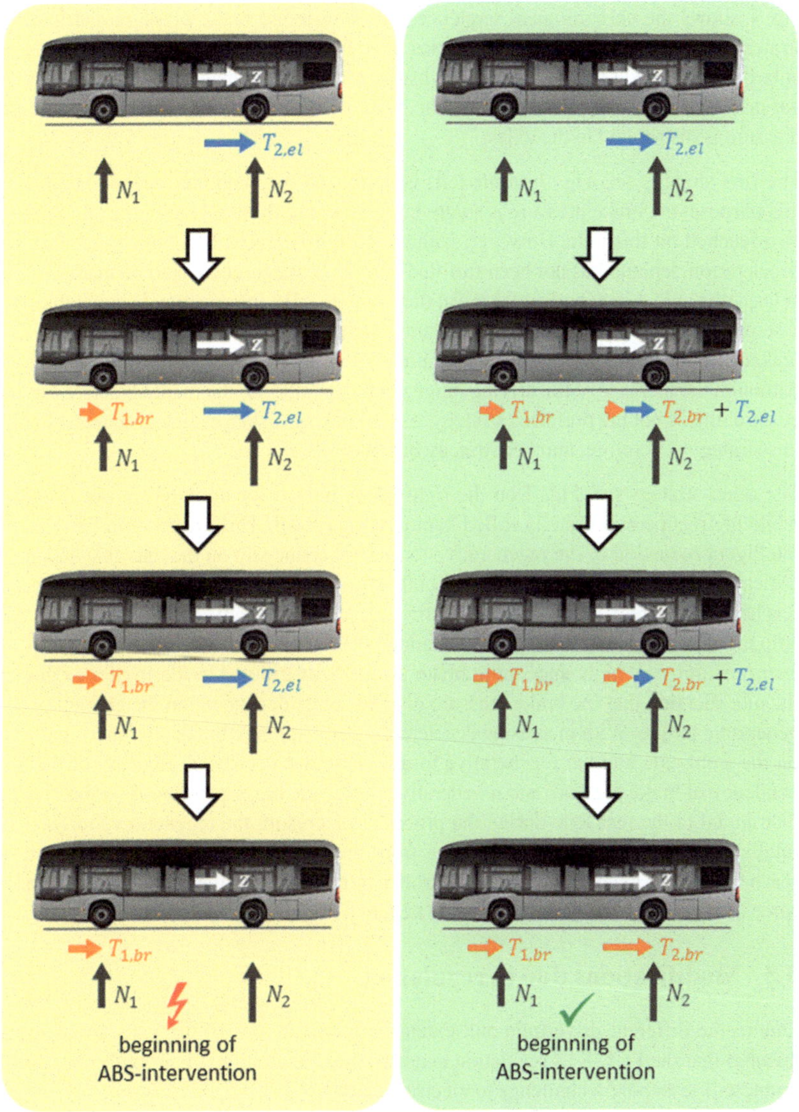

Figure 7: Brake blending with different brake pressure distribution strategies on low friction
Left: Pressure build-up solely on the front axle
Right: Reduction of the regenerative torque during pressure build-up on all axles

Fig. 7 shows the chronological sequence of two different brake pressure distribution strategies on low friction. Both strategies have in common that at first the vehicle is only decelerated via recuperation. This happens as long as the slip on the rear axle is not critical. Once the slip limit is reached, the strategies differ in the way the brake pressure is distributed to the axles.

The first strategy (cf. Fig. 7 on the left) is optimized for maximum recuperation. For this purpose, the brake pressure is solely built up on the front axle until the slip limit is also reached on this axle. However, if the slip limit is reached on both axles, but the deceleration demand has not been fulfilled yet, the brake torque has to be increased on at least one axle. Since both axles are at the limit, an ABS intervention is inevitable. As a result, the regenerative torque is instantly reduced to zero and the rear axle is not braked for a few milliseconds until the brake pressure has been built up on that axle. Although the increase of the driving range via recuperation is the main objective for the eCitaro, this is not the preferred strategy, as the braking distance slightly increases with this strategy. Therefore, another strategy has been developed.

The other strategy (cf. Fig. 7 on the right) is optimized for minimum brake distance while the recuperation rate is stilled kept at a high level. This is accomplished via an intelligent reduction of the regenerative torque. Once the slip on the rear axle becomes critical, brake pressure is distributed to both axles. Since the slip limit has already been reached on the rear axle, the regenerative torque has exactly to be reduced by the amount of brake torque which has been added via the service brake. At first, the loss of energy is not very high since more brake torque is distributed to the front axle due to the modification that the brake pressure distribution is dependent on the amount of regenerative torque. With increasing deceleration demand more brake pressure is built up on the front axle and the regenerative torque is further decreased. Because of the dependency of brake pressure and regenerative torque the brake pressure is automatically distributed to the rear axle during the process. As a result, the regenerative torque has totally been replaced with brake torque from the service brake once the slip limit is reached on both axles. If the ABS controller interferes now, there is no loss of brake force and therefore the brake distance is kept to a minimum.

3.3 Modifications due to regulation

Due to the different drive train and extensive modifications of the EBS, it has to be ensured that the ECE-R13 regulation is not violated at any time. Especially the type-0 and test-II tests pose a challenge to electric vehicles.

3.3.1 Type-0 test

The type-0 test states that a maximum deceleration of 5 m/s^2 has to be reached with the service brake at any time [9]. For conventional vehicles, for which the service brake is constantly used, this test is not a big challenge if the brake layout has been designed correctly. However, since the electric motors in purely electric vehicles have a lot of power and their maximum torque at low speeds (cf. chapter 2), the service brake is only used shortly before standstill when the regenerative torque is ramped out because of efficiency reasons. In those cases, the braking power is very low because of the low velocity. As a consequence, the brake pads do not heat up which leads to a decrease of the friction coefficient over time.

There are two solutions for this problem, a hardware and a software one. In case of the hardware solution, the brake force loss is compensated via bigger brake cylinders and special brake pads which have a higher friction coefficient. In contrast, the software prevents a loss of brake force by using the service brake at high speeds from time to time. As a result, the recuperation rate is a lot lower with this solution. Therefore, the hardware solution has been chosen for the eCitaro as the increase of the driving range is the main objective of the project.

Figure 8: Tests in high mountains

3.3.2 Type-II test

The type-II test states that a maximum deceleration of 5 m/s^2 has to be reached with the service brake after the vehicle drove a 6 km long slope with a 6% declination at a constant speed of 30 km/h [9]. To prevent the service brake from overheating, the endurance brake is normally used while driving downhill. In contrast to the brake torque of the endurance brake, the regenerative torque is not constantly available at a speed of 30 km/h. If the battery is fully loaded and no brake resistors are available. In case of the eCitaro, the problem is avoided as a brake resistor is needed for the air conditioning anyway.

3.4 Additional modifications

Apart from the already mentioned modifications, there are a few more topics which are less complex, but also important for the control of electric vehicles.

The main controlled variable of the EBS is the deceleration of the vehicle and the main manipulated variables are the brake pressures and the endurance brake or regenerative torque. For this purpose, the deceleration is converted into brake pressures and torques and the other way round by means of the learned vehicle mass. Since the learned vehicle mass can deviate from the real one, it is advantageous to learn the actual ratio between the deceleration and the regenerative torque.

In contrast to combustion engines, electric motors do not have a drag torque. Although this can be simulated by the EBS, the eCitaro currently does not have such a pseudo drag torque behavior. The recuperation does not begin until the brake pedal has been applied. This way, the driver can decide whether he wants to decelerate or not.

4 Conclusion

In this paper it is described, how an electronic brake system (EBS) has to be modified for the use in an electric bus. Most of the modifications result from the different characteristics from endurances brakes and electric motors. Electric motors have a better response time, a higher maximum torque and more power than conventional endurance brakes. Additionally, they can be controlled more precisely and their maximum torque is at low speeds, which is why electric vehicles can brake fully electrically nearly until standstill.

The different dynamics of conventional endurance brakes and electric motors require the biggest modifications in the EBS. In contrast to endurances brakes, electric motors need be controlled by an open-loop system if the full dynamic shall be exploited. Furthermore, the different dynamics have to be taken into account, when the brake torque is shifted from the electric motors to the service brake.

As the increase of the driving range is the primary objective, the recuperation shall be maximized by holding back the service brake as much as possible. The higher maximum torque and the better controllability of the electric motors give the EBS more opportunities to do so. Nevertheless, in situations where the service brake cannot be hold back, the EBS has to ensure via an intelligent brake pressure distribution that the driving stability is not negatively affected by the use of the service brake.

Besides all those software modification, also hardware changes are necessary, in order to avoid any violations of the ECE-R13 regulation. Due to the low usage of the service brake, the friction coefficient of the brake pads decreases over time. The brake force loss is compensated via bigger brake cylinders and special brake pads which have a higher friction coefficient.

References

[1] Daimler AG: *Anti-lock braking system for commercial vehicles*. Stuttgart, 26.12.2005

[2] Toepfer, Bernhard; Reiner, Michael; Daimler-Benz AG: *Verfahren zur lastabhängigen Regelung des Bremsdruckes an Fahrzeugen und Vorrichtung zur Durchführung des Verfahrens*. German patent DE3829951C2, 20.08.92

[3] Reiner, Michael; Mercedes-Benz AG: *Verfahren zur Steuerung einer Dauerbremse eines Kraftfahrzeuges*. European patent EP0715999A1, 12.06.1996

[4] Reif, Konrad: *Brakes, Brake Controland Driver Assistance Systems: Function, Regulation and Components*. Springer Vieweg, Wiesbaden, 2014

[5] Hilgers, Michael: *Elektrik und Mechatronik* in *Nutzfahrzeugtechnik lernen*. Springer Vieweg, Wiesbaden, 2016

[6] Daimler AG: *Lokal emissionsfrei und fast geräuschlos. Weltpremiere des Mercedes-Benz eCitaro*. Accessed on: May 17th, 2019

[7] Pischinger, Stefan; Seiffert, Ulrich: *Vieweg Handbuch Kraftfahrzeugtechnik*. Springer Vieweg, 8. Auflage, Wiesbaden, 2016

[8] Fuchs, Alexander; Klepser, Thorsten; Daimler AG: *Verfahren zum Betreiben eines Kraftwagens*. German patent DE102009039347A1, 03.03.2011

[9] Economic Commission for Europe of the United Nations: *Regulation No 13*, 18.02.2016

TIRE.WHEEL.TECH SECTION

WHEEL TECHNOLOGIES AND REQUIREMENTS

Contribution of tires and wheels to the WLTP

Dr. Günter Leister,
Daimler AG

1 Introduction

Starting in 1997, all fuel consumption and emission values in Europe were based on the so-called "New European Driving Cycle (NEDC)". Although this cycle did not represent the typical vehicle-driver related fuel consumption values, it was very suitable for comparing cars and engines.

In 2007, the UN in Geneva decided to develop a worldwide-harmonized test cycle WLTP (Worldwide Harmonized Light Duty Test Procedure) as a new basis for fuel consumption and emission homologation.

Second goal was to have more realistic homologation values (higher values than today but closer to typical vehicle-driver values), decided by the EU commission and strongly supported by the car industry.

2 NEDC, WLTC and WLTP

There was a strong timeline (wenn im Sinn von "angespannte Zeitschiene", könnte man auch "tight timeframe" verwenden) for the change from NEDC to the WLTP coupled with new targets for the CO2 emission, Fig 1. Part of the WLT Procedure is the so-called WLT Cycle.

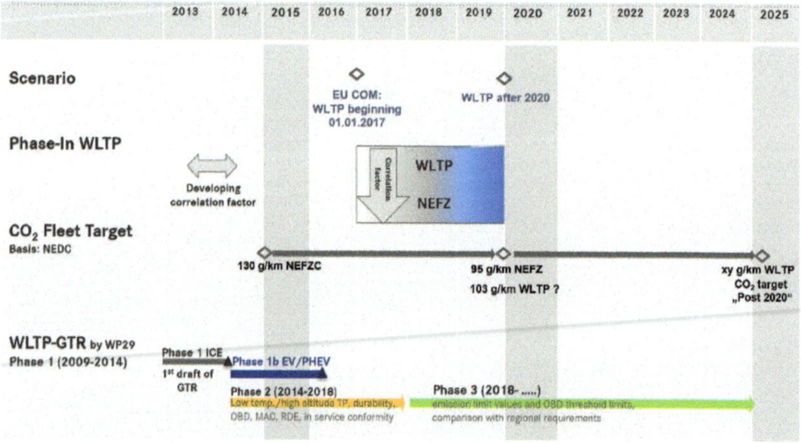

Fig.1: Timeline of the WLTP

2.1 WLTC (Worldwide Harmonized Light Duty Test Cycle)

The new WLT Cycle consists of four sub cycles with so-called Low Speed / Mid Speed / High Speed and Extra High Speed cycles. Looking into the details, the dynamics and the average of the driving velocity of the WLTP is 12 % higher, which leads to approx. 2 – 5 % more fuel consumption, compared to the old NED Cycle, Fig 2.

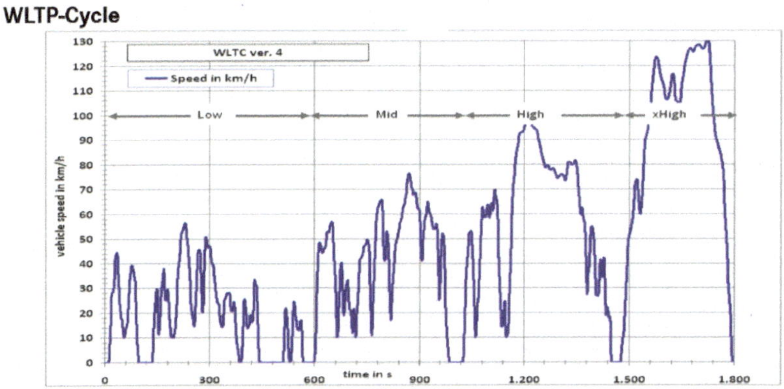

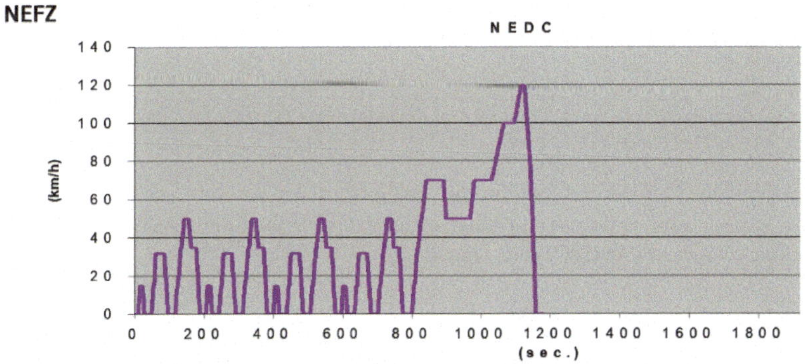

Fig 2: WLTP Cycle and NED Cycle

2.2 WLTP (Worldwide Harmonized Light Duty Test Procedure)

The change in conditions of the procedure is more relevant than the cycle itself. The conditions lead to an increase of CO2 values of approx. 10 – 25 %. Smaller effects are the reduction of tolerance subtraction, more precise temperature conditions and further detailing of the described preconditioning cycle. The most important change is the consideration of the real vehicle weight after production, the tire rolling resistance label and the aerodynamics for each individual car.

For the evaluation of the CO2 value, an interpolation method is used. Car manufacturers have to evaluate the CO2 value for the best-case and worst-case vehicle by testing the car on a track. For the best- and worst-case vehicles the parameters weight, aerodynamics and rolling resistance values of the tires have to be evaluated. For each individual vehicle, the real CO2 emission value has now to be computed by interpolation methods, Fig. 3.

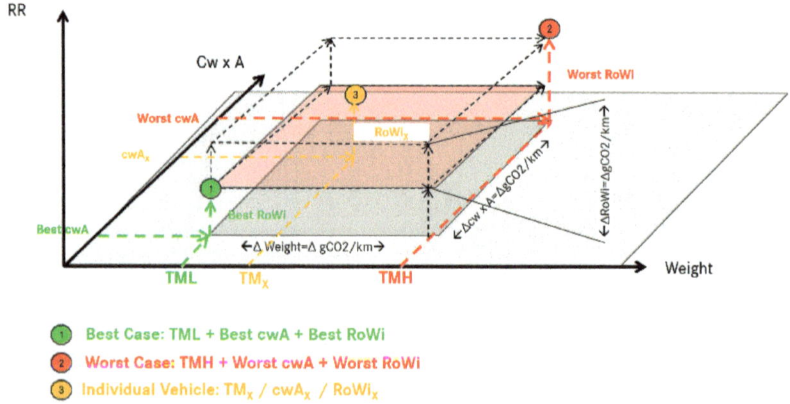

Fig 3: Interpolation cube for determination of the individual CO2 value

Not the individual value is used for the rolling resistance but the so-called class value. The "class" value is the average value of the range of the rolling resistance class according to Virtual Lab (ISO) rules, Fig 4.

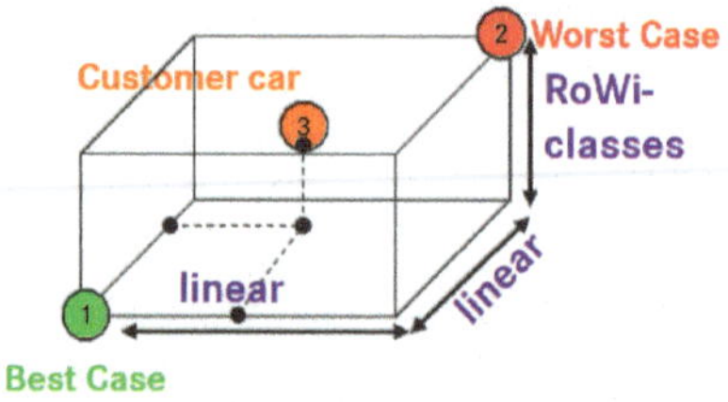

Class	C1 range	C1 class value
1	RRC ≤ 6.5	⇨ 5.9
2	6.5 < RRC ≤ 7.7	⇨ 7.1
3	7.7 < RRC ≤ 9.0	⇨ 8.4
4	9.0 < RRC ≤ 10.5	⇨ 9.8
5	10.5 < RRC ≤ 12.0	⇨ 11.3
6	RRC > 12.0	⇨ 12.9

Fig 4: Virtual Lab rules for rolling resistance classes

2.3 Sensitivity Vehicle from NEDC to WLTC

Analyzing the three effects regarding the WLT Cycle weight, aerodynamics and rolling resistance, it can be shown that the aerodynamic effects are now more important compared to the NEDC. The change in the effect of tire rolling resistance and weight are slightly small, Fig. 5.

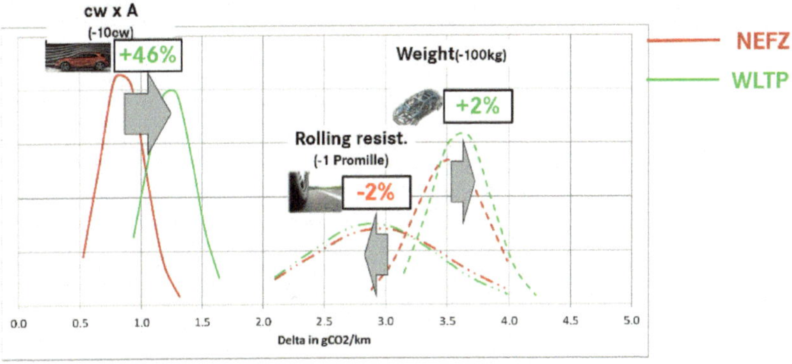

Fig 5: Average influence change of weight, rolling resistance and aerodynamics

3 New COC Document

The new legislation leads to COC documents with new contents: WLTC (low, medium, high, extra high, combined, in gram CO2), actual mass, front area, class of rolling resistance, individual rolling curve: f0, f1, f2 (which depends on the aerodynamics of the individual wheel design).

Plugins have additional values like energy consumption in Wh / km, electric driving distance in km and electric driving distance city in km, Fig 6.

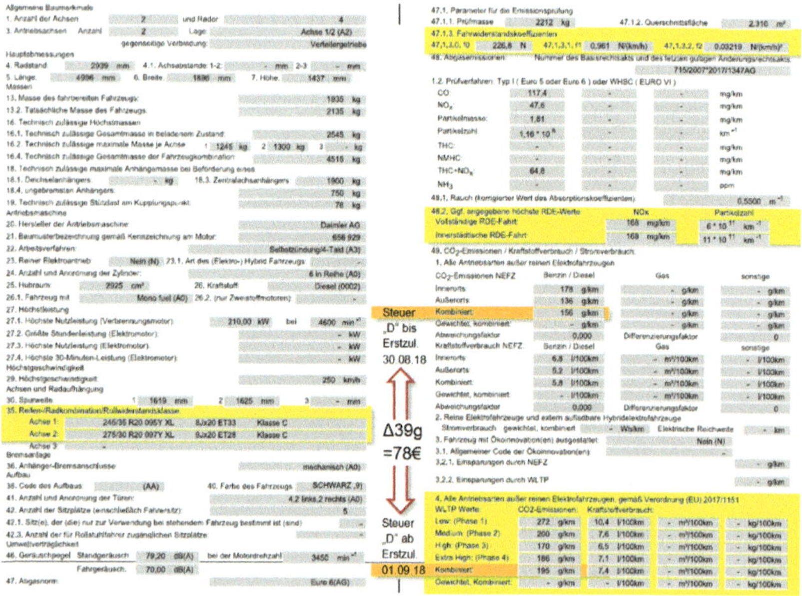

Fig 6: COC Document containing WLPT values

For a car manufacturer, the influence of the individual customer options on the CO2 value is very important, Fig 7. Indeed, the influence of the tire dimension (including aerodynamics, weight and rolling resistance) is dominant. This may lead to completely different take rates of large wheels, depending on individual country taxation.

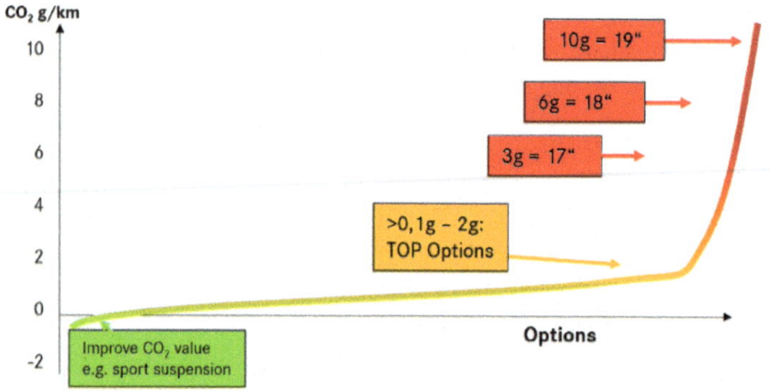

Fig 7: Influence of options on CO2 values

The current situation is that a new vehicle sold by a car manufacturer must be equipped with original tires and wheels when leaving the dealer. Afterwards, the customer can change tires and wheels without consequences. This may also influence the customer's behavior.

4 Method to Evaluate Influences of Tires and Wheels

To have more transparency of the WLTP influences of tires and wheels, it helps to work with a cost method. Looking on the regulations: if a car manufacturer is not able to fulfill the target of the legislation in the overall fleet, 95 Euro per car and per gram CO2 tax has to be paid. To avoid this cost, technical improvements in relation to aerodynamics, weight, rolling resistance (also influenced by tire pressure) are possible, Fig 8.

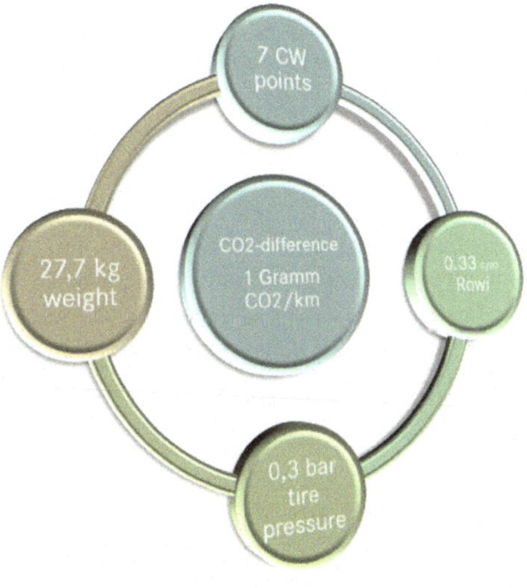

WLTP cost equivalence 95,- €/ Gramm CO2

Fig. 8: Influence of tires and wheels on CO2

4.1 Tire Influence

The main tire influence factors are the rolling resistance and the tire pressure. Nevertheless, these will lead to some disadvantages which have to be compensated by other chassis components. Without doing anything, the change of rolling resistance will lead to effects described in Fig. 9.

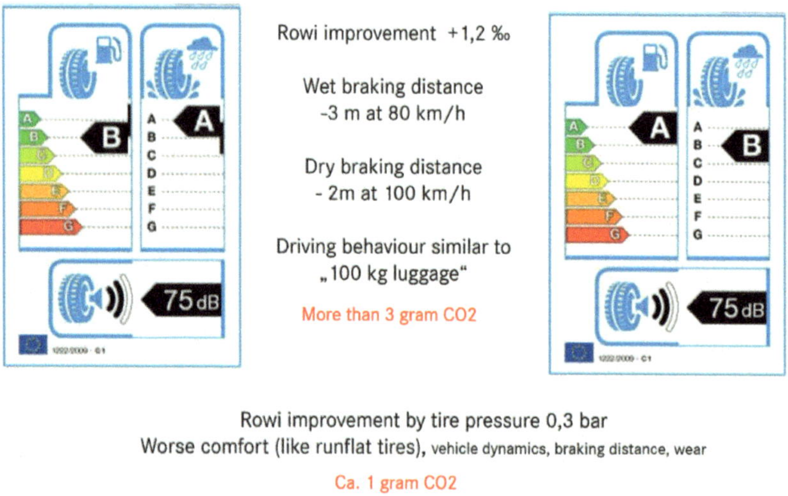

Fig 9: Interaction of tire properties

There is also an influence of the aerodynamics of the tire, Fig 10. Up to now, it is hard to precisely determine this influence in advance because so many measurements are necessary after the tire sidewall shape is finished. Nevertheless, there are many rules in the book of requirements to avoid very bad aerodynamic values for the tires, [1].

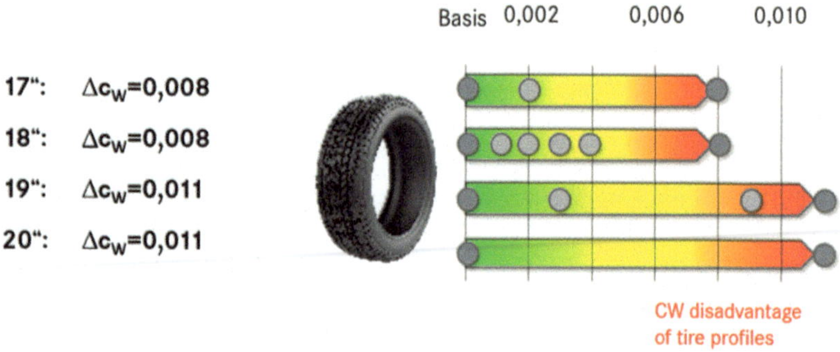

Fig 10: Influence of tire profile on the CW value

4.2 Wheel Influence

In addition, the wheel concept plays an important role. Looking on the cw influence and the weight of a wheel from a technical side, a very clear recommendation for the wheels can be given. Yet, the design of "ugly" wheels which no customer will want to buy is not an option, even in times of hard CO2 regulation, Fig 11.

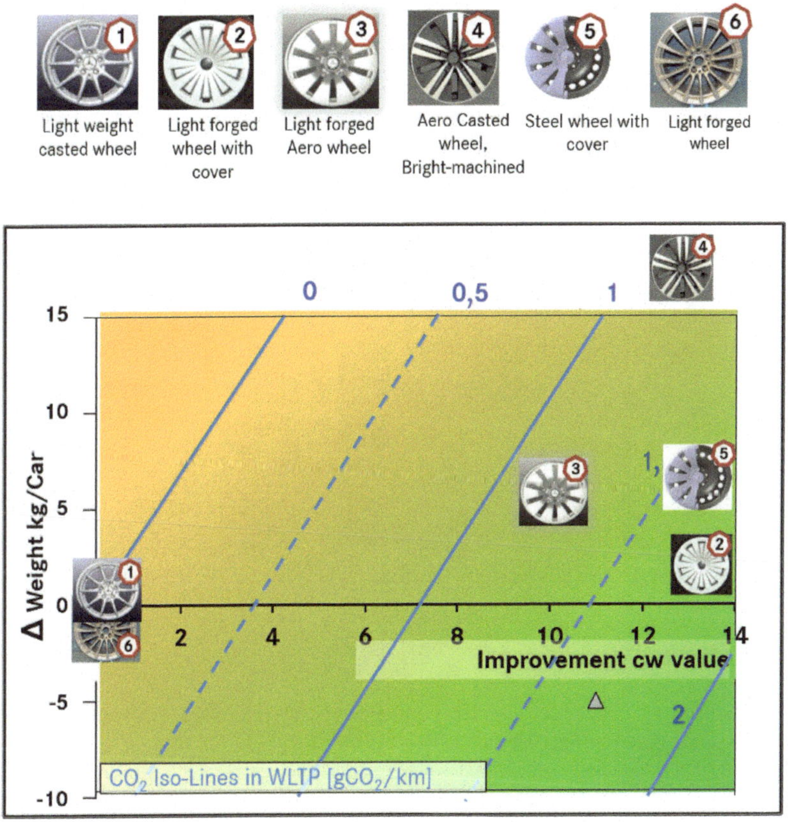

Fig 11.: Influence of wheel weight and Aerodynamics on CO2 values

5 Result

Linking tire and wheel influence to actual effects on CO2 values of individual wheel options is possible. Looking at the current E-Class after optimization of rolling resistance, aerodynamics and tire pressure, the influence of wheels and tires on the CO2 emission is shown in figure 12. For example, a customer buying an Avantgarde Line car fitted with 19-inch wheels instead of 17 inches will have a 3-gram higher CO2 value in his COC paper.

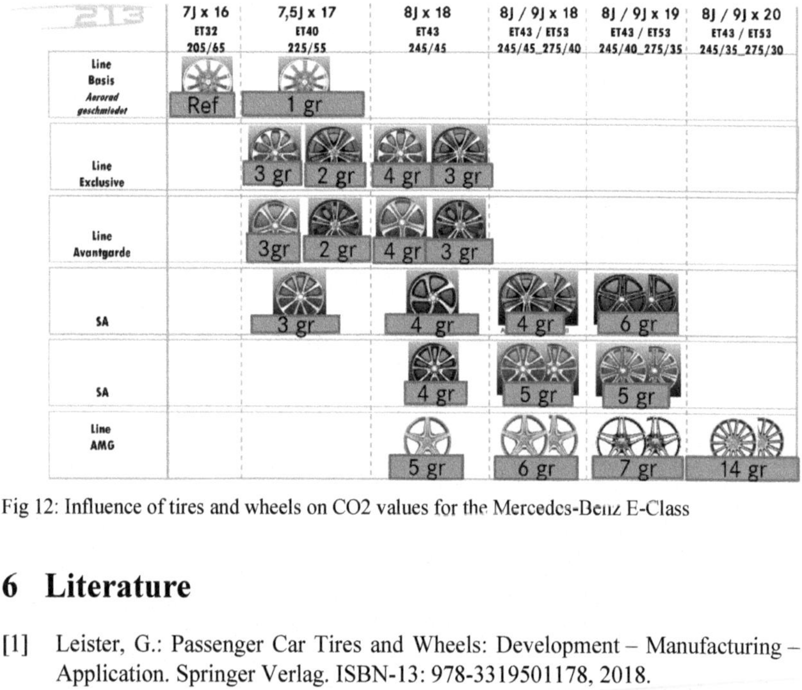

Fig 12: Influence of tires and wheels on CO2 values for the Mercedes-Benz E-Class

6 Literature

[1] Leister, G.: Passenger Car Tires and Wheels: Development – Manufacturing – Application. Springer Verlag. ISBN-13: 978-3319501178, 2018.

Wheel casting technology in the next generation

Dipl.-Ing. (FH) Ralf Bux

ENTEC-STRACON GMBH

1 Introduction

The basis for driving safety and driving comfort in light-alloy wheels for cars are the unspring masses, whereby it is crucial that the weight of the wheels is as low as possible. Due to the inertia and the rotational moment one is endeavored to use light wheels. For this reason, on the one hand, attempts are being made to structurally implement lightweight construction on wheels. On the other hand, there are efforts to reduce weight through the selection of materials.

The prior art is currently cast or forged wheels of aluminum or magnesium alloys which are made to a very high percentage by the low pressure die casting process, Fig 1.

Fig 1: casting and forged wheel

In addition to these driving dynamics demands increasingly aerodynamic or crash relevant interpretations of the wheels play a major role. As the aerodynamic properties of the wheels are directly related to fuel consumption and CO2 emissions, there was also an increased need for action on the part of the legislature, Fig 2.

Fig 2: aero casted wheel

Homologation requirements within the overall registration of passenger cars, in particular the WVTA (Whole Vehicle Type Approval) in combination with the WLTP (Worldwide Harmonized Light Vehicles Test Procedure), aggravate these requirements, so that the basic equipment of the entire vehicle is no longer within the approval procedure (WVTA) describes the vehicle, but all equipment variants. This change in the type test by the globally standardized light vehicle test procedure (WLTP) calls for a rethink in the design of vehicle parts in terms of aerodynamics and lightweight construction.

In addition, these lightweight construction and aerodynamic requirements are underpinned by the increasing use of electromobility under the slogan "Lightweight increased range".

Depending on the vehicle type, different wheel dimensions are used, which are aerodynamically worse and cause a higher block function in the frontal and offset crash, which worsens the classification result of the entire car series, Fig 3.

Fig 3: Mercedes EQ with aero casted wheels

Fig 4: Mercedes EQ offset crash result

These increasing demands, consisting of lightweight construction, aerodynamics and crash require a process change in the production of vehicle wheels, since the standard casting processes such as low-pressure die casting processes cannot meet these requirements procedurally optimal, Fig 4.

2 Knowledge community with a holistic approach

Our project team, consisting of material science and casting technology scientists, wheel experts and machine manufacturers, have developed a system in conjunction with a process that will provide a solution to these changes, Fig 5.

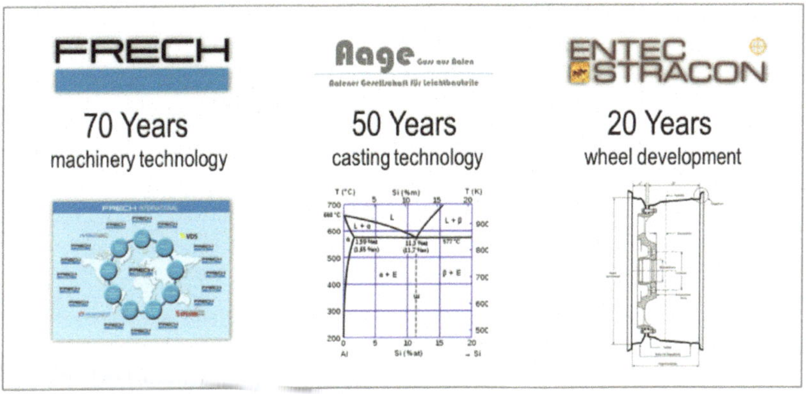

Fig 5: Experts Community TURBU-PRESSURE-CASTING

A holistic approach to product-specific requirements and a positive energy balance within the manufacturing process in conjunction with the changing demands of the automotive industry was addressed. In addition, resource-saving options were consistently defined and implemented.

All in all, we have combined the positive components of existing technologies such as low pressure die casting or die casting in the mold, which makes it possible to manufacture light metal wheels very efficiently and flexibly despite very different cross sections.

3 Casting process and its advantages

Our TURBU-PRESSURE-CASTING, in which despite high casting velocity without "TURBU-lence" - "PRESSUR-ized" is poured, in connection with targeted partial tool temperature is able to produce the target wheel geometry almost finished, that means without extensive reworking and machining, Fig 6.

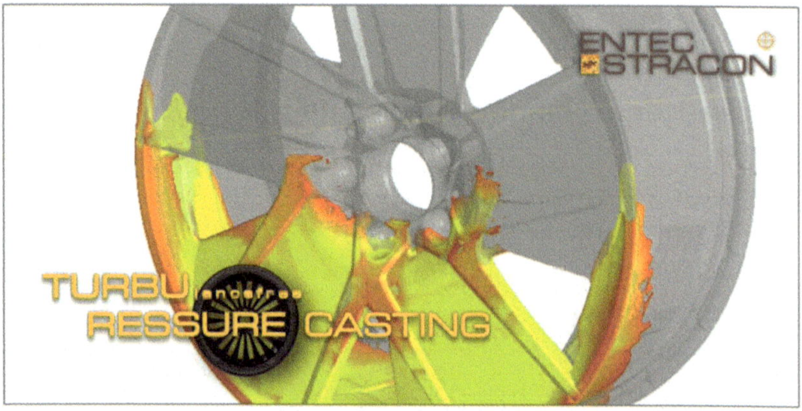

Fig 6: Technology outlook TURBU-PRESSURE-CASTING

The light alloy wheel produced by our patented process already has its required strength right after casting, so that additional energy-intensive manufacturing processes such as cutting, flow forming or head treatment (T6) are unnecessary.

In order to comply with the CO2 limits, more expensive additional plastic cover is increasingly being used. By means of our method it is possible to produce aerodynamic-flashes by casting and thus to dispense with such additional elements. In addition, bolt hole or stiffness structures are possible, which could be produced by the standard method only by mechanical processing or not at all.

All in all, our process offers advantages in all areas of design and construction, production (cycle time and process) or directly on the vehicle (lightweight construction, aerodynamics), so that ecology and future-proofing are just as feasible as cost-effectiveness combined with high quality.

4 Our casting cell as a complete system enables smooth system changes

Together with our cooperation partner, we have developed a casting cell based on a special cold chamber die casting machine, which makes it possible to prepare the alloy wheel immediately after the casting so that only "good parts" leave the cell, which only have to be slightly reworked in the tire seat and in the wheel center. Thus, high value creation losses within the production chain are not to be expected, Fig 7.

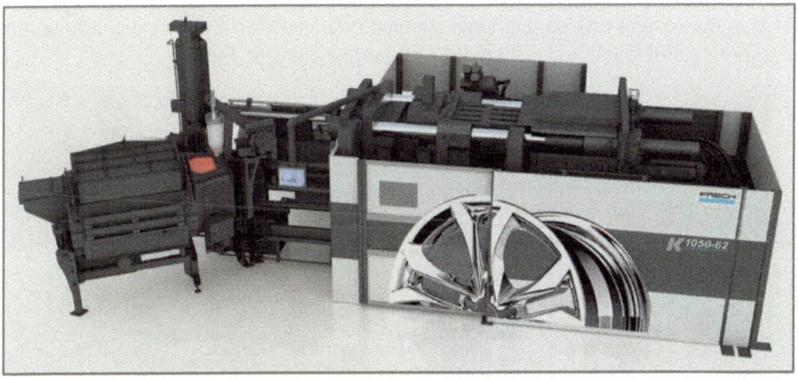

Fig 7: Complete casting cell for TURBU-PRESSURE-CASTING

Furthermore, due to its high level of automation, the casting cell has all the quality requirements, also in relation to Industry 4.0, and enables fully automatic tool changes of less than 10 minutes. With a cycle time of less than 2 minutes, the cell sets new standards in production speed, Fig 8.

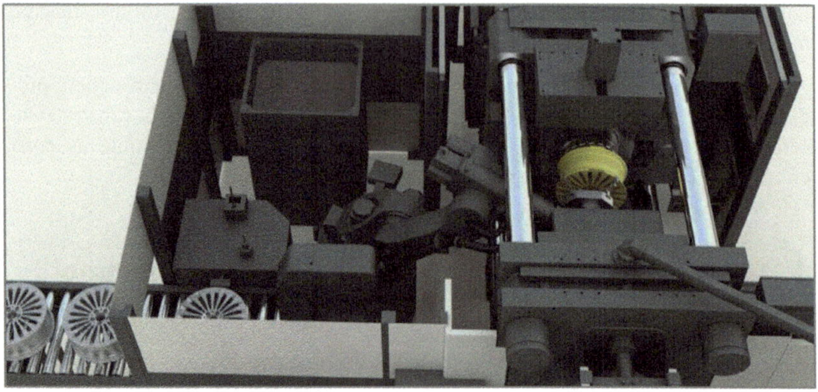

Fig 8: Details for processing within the casting cell

Low-pressure die casting has the advantage over the forging process that the tools are very cost-effective. Especially for wheels with lower production volume, the impact on the parts price is much lower than forged wheels. This advantage is also incorporated in our casting cell, so that tools are much cheaper than a classic die-casting tool or a forging tool thanks to a split molding / guiding concept, Fig 9.

Fig 9: Tool change without manual working

5 Result and Outlook for the series introduction

In summary, 80% of the casting cell is the production of a wheel, so that a few additional steps create a possibility for people and the environment to do goods.

With the company CMS we have a TIER-1 wheel manufacturer and partner who exclusively implements the market launch with us. It is planned in 2019, a first series release.

If you have any questions, feel free to contact both CMS or ENTEC-STRACON:

CMS Jant ve Makina San. A.Ş.

(Çiğli Factory)

10006/1 Sok. No.12 35620 A.O.S.B

Çiğli, İzmir - TÜRKİYE

Tel: (+90) 232 399 1100 Pbx

Fax: (+90) 232 399 1111

https://www.cms.com.tr/en/

ENTEC-STRACON GMBH

(Development center)

Robert-Bosch-Straße 80

Aalen - GERMANY

Tel: (+49) 7361 880939 0

Fax: (+49) 7361 880939 1

https://www.entec-stracon.com/en/

6 Literatures

[1] Daimler AG, Global Media Site, Mercedes pictures

[2] ENTEC-STRACON GMBH, Published patent content

[3] Oskar Frech GmbH + Co. KG, Excerpt from press releases

Michelin proposal (XL+) for high-load capacity tires needs

Stephane Bertoldi
Pierre Chandezon
Bruno Guimard

Michelin

MICHELIN
MICHELIN PROPOSAL (XL+)
FOR HIGH LOAD CAPACITY TIRES NEEDS
Stephane Bertoldi
Pierre Chandezon
Bruno Guimard
chassis.tech plus 2019 - High load capacity tires
MICHELIN
Creation/Date : 13/05/2019
Michelin Confidential D4

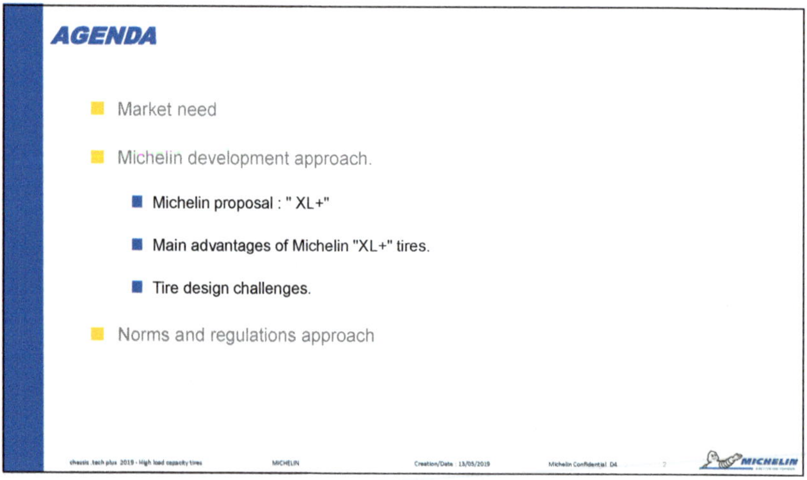
AGENDA
Market need
Michelin development approach.
Michelin proposal : " XL+"
Main advantages of Michelin "XL+" tires.
Tire design challenges.
Norms and regulations approach
chassis.tech plus 2019 - High load capacity tires
MICHELIN
Creation/Date : 13/05/2019
Michelin Confidential D4
2
MICHELIN

BACKGROUND

- Vehicle mass has been increasing significantly over the last years (additional equipment, more versions, larger cars for a given segment) despite introduction of lightweight technologies.
- The introduction of BEVs (Battery mass) leads to an acceleration of this tendency.
- Using the current tire technology, mass increase leads to :
  - Bigger wheel diameters
  - Larger wheelhouse
  - Higher vehicle and center of gravity.
- SUVs are less concerned than sedans.
- The ACEA has made an official request to ETRTO for "High Load Capacity tires" that carry more load using the same tire size.

NEED FOR HIGH LOAD CAPACITY TYRES

- For a given vehicle, the outside diameter is identical for all tire sizes & vehicle versions.
- Load capacity is more critical for largest wheel diameter (lower tire sidewall height) even when using the maximum width allowed by chassis packaging.

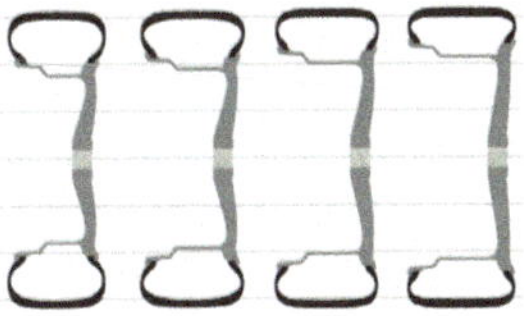

Width	H/B	Seat	SL	XL
225	50	17	94	98
245	40	18	93	97
255	35	19	92	96
265	30	20	90	94

In this case, even XL for the 20 inch wheel may be insufficient for all vehicle versions.

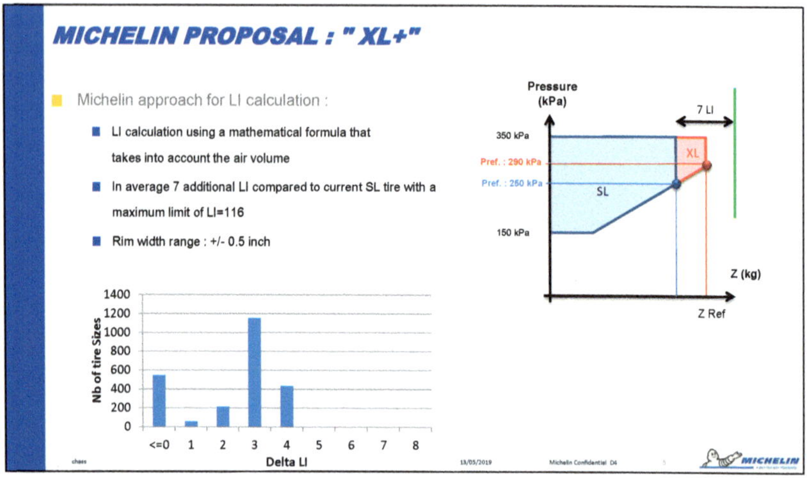
MICHELIN PROPOSAL : " XL+"
Michelin approach for LI calculation :
LI calculation using a mathematical formula that takes into account the air volume
In average 7 additional LI compared to current SL tire with a maximum limit of LI=116
Rim width range : +/- 0.5 inch
Pressure (kPa)
7 LI
350 kPa
Pref. : 290 kPa
Pref. : 250 kPa
150 kPa
XL
SL
Z (kg)
Z Ref
1400
1200
1000
800
600
400
200
0
Nb of tire Sizes
<=0
1
2
3
4
5
6
7
8
Delta LI
13/05/2019
Michelin Confidential D4
MICHELIN

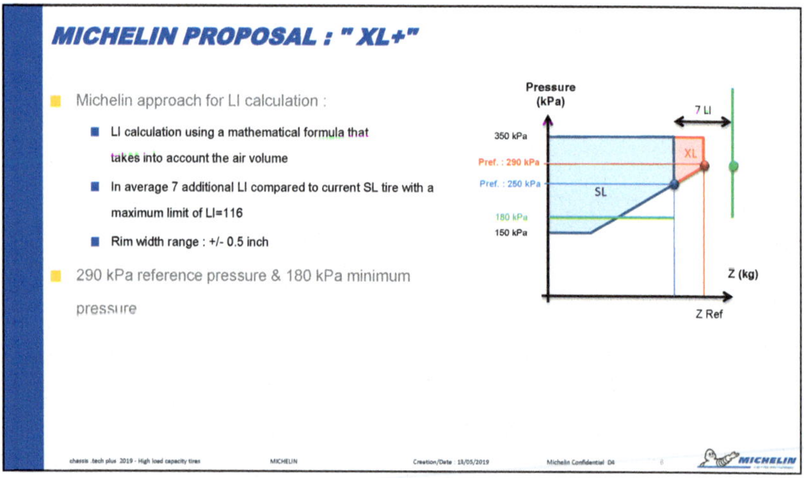
MICHELIN PROPOSAL : " XL+"
Michelin approach for LI calculation :
LI calculation using a mathematical formula that takes into account the air volume
In average 7 additional LI compared to current SL tire with a maximum limit of LI=116
Rim width range : +/- 0.5 inch
290 kPa reference pressure & 180 kPa minimum pressure
Pressure (kPa)
7 LI
350 kPa
Pref. : 290 kPa
Pref. : 250 kPa
180 kPa
150 kPa
XL
SL
Z (kg)
Z Ref
chassis .tech plus 2019 - High load capacity tires
MICHELIN
Creation/Date : 13/05/2019
Michelin Confidential D4
MICHELIN

MICHELIN PROPOSAL : " XL+"

- Michelin approach for LI calculation :
  - LI calculation using a mathematical formula that takes into account the air volume
  - In average 7 additional LI compared to current SL tire with a maximum limit of LI=116
  - Rim width range : +/- 0.5 inch
- 290 kPa reference pressure & 180 kPa minimum pressure.
- Adapted slope to offer more load at high pressure
- Same denomination as current tire (i.e. 205/55 R16 97Y) with a specific "XL+" suffix marked on the tire

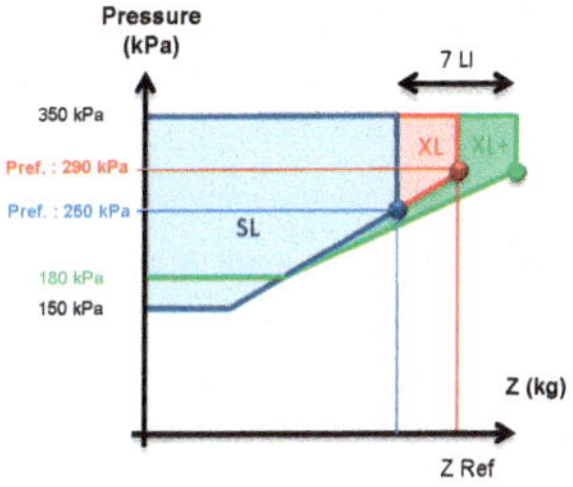

MAIN ADVANTAGES OF MICHELIN XL+ TIRES

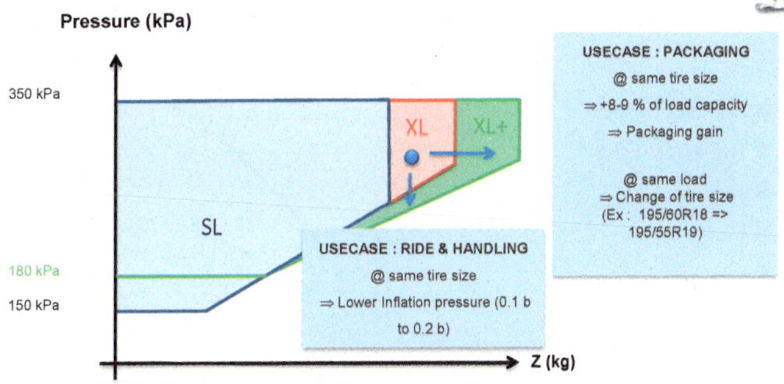

TIRE DESIGN CHALLENGES

- Implicit : Product integrity is a prerequisite for the customer
  - Overall durability and high speed endurance
  - Road hazard vs. usage
  - No bead unseating
  - ....

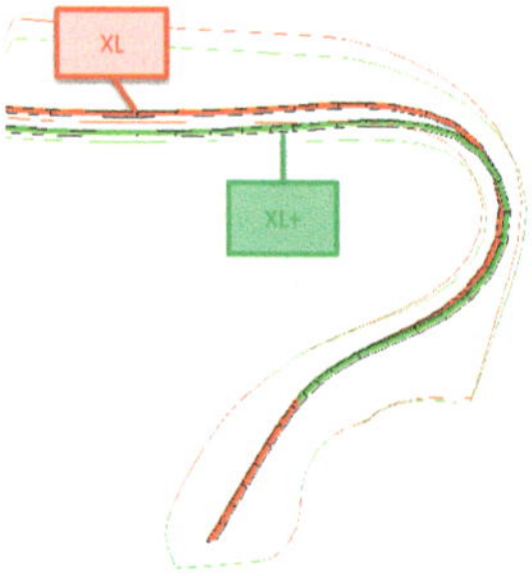

Impact of tire load on tire deflection

TIRE PERFORMANCES : EXPLICIT

- Explicit : Expected by the customer (OEM and consumer)
  - Handling / safety
  - Rolling resistance
  - NVH
  - Wear
  - ...

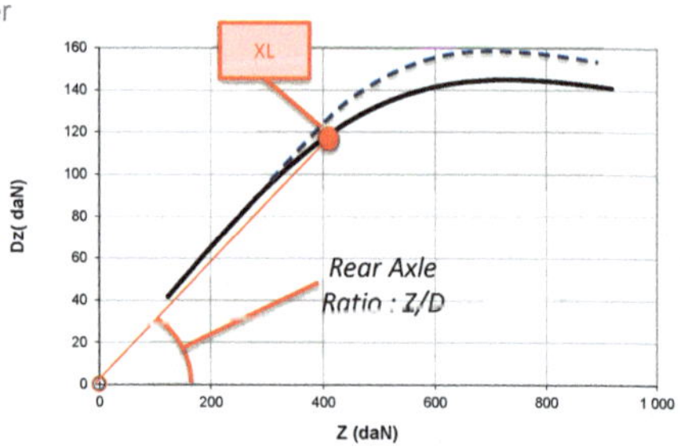

TIRE PERFORMANCES : EXPLICIT

- Explicit : Expected by the customer (OEM and consumer)
  - Handling / safety
  - Rolling resistance
  - NVH
  - Wear
  - ...

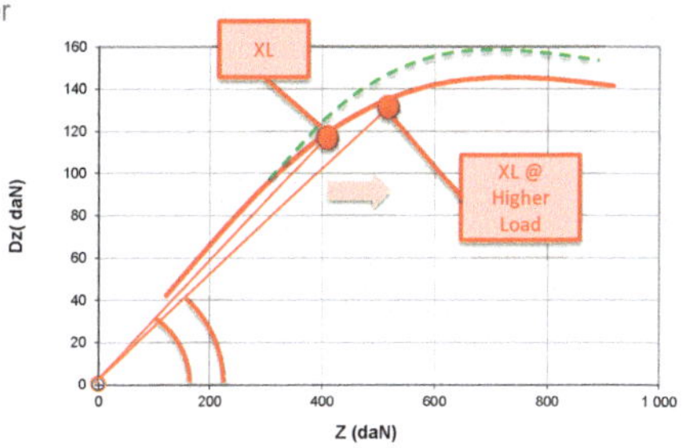

chassis.tech plus 2019 - High load capacity tires MICHELIN Creation/Date : 13/05/2019 Michelin Confidential D4 11

TIRE PERFORMANCES : EXPLICIT

- Explicit : Expected by the customer (OEM and consumer)
  - Handling / safety
  - Rolling resistance
  - NVH
  - Wear
  - ...

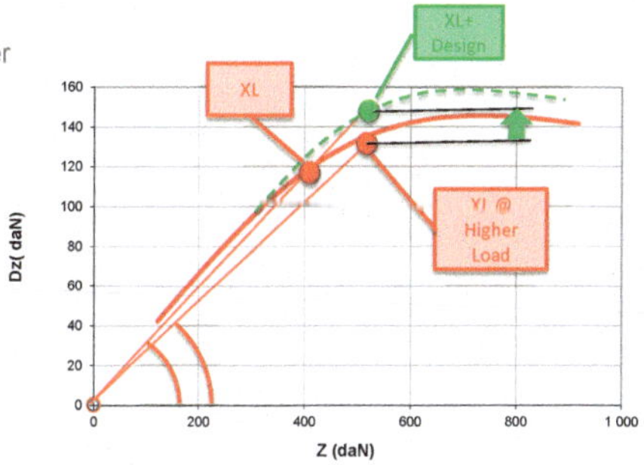

chassis.tech plus 2019 - High load capacity tires MICHELIN Creation/Date : 13/05/2019 Michelin Confidential D4 12

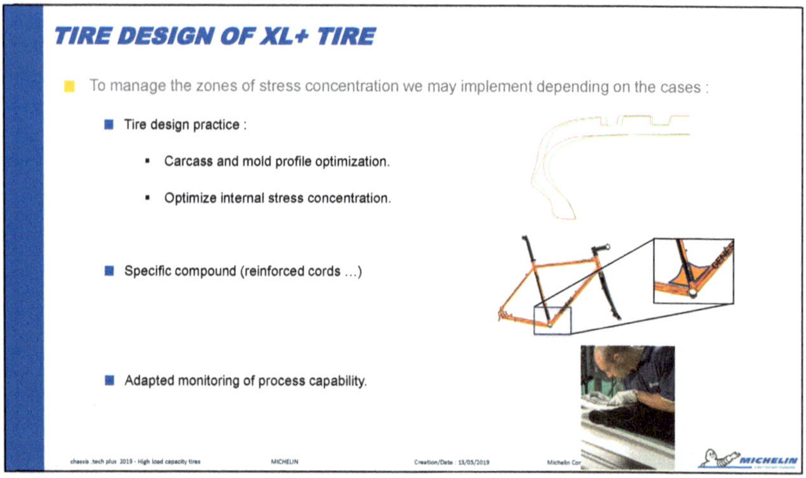
TIRE DESIGN OF XL+ TIRE
To manage the zones of stress concentration we may implement depending on the cases :
Tire design practice :
Carcass and mold profile optimization.
Optimize internal stress concentration.
Specific compound (reinforced cords …)
Adapted monitoring of process capability.
chassis .tech plus 2019 - High load capacity tires
MICHELIN
Creation/Date : 13/05/2019
MICHELIN

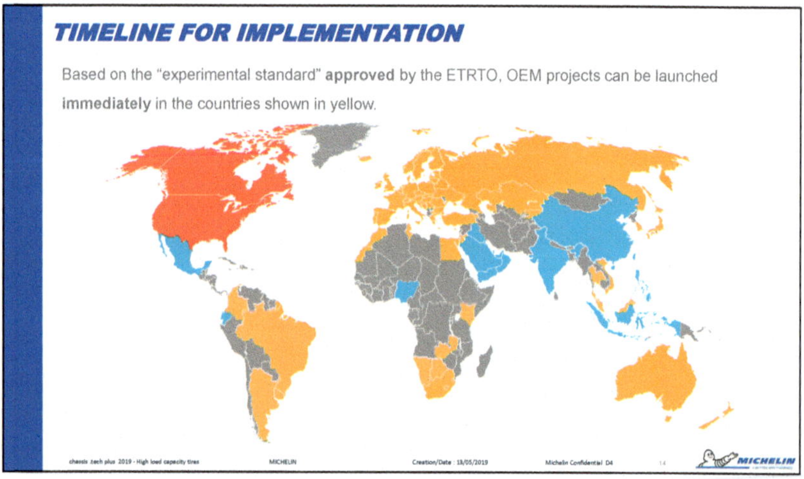
TIMELINE FOR IMPLEMENTATION
Based on the "experimental standard" approved by the ETRTO, OEM projects can be launched immediately in the countries shown in yellow.
chassis .tech plus 2019 - High load capacity tires
MICHELIN
Creation/Date : 13/05/2019
Michelin Confidential D4
14
MICHELIN

TIMELINE FOR IMPLEMENTATION

- USA :
  - Each dealer and the NHTSA must be informed about the tire to rim fitting information for tires that are not listed in publications of TRA, ETRTO, JATMA, TRAA, ALAPA or SABS
  - Electronic publication may be sufficient, whereas the Confidential EDI and/or an Experimental Standard do not qualify as publications
- China :
  - Certification body is in charge
  - Certification body may request a proof that a tire design follows consistent rules. The listing of tire sizes in the application could possibly be sufficient
- Brazil :
  - Certification bodies typically require that a tire is listed in a publication of one of the major standardisation organizations

chassis.tech plus 2019 - High load capacity tires MICHELIN Creation/Date : 13/05/2019 Michelin Confidential D4 15 MICHELIN

CONCLUSIONS

- XL+ is a tire solution for future vehicles through:
  - More Load Capacity
  - Inflation pressure degree of freedom.
- XL+ tires are designed specifically to fit implicit (endurance) and explicit (handling, RR…) tire requirements
- Egg and chicken kind of issue:
  - Without a published standard, application is limited to countries covered by UNE ECE reg
  - Some regulations need publication in a standard.
  - To get a standard approved, one needs to prove the experimental standard is used on the market and works

chassis.tech plus 2019 - High load capacity tires MICHELIN Creation/Date : 13/05/2019 Michelin Confidential D4 16 MICHELIN

MICHELIN
THANK YOU
chassis .tech plus 2019 - High load capacity tires
MICHELIN
Creation/Date : 13/05/2019
Michelin Confidential D4
17

TIRE TESTS AND SIMULATION

Stopping distance variation with road surface temperature and tire wear

Bissoli Alberto Giovanni, FCA Italy

Murgia Stefano, FCA Italy

Tambellini Marco, Università di Pisa

1 Abstract

Stopping distance measurements are affected by a large quantity of noise factors that contribute to make them not comparable when measured in particular conditions.

For automotive companies and tire manufacturers is not feasible, in terms of costs and available time, to control such noise factors, of which ones of the most important are:

- Road surface temperature during tests;
- Tire wear.

As regards road surface temperature, currently the state of the art of tire standards that take it into account are SAE J2909 [1] and ISO 21994:2007 [2].

Regarding testing temperature, the ISO standard states that:

- The ambient temperature should stay between +5 and +35°C, and its variation should be less than 10°C during a testing session.
- The road surface temperature should stay between +10 and +40°C, and its variation during a sequence of measurements should be less than 10°C.
- Measurement performed within acceptable range can be compared only if temperature difference between one another is less than 10°C.

As previously stated automotive and tier one companies cannot wait such a range of environmental conditions to perform and compare their tests (and, in addition, the need to compare tests from months, or even years apart can rise).

On the other hand, SAE J2909 gives a polynomial formulation that, starting from the adhesion coefficient of the track used, gives a polynomial corrective formula to adjust measured stopping distances to the desired road surface temperature.

However, this implies the need to measure the road surface friction coefficient, with a standard reference tire and a standard method (ASTM E 1136 - 1137 [3]) every time before tests.

Moreover, both ISO and SAE standards do not take into account that road surface temperature can have different influence on different types of tire for what regards compound and tread pattern (e.g. on summer, all season and winter tire).

It is very important, for a testing proving ground, to have the opportunity to compare test performed in different environmental conditions or to estimate tire performance at different road surface temperature.

In addition, also tire wear should be taken into account while performing stopping distance tests. In particular, experimental evidence shows that measured stopping distances are shorter after a certain number of braking maneuvers performed on the same tire.

For this reason it is important to understand when stopping distance measurements give more repeatable results on a single tire during a brake test session, and which burnishing procedure to perform in order to save time and improve measurement repeatability.

In this paper conclusion of six months experimentation and measurement on five different summer tires and one all season are reported.

From data analysis, a corrective formula for stopping distance measurement with road surface temperature is proposed.

This formula is compared to the other ones present in literature and will be validated for future use.

Furthermore some important considerations on the effect of tire wear, intended as the number of braking maneuvers performed on the same tire, are done.

2 Introduction

The activity presented in this paper is placed in a broader context of a collaboration between FCA Italy and the *Università di Pisa* on the characterization of the longitudinal tire's behavior [4], [5]. The results so far obtained from this collaboration are the characterization of the tire longitudinal slip curve from vehicle road tests and an assessment of stopping distance variation with road surface temperature and tire wear.

The reason for this study is that stopping distances are a very important indicator of a car, equipped with its own tires, performance and they are measured by car manufactures and specialized magazines. It is very important to understand how they can vary under different testing conditions, in order to be able to evaluate results in a more objective way and to shorten testing and data analysis time.

In particular this paper will focus on evaluating the effect of the solely road surface temperature on the value of measured stopping distances. In addition, also tire wear, intended as the number of ABS braking performed on the tire, is evaluated.

For this purpose different summer tires and one all season tire are measured in a six months' time span, with different conditions of road surface temperature, keeping all other noise factors as constant as possible.

The vehicle used for tests comes from the activity performed to evaluate longitudinal tire slip curves with on vehicle tests, reported in [6]. For this purpose, the vehicle was equipped only with front brakes (rear brakes were removed for technical reasons). This is the reason why absolute values of stopping distance reported in this study are relatively high. The measurement system equipped in the vehicle and the calculation of stopping distances were developed to have the highest simplicity maintaining at the same time high data reliability. For this reason, stopping distances measured with the developed system and methodology were compared to those measured with a measurement system compliant with ISO 21994:2007 standard.

The test plan was developed in order to allow the evaluation of different road surface temperatures for each tire, keeping the wear as low as possible between test sessions at different temperatures.

The choice to study the effects of road surface temperature mainly on summer tires comes from different reasons:

- Summer tires are the most used in European market;
- In respect to winter and all seasons tires, variation of stopping distances with tire wear is limited. This allows to evaluate more easily the influence of the solely temperature on stopping distances. For a comparison of the wear effect on tests results in a summer and in an all season tire refer to figure 1 and 2.

With obtained data, a corrective formula for road surface temperature is determined and compared with those present in literature.

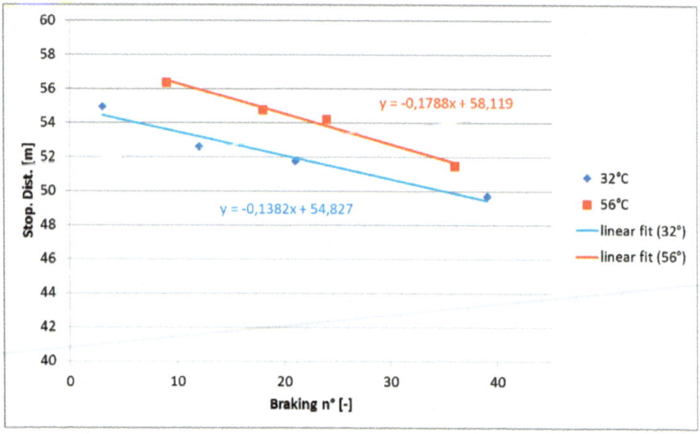

Figure 1: effect of tire wear on stop. dist. of an all season tire for tests at low and high road surface temperature

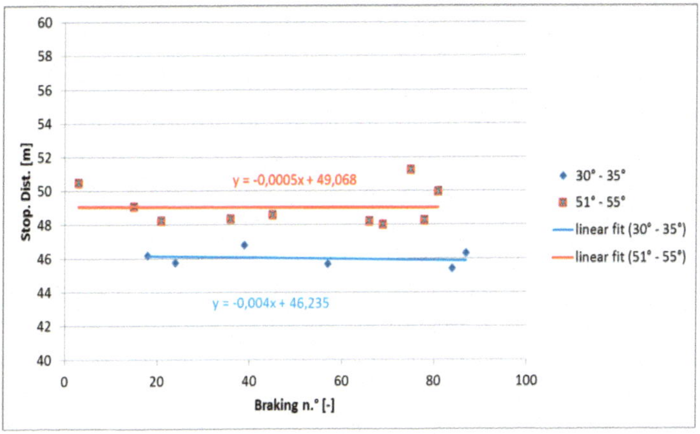

Figure 2: effect of tire wear on stop. dist. of a summer tire for tests at low and high road surface temperature

3 Measurement Setup

As previously stated, the vehicle used to perform the activity is the same used to obtain longitudinal tire slip curves from on vehicle road tests. For this reason rear brakes where removed, in order to make possible to measure vehicle longitudinal speed with on-board equipped wheel speed sensors, without the need to use an external GPS system.

As showed in next paragraphs, this particular setup does not have an effect on the results of this work.

The vehicle was equipped with a CAN network logger programmed to acquire the following signals:

- Vehicle speed [kph];
- Wheel speeds [kph];
- Longitudinal acceleration [m/s^2];
- Lateral acceleration [m/s^2];
- Master Cylinder pressure [bar];
- Steering wheel angle [deg].

To calculate stopping distances the signals of interest are:

- Wheel speeds, in particular the rear ones; having removed rear brakes, rear wheels are, with excellent approximation, in pure rolling conditions (free wheeling) during the ABS braking (neglecting the rotational inertia of the wheels). This allows to integrate these signals in order to obtain the distance covered by the vehicle during the braking maneuver;
- Master Cylinder pressure: this signal is used to determine the starting point of the ABS braking.

All other signals where used as a check for the quality of the maneuver, which has to be performed starting the ABS braking with longitudinal and lateral accelerations near to zero (i.e. starting from constant speed and with steering wheel angle equal to zero).

Stopping distance results obtained with the previously presented measurement setup were compared to result obtained with an instrumentation compliant with ISO 21994:2007 requirements, showing no major differences between them. For this reason the more agile CAN network logger was used during tests. Results of this comparison can be seen in figure 3.

Road surface temperature was measured for each test in two different ways:

- Using an infrared thermometer directly on tarmac surface;
- Using a temperature sensor installed below the tarmac surface on the test track.

These two methods were compared in order to obtain the most reliable results. Conclusions show that the infrared thermometer provides sufficiently good measurements and thus was the only method used in the later part of the activity.

Vehicle weight was monitored during all the tests. Fuel tank was refilled each time before tests in order to not introduce any additional variability source.

4 Test Plan

For this particular analysis five brand new summer tires and one all season tire from different suppliers where selected.

For each tire type the following tests were performed:

- 10 ABS braking from 100 to 0 kph (without any burnishing procedure) were performed at constant road surface temperature and stopping distance were measured. The purpose was to identify the wear effect at constant temperature on stopping distances at the beginning of the tire life, in order to identify the most effective burnishing procedure before performing stopping distance test sessions (or to compare results from tires with different wear status);

- After the first burnishing brake maneuvers, test sessions at constant road surface temperature were performed. Each test session consisted in three ABS braking respectively from 80, 100 and 120 kph;
- For each test session (consisting of three ABS braking) stopping distance value from 100 kph to 0 was extrapolated and road surface temperature registered;
- For each tire, test sessions were performed in a range of road surface temperature going from about 25°C to about 60°C.

At the end of testing plan, each tire was tested with about 80 - 100 ABS braking, that is until tire wear was within the limits to have significant test results.

5 Stopping Distance Measurement

ABS braking maneuvers were performed in a dry and flat track surface, with adhesion coefficient higher than 0.9.

Each braking maneuver was performed as following described:

- Vehicle was driven at the nominal starting speed V_0 (i.e. 80, 100 and 120 kph);
- The highest gear compatible with vehicle speed was selected;
- The longitudinal acceleration was kept equal to zero (i.e. constant speed) for at least 2 seconds before applying the brake;
- The lateral acceleration was kept equal to zero (i.e. straight driving) for at least 2 seconds before applying the brake;
- A step application was imposed on the brake pedal, moving from gas pedal to brake pedal as fast and as strong as possible to get full ABS condition (i.e. panic braking);
- ABS was kept active for all the braking maneuver;
- The clutch pedal was pressed only to avoid engine stall.

During a test session brake system temperatures were kept between 80 and 120°C.

As regards data analysis, the trigger event for stopping measurement's start was defined as the instant of first foot contact with the brake pedal. Since only CAN signals were acquired, trigger signal was defined as the moment in which master cylinder pressure was equal or higher than 2 bars.

Since acceleration and vehicle speed present pitch oscillations around the complete stop of the vehicle, the end condition V_f was defined as vehicle speed equal or lower to 2 kph.

Stopping distance was calculated, for each braking, integrating the mean of the signals taken from rear wheel speed sensors between these two trigger conditions V_0 and V_f.

The distance travelled is then defined as:

$$\int_{t_0}^{t_f} V(t)dt \tag{1}$$

where, as already stated, in order to evaluate the absolute speed $V(t)$ of the vehicle the signals coming from rear wheel speed sensors where used:

$$V(t) = R_r \frac{w_{r1} + w_{r2}}{2} \tag{2}$$

As stated in previous paragraphs the master cylinder pressure signal was used to identify the brake starting trigger (and thus the starting speed V_0). The end speed was defined as $V_f = 2\ kph$.

For stopping distance calculation a *for* cycle was used applying the trapezoid method as follows:

$$s_{(k)} = s_{(k-1)} + \frac{V_{(k)} + V_{(k-1)}}{2}\left(t_{(k)} - t_{(k-1)}\right) \tag{3}$$

from $k = 2$ to $k = length(V)$, where V is the vehicle speed array and t is the elapsed time.

The stopping distance s_{trigg} is the last element of the array s obtained as in Eq. 3.

In order to provide a result for a certain value of start condition speed (in this work 100 kph) data of the three stopping distances from the three values of V_0 were fitted with a polynomial curve of the type:

$$s = p_1 V_0^2 + p_2 V_0 \tag{4}$$

In this way is it possible to have a realistic value of stopping distance s even if the initial trigger speed V_0 for each brake maneuver is slightly different from the standards.

In figure 3, the difference of stopping distance results obtained integrating rear wheel speed signals from CAN and using GPS is shown. Although at a first sight the results are not matching, the percentage error committed is quite negligible (*0,1%*). This difference is due to an offset present on the vehicle speed signal (the one measured with GPS system is higher than the one measured through CAN network).

The robustness and reliability of the integration method has been verified for different tests and results confirmed its goodness.

It should be remembered that stopping distance results presented in this work are relatively high because they were performed in a vehicle equipped only with front brakes.

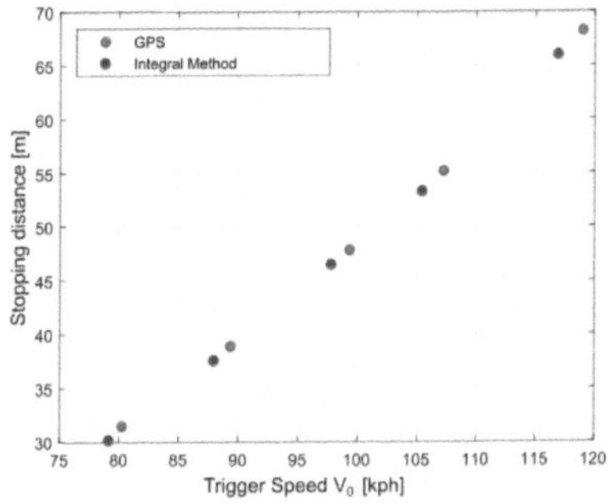

Figure 3: Comparison between GPS and integral method

6 Data Analysis

One of the first observations coming from this work is that, although a number of sources refers to atmospheric temperature, this is not the right parameter to consider when evaluating tire's performances, since it is not directly linked to road surface temperature. This result can be clearly analyzed from results reported in Table 1.

Table 1: Comparison between atm. temp. and road surface temp.

Test n. [-]	Atm. Temp. [°C]	Road Surf. Temp. [°C]	Stop. Dist. [m]
5	25	54	49,1
10	25	41	47,1
11	25	44	47,6
14	25	47	48,0
16	25	38	45,9

From the table is possible to see how, testing at the same atmospheric temperature, does not lead to have tests at the same road surface temperature. Plotting these points in a graph it is than clear that the right parameter to refer when performing stopping distance tests is road surface temperature. In figure 4 it is possible to see how stopping distance measurements performed at the same atmospheric temperature show a trend clearly linked to road surface temperature.

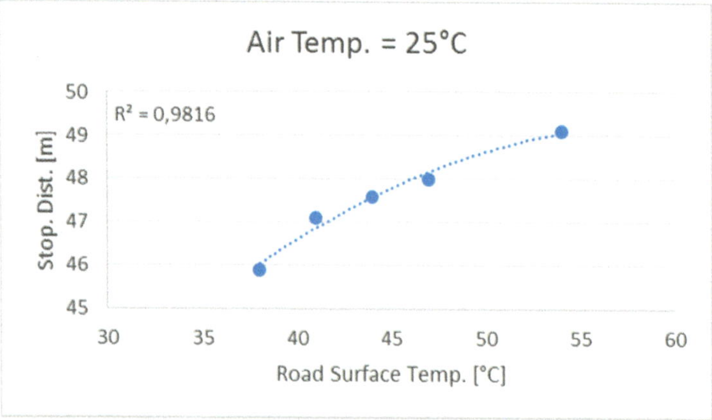

Figure 4: Stopping distance results at constant atm. temp.

As stated in previous paragraphs, tires where tested at different road surface temperature conditions and stopping distances were measured. An example of the complete testing performed on each tire type could be seen in Figure 5. Red squares represent result of maneuver performed with tarmac temperature higher than *50°C*, while blue dots represent results for tarmac temperature lower than *35°C*. It could be soon noticed that tests performed at higher tarmac temperatures present also higher stopping distance values in all wearing condition of the tire itself. In general, the difference is higher at the beginning of tire life (more than 4 m difference, i.e. about 7%), and tends to lower down at the end of tire's useful life (about 1 m of difference). In other few cases the difference stays constant during all the tire's life span.

It can also be seen that tire wear has a higher influence for tests at higher tarmac temperature. To see more clearly these two phenomena, refer to Figure 6. The higher fitting line refers to tests performed on the new tire, the lower one to tests performed with the same tire, but after several tests (worn tire). The dependence to road surface temperature is confirmed.

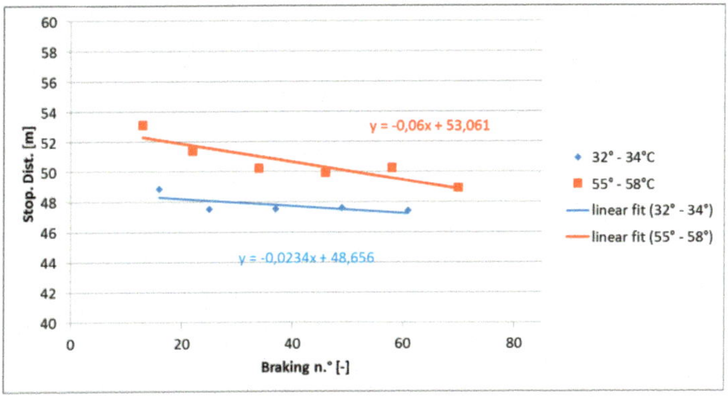

Figure 5: Stop. dist. behaviour at high and low road surface temp. in the whole tire life span

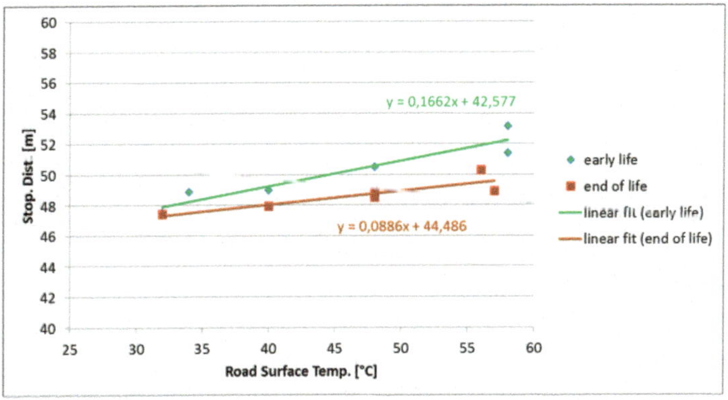

Figure 6: Stop. dist. behaviour with road surf. temp. at the beginning and at the end of tire life span

These type of considerations can be done for all the different summer tires tested.

From tests, tendencies can be extrapolated and a general corrective formula for stopping distances, measured in different road surface temperatures, can be summarized.

As regards tire wear, particular considerations can be done for the first braking, done at the earliest stage of the tire life. As it can be seen in Figure 7, first ABS brake maneuvers show higher stopping distance values in respect to the following. Stopping distance results tend to stabilize after a certain number of braking maneuvers, different for each specific tire. However, it can be seen that a number between 15 and 20 braking brings the tire at its mean stopping distance performance. This means that a good burnishing procedure will provide more reliable and stable results.

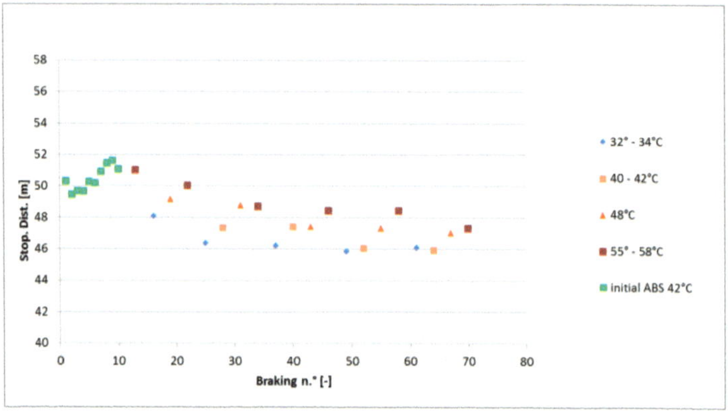

Figure 7: Testing results with focus on first 10 ABS braking at constant road surface temp.

7 Corrective Formula Identification

In Table 2 and Table 3 a resume of testing results for the five summer tires can be seen and analyzed. In addition the same considerations are done for one all season tire (tire 6). Summer tires were chosen considering a wide span of different tire types: from the sportier, to low rolling resistance tires.

Tires were tested in a temperature range that goes from approximately 30°C to about 60°C and a maximum number of brake maneuvers from 70 to 100 for summer tires and about 40 for the all season tire, which had worn out more easily.

Stopping distance variation range is different from tire to tire and goes from a minimum of 2,2 m (tire 5) to a maximum of about 7 m (tire 6, the all season one). Considering only summer tires the mean variation of stopping distances during tests was of approximately 4,2 m (7,8% of total variation).

Considering the results of consecutive brake test sessions performed at different road surface temperature, the stopping distance variation due to road surface temperature was calculated. Having then clear the temperature range in which such stopping distance variations were measured it was possible to calculate the stopping distance versus road surface temperature gradient for each tire [m/°C]. For this calculation were considered, for each tire, brake maneuvers performed in the early stage of tire life (i.e. after the burnishing procedure and until approximately 30 braking maneuvers).

With this value the stopping distance at the reference road surface temperature was calculated. The reference road surface temperature chosen was selected as the average one measured in Balocco Proving Ground in a one year time span, equal to 25°C.

Considering the high absolute value of stopping distance for the test vehicle (equipped with only front brakes), the percentage value of stopping distance versus road surface temperature was calculated (Table 2).

The hypothesis behind this data analysis methodology is that tire wear variation during the considered test sessions is negligible. To validate this hypothesis, as already said, a large number of brake maneuvers were performed on each tire and the variation of stopping distances with tire wear at constant road surface temperature was calculated.

As it is possible to see from table 3, the wear gradient for summer tires is always smaller than the temperature gradient (in most cases one order of magnitude smaller), so it is possible to neglect tire wear influence committing only a marginal error.

For the only all season tire tested the situation is different: in this case the wear gradient is higher than the temperature one (as it could be expected, since all seasons tires are designed to guarantee acceptable performance in a wider span of weather conditions). For this reason, the decision was to deeper the study on all seasons tires in a different work.

As regards stopping distance variation with road surface temperature for summer tires, looking at these conclusions appears difficult to synthetize a general corrective formula without losing any information about the test, in particular if the type of tested tire is not well known. The decision on how to proceed in this case can be done depending on the necessities: an option could be to select a correction coefficient depending on tire type (i.e. summer with low rolling resistance, summer with sporty features, normal summer). Another option could be to synthetize a correction coefficient for all summer tire types. This can be the simplest way to proceed, but it will inevitably end with loss of information. The choice will depend on the precision needed in each particular case and from available data (i.e. number of summer tire types previously analyzed with this methodology).

The average of these values for summer tires is equal to 0,0956 m/°C, equivalent to a 0,211% variation for °C of road surface temperature.

With this value a corrective formula can be synthetized. It provides a method to bring the measured stopping distance value to the one that would be measured at a reference value:

$$S_c = S_m - \left\{ S_m \cdot \left[k \cdot \frac{(T_t - T_{ref})}{100} \right] \right\} \tag{5}$$

where S_c is the corrected stopping distance in meters, S_m is the measured one, k is the corrective coefficient (as said equal to 0,211 for summer tires, based on data acquired in this work), T_t is the measured road surface temperature and T_{ref} is the reference one (the one at which test engineer want to compare results, in this case equal to 25°C).

Table 2: Testing results – road surface temperature

	Tire 1	Tire 2	Tire 3	Tire 4	Tire 5	Tire 6 (AS)
Stp. Dist. range [m]	45,4 ÷ 51,3	47,4 ÷ 53,1	44,5 ÷ 48,7	43,1 ÷ 46,0	42,9 ÷ 45,2	50,6 ÷ 57,4
Tot. Stp. Dist. Δ [m]	5,9	5,7	4,2	2,9	2,3	6,8
Stp. Dist. Δ due to temp. variations [m]	3,4	3,5	2,2	1,4	1,0	3,8
Temp. range [°C]	27 ÷ 58	32 ÷ 58	32 ÷ 56	25 ÷ 48	15 ÷ 57	31 ÷ 56
Rate [m/°C]	0,131	0,166	0,087	0,049	0,045	0,083
Rate (Δ Stop. Dist. / °C) [%]	0,292	0,356	0,191	0,113	0,102	0,163

Table 3: Testing results – wear

	Tire 1	Tire 2	Tire 3	Tire 4	Tire 5	Tire 6 (AS)
Stp. Dist. Δ due to wear (n° ABS braking) [m]	0,5	3,2	2,5	2,0	1,5	5,1
Total ABS braking per-formed [-]	93	70	82	88	79	39
Rate [m/n° ABS]	0,005	0,048	0,021	0,040	0,022	0,179
Rate (Δ Stop. Dist / n° ABS) [%]	0,011	0,103	0,046	0,092	0,050	0,351

With this formula and the calculated coefficient, it is possible to compare tests done at different road surface temperature, or also estimate, from one single test session, the range of possible performances of the specific tires in different testing conditions.

8 Comparison with state of the art procedures

As stated in previous paragraphs the main international procedures which give indications about stopping distance measurements are ISO 21994 and SAE J2909.

As regards the ISO, it states only that stopping distances can be measured with atmospheric temperature between +5 and +35°C, that road surface temperature should stay between +10 and +40°C and that comparison between two measures can be done only if the atmospheric temperature is equal or lower than 10°C. As shown in paragraph 6, atmospheric temperature is not directly linked with road surface temperature (which is the right parameter to be checked and took under control), and, in addition, 10°C of road surface temperature difference could mean to have, with all others noise factors kept constant, a difference in measured stopping distances of about 4% (meaning that they are comparable only committing an error). For this reason, the corrective formula presented in this paper will give to test engineers additional information in respect to what presented in ISO 21994.

The other international standard involved is the SAE J2909. This standard gives a polynomial correction related to road surface temperature and it is optimized for all season tires types (North America market). It implies to measure the friction coefficient for each test (or test track) in order to be able to correct that value for road surface temperature. It is a good method indeed, but measuring the adhesion coefficient for each test is, of course, time consuming.

A comparison between this method and the corrective empirical formula is done and results are shown in figure 8, in which blue dots represent experimental data acquired through measurements at different road surface temperature.

Supposing to perform only the measure at the lower temperature (33°C) and to report this measured "true" value to other temperatures (higher in this case), it can be seen that the error committed using the corrective formula obtained in previous paragraph is less than the one committed using the corrective formula given by SAE document. The reason for this could be that the corrective formula is specific for summer types of tires, while SAE formula is more linked to all seasons tires.

In the presented case, the difference between measured values and those obtained with the corrective formula is of about 1,1 m (2,2%), while the error committed with SAE method is 2,9 m (5,9%).

For the studied corrective formula, this gap can be further reduced using a coefficient specific for the type of summer tire used: for example, being the tested tire a low rolling resistance one, the coefficient calculated on the same type of summer tire tested can be used instead of the average one.

As mentioned before, choosing which coefficient to use depends firstly on available data (about which type of tire is under testing and previous measurements on that specific tire type) and on the precision level that the test engineers would like to reach.

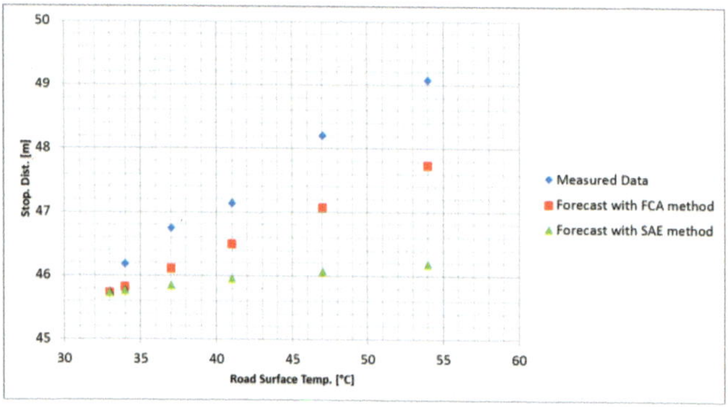

Figure 8: Comparison between measured data and corrections by FCA (general coefficient) and SAE

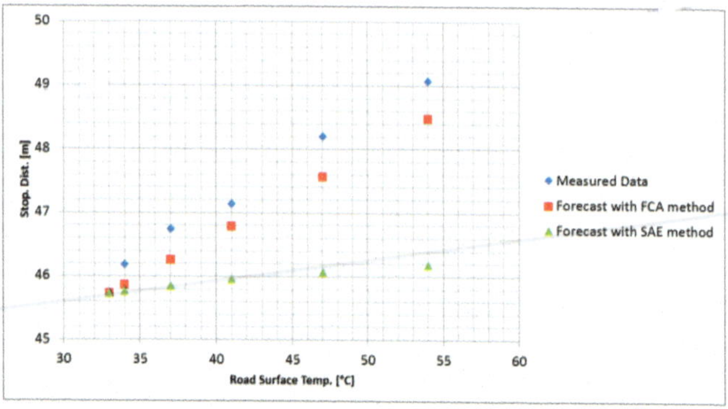

Figure 9: Comparison between measured data and corrections by FCA (specific coefficient) and SAE

9 Conclusions and next steps

For the first time in literature a specific type of tires (summer) was studied, measuring its performance of stopping distance on different conditions of road surface temperature.

Five summer tires were tested in a six months time span. From acquired data a formula was studied which allows to correct results, and thus to compare tests performed in different road surface temperature conditions.

Results of correcting stopping distance measurements for road surface temperature for summer tires are presented and compared to SAE J2909 correction. They show that the new formulation can produce more accurate results.

The precision of the coefficient to be used in the corrective formula can even increase increasing the number of tires analyzed with the method presented in this paper.

The same type of formulation can be used also for other families of tires (i.e. winter or all seasons), but specific test will be done in the future to obtain more specific results. As highlighted from first tests, the major limit to perform this kind of study with all seasons and winter tires is that their performances are more affected by wear, so they cannot be tested with a high number of braking without introduce a variability due not to road surface temperature, but to the wear itself.

In addition, the effect of the first braking on brand new tires was studied, highlighting its influence on stopping distances. Although is clear that a number between 15 – 20 braking from 100 kph is needed to bring the tire at its mean performance in terms of stopping distances, a more accurate study on this phenomena is needed to understand which are the reasons behind it.

10 References

[1] SAE J290, Light Vehicle Dry Stopping Distance, SAE International, 2010.

[2] ISO 21994:2007, Passenger cars -- Stopping distance at straight-line braking with ABS -- Open-loop test method, 2007.

[3] ASTM E1136-17, Standard Specification for P195/75R14 Radial Standard Reference Test Tire, ASTM International, West Conshohocken, PA, 2017, www.astm.org.

[4] Y. Gaspari (2018), "Influence of tire wear and asphalt temperature on vehicle stopping distance", Master's Thesis, Università di Pisa, 2018

[5] L. Ceccarini, “Identification of tires longitudinal slip curve from experimental data in order to improve effectiveness and robustness of ABS tuning”, Master’s Thesis, Università di Pisa, 2017.

[6] A.G. Bissoli, L. Ceccarini, Y. Gaspari, S. Murgia, “Longitudinal tire slip curve identification from vehicle road tests”, 9th International Munich Chassis Symposium, pp. 877-900, 2018.

Exploring simulator technology for tire development

Francesco Calabrese, Dr. Manfred Bäcker, Axel Gallrein
Fraunhofer ITWM

Introduction

The choice of the optimal tire setup for a given vehicle is not a trivial task. Nowadays car manufacturers often collaborate with the tire manufacturers during the development phase of new vehicles in order to improve the quality of their final products. At the beginning of a new development project, vehicle engineers define the so called vehicle key performance indicators (KPIs) and respective targets that the vehicle has to achieve. Regarding tires, the most crucial point is to answer the following key questions:

1. Which tire and rim size to choose?
2. What is the optimal tire pressure setup?
3. How close can we come to the desired key performance indicators by improving the tire and how much must be done on the vehicle side?

The authors, with the help of a premium car brand partner, proposed and validated a method based on off-line MBS full vehicle simulations (see [1]). This methodology uses CDTire/3D for the tire simulation technology, this model can successfully predict KPI variations for variations the tire size: starting from the parametrized predecessor tire model, the effect of a different tire/rim size or inflation pressure on the tire characteristics can be estimated, even if the varied tire is not physically available ([1] [2]). However, this method is not fully satisfactory, because the 'virtual tire' candidates cannot be evaluated by a professional driver in order to understand if the vehicle is 'drivable' and satisfies the expectation of the final user – it must still be prototypically manufactured.

The main reason why driving simulators and driver-in-the-loop (DIL) simulations become increasingly attractive is that they allow professional drivers to subjectively asses the tire-vehicle interaction during the development phase without a physical prototype. In this way, changes to the vehicle can be assessed in a few hours in the simulator environment. Moreover, the professional driver can also communicate his assessment to the tire engineers.

To close the virtual predevelopment loop, the communication between the professional driver and the tire engineer would ideally lead to adjustments of the virtual tire prototype within the same 'simulator testing session', allowing the driver to assess also the tire development. In this paper, the authors propose a methodology for 'exploring the driving simulator technology for tire development 'to make exactly that possible.

Introduction of the general methodology

This paragraph describes the general methodology.

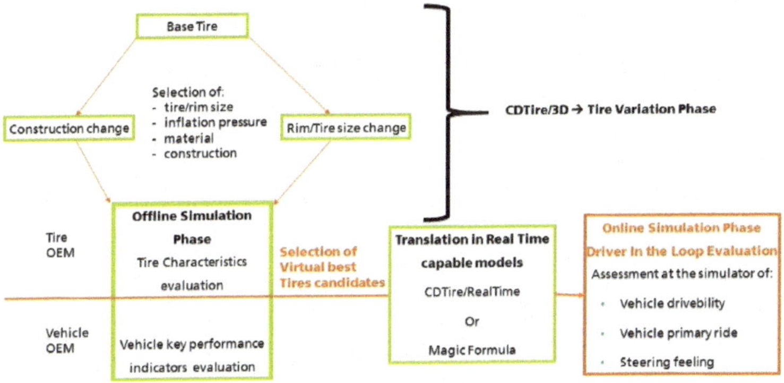

Figure 1: Description of the general methodology

Figure 1 depicts the general idea of the process starting from an existing tire that is the base tire of the tire development process. This tire could be the 'predecessor tire' that was designed for the previous version of the vehicle or an existing tire that is available on the market. For the 'base tire', all the measurements necessary to parametrize a CDTire/3D model must be available and the normal parametrization process then creates a base virtual model.

Once that the virtual tire model of the base tire is available, the 'variation phase' starts. During the variation phase, tire engineers can generate several instances of the virtual base tire by changing:

- Tire size
- Rim size
- Materials (for example rubber shear stiffness of tread, carcass stiffness etc.)
- Inflation pressure

In the next paragraph, we show an example of the results of such a tire variation phase performed by using CDTire/3D and the tools available in the CDTire model suite.

The variations of virtual tires produced can then be directly used in off-line simulations. In this phase, virtual standalone tire characteristics (like static stiffness, cornering stiffness, aligning toque, dynamics behavior etc.) can be collected and evaluated. This phase is important to have a realistic forecast of tire improvements and knowledge of the tire's performance limit.

The part that regards the 'tire variation' and generation of 'virtual tire instances' should be under the responsibility of a tire maker, given that the tire maker knows if a certain size of a tire can be manufactured and if certain variations of material (for example different rubber compounds) are realizable in production. The main advantage of using CDTire/3D in this phase over using an FEM approach are:

- The prediction of the tire characteristic variation due to modification of the base tire can be available in few minutes of simulation
- The CDTire/3D virtual prototype generated can be directly used in full MBS simulations

Off-line full vehicle simulation are carried out in order to understand the effect of these variations on the objective KPIs (on-center behavior, understeering gradient etc.) and a selection of the best virtual tire candidates can be done by the vehicle maker.

However, CDTire/3D is not real time capable, and as such does not directly support the respective simulator scenarios. Instead, the authors propose to use the CDTire/Realtime or a Magic Formula (PAC2002) tire model that must be derived from the virtual tire prototype. These models can then directly be used in DIL simulations at driving simulator and allow professional drivers to subjectively rate the drivability of the new vehicle prototypes. The chosen 'optimal' tires can then be engineered, manufactured and tested with the vehicle prototype on the proving ground for the final approval.

In the next paragraph, the authors will:

- Shortly explain some modeling aspects of the CDTire/3D model and its resizing capability
- Show simulation results of the variation phase on the tire characteristics

The last paragraph will analyze the translation to CDTire/Realtime and Magic Formula. In particular the comparison of the capability of these two modelling approach will be described, as well as the respective advantages/disadvantages in terms of fidelity of simulation, feeling at the simulator and efficiency impact on the methodology.

CDTire/3D

The model

Fraunhofer ITWM has developed the structural tire model CDTire/3D [4,5,7] used in the vehicle industry for comfort, durability and advanced handling scenarios. This model is based on a spatial finite difference (FD) formulation of the tire modeled as a shell. The functional layers of a tire (like cap ply, belt plies and carcass/body plies) are accumulated into the shell properties during tire initialization, but are accessible through their respective material parameters in the parameter file. The modelling of each cord-reinforced layer includes a non-linear part in the elastic component of the material description due to different behaviour under compression and tension regimes. The geometric formulation of the material behaviour allows for very large deformations. The dissipative parts of the material description combine viscous-elastic and inner friction behaviour. The tread is modelled by a brush-type contact formulation, allowing for local stick-slip effects. Figure 2 sketches the principle functional components of this structural MBD tire model:

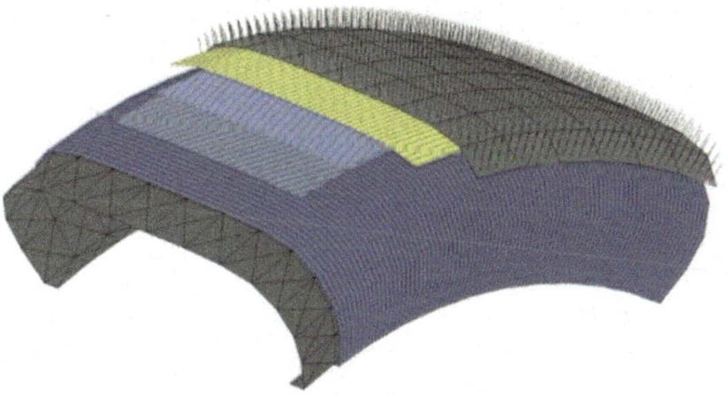

Figure 2. Functional layers of tire in CDTire

The model features a strict separation between material and geometric properties with the inflation pressure applied correctly onto the inner liner surface of the tire.

The model is set up with the help of a Construction Assistant.

The Construction Assistant

The Construction Assistant utilizes the inflated cross section geometry to parameterize the tire based on its constructional properties, such as the properties of the functional layers of tread, cap ply, belt and carcass layers – as well as geometric properties, such as position of the functional layers. See Figure 3 for a layout of the Construction assistant.

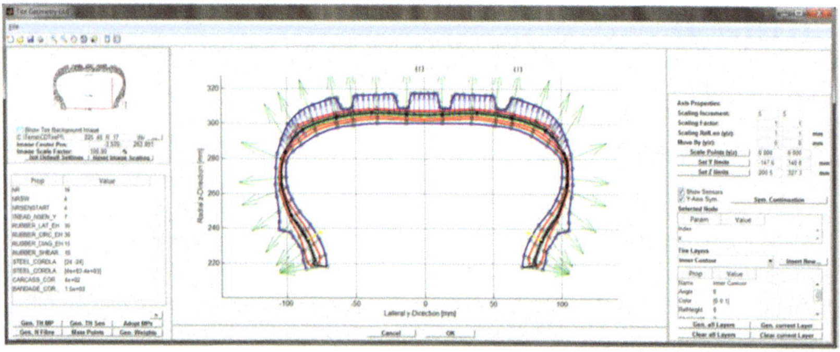

Figure 3. Construction assistant

From the material properties of the functional layers and its geometric dimensions, a discrete shell model is derived.

Example of 245/30 R20

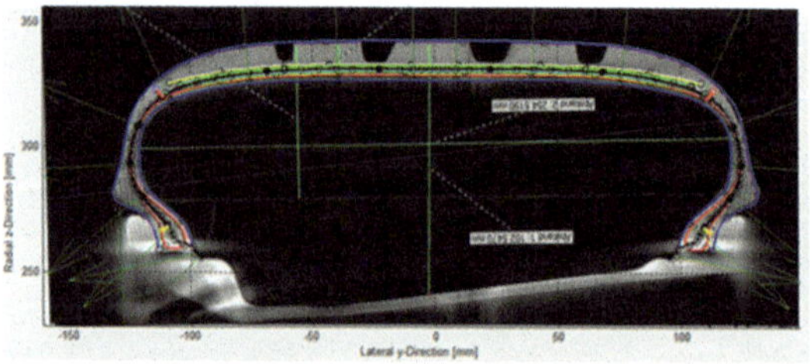

Geometrical properties :

- Cap Ply Geometry
- Steel Coord Geometry
- Steel Angle
- Carcass Geometry
- Inner Section contour
- Outer Section Contour

Material properties :

- Bead Rubber stiffness
- Tread Rubber stiffness
- Cap Ply elastic properties
- Cap Ply Prestrain
- Steel elastic properties
- Steel Prestrain
- Carcass elastic properties
- Carcass Prestrain

Figure 4. Constructional, geometrical and material properties that are input in the CDTire model

As showed in the Figure 4 all the constructional and functional entities and material are represented though a geometrical and material representation in the tire model. This allows expert users to do modification to the tire construction in order to understand the effect on the tire characteristics. An example of construction variation will be shown in the next paragraphs.

The Morphing algorithm for tire resizing

As the model features a strict separation between material end geometric properties, Fraunhofer ITWM developed a method to morph an existing tire based on nominal size specifications

- Tire nominal width (e.g. 225 mm)
- Tire aspect ratio (e.g. 45 %)
- Rim nominal diameter (e.g. 17'')
- Rim nominal width (e.g. 7.5')

The morphing algorithm adapts (morphs) the geometric description of the reference configuration of the cross section of the tire according to changes to any of these 4 nominal descriptive parameters without modifying the material description. It also adjusts the mass distribution according to the change in dimension. In short, the tire can be morphed e.g. from a 225/45 R 17 (x7.5) to a 235/40 R 18 (x8). Figure 4 shows this morphing in principle:

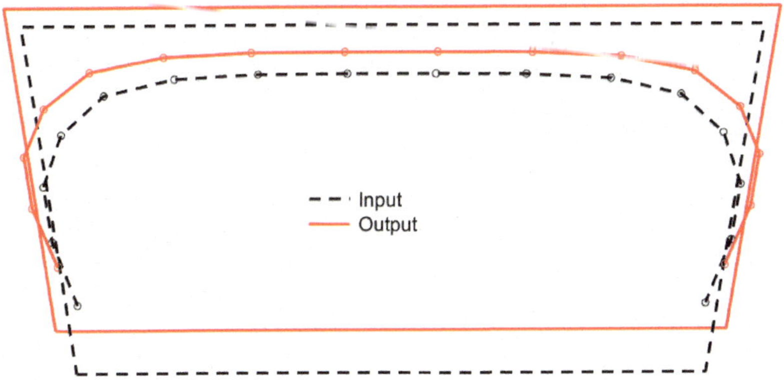

Figure 5. Tire morphing

In figure 5, the trapezes are made up of the nominal size parameters and shows the effect on the shell mid-plane geometry of the base and morphed model.

Tire/Rim resizing

Now the question that we want to answer in this chapter is: given a parametrized CDTire model for the base tire (for example a 235/55 R19 8J), is it then possible to re-size this tire to a different tire/rim size (for example 255/50 R19 9J)?

The general meaning of the morphing capability is: starting from a parametrized tire model called 'predecessor tire' or 'base tire':

a) To have a tire model that allows to predict the qualitative behavior of a tire that does not exist yet in the market by re-sizing the 'base tire' to the 'target size'
b) To predict the qualitative difference between tires of different sizes
c) To have the possibility of changing the rim width

CDTire/3D is predictive in 'tire resizing' because the model features a clear separation of mass, geometrical and material property.

The meaning of the morphing technique used in this article is the following:

- Starting from the 'base tire' to obtain a 'morphed tire' that keep the same material properties but update geometry and mass to the new size. Now the 'morphed tire' will have the same quantitative and quantitative behavior of an existing real tire if this has also the same material of the 'base tire'. The validation of this method has been proposed in the article [1].

From the application point of view this approach can be used in the early stage of the tire design with the following strategies or final aim:

d) Starting from the 'predecessor tire' to create a model of a 'virtual tire' with a different tire/rim size that does not exist yet in the market or it has never been equipped on this vehicle before. This is important because the vehicle maker has to have an idea of the qualitative behaviors on the new 'target tire'.
e) Understand the qualitative spread of performance between different tires of different sizes: for obvious cost reasons it is not possible in the early stage of development to make testing on vehicle of all the tire size combinations. For the same reason it is not possible to parametrize all the possible tire and make this tests virtually. The idea is to predict the spread of performance qualitatively by using the morphing techniques and evaluate his effect on the vehicle performances via simulation.

In the next paragraph some result of the re-sizing feature of CDTire/3D are shown. The trends showed in the results are confirmed from literature by experience.

Rim width effect

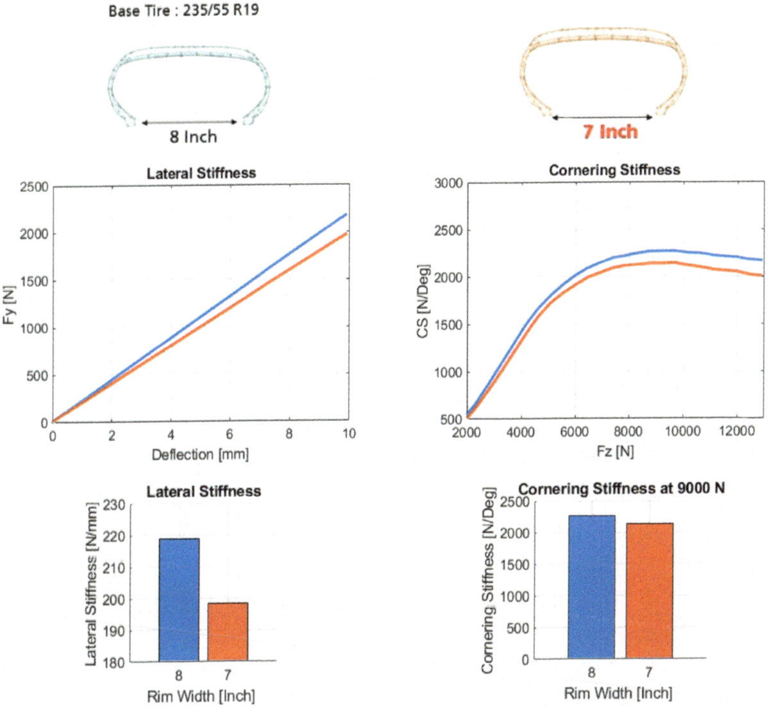

Figure 6. Resizing the rim width. Base tire: blue lines, tire variation in orange.

In the figure 6 it is shown how by having larger the rim width (without changing the tire) the model predict a larger of lateral stiffness, as well larger cornering stiffness. Since the percentage increase of lateral stiffness is bigger respect the percentage increase of cornering stiffness the model will predict a shorter relaxation length by increasing the rim width.

Tire size effect

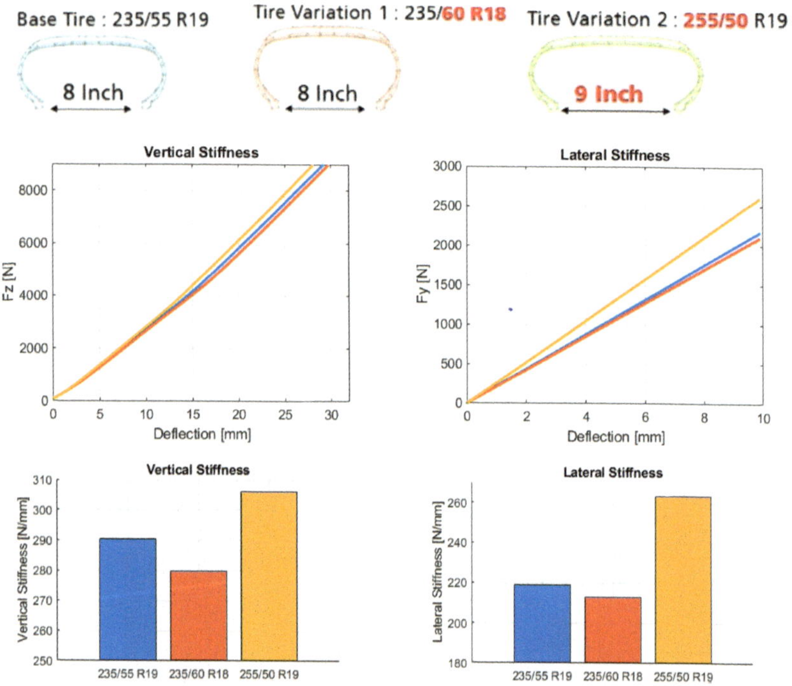

Figure 7. Static stiffness variation: In blue the base tire, Resizing of the rim radius from R19 to R18 in orange. Resizing of the tire width from 235 mm to 255 mm in yellow.

In the figure 7 an example of tire resizing is presented. In particular the model shows how by reducing the rim radius tire becomes 'softer'. The general tire stiffness increase by increasing the tire width.

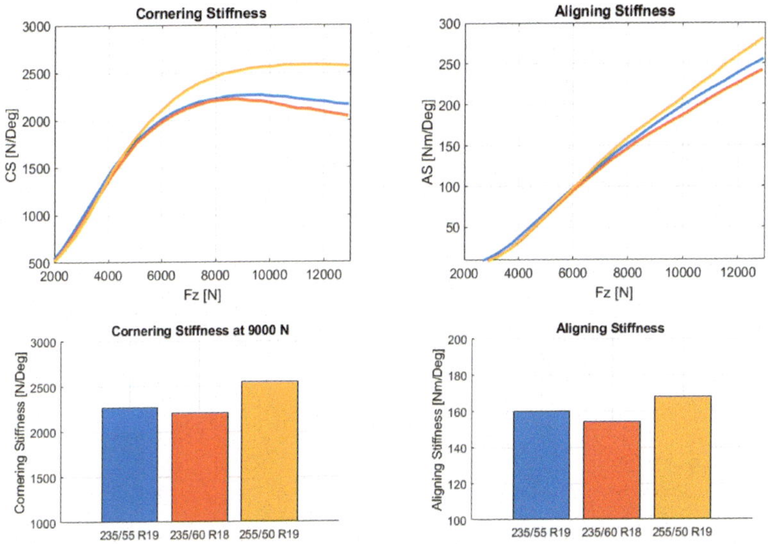

Figure 8. Handling characteristics: In blue the base tire, Resizing of the rim radius from R19 to R18 in orange. Resizing the tire width from 235 mm to 255 mm in yellow

Figure 8 shows how the model predict the fact that tire with wider belt can produce more cornering stiffness at high preload Fz because of the stiffer belt.

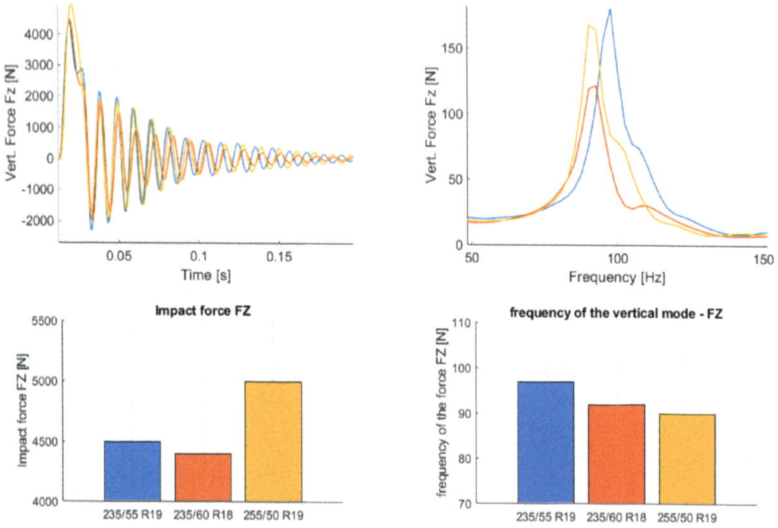

Figure 9. Cleat test and typical tire frequency: In blue the base tire, Resizing of the rim radius from R19 to R18 in orange. Resizing the tire width from 235 mm to 255 mm in yellow

Last example of simulation proposed in figure 9 shows how the tire size affect the vibrational behavior of a tire. In particular by increasing the width (tire variation represented in yellow) the vertical stiffness of the tire increase as showed in figure 7, at the meantime also the mass of tire will increase. In this case the mass variation is dominant, as result the typical frequency will decrease. Also in the case of rim radius reduction the typical frequency will decrease, but in this case this is due to the vertical stiffness reduction since the mass is almost the same of the base tire.

Variation of the tire tread material

As described in the previous paragraph in CDTire/3D material and geometrical properties of the tire construction are represented. Due to these features tire makers can also vary these properties in order to create virtual prototypes that predict the effect of these changes. In the following figure it is shown how a stiffening of the tread compound affect the cornering stiffness and the aligning stiffness.

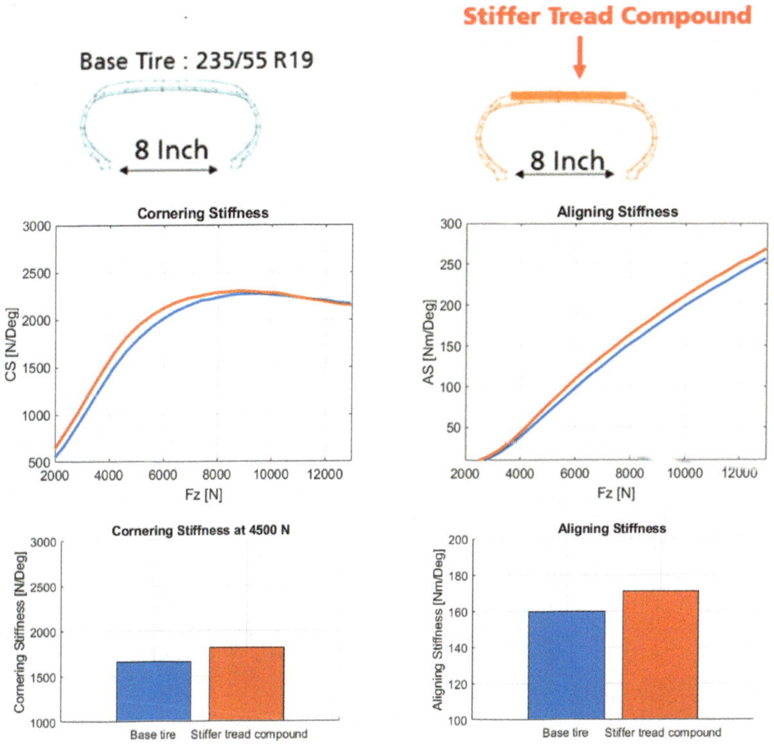

Figure 10. Variation of the handling characteristics when changing the tread rubber elastic properties: In blue the base tire, In red the same tire with a variation of 10% of the tread shear stiffness

Variation of the bead geometry and material

Last example of tire variation shows how the bead construction strongly affect the tire properties. In particular by using stiffer rubber in the bead area (so called apex rubber) it is possible to increase the longitudinal tire stiffness. Longer carcass loop in the bead area increases the tire lateral stiffness and consequently reduce the relaxation length. In figure 11 the results of the model prediction are shown.

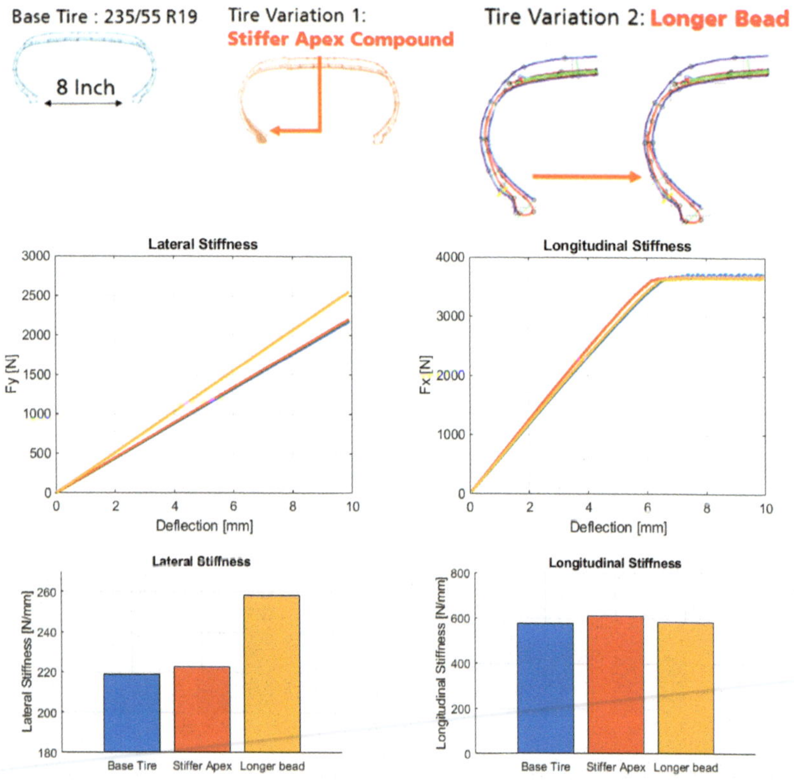

Figure 11. Base tire blue line. Effect of the increase of the apex rubber (red line) or of longer bead (yellow line)

Results of the experience at the simulator

From the variation phase described in the previous paragraph and after the evaluation of the KPIs variation through full MBS simulation has been done, the selection and translation of the CDTire/3D to a real time capable tire model can be performed. The process of derivation of CDTire/Realtime or Magic formula (PAC2002) models are realizable through dedicated tool developed by Fraunhofer. In both cases, virtual measurements of tire characteristics of the virtual prototypes are used to derive the real time models.

The typical work flow of a joint tire development loop is depicted in Figure 12:

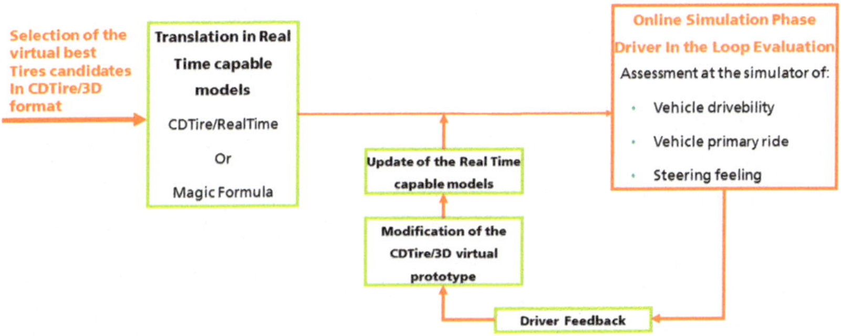

Figure 12. Typical work flow during the DIL simulation at a driving simulators.

The tire development loop consists of the following steps:

- The vehicle with the derived real time tire models is tested by the driver.
- The driver gives his feedback by rating the subjective KPIs.
- The tire engineer can react to the feedback by modifying inflation pressure, wheel size or material properties of the virtual tire prototype.
- Updated real time models are derived and the driver can re-do the tests.

This loop can continue until the driver is satisfied and the KPIs targets are reached. One single iteration of the loop in Figure 12 takes approximately 2 hour. Fraunhofer successfully tested this process together with a tire manufacturer and a vehicle manufacturer in a common tire development project.

In this project, each development step was always checked using both derived CDTire/Realtime and derived Magic Formula tire models. The professional drivers also assessed their experiences of this activity:

- In case of CDTire/Realtime, the drivers could consistently identify the respective tire design change performed by the tire engineer.
- The drivers consistently described the subjective experience of using the derived CDTire/Realtime models as more realistic, compared to the Magic Formula tire models. The next paragraph makes an effort to explain this phenomenon.

Within the scope of one simulator session, the driver modification requests on the tires are more easily performed by a translation from the virtual tire prototype to the CDTire/Realtime model. This is because in order to derive a new Magic Formula model several hours are needed, due to the amount of time needed to simulate all the virtual experiments and use a regression algorithm to identify the new Magic Formula tire model. In case of CDTire/Realtime, the time needed to apply a variation to the new tire model is practically less than one hour. This is mainly because the parameters of the CDTire/3D and CDTire/Realtime models are of the same nature and have physical meaning and change on the CDTire/3D parameter set can be easily translated into a change of the CDTire/Realtime parameters.

In the last paragraph the authors will try to explain which are the main modelling reason for which a professional driver feels a vehicle equipped with CDTire/Realtime 'feels' more real than the Magic Formula tire model.

Modelling aspects that make CDTire/Realtime more realistic in terms of the driving feeling at the simulator

As described in the previous paragraph, the professional drivers described the experience using CDTire/Realtime at the driving simulator as more realistic compared to the Magic Formula tire models.

The authors think that the main differences that generate a better realism are the following:

a. CDTire/Realtime is a model with a flexible belt which allows a physical 'enveloping' of the 3D road profile. This should lead to better fidelity of the simulator behavior respect to the vehicle primary ride and body movements.
b. CDTire/Realtime can give a physical answer of the dynamic vibrations generated from the tire interacting with the road. For example hitting a curb would generate the correct dynamic response.
c. The self-aligning torque in driving and braking conditions generated by CDTire/Realtime is closer to the actual tire behavior. This is shown in the combined slip comparison in Figure 13, where the self-aligning torque clearly is affected by the longitudinal force, due to the lateral belt flexibility. In the Magic Formula model this effect is not present. The self-aligning torque will change the torque feedback at the steering wheel and will give an appropriate feeling to the driver.

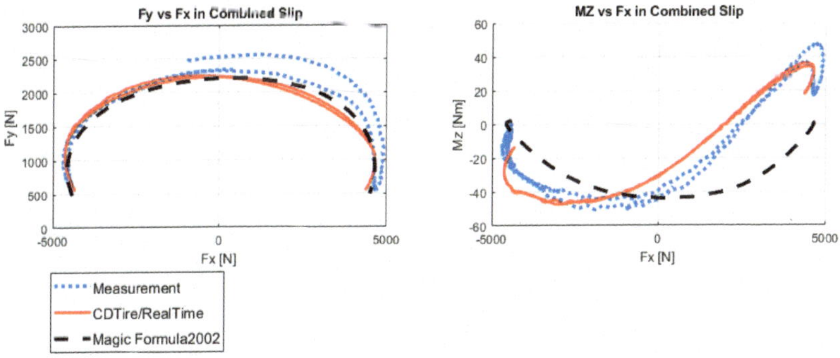

Figure 13. Self-aligning toque during combined slip experiment. Comparison between measurements, Magic Formula and CDTire/Realtime.

d. The Magic Formula model does not have a real velocity dependency. It is valid only for one velocity, the nominal velocity for which the measurement are performed. This means that all the velocity effects like stiffening of the tire due to centrifugal forces and especially decreasing friction by higher sliding velocities are neglected.

Those effects are present in CDTire/Realtime. In Figure 14, a comparison between measurements (right side) and simulations is shown. From the results it is clear that the Magic Formula model is not having a correct behavior in the friction-velocity dependency. Furthermore in Figure 15 it is shown how the lateral force characteristic is affected by the longitudinal velocity in the CDTire/Realtime case while for the Magic Formula model stay constant.

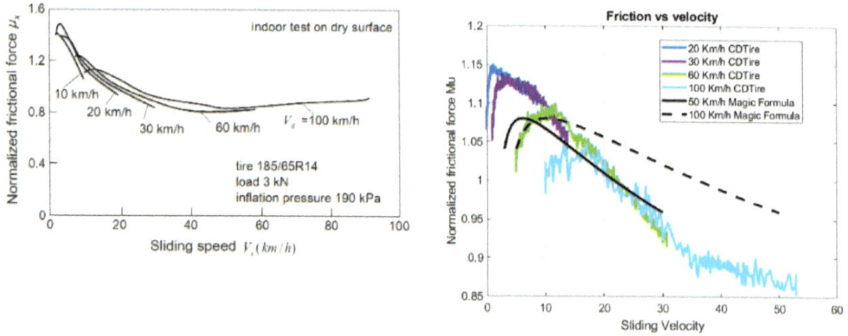

Figure 14. Velocity dependency of the friction coefficient. Left side measurement published in [8]. Right side comparison between CDTire/Realtime and Magic Formula.

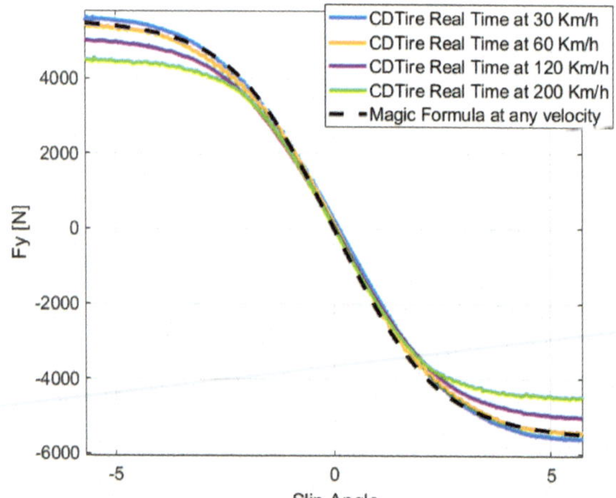

Figure 15: Lateral force Fy for different velocities calculated with CDTire/Realtime and comparison with Magic formula.

An example of how much the tire force-velocity dependency can affect the results of a full vehicle simulation is shown in figure 16. In this figure vehicle ramp steer maneuver simulations are shown for 2 different velocities (60 Km/h and 150 Km/h) by using MF model and CDTire/Realtime. The max lateral acceleration provided by the magic formula model does not change by increasing the velocity. This mean that the driver would not feel a 'velocity' effect and a consistent 'tire limit' at the simulator.

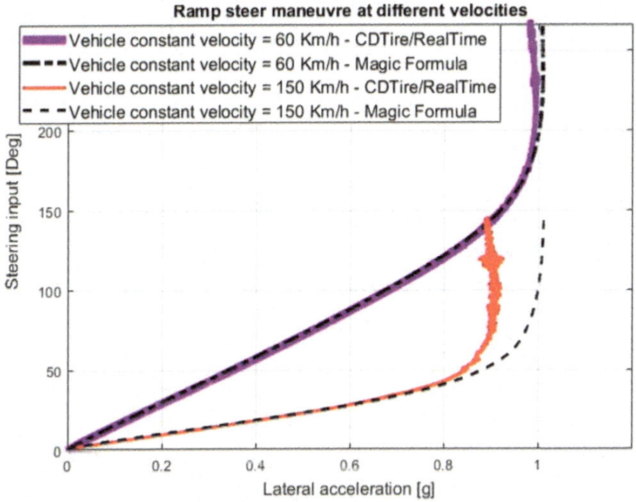

Figure 16. Comparison of result of ramp steer at different velocities with CDTire/Realtime and Magic Formula

Conclusion

In this paper the authors proposed a methodology that describes how to use the CDTire model suite and the vehicle driving simulator technology in order to improve the modern virtual tire development process. In particular this methodology allows professional drivers to test and subjectively judge the drivability of virtual tires prototypes of different tire/rim size or construction in the early stage of the vehicle development. This evaluation would help tire and vehicle engineers to do the right choices regarding the tire size and construction at the beginning of the development phase by saving hours of expensive test at the proving ground.

The authors showed how the key ingredient of the methodology is the usage of a physical 3D structural tire model like CDTire/3D that allows to predict, starting from a virtual model of the 'predecessor tire', the effect of the variation of the tire/rim size, changes in tire construction (material properties or geometric properties) and the inflation pressure. The prediction capabilities are shown in the first chapter by meaning of tire characteristic variation. By using this approach, it is possible to predict the limit performance of a tire prototype in a physical way.

Given that CDTire/3D is not real time capable, it must be converted into an equivalent CDTire/Realtime or Magic Formula model in order to make the driving simulator session feasible. Deriving the CDTire/Realtime for the various virtual CDTire/3D prototypes results quite easy since only few virtual tire characteristics are needed and the structural parameters between the two models are quite similar in terms of physical meaning, so variations on the CDTire/3D model can be easily transferred to CDTire/Realtime model type.

On the other hand the derivation of the Magic Formula models needs a quite large number of virtual experiment calculated by using the CDTire/3D in order to identify through a regression algorithm this semi-empirical model. The time cost of this process is quite high. This is the practical reason for which the CDTire/Realtime model is preferred: changes of tire virtual prototypes requested during the driving simulator session can be realized in a reasonable time. This allows to re-test the tire virtual prototype in the same session.

Furthermore practical experience shows that the professional drivers claim a more realistic driving feelings by using CDTire/Realtime respect to the Magic Formula. The reasons of the better performance of CDTire/Realtime are due to the physicality of its modeling approach. In fact in the last chapter it is shown how the Magic Formula cannot produce realistic dynamic response due to an interaction with a 3D road, cannot replicate a consistent 'velocity effect' as well as realistic self-aligning torque MZ in driving or braking conditions. Those differences are felt from the driver in terms of body movements and acceleration and steering feedback torque.

References

1. Calabrese F., Dusini, L., Bäcker M and Gallrein, Axel, "Managing the variety of potential tire / wheel sizes in the early vehicle development process",9th International Munich Chassis Symposium 2018",Springer Fachmedien Wiesbaden,978-3-658-22050-1

2. Bäcker M., Gallrein A., Calabrese F., Leister G., "Simulation of a tire blow-out in a full vehicle scenario. " In: Pfeffer P. (eds) 7th International Munich Chassis Symposium 2016. Proceedings. Springer Vieweg, Wiesbaden

3. Koutny, F., "A Method for Computing the Radial Deformation Characteristics of Belted Tires," Tire Science and Technology, TSTCA, Vol. 4, No. 3, Aug. 1976, pp. 190-212

4. Gallrein, A., Baecker, M., Gizatullin, A.: "Structural MBD Tire Models: Closing the Gap to Structural Analysis – History and Future of Parameter Identification", SAE Technical Paper 2013-01-0630, 2013, doi:10.4271/2013-01-0630.

5. Baecker, M., Gallrein, A. and Haga, H. "A Tire Model for Very Large Tire Deformations and its Application in Very Severe Events" SAE Int. J. Mater. Manuf. 3(1): 142-151, 2010

6. Leister G., Fahrzeugreifen und Fahrwerkentwicklung: Strategie, Methoden, Tools (ATZ/MTZ-Fachbuch) Gebundene Ausgabe – 11. Dezember 2008, ISBN-10: 3834806714

7. A. Gallrein, J. DeCuyper, W. Dehandschutter, M. Baecker: "Parameter Identification for LMS CDTire", 3rd Int. Tyre Colloquium, Tyre Models for Vehicle Dynamics Analysis Proceedings, Vehicle System Dynamics Vol.43, Supplement, 2005, pp.444-456.

8. Nakajima, Yukio. (2019). Advanced Tire Mechanics. 10.1007/978-981-13-5799-2.

Contact Information

Dr. Manfred Bäcker, Axel Gallrein, Francesco Calabrese
Fraunhofer ITWM
Fraunhofer-Platz 1
67663 Kaiserslautern
Germany

http://www.itwm.fraunhofer.de

manfred.baecker@itwm.fraunhofer.de
axel.gallrein@itwm.fraunhofer.de
francesco.calabrese@itwm.fraunhofer.de

Model-based development of iTPMS (indirect Tire Pressure Monitoring System)

Dipl.-Ing. Wenrui Han, Prof. Dr.-Ing. Günther Prokop
Lehrstuhl für Kraftfahrzeugtechnik, Technische Universität Dresden

Dr. Thomas Roscher
AUDI AG

1 Introduction of iTPMS

The tire as the only contact between vehicle and road surface has a significant influence on vehicle safety, ride comfort, driving dynamics and fuel economy. A tire can reach the optimum performance, only when an appropriate tire pressure has been adjusted. Due to the importance of tire pressure, the USA has published a legal regulation (FMVSS 138) to require that tire pressure monitoring systems have to be installed in light vehicles since 2007. A similar regulation in Europa has been released since 2012. Likewise, all vehicles in China must be equipped with TPMS from 2020.

There are mainly two types of tire pressure monitoring systems: direct and indirect measuring systems. The direct TPMS employs pressure sensor on each wheel to measure tire pressure. The indirect TPMS uses the tire itself as a sensor, where the pressure-dependent properties of wheel angular speed signal can be used to detect a tire pressure loss.

The two pressure-dependent phenomena of a wheel during pressure loss are the change of wheel rolling radius and the change of natural frequencies contained in the wheel angular speed. As figure 1 (a) shows, in a one-wheel pressure loss scenario (puncture) the wheel angular speed of a deflated wheel increases in comparison with the other wheels, accordingly, the rolling radius of the wheel decreases. Besides the change of rolling radius, the other phenomenon is a shift of natural frequencies contained in the wheel angular speed. Wheel angular speed signal contains vibration information of tire torsional vibration. The natural frequencies of tire torsional vibration are strongly dependent on the tire pressure. These two phenomena can be used for the detection of 1-3 wheels press loss scenario. However, because of the simultaneous change of the rolling radius of 4 wheels, the first phenomenon can not be applied in 4 wheel pressure loss scenario (see figure 1 (b)). In this case, only the change of natural frequencies contained in the wheel angular speed can be implemented to detect a four wheels pressure loss.

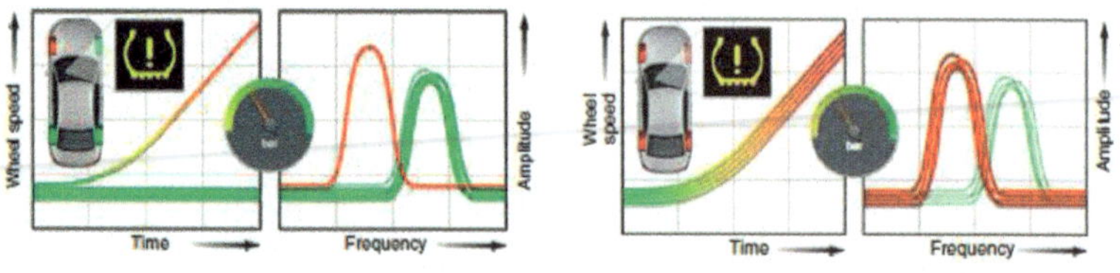

Figure 1: (a) One-wheel puncture scenario (left) (b) Four-wheels diffusion scenario (right) [1]

The above mentioned pressure-dependent properties of tires can not just be affected by tire pressure. Especially the tire natural frequencies can be influenced by tire construction, rubber compound, drivetrain, axle elastic kinematics etc. In the current development of iTPMS, these influencing factors can be identified with different kinds of tests and therefore the vehicle test plays an important role for the application of iTPMS. In order to reduce the test effort and realize a predication of iTPMS functionality in the early phase of the whole vehicle development process, diverse simulation methods can be used to replace the vehicle test and to investigate the influencing factors on this system.

2 Overview of virtual development of iTPMS

As mentioned in the first section, the current development of iTPMS is based on vehicle tests under different conditions. Figure 2 shows the conventional development process of iTPMS. If the vehicle tests can be replaced by simulation or rig tests, the expenditure on the iTPMS development can be significantly reduced. The virtual tests and rig tests can be executed at an early phase of development process, in which the available prototype vehicles are limited.

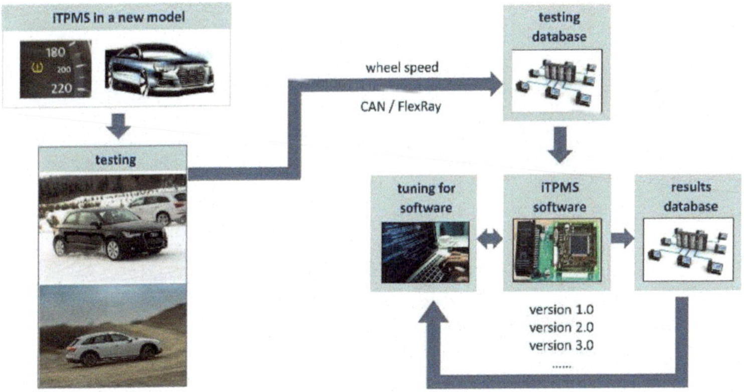

Figure 2: Development process of iTPMS

For a virtual development of iTPMS, some main questions need to be answered before the virtual tests are executed. These questions are e.g. which tire model is suitable for the virtual tests? What are the influencing factors on the tire vibration behavior? Can a simulation model represents the tire and vehicle behavior like a real-world test? In order to approach these questions systematically, the virtual methods can and will be classified in three levels: component level, subsystem level and full vehicle level.

Figure 3 shows the tools for the virtual development of iTPMS. Generally, two kinds of models have been used. A simplified model, which is described by the equations of motion, can improve the understanding of tire vibration behavior and its correlation with some tire construction parameters. The small number of needed parameters is an obvious advantage in comparison with other complex models. However, a complex model such as a multibody model can not be totally replaced by a simplified model. This complex model can better describe some dynamic properties like the dynamic stiffness and non-linearity of rubber components. The parameterization and validation are two important parts for a reliable model. For this reason, the simulation results need to be compared with the data from test rig and vehicle test.

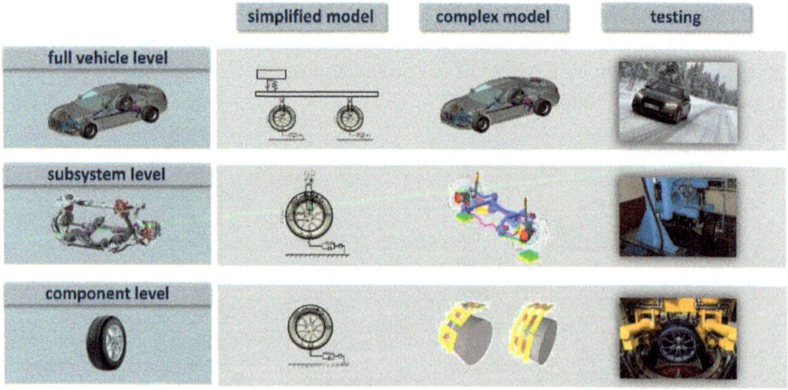

Figure 3: Tools for a virtual development of iTPMS [2]

3 Development in three levels

The virtual development has been divided into 3 levels: component level, subsystem level and full vehicle level. In this section, the models and results in different levels will be introduced.

3.1 Component level: tire model

The tire is the most important factor for virtual development of iTPMS. Different characteristics of a tire such as tire construction, rubber compound etc. can lead to different tire performance. If a pressure loss need to be detected in simulation, a tire model should be able to represent the typical tire behavior of different kinds of tires and the behavior of tire model should be comparable to the tire behavior which can be measured in the real-world test. After a comparison of different tire models, the Ftire model and the rigid ring tire model have been selected to investigate the tire behavior at pressure loss. Both models and the reason of the selection will be introduced in the following sections.

3.1.1 Ftire model

The Ftire model is a widespread tire model for full vehicle simulation. The structure of Ftire can be seen in figure 4. In the Ftire model the tire belt is described as an extensible and flexible ring. This ring consists of a finite number of belt elements. These belt elements are coupled with each other by stiff springs both in-plane and out-of-plane. To every belt element, a number of mass-less tread blocks are appended. These blocks carry nonlinear stiffness and damping properties in radial, tangential and lateral direction. [3]

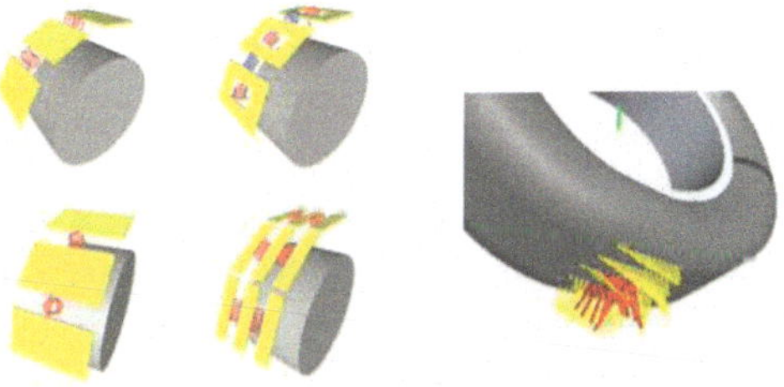

Figure 4: Coupling of belt elements (left) and tread elements (right) [3]

The Ftire model is able to describe the pressure-dependent tire vibration behavior. Figure 5 shows a comparison of frequency spectrum of wheel angular speed at three pressures between rig test and simulation. This result is from a cleat test, which has been executed at 30 km/h on an external drum tire test rig and on a corresponding virtual tire test rig. From both plots a typical natural frequency of a tire between 30 and 40 Hz can be found, which is used in iTPMS to detect a pressure loss. The pressure dependency of this natural frequency can be performed with the Ftire model in general, though the frequency shifts of the Ftire at three pressures show some deviations from the measurement, because the wheel angular speed vibration has not been taken into account in the current standard process of Ftire parameterization.

Generally, the Ftire model is suitable for the further simulation of iTPMS. However, due to the high complexity of Ftire model a long simulation time can not be ignored. Otherwise, a large numbers of parameters mean a complex procedure for parameterization.

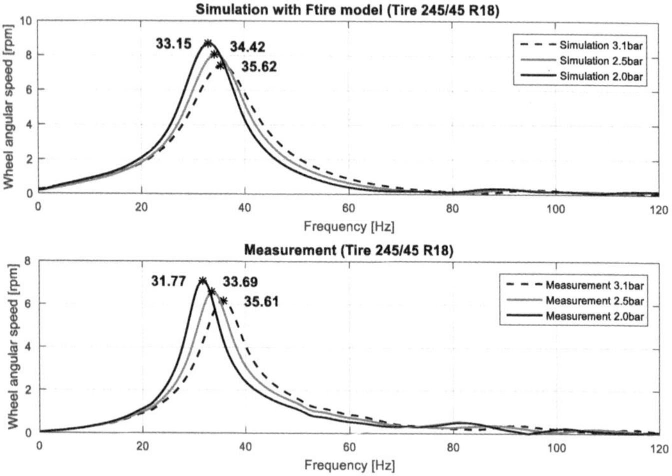

Figure 5: Frequency spectrum of wheel angular speed from test and Ftire simulation

In order to reduce the computing time and simplify the parameterization process, a rigid ring model has been selected to represent the tire vibration behavior at pressure loss.

3.1.2 Rigid ring tire model

The rigid ring tire model was created by ZEGELAAR [4] in 1996. In the next sections, the structure, modification and parameterization of the rigid ring tire model and a series of analyses with the model will be introduced.

3.1.2.1 Structure of the rigid ring tire model

The rigid ring tire model can be seen in figure 6. The tire belt is considered as a rigid ring. The rim and the rigid ring are coupled by springs and dampers in radial direction and torsional direction. The contact between rigid ring and road surface is described with a contact model. The rigid ring tire model in figure 6 has four degrees of freedom: rim rotation (φ_F) and belt rotation (φ_G), translation of belt in longitudinal and vertical direction (x_G, z_G).

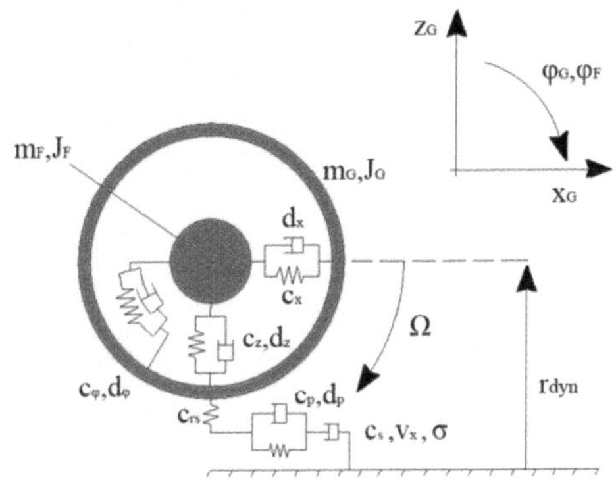

Figure 6: Rigid ring tire model [2]

Even though the rigid ring tire model is simple, it can also describe the tire natural frequency and its pressure dependency, which is relevant to the functionality of iTPMS. Figure 7 shows a comparison between the rig test and the simulation with the rigid ring tire model. The frequency shifts of a tire at three pressures can be well represented with the rigid ring tire model, because the parameterization of the rigid ring model has been run with the focus on the wheel angular speed behavior and its pressure dependency.

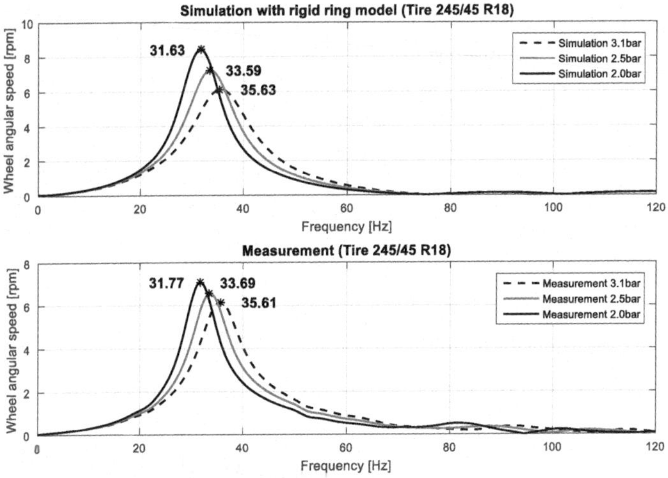

Figure 7: Frequency spectrum of wheel angular speed from test and rigid ring simulation

3.1.2.2 Modification of the rigid ring tire model

The rigid ring tire model is a single point contact model. This model is theoretically valid for long wavelength (longer than 3 meter [4]) road irregularities and can be used on surfaces with random unevennesses. However, the iTPMS should be able to detect a tire pressure loss on any road surfaces. In order to implement the rigid ring model in the simulation with different kinds of excitations, the contact model needs to be modified. The tire enveloping property, which can not be represented by rigid ring tire model, is essential for a simulation with short wavelength road unevennesses such as a cleat. For this reason, a simplified flexible ring has been used as the contact model to describe the tire enveloping property in the contact area.

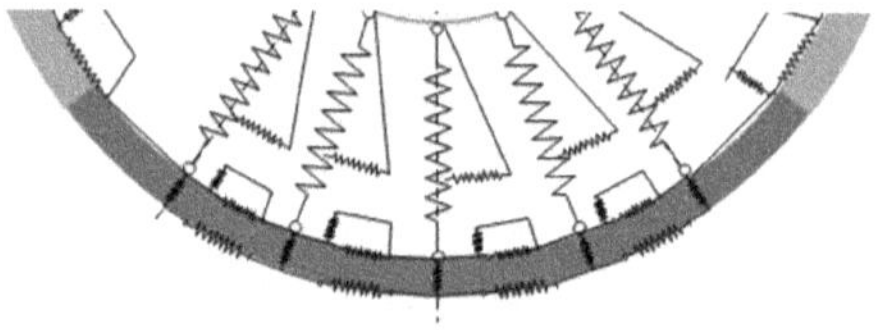

Figure 8: Flexible ring as a contact model for rigid ring tire model

Figure 8 shows the structure of the contact model. The contact area consists of a defined number of belt elements. The belt elements are coupled by springs in tangential and radial direction. These springs can be understood as the stiffness of tread in tangential and radial direction. The enveloping property can be performed by the numerous deformable belt elements in the contact area. Due to the deformation of each spring element, in the contact area occur resultant forces in tangential and radial direction. The forces can be used as an input for the rigid ring tire model. The modified contact model enables the simulation with the rigid ring tire model on any kinds of road surfaces. After a preprocessing of the contact model, a displacement excitation can be transformed into a force excitation. The vibration of a rigid ring tire model can be excited by the force excitation. Figure 9 shows a simulation at a speed of 4 km/h. The enveloping property of this rigid ring tire model can be seen in the vertical force (a small bump) in the time domain.

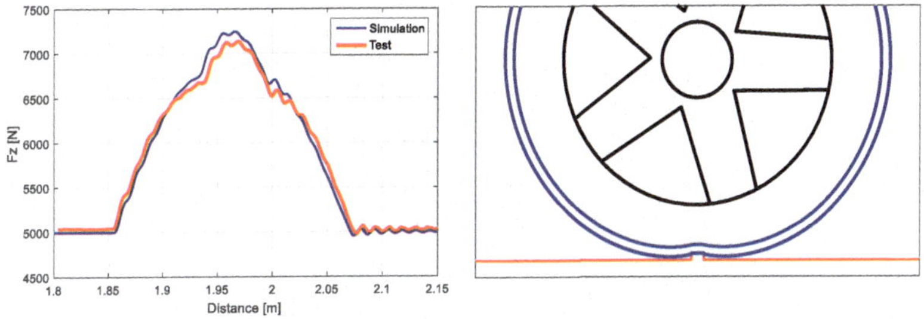

Figure 9: Enveloping property described by contact model: vertical force (cleat test at 4 km/h) between test and simulation (left), visualization of enveloping property of contact model (right)

3.1.2.3 Parameterization of the rigid ring tire model

The parameters are crucial to the quality of a model. A parameterization process is often a challenge for a high-quality model. The complexity of a parameterization process for a model and the accuracy of the model are points, which need to be considered for the development of a practical and pragmatic parameterization process.

As figure 6 shows, the rigid ring model consists of following parameters: mass of tire belt (m_G), moment of inertia of rim (J_F) and of tire belt (J_G), tire longitudinal stiffness and damping (c_x, d_x), tire vertical stiffness and damping (c_z, d_z), tire torsional stiffness and damping (c_φ, d_φ). It should be noted that the tire stiffness and damping in longitudinal and vertical direction are considered to be equal. The basic contact model has following parameters: longitudinal tread stiffness and damping (c_p, d_p), slip stiffness (c_s), longitudinal slip (s_x) and relaxation length (σ). v_x is the longitudinal velocity and r_{dyn} is the dynamic rolling radius.

Theoretically, all parameters can be determined by different kinds of rig tests. However, the deviation and tolerance of each parameter can accumulate. The final simulation result of the tire model may well not be perfect. In reality, the change of frequency at pressure loss is usually between 0.5 and 2 Hz. Due to the high requirement on the accuracy of the tire model, the parameterization on test rig is not sufficient. The other pure mathematical methods such as fitting and optimization can make the simulation result very close to the test result. However, the physical meaning of each parameter can not be guaranteed with this method. For this reason, a hybrid method, namely a combination of both methods could be the best way to parameterize the described rigid ring tire model.

Figure 10: Tire test rig in TU Dresden

First, the most important parameters on the iTPMS-related vibration will be identified by a sensitivity analysis. These important parameters need to be defined with measurement. The result of sensitivity analysis can be seen in figure 12. The mass and inertia are the parameters, which can be easily measured. The dynamic rolling radius, radial stiffness can be defined on a tire test rig (see figure 10) [5] with a rolling test. The other parameters can be determined by a fitting method. With this hybrid method, on one hand, the complexity of the parameterization process can be reduced. On the other hand, the accuracy of the model is guaranteed. The measurement data at 2.0, 2.5 and 3.1 bar are used for the parameterization of the model. After the parameterization, the model will be validated at three other pressures, so that the interpolation and extrapolation capability of the model can be verified. The final results can be seen in figure 7, the deviation of the peak frequencies at different pressures between simulation and test are below 1%.

3.1.2.4 Application of the rigid ring tire model

The rigid ring tire model needs to be used in the full vehicle simulation to realize a virtual test of iTPMS. Besides, this model is also a tool for the investigation on tire vibration behavior and its correlation to tire parameters.

Compared to a complex model like Ftire model, the rigid ring tire model can be better analyzed with mathematical methods. The model is described by equations of motion. The eigenvalues and eigenvectors can be calculated by solving the equations. This modal analysis improves the understanding of tire vibration behavior. The absolute value and phase information show the status of each degree of freedom at each eigenmode [6]. Figure 11 shows the absolute value and phase information of 4 degrees of freedom at the first natural frequency. It can be seen from the calculation that the longitudinal and torsional vibration of tire belt are participant in the first natural frequency and they are moving in phase. This vibration is the so called coupled longitudinal and torsional mode of a tire. iTPMS uses this mode to detect the pressure loss, because this natural frequency is strongly pressure dependent.

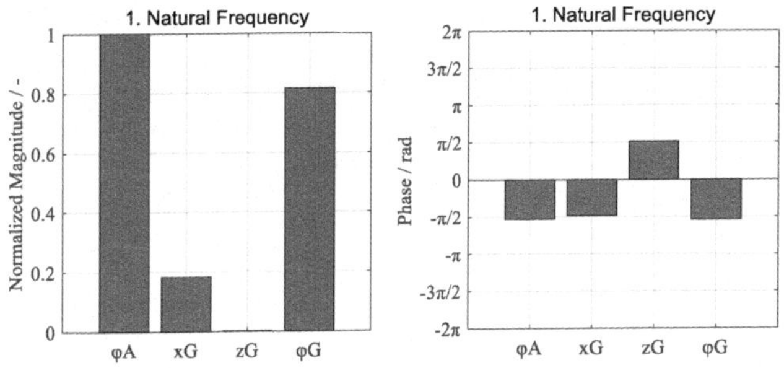

Figure 11: Modal analysis of rigid ring tire model [7]

Due to the limited number of parameters, a sensitivity analysis of this model is possible. Figure 12 shows the influences of all model parameters on the first tire natural frequency that has been introduced in figure 11. The theoretical analysis can indicate the important parameters for tire natural frequency. Under the consideration with the change rate of each parameter in reality, the important influencing factors can be identified. For example, the change of rolling radius by 30% is just for a theoretical analysis. In reality, the change of rolling radius at 20% pressure loss is just 0-2%. In this case, the influence of rolling radius on the tire natural frequency is less critical than the result from the theoretical analysis in figure 12.

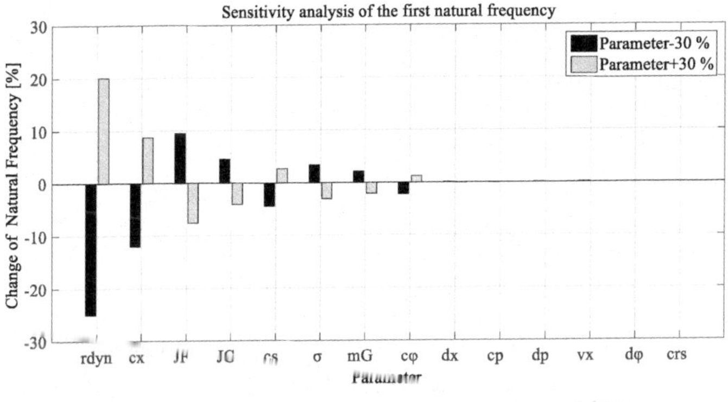

Figure 12: Sensitivity analysis of rigid ring tire model [2]

3.2 Subsystem level: axle model

Axle constructions of modern vehicles become more and more complex due to the requirement on better performance of driving dynamic and ride comfort. The axle kinematics and compliance can influence the tire vibration behavior. Especially the axle stiffness can change the tire frequency behavior. In order to ensure the functionality of iTPMS in a new vehicle concept phase, the influences from axle need to be investigated. For the investigation of the correlation between axle and tire, two types of models are used: multibody model and simplified concept model.

3.2.1 Multibody model

Multibody model is a complex model, which can emulate the movement and characteristic of every axle component. For example, the stiffness and damping of suspension rubber bushing in three direction can be described with a suitable model.

Figure 13 shows a simulation with a rear axle model. The tire natural frequencies can be gained from a cleat test. All axle components, which have been marked on the right diagram, could potentially change the tire frequency behavior. For this reason, a sensitivity analysis of the stiffness and damping of these suspension components in every direction has been executed. The influences of the elastic components on the first tire natural frequency by the change of the stiffness by 25% can be compared with the change of the first tire natural frequency with 20% pressure loss.

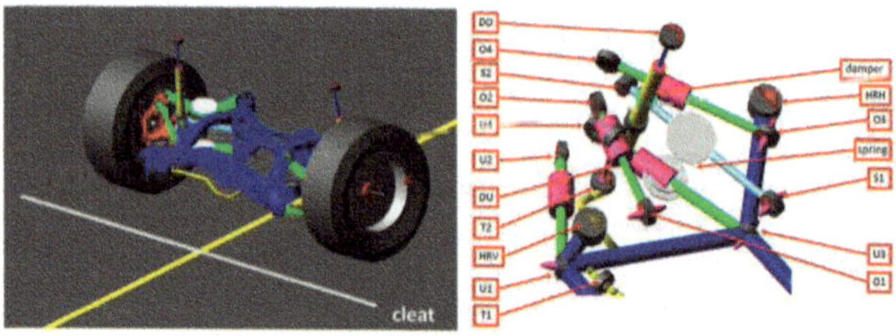

O1-O4 U1-U4	Suspension bushing (O: top; U: bottom)
DO & DU	Strut bearing (O: top; U: bottom)
S1 & S2	Toe link bushing
T1 & T2	Anti-roll bar bushing
HRV	Front subframe bushing (rear axle)
HRH	Rear subframe bushing (rear axle)

Figure 13: Simulation with rear axle model

In figure 14 it can be seen that the change of the stiffness of axle components does not influence the tire frequency behavior significantly. The investigation is just for the static stiffness of all components, the dynamic stiffness has been investigated with another simulation in which the simple bushing models which can just describe the static stiffness have been replaced by the complex bushing models. Otherwise, the elasticity of rigid components such as links, subframe etc. also have been researched and investigated [9]. In these investigations, the subframe bushing and subframe itself show stronger influences than the other components. The influence of subframe bushing on the wheel angular speed frequency behavior was also detected from a vehicle test by SÜNDER [8]. Both components need to be observed in the early concept phase of a vehicle, in order to ensure the functionality of iTPMS.

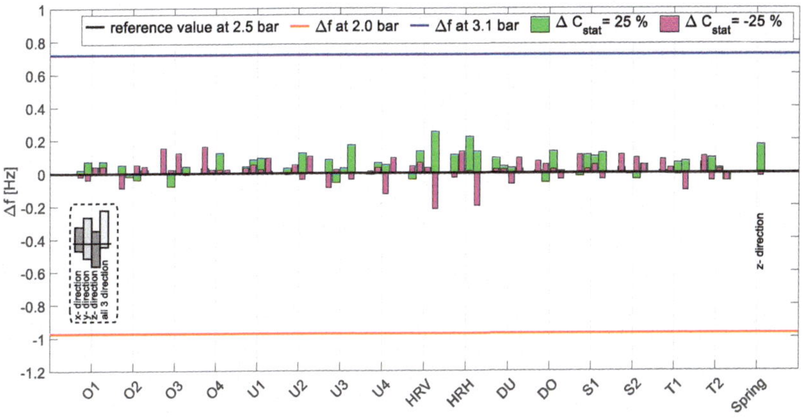

Figure 14: Sensitivity analysis of axle components [9]

3.2.2 Simplified model

The simplified models of front and rear axles have been created for the investigation of axle influences. The models can be seen in figure 15. The modeling of front and rear axles are different, because subframe of rear axle is elastically connected to the vehicle body with bushings. Similar to the rigid ring tire model that these models can be described with differential equations. The modal analysis and sensitivity analysis can be executed to determine the eigenmodes of both axles and to define the important parameters. The parameterization of both models are carried out on the axle test rig [10] in TU Dresden.

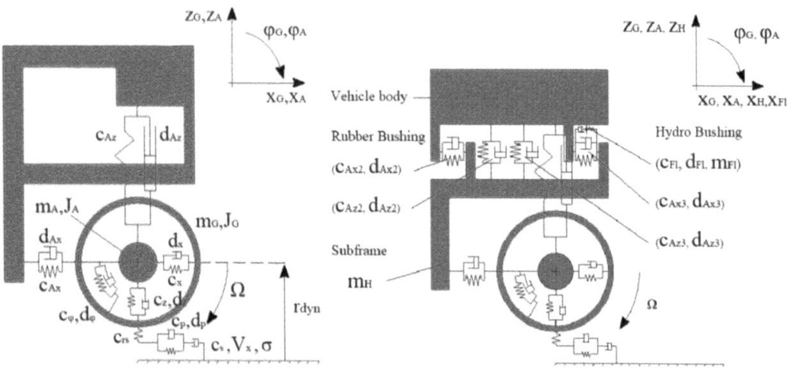

Figure 15. Front axle (left) and rear axle (right) based on rigid ring tire model [2]

Figure 16 shows the frequency spectra of wheel angular speed of front and rear axles. Both axle models can represent the vibration from axle (10-20 Hz) and the vibration from tire (30-50 Hz). Figure 16 shows the change of frequency spectrum after the changing of tire radial stiffness. A reduction of tire radial stiffness is a phenomenon of pressure loss. With a decreasing of tire radial stiffness shifts the tire natural frequency to lower values. This is a typical phenomenon which iTPMS uses for pressure loss detection. The change of frequency spectrum by pressure loss (reduction of radial stiffness) is different between front and rear axles. Due to the different modeling of both axles, the shifts of the tire natural frequency look different. The shift of rear axle is inconsistent compared to the front axle. The peak of natural frequency shifts from about 41 Hz to 37 Hz while changing of tire radial stiffness in a certain area. This change of the peak of natural frequency can possibly influence the pressure detection. This phenomenon can be explained after a modal analysis: in the rear axle analysis exist two vibration modes between 30 Hz and 50 Hz. In most instances, both peaks seem like one vibration mode. The influencing factors of both vibration modes are different. One mode can be strongly influenced by tire radial stiffness and the other mode can be strongly influenced by the mass of subframe. In order to guarantee the high quality of iTPMS regarding detection performance and robustness, a deeper understanding of both vibration modes is essential. The rear axle layout in terms of an optimization of iTPMS might be considered in the vehicle development process.

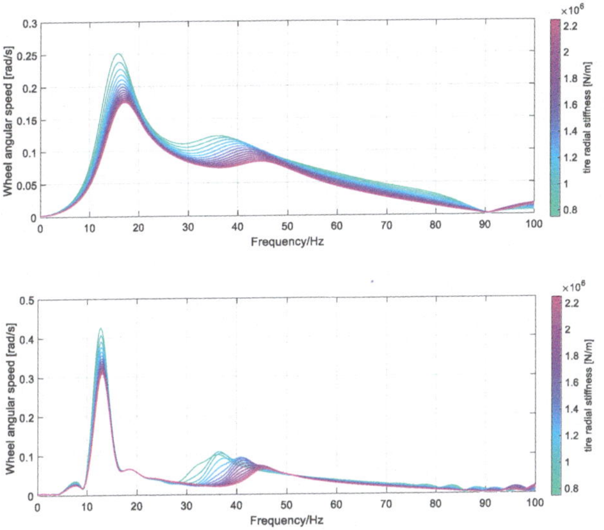

Figure 16: Analysis of the influence of tire radial stiffness on the wheel angular speed spectrum of front axle (top) and rear axle (bottom)

3.3 Full vehicle level: vehicle model

The goal of the full vehicle simulation is to emulate the real-world vehicle test in a simulation environment. It means that the road, tire, suspension and vehicle should be set up as in the real-world vehicle test. As the first step, the virtual test is executed with a complex vehicle model so that the simulation is as close as possible to the vehicle test.

Tire model and vehicle model are two main parts of a simulation. In the section 3.1.2.3 it has been introduced, how a tire model can be parameterized. The tire model performance directly influence the accuracy of a simulation. A comparison between test and simulation with Ftire model and rigid ring tire model has been shown in the section 3.1. The parameterization of a multibody vehicle model is a highly complex topic and will not be described in detail in this paper.

Besides the tire model and the vehicle model, the road on which the vehicle tests take place should be transformed to a digital road as an input for the simulation. Figure 17 shows the proving ground, on which the vehicle test takes place. With 3D scanning technology the road information (incl. the wide, height, unevenness etc.) can be collected and saved in a CRG road file. This road file can be used in the simulation.

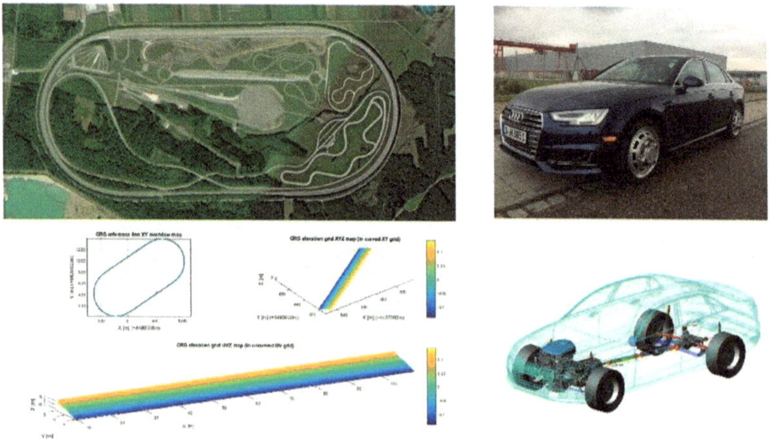

Figure 17: Proving ground and test vehicle (top) vs. road model and vehicle model (bottom)

In order to ensure the iTPMS compatibility with every vehicle-tire combination, diverse tires from different tire manufactures need to be tested in vehicle tests. Different types of tests will be executed to guarantee a failure-free functionality of iTPMS. In this section, a simple test will be introduced. The test can be mainly divided into two parts: calibration phase and detection phase. During the calibration phase, all tires are inflated according to the nominal tire pressure. A vehicle runs on the proving ground with different speeds and loads. In this phase, the tire behavior can be recorded by a logging equipment. In the detection phase, the vehicle runs on the same route with under-inflated tires (e.g. 20%), a warning of iTPMS within a certain time is expected. The warning is based on the two algorithms, which have been introduced in the section 1: the change of rolling radius and the change of natural frequency contained in the wheel angular speed. The change of rolling radius can be easily represented in the simulation. The simulation results about rolling radius will not be shown in this part. The focus of the analysis on full vehicle level in this paper is the tire frequency behavior at pressure loss. In this section, the frequency spectrum of wheel angular speed will be mainly described. In figure 18 is a result of front axle at two pressure levels at 45 km/h on the proving ground.

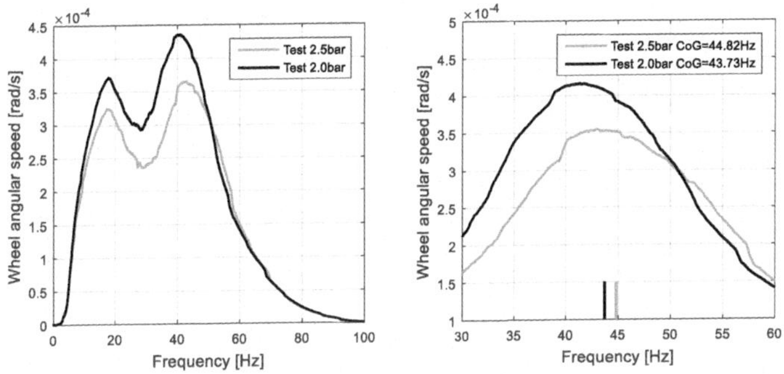

Figure 18: Frequency spectrum of wheel angular speed from vehicle test at two pressures: (left) spectrum 0-100 Hz with a 10-60 Hz filter (right) spectrum 30-60 Hz with a 30-60 Hz filter

From the vehicle test results it can be seen that the peak of the spectrum from 30 to 60 Hz can be hard to identify, because the excitation from the road surface is more complex than the excitation on a test rig. In this case, another parameter is always used for a detection of the frequency shift. This parameter is the "Center of Gravity (CoG)" of the spectrum between 30 and 60 Hz. With the help of this parameter, the change of spectrum due to pressure loss can be detected. According to test results from tire test rig, this parameter has a high correlation with the parameter "peak of natural frequency". The definition of CoG can be seen in this equation [11]:

$$\text{CoG} = \frac{\sum_{i=30Hz}^{60Hz} A_i \cdot f_i}{\sum_{i=30Hz}^{60Hz} A_i}$$

Figure 18 shows that the spectrum has about 1.09 Hz CoG shift due to the 20% pressure loss. This CoG shift can also be gained from simulation. Figure 19 shows results of the multibody simulation with Ftire model and rigid ring tire model. The virtual tests are executed at two pressures (normal pressure and 20% under-inflation) on the proving ground model.

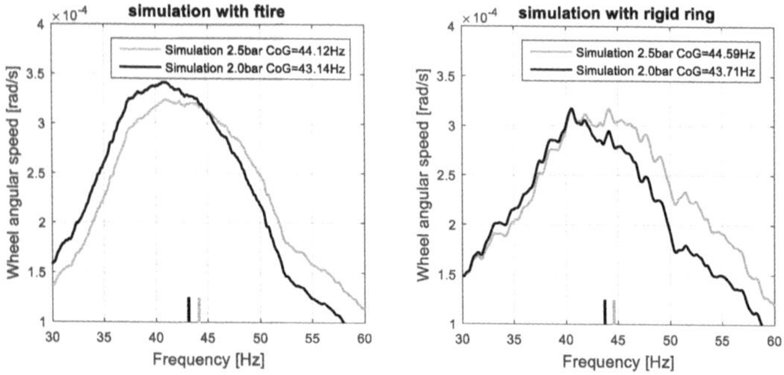

Figure 19: Frequency spectrum of wheel angular speed from simulation at two pressures: (left) simulation with Ftire model (right) simulation with rigid ring tire model

It can be seen from figure 19 that the pressure loss can be detected in the simulation not only with the Ftire model, but also with the rigid ring tire model. The CoG shifts from vehicle test and simulation are about 1 Hz. However, there are small deviations between tests and simulations. The deviation can be strongly influenced by the model parameterization. Otherwise, the processing of raw signal such as filtering and smoothing can also influence the results. These influencing factors need to be considered in order to ensure high accuracy of a simulation.

4 Conclusion and outlook

The aim of this paper is to introduce a model-based development of iTPMS. The investigations in this paper are mainly divided into three levels: component level, subsystem level and full vehicle level. The purpose of the virtual development of iTPMS is on one hand to deepen the understanding of iTPMS and its influencing factors and on the other hand to realize virtual tests to partly replace the real-world vehicle tests.

On component level, the Ftire model and the rigid ring tire model have been presented. Especially for the rigid ring tire model, the modification, parameterization and application of the model have been described.

On subsystem level, a result of an analysis about the influences of axle components has been shown. Otherwise, a simplified axle model has been created. This simplified model can be used to get a deeper understanding of the influences from axle.

In the last section, a comparison between vehicle test and vehicle simulation has been presented. The phenomenon that occurs in vehicle test during pressure loss can also be represented in the mutibody simulation with Ftire model and rigid ring tire model. This comparison shows that the vehicle test can generally be replaced by simulation. However, the accuracy of the simulation is strongly dependent on the tire model. A systematic parameterization of the tire model, which has been introduced on component level, is essential for a virtual test with high accuracy.

In order to implement the simulation into the iTPMS development process, the tire model needs not only to reach a high accuracy, but also to be validated with different types of tires. With the help of a parameterization and validation process, a virtual test with different tires can be realized. An overall process from tire model parameterization to the analysis of vehicle simulation result should be summarized into a process and used into the development process of iTPMS.

References

1. UNDERBERG, V., ROSCHER, T., JENNE, F., PUCAR, P., STURMHOEBEL, J.: Tire Pressure Monitoring Systems, Encyclopedia of Automotive Engineering, 2014
2. HAN, W., GUO, Y., ROSCHER, T., PROKOP, G.: Simulative research on the tire torsional vibration and its vehicle relevant influencing factors, 19th Stuttgart International Symposium, Stuttgart, 2018
3. GIPSER, M: FTire - Flexible Structure Tire Model: Modelization and Parameter, 2017
4. ZEGELAAR, P.: The dynamic response of tires to brake torque variations and road unevenesses, PhD Thesis, Delft University of Technology, 1998
5. SARKISOV, P., KUBENZ, J., VAN PUTTEN, S., PROKOP, G.: Steifer Reifenprüfstand für höhere Messgenauigkeit, Automobiltechnische Zeitschrift (ATZ) 04/2015 S. 46-51
6. SÜNDER, R., PROKOP, G., ROSCHER, T.: Comparative Analysis of Tire Evaluation Methods for an Indirect Tire Pressure Monitoring System (iTPMS)" – SAE Int. J. Passeng. Cars - Mech. Syst., 2015
7. HAN, W., FÜLLGRABE, F., ROSCHER, T., PROKOP, G.: Next steps towards virtual development of indirect Tire Pressure Monitoring System (iTPMS), 14th Intelligent Tire Technology Conference, Dresden, 2018
8. SÜNDER, R.: Entwicklung eines optimierten Verfahrens zur Reifenabsicherung für indirekte Reifendruckkontrollsystem, Dissertation, Technische Universität Dresden, 2017
9. HAN, W., ROSCHER, T., PROKOP, G.: Untersuchung des Einflusses der Elastizität von Fahrwerkskomponenten auf das Reifenschwingverhalten mittels MKS, WKM Symposium, Darmstadt, 2018
10. BINDAUF, A., ANGRICK, C., PROKOP, G.: Fahrwerkscharakterisierung an einem hochdynamischen Achsprüfstand, Automobiltechnische Zeitschrift (ATZ), 12/2014, S. 76-81.
11. SÜNDER, R., PROKOP, G.: Rig based Tire Evaluation for an indirect Tire Pressure Monitoring System (iTPMS), 14th Mini Conference on Vehicle System Dynamics, Budapest, November 2014

VEHICLE DYNAMICS AND TIRE DEVELOPMENT

Influence of rolling-resistance-optimized tires on the rollover stability of battery electric SUVs

M. Sc. Fan Chang[1], Dipl.-Ing. Florian Birnbaum[1], Dr.-Ing. Sebastiaan van Putten[2], Dipl.-Ing. Jan Kubenz[1], Prof. Dr.-Ing. Günther Prokop[1]

[1] Institute of Automotive Technologies Dresden – IAD, Chair of Automotive Engineering, Technische Universität Dresden, 01069 Dresden

[2] Virtual Chassis, Concept Attributes and Functions, AUDI AG, 85045 Ingolstadt

1 Motivation and introduction

Due to the high mass and required volume for a high capacity battery, Sports Utility Vehicles (SUVs) with their large dimensions and high dead load of structure currently appear to a suitable vehicle concept for electrification. Mounting the battery under the bottom of chassis leads to larger moments of inertia and lower center of gravity, whose influence on dynamic rollover behavior has not been investigated sufficiently. In addition, the influence of tire characteristics on nonlinear driving dynamic behavior i.e. rollover performance has been studied only partially. Under the constraints that battery electric SUVs are mostly equipped with rolling resistance optimized tires in order to extend the electric range, the issue has gained attention again.

Thus, the combination of the new vehicle concept parameters and different tire characteristics requires a new consideration for rollover stability. In order to increase efficiency, the interaction of characteristics of rolling resistance optimized tires with rollover stability has to be assessed in the early phase of chassis development, within a virtual environment.

This paper aims for investigation on influence of tire parameters on vehicle dynamics and especially rollover stability to gain a novel virtual assessment of tire characteristics for electric SUV. It is structured as follows. Chapter 2 presents a summary of the state of the art and previous works to give an overview of the rolling optimized tires as well as the existing rollover analysis methodologies. Subsequently, the differences in tire characteristics between performance orientated and rolling resistance optimized tires of electric SUVs are described in chapter 3. Based on this, the performance orientated tire is reconstructed towards rolling resistance optimized tires with a reasonable approach in chapter 4, while the chapter 5 depicts rollover behavior of the two tires through different analysis methodologies. Based on above-mentioned preparations, the rollover behavior is simulated and analyzed gradually at the full vehicle level, not only through Fishhook maneuver but also through effect chain analysis in the nonlinear dynamic range. Moreover, the influential parameters are further evaluated in chapter 6. In conclusion, a summary over the influence of tire parameters on the rollover behavior is executed in chapter 7.

2 State of the art

According to the outline described in the introductory part, this chapter thematizes the overview of the current state of the art. At the beginning, basic dependencies of tires optimized for rolling resistance are discussed and the interactions with the vehicle dynamics in the linear and limit range are elaborated. Subsequently, the methodologies for analyzing the rollover stability, already known from previous publications, and the used simulation tools will be shortly summarized. Based on this, the derived research gaps for this paper can be defined.

Rolling resistance optimized tires

The efforts on the part of the OEMs (Original Equipment Manufacturers) to reduce rolling resistance have been known at least since the mid-1970s [1]. This can be comprehended by the significant proportion of the tires in the total energy consumption of the vehicle. In particular, through increased efforts to reduce CO_2 emissions, the energy efficiency of tires is increasingly being brought into the focus of current research topics. Confirming with [2], the percentage of CO_2 emissions being attributable to the tire influence is comparatively independent from the underlying driving cycle i.e. NEDC (New European Driving Cycle) and WLTP (Worldwide harmonized Light vehicles Test Procedure).

A comprehensive overview of the measures to reduce rolling resistance is already available in [3]. As a consequence, the rolling resistance coefficient can be reduced by three basic actions:

1. Reduction of the dissipative material volume
2. Reduction of the deformation or of the occurring strains ε
3. Reduction of the loss factor $\tan \delta$

However, the challenge tire manufacturers are facing is not the design of tires with a low rolling resistance, but the development of tires that still meet the OEM's requirements in terms of wear, driving dynamics, and acoustics as well as mechanical comfort. Due to physical interactions between the various tire properties considered, this results in a variety of trade-offs, which should ideally be resolved, depending on the specific application. In particular, the third mentioned possibility of rolling resistance reduction eventuates in a target conflict, which is decisive in the linear range as well as the limit range of vehicle dynamics. This conflict of objectives is adequate to the fact, that the tread is on the one hand the only part of the tire, which is in direct contact with the road and therefore solely responsible for the force transmission. On the other hand, the tread compound generates the relevant part of hysteresis related energy losses. Consequently, there is a significant interaction in material design because an essential friction mechanism bases on material hysteresis. Here, a proportionality between the maximum friction coefficient and the loss modulus can be assumed. As already explained, a significant proportion of the energy losses can therefore be attributed to exactly the tire component, which is relevant for the force transmission. By observing this effect over the frequency domain, it becomes apparent that the existing conflict of objectives is resolvable through to the different excitation mechanisms. As illustrated in Figure 1 the excitation frequencies of 10^3-10^7 Hz are relevant for the friction mechanisms, whereas the rolling resistance correlates with the wheel frequency 10^1-10^2 [4]. [5]

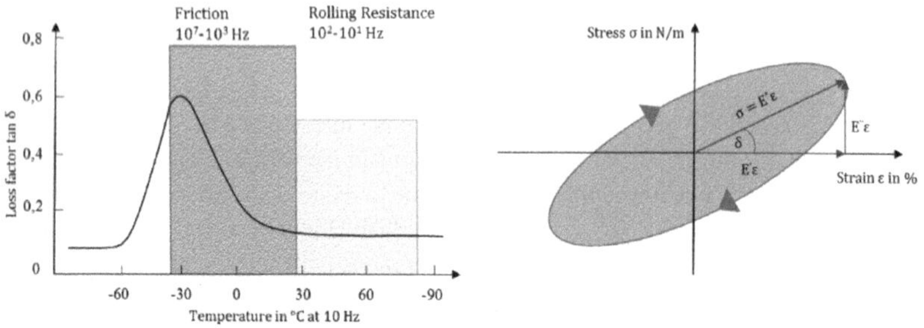

Figure 1 Exemplary loss factor over frequency and hysteresis loop according to [4]

Several studies known from the literature conducted only the effects of rolling resistance optimization on the component level. Moreover, the interactions between tire properties and vehicle dynamics have been for instance extensively analyzed in [1], [6] and [7]. From the linear range to limit range, the influence of tire properties on rollover behavior is discussed in [8]. However, this work is limited to the steady state range. Further investigations of the tire influence on dynamic limit range i.e. on rollover behavior cannot be found in the existing literatures.

Rollover and tire analysis methodology

On the other hand however, a sound foundation for the rollover modeling and its analysis methodology as full vehicle evaluation have been established in previous works. In order to simulate the untripped rollover behavior, a nonlinear two-track model with full axle KnC (Kinematics and Compliance), a dynamic steering and a nonlinear tire model [6] was firstly introduced and this model has been verified to be able to roll over [9]. In order to parameterize the model for this work, the inertial parameters of a battery electric SUV in middle class were identified by a pendulum test. The KnC characteristics were measured through a quasi-static KnC test bench. The utilized nonlinear MFT (Magic Formula Tire) model parameters were assigned through a highly precise hybrid method [10].

Based on aforementioned vehicle model, the individual effects and cumulative influence of concept parameters, which alter from conventional SUVs to battery electric SUVs, on rollover stability have been investigated in [11]. However, only the Fishhook [12] behavior has been discussed. In order to establish a systematical analysis methodology for rollover behavior, an effect chains analysis in the nonlinear dynamic range was proposed in [13]. With the help of this effect chain analysis, the interaction and operation principle of roll and yaw eigenfrequencies, which vary severely due to the battery under the bottom of chassis, have been demasked in [14]. On the basis of previous findings, not only the Fishhook maneuver but also the discrete single sine maneuvers have to be simulated to achieve a profound consideration in the complete nonlinear dynamic range. Thus, these two maneuvers will be utilized to evaluate the influence of tire characteristics on rollover behavior in chapter 5.

From the current state of knowledge, the following main areas of investigation for the reduction of existing research gaps can be deduced. Thereby, it should be clarified in the further course of this paper, whether rolling resistance optimized tires tend to understeer owing to their lower friction coefficients and therefore are reasonable for rollover. Besides, it should be studied, whether this type of tire tends to rollover due to the lower cornering stiffness, which leads to a more unstable rear axle. Moreover, wheel load and camber angle dependencies will be also analyzed.

3 Differences in tire characteristics between performance orientated and rolling resistance optimized tire of electric SUVs

At the beginning of this chapter, objective tire characteristics of two exemplarily selected tires of the same dimension and tread design are compared. These are a performance orientated and a rolling resistance optimized tire.

In the following section, examples of an ordinary and a rolling resistance optimized tire in the dimension 255/50 R20 are analyzed by means of a standard objective methodology for the utilized battery electric SUV. Figure 2 compares selected tire characteristics of the two considered tires versus a reference tire. In addition to the expected difference in rolling resistance coefficient, cornering stiffness, longitudinal and lateral relaxation length as well as lateral friction coefficients show significant differences between the compared specifications. However, a differentiation with respect to other parameters such as vertical stiffness c_z is hardly possible. It is assumed that the constructive and material parameters of the compared tires have a subordinate impact on the vertical stiffness and that the inflation pressure is the dominant influence parameter. In contrast, the tire enveloping stiffness c_{TES}, as a measure of the tire's absorbing performance in case of road unevenness, differences resulting from constructive changes are observed. However, this is for the rollover stability of the entire vehicle of minor importance. Because this characteristic value is primarily relevant for driving over singular road obstacles and in this way for ride comfort. Such kind of road unevenness do not appear during the considered rollover maneuvers.

The following section analyzes the existing differences between the tires in more detail to highlight the impacts of low rolling resistance tires on rollover criticality. As aforementioned, especially the cornering stiffness and the friction coefficients are crucial for vehicle dynamics. For this reason, these tire characteristics are focalized in the following

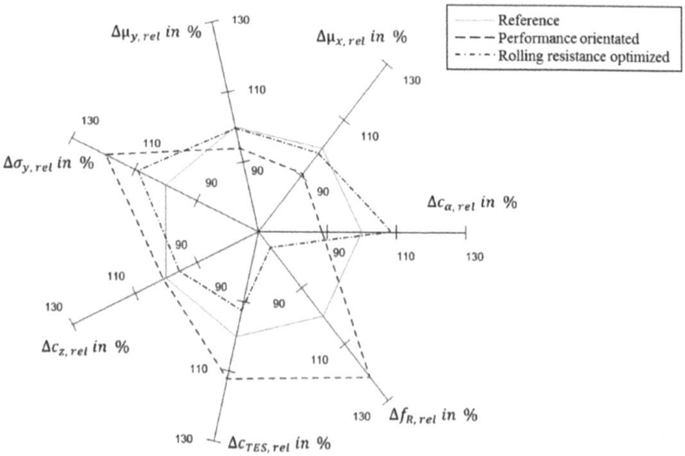

Figure 2 Distinctions in objective evaluation of specific tire properties

In addition to the diverging carcass stiffness, the differences in the cornering stiffness result from variations in the profile stiffness of the corresponding tread compounds. Under the premise of ideal adherence, rigid carcasses and approximately the same contact lengths, the lateral deflection of the profile treads only depends on the kinematic mechanism of the rolling tire. Consequently, identical deflections of the tread elements would result for both tires under identical small slip angles. For this reason, the cornering stiffness generated in this operating condition is solely dependent on the profile stiffness. Thus, it can be concluded that the rolling resistance optimized tire has a lower profile stiffness, whereby, the deformation work is less due to the lower force. Assuming material isotropy in the horizontal direction, the energy dissipation is in accordance with the rolling kinematics also lower. Similar findings can be withdrawn from [3]. The lower tread and consequently cornering stiffness of the in terms of energy efficiency optimized tire is explained by a lower complex modulus of elasticity of the dedicated tread.

Based on the lateral relaxation length, different factors can be investigated in accordance with equation 1. Both, the conventional and the rolling resistance optimized tires have a shortened lateral relaxation length in relation to the reference.

$$\sigma_y = \frac{c_\alpha}{c_y} \tag{1}$$

However, the change in the relaxation length of the conventional tire due to the greater cornering stiffness results from a stiffened carcass. In contrast, the quicker lateral force built up of the rolling resistance tire is caused by the significantly degraded cornering stiffness.

The significant distinctions in friction coefficients are a consequence of the reduced loss factor, which is forced by the rolling resistance optimization. As already described, the dependency of the loss factor on the rolling resistance and the friction coefficient on the frequency can be decrypted. However, the course of the loss factor cannot be manipulated freely over the frequency, so that, despite the existing findings on the use of silica, the rolling resistance tire must have a lower loss factor over the entire area of interest. This behavior can be comprehended by means of Figure 1.

4 Stepwise reconstruction from conventional tires to rolling optimized tires

In this section, the procedure for the transformation of the previously introduced conventional tire towards the tire with a reduced rolling resistance coefficient is derived. Since the used MFT model is an empirical approach, it is possible to manipulate a single tire characteristic, thereby allowing analysis of single effects on full vehicle behavior. Nevertheless, appropriate procedure is required for the targeted analysis of tire properties, which is set out below.

In principle, regarding the characteristics of the horizontal dynamics, the longitudinal characteristics are principally adapted before the lateral ones when converting the conventional tire to the rolling resistance tire. At the beginning, the longitudinal and lateral slip stiffness that influence the driving dynamics in the linear range of vehicle dynamics are adjusted. Subsequently, the transient behavior is adapted by the transformation of the relaxation lengths. Next, the vertical stiffness is transferred to the limit range, since it is assumed that the roll model of the vehicle is only activated in the rollover-relevant area at larger slip angles respectively outside the linear force range.

In the third step, the friction coefficients are degraded to the level of the rolling resistance tire. The coefficients of friction in turn decisively influence the driving dynamics of the entire vehicle in the limit range. Subsequently, the nonlinear range is considered, that is the transition range between the linear range and the maximum force level above the slip values. Then the curvature factors (decreasing course of the longitudinal and lateral force) are fitted after exceeding the maximum. Finally, the remaining tire properties are aligned, so that the final rolling resistance optimized tire is available.

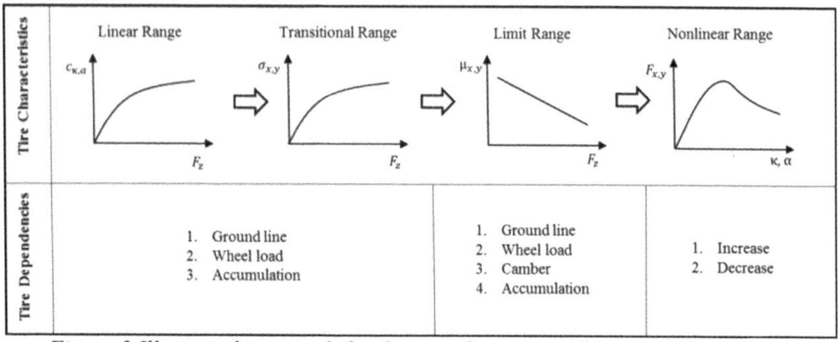

Figure 3 Illustrated approach for the transformation of the experimental tires

The upper part of Figure 3 summarizes the above-described procedure and shows in the lower part that the transformation of the slip stiffness, relaxation lengths and friction coefficients are divided into the single effects of the basic level, the wheel load dependency and the accumulation of both effects. It should be noted, that for the interpretation of the results at the component level, the operating range of tires during dedicated vehicle manoeuvres has to be considered. Because of the high wheel loads and the high sideslip values during this kind of vehicle manoeuvres, no component tests are feasible. This leads to the extrapolation of MFT model in this area.

5 Different rollover analysis methodologies

Utilizing the two rollover analysis methodologies described in chapter 3, this chapter firstly compares the cumulative effects of the two tires (performance orientated and rolling resistance optimized) for rollover behavior. A further investigation of single tire parameters will be conducted exemplarily in the next chapter.

The routine standardized method, Fishhook maneuver comes into view firstly. This maneuver was characterized as the most rollover critical test because of the semi-closed loop condition driven by a steering robot. At the time at which the maximum roll angle i.e. zero roll angle velocity is reached, a steer machine initiates the countersteer action. By this means, Fishhook maneuver shows the dynamical overshoot of the roll model. In this work, longitudinal vehicle velocity was slightly reduced from the typical 80 km/h and the steering velocity has been modified to avoid a total spin out in the yaw motion. To objectify the rollover criticality of Fishhook maneuver, roll angle, sideslip angle at the rear axle and wheel lift of respective wheels are selected as characteristic values.

These three values correspond to the roll motion, yaw motion and rollover performance respectively.

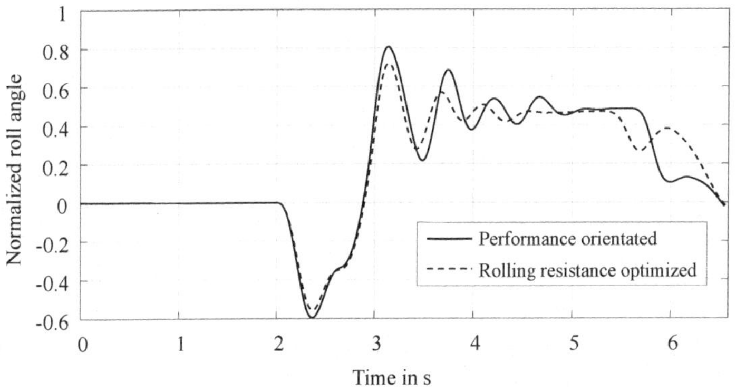

Figure 4 Normalized roll angle

Figure 4 depicts the roll angle behavior with the two different tires. Both vehicles begin with a straight moving and initiate the maneuver after 2 seconds. The first magnitude in the negative range indicates that the vehicles roll around their roll axis in the counterclockwise direction. After the counter steering at the maximum roll angle, the vehicles roll in the other direction and the maximum roll angle magnitude indicates the rollover criticality. In comparison to vehicle with rolling resistance optimized tires, the one with performance orientated tires demonstrates a larger roll angle and oscillation process due to the higher friction coefficient, which means larger maximum lateral acceleration.

From the full vehicle level to component level, the wheel lifts are the most descriptive value to objectify the rollover stability. The left diagram of Figure 5 shows the wheel lift behavior of the front left wheel. It begins with a negative value, which highlights the tire deflection during driving straight ahead. Subsequently, the tires deflects more because of lateral shift of CoG (Center of Gravity) in direction of outer curve. It can be clearly observed, that the vehicle with conventional tire is subject to the larger maximum wheel lift and the wheel tips up longer off the ground. Besides, the second and the third wheel lift peaks can be noticed while the vehicle with rolling resistance optimized tires becomes stable after the first amplitude. In comparison to the front, the rear axle shows more differences between the two tires, in the right diagram of Figure 5Figure 5 Normalized wheel lift at front and rear axle. The performance orientated tire exhibits in that case a rollover risk while the rolling resistance optimized tire demonstrates barely any wheel lift.

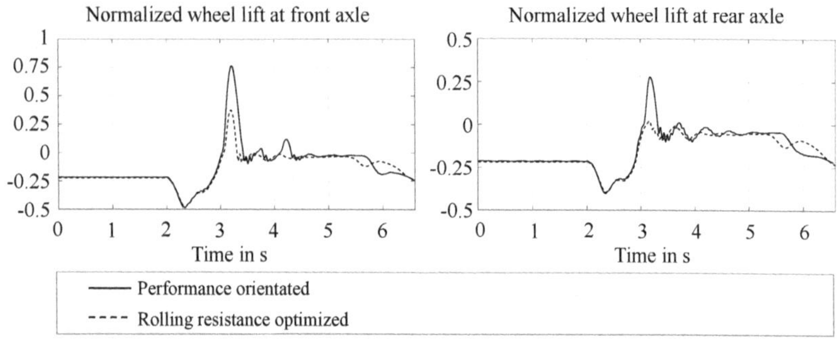

Figure 5 Normalized wheel lift at front and rear axle

Summarizing the results from fishhook maneuver, the rolling resistance optimized tires prove advantageous in consideration of rollover stability compared with conventional tires, which can be mainly explained by the lower friction coefficient, induced through the lower loss factor.

The other analysis method to investigate the rollover behavior is the aforementioned discrete sinusoidal weave maneuver with different steering input frequencies. From this maneuver, the effect chain analysis can be accomplished. Figure 6 shows the nonlinear effect chain from the steering angle via yaw rate and lateral acceleration to roll angle. The same lateral acceleration for the two configurations have been targeted, so that the excitation for the roll model is identical. The partition of the overall transfer behavior into separate transfer links allows a systematical approach to analyze the tire properties. The physical background of this effect chain has been more precisely interpreted in the previous work [13].

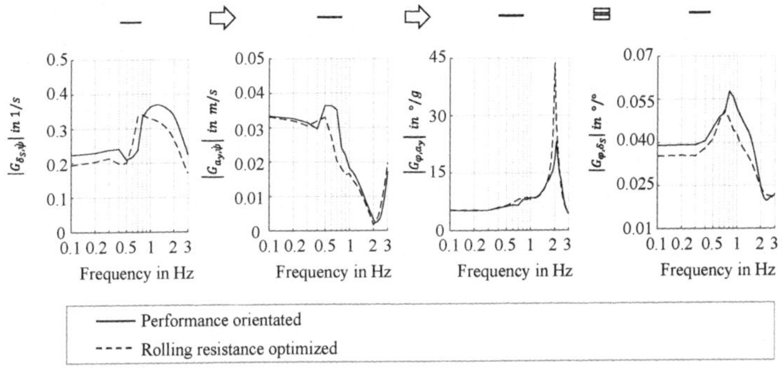

Figure 6 Effect chain analysis comparing the two selected tires

In Figure 6, it can be revealed, that the vehicle equipped with rolling resistance optimized tires shows a lower yaw resonance frequency in the transfer behavior from steering angle to yaw rate. This is reasonable through the lower cornering stiffness of rolling resistance tires. In order to reach the same target lateral acceleration however, the greater sideslip angle at rear axle i.e. effective axle sideslip angle must be built up in Figure 7. In this case, the tire sideslip angle demonstrates a nonlinearity strongly due to the gradient degression of lateral force characteristic curves.

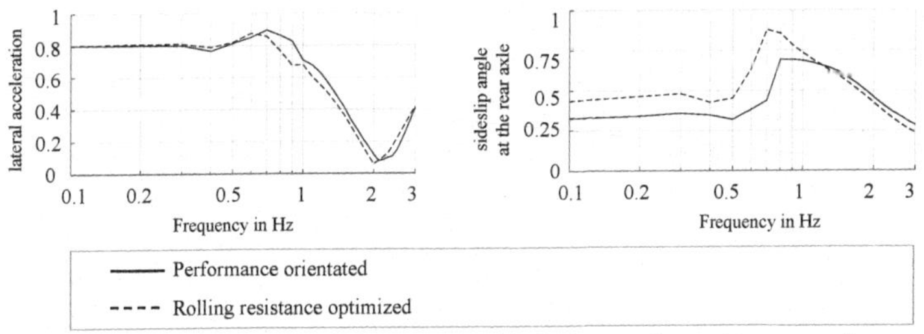

Figure 7 Normalized lateral acceleration and sideslip angle at the rear axle comparing the two selected tires

It is noticeable that the both vehicles have the identical maximum absolute lateral acceleration in the range from 0.1 Hz to 0.7 Hz in the left diagram of Figure 7. However, the maximum absolute roll angle behavior demonstrates a remarkable difference in Figure 8, which is caused by the nonlinear distortion. Through this phenomenon, the third odd harmonic of the input frequency is excited, which additionally excites the roll resonance frequency. Around the frequency of 0.7 Hz, the vehicle with rolling resistance tires is more rollover critical than the one with conventional tires. Further extensive explanation for this nonlinear phenomenon can be found in [14]. However, from the frequency of 0.8 Hz on, the vehicle with performance orientated becomes more critical because of greater absolute lateral acceleration, which can be observed in the left diagram of Figure 7.

By comparing performance orientated and rolling resistance optimized tires with respective analysis methods, it can be concluded that the tire properties have different influence on the rollover behavior in different maneuvers. On the one hand, rolling resistance optimized tires typically show a lower lateral friction coefficient, which reduces the risk of rollover in Fishhook maneuver. On the other hand, cornering stiffness of rolling resistance tires is typically lower, which reduces however rear axle stability and excites more nonlinear distortion. The additionally excited third odd harmonics can generate a greater roll angle, which leads to a rollover critical situation. Combing the two different analysis methods, a more cohesive and systematical consideration of rollover behavior can be conducted.

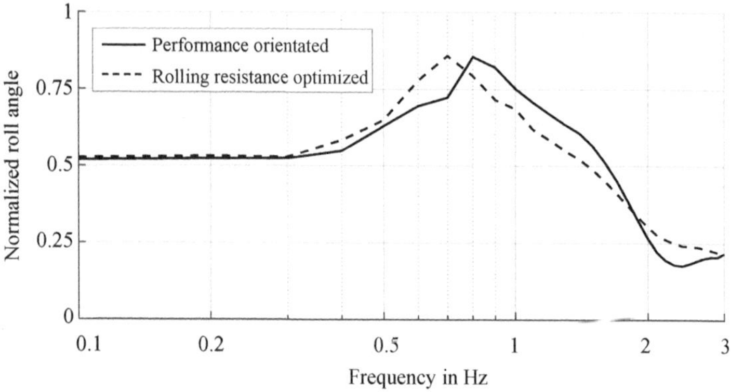

Figure 8 Normalized roll angle comparing the two selected tires

6 Further investigation of the essential parameters in consideration of rollover

In this chapter, the stepwise-reconstructed tires will be simulated gradually through these two maneuvers, in order to investigate the essential tire parameters for rollover further. For Fishhook maneuver, maximum wheel lift has been selected as the characteristic value to objectify the rollover criticality. Summarizing all the single results together, a cascading diagram can be drawn up in Figure 9Figure 8 to gain an overview of the influence for the gradually varied tire properties. The numbers correspond the described procedure in Figure 3.

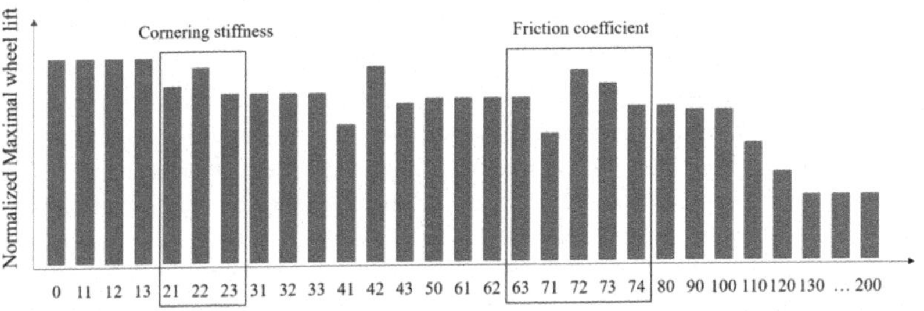

Figure 9 Waterfall diagram of maximum wheel lift for the stepwise reconstruction

Reading from left, the maximum wheel lift of the vehicle with performance orientated tires is the most rollover critical, whose absolute value is identical with the maximum wheel lift in Figure 5. The next three bars (No.11, 12 and 13) show that the changes of slip stiffness in dependence on wheel load and curvature factor have barely any influence on wheel lift behavior. Afterwards, the adaption of cornering stiffness from conventional tires to rolling resistance optimized tires demonstrate the most severe differences. The lower basic level (No.21) of cornering stiffness in linear range leads to a reduction of wheel lift, while wheel load dependency (No.22) of cornering stiffness in turn impairs the rollover stability. It can be seen that the latter effect dominates, so that the rollover stability is improved cumulatively (No.23). After the changes in linear range, the transient adaption has been conducted. Based on No.31, 32 and 33, the longitudinal transient characteristics appear to have no effect on the rollover safety. In contrast, the lateral transient characteristics show noticeable distinction. The lower basis level (No.41) of lateral relaxation length indicate decreases in the wheel lift, while the wheel load dependency (No.42) increases the rollover criticality. The cumulative combination (No.43) of the both effects reveals that the wheel load dependency outweighs. Subsequently, the slightly reduced vertical stiffness (No.50) shows consequently negligible increased wheel lift.

Since the Fishhook maneuver is a quasi-pure transversal dynamic maneuver, the longitudinal friction coefficient (No.61, 62, 63) demonstrates consequently no influence on rollover behavior. In terms of lateral friction coefficient (No.71, 72, 73 and 74), noticeable differences can be observed. Firstly, the lower basic level of friction contributes to the rollover stability due to the limitation of lateral acceleration. In contrast, the effect of load dependency (No.72) of friction coefficient is unfavorable for rollover stability. It is worth mentioning, that dependency of camber has a positive influence. The compensation of the above three properties shows that the lower grip level of rolling resistance optimized tire improves the rollover stability. Afterwards, the curvature factor (No.90) of lateral force and shape factor (Nr.110) of lateral force have certain influence because of the local cornering stiffness. At the end of the diagram, the No.200 shows the maximum wheel lift of vehicle equipped with rolling resistance optimized tires, which corresponds Figure 5.

However, the Fishhook behavior depends primarily on the current maneuver inputs such as vehicle and steering velocity. The trend of the variation of tire parameters can only be explained partially. On the other hand however, the effect chain analysis offers the possibility to evaluate the influence of tire characteristics on the different transfer behaviors separately and systematically. Therefore, the effect chain analysis shall be conducted in order to gain a cohesive understanding. In this paper, emphasis will be given to investigating the influence of cornering stiffness and lateral grip level on the nonlinear effect chain. The other tire parameters can be analogously analyzed.

Figure 10 depicts the comparison of effect chain from performance orientated tires via No.21, 22, to No.23 in order to study the cornering stiffens in detail.

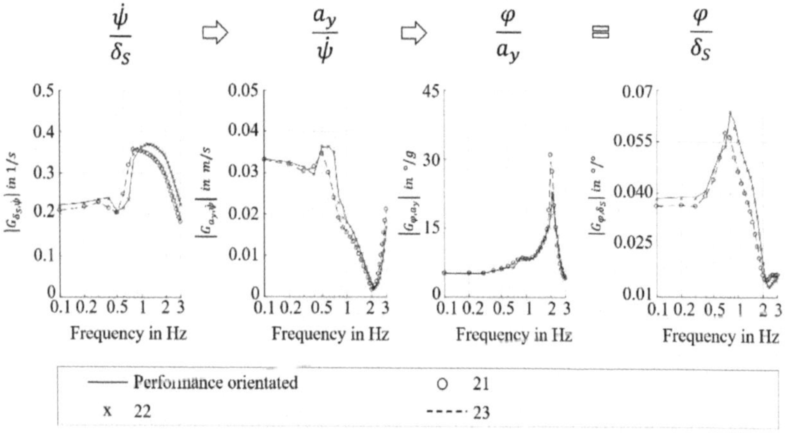

Figure 10 Effect chain analysis for conventional tire and step No.21, 22 and 23

From the first transfer behavior, it can be withdrawn that the reduction of cornering stiffness (No.21) leads to a reduction of yaw eigenfrequency. Meanwhile, dynamic gain increases in the transfer behavior from lateral acceleration to roll angle around 0.6 Hz, which indicates the reinforced nonlinear distortion.

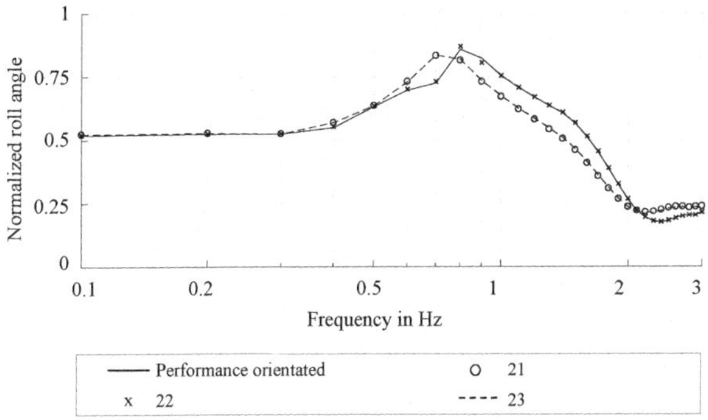

Figure 11 Normalized roll angle for conventional tire and step No.21, 22 and 23

In this case, the roll model will be additionally excited according to Figure 11 and the rollover criticality increases. In comparison to Fishhook evaluation, the decrease of cornering stiffness shows an opposite effect around the frequency of 0.7 Hz in Figure 11. In the next step, the load dependency (No.22) has negligible influence on the rollover behavior and the curves of No.22 overlap with performance orientated curves in Figure 10 and Figure 11. This result differs from the statement of fishhook maneuver. Furthermore, it can be seen that the curves of No.23 overlap with the ones of No.21 i.e. the influence of the basic level of cornering stiffness dominates. The vehicle has become under this circumstances more rollover critical.

Another significant parameter is the lateral friction coefficient. The dependency of basic level, wheel load and camber have been stepwise simulated. Figure 12 presents the simulation results and depicts that the decrease of lateral friction coefficient (No.71) leads to a greater dynamic yaw gain. At the same time, greater sideslip angles are built up to reach the target lateral acceleration. The vehicle is totally in the nonlinear dynamic range and consequently subject to the additional excitation of third odd input frequency.

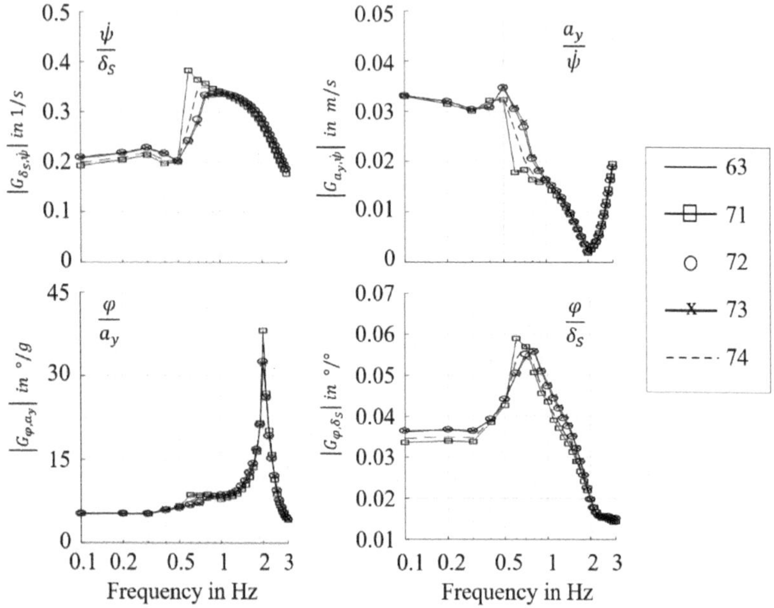

Figure 12 Effect chain analysis for step No.63, 71, 72, 73 and 74

According to Figure 13, the decrease of basic level of lateral friction coefficient has thereby an opposing effect in in comparison to the results of Fishhook. Moreover, the wheel load dependency (No.72) demonstrates barely any influence on the roll angle and the curve No.72 overlaps with No.63 in Figure 13, while it is not favorable in the Fishhook evaluation. Similarly, the camber dependency impairs the wheel lift behavior in Fishhook and however, shows fewer influence the effect chain. Combing the three effects together, No.74 reveals that the basic level is the dominate factor and the vehicle will be more rollover critical though the adaption of lateral friction coefficient in this maneuver.

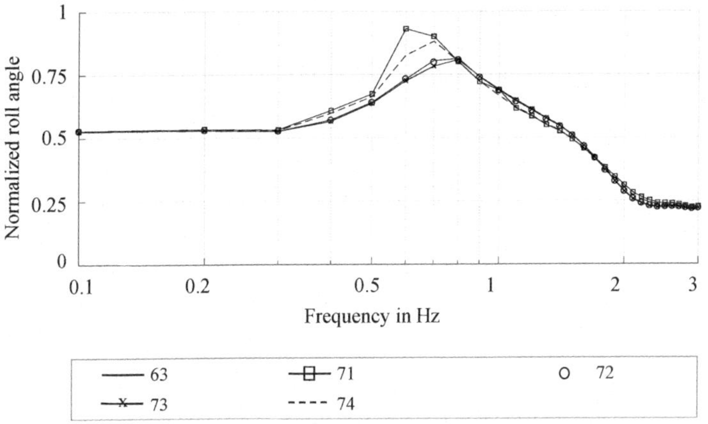

Figure 13 Frequency spectrum of normalized maximum roll angle from different dependences of friction coefficient

7 Summary

After a short introduction, this paper sets its objective to investigate the repercussions of rolling resistance optimized tires on rollover criticality of electric SUVs. Firstly, fundamental issues in the enhancement of the energy efficiency of tires are elaborated. Subsequently, a short introduction of current acknowledgements indicates that influence factors of tires on typical vehicle dynamics are extensively investigated, except of the full extent to the nonlinear range of vehicle dynamics. For this purpose, the utilized vehicle is described. Whereby the objective characterization of tire friction properties on real roads provides an enormous advantage towards existing studies. On the basis of these and moreover the different test maneuvers known from previous work, research gaps defined at the outset could be contracted. After a systematical reconstruction from a performance orientated tire towards to rolling resistance tires, two rollover analysis methods have been carried out. Based on the simulation results, lower roll angles and lower wheel lifts can be observed during the Fishhook maneuver due to lower friction coefficients of rolling resistance optimized tires. This leads to the conclusion, that lower friction coefficients are affirmative regarding rollover stability. Otherwise the low rolling resistance tire results in the course of the discrete sinusoidal weave maneuver to degradation of the rollover behavior. This is adjusted by the reduced cornering stiffness of the low rolling resistance tire. In summary, the tire properties have different effects on different rollover critical maneuvers, which should be taken into consideration in the early stage of chassis development, in order to gain a better design compromise.

8 Outlook

As summarized in the last chapter, the influence of two substantial tire parameters on the rollover behavior is explained emphatically. The next step has to be focused on the other tire parameters such as curvature factor and shape factor of lateral forces characteristic curve.

Besides these parameters, the challenge lies in the tire measurement at the component level. Because of the current state of testing technology, the measurements of tire conditions especially with regard to the high vertical forces in combination with consequently high resulting forces are classified as difficult. Furthermore, overturning moment is worth being investigated, which can be proven as a significant influence on the rollover behavior according to literature [6]. In particular, the mechanisms of the overturning moment have to be explained by means of experimental and simulation-based methods. However, testing technology poses a challenge yet again in this aspect. Moreover, studies with more complex models such as SWIFT (Short Wavelength Intermediate Frequency Tire model) and F-Tire, which can depict physical interactions, allow more possibilities to consider the physical process and offer more possibilities of extrapolation.

Lastly, the connection and relationship between fishhook and discrete single weave maneuver requires further studies. Through this, the investigation of rollover behavior can be more systematical and comprehensible.

Bibliography

[1] S. CLARK, J. WALTER and T. LACLAIR, "The Pneumatic Tire," Washington D.C., U. S. Department of Transportation, 2006, pp. 475-612.

[2] J. PAVLOVIC, A. MAROTTA and B. CIUFFO, "CO2 emissions and energy demands of vehicles tested under the NEDC and the new WLTP type approval test procedures," *Applied Energy,* vol. 177, p. 666, 2016.

[3] M. VENNEBÖRGER, C. STÜBEL, B. WIES and K. WIESE, "Leichtlaufreifen für PKW mit niedrigem CO2-Austoss," *ATZ,* vol. 115, pp. 572-577, 2013.

[4] M. ERSOY and S. GIES, Fahrwerkhandbuch : Grundlagen – Fahrdynamik – Fahrverhalten – Komponenten – Elektronische Systeme – Fahrerassistenz – Autonomes Fahren– Perspektiven, Wiesbaden: Springer Vieweg, 2017.

[5] T. SCHULZE, G. BOLZ, C. STÜBEL and B. WIES, "Reifen im Zielkonflikt zwischen Rollwiderstand und Nassgriff," *ATZ,* vol. 112, pp. 516-523, 2010.

[6] H. B. PACEJKA, Tyre and Vehicle Dynamics, Oxford: Elsevier, 2006.

[7] U. PECKELSEN, Objective Tyre Development - Definition and Analysis of Tyre Charakteristics and Quantification of their Conflicts, Karlsruhe: KIT Scientific Publishing, 2017.

[8] Y. LI, W. SUN, J. HUANG, L. ZHENG and W. Yanyang, "Effect of vertical an lateral coupling between tyre and road on vehicle rollover," *Vehicle System Dynamics,* vol. 51, no. 8, pp. 1216-1241, 2013.

[9] F. CHANG, S. VAN PUTTEN and G. PROKOP, "Analysis of rollover behavior of SUVs in the early phase of chassis development," *18th Stuttgart International Symposium,* 03 2018.

[10] S. VAN PUTTEN, Eine hybride Methode zur objektiven Beschreibung von Reifencharakteristika, Dissertation, Technische Universität Dresden, Göttingen: Cuvillier, 2017.

[11] F. CHANG, S. VAN PUTTEN and G. PROKOP, "Chassis Design of Electric SUVs in Consideration of Rollover," *27th Aachen Colloquium Automobile and Engine Technology,* 10 2018.

[12] Ford, Consumer Information Regulations: Rollover Resistance, NHTSA-2001-9663-24, 2001.

[13] F. CHANG, K. KRAUTER, S. VAN PUTTEN, J. KUBENZ, A. OHLETZ and G. PROKOP, "Cause and effect chains analysis of rollover behavior with respect to chassis design," *19. Internationales Stuttgarter Symposium,* March 2019.

[14] F. CHANG, K. KRAUTER, S. VAN PUTTEN and G. PROKOP, "Analyzing the rollover stability of SUVs," *ATZ Worldwide,* April 2019.

Methodology for the measurement of tire wear particles

Prof. Dr.-Ing. K. Augsburg – Technische Universität Ilmenau

M. Sc. Felix Wenzel – Technische Universität Ilmenau

Dr.-Ing. S. Gramstat – AUDI AG, Ingolstadt

1 Introduction

Non-exhaust emissions from on-road vehicles are gaining increasingly public attention. Besides the disk brake also the tires represents a significant source of abrasion particles of road traffic. The tyre tread surface wears out due to frictional contact with the road surface. To this effect, a new measurement facility, unique in its kind, was designed and developed at the Technische Unversität Ilmenau to assess tyre particulate emissions.

Because of numerous overlapping environmental factors, measuring tyre particles in real road traffic represents an arduous task. Therefore, a test setup for measurement under laboratory conditions on a corner module test bench is developed. First, the requirements for the experimental setup are described. Furthermore, a simple test cycle is defined and the results are analysed. The authors also attempted to identify the relevant influential parameters. The main feature of this measurement system lies in the physical separation of the particle streams coming from tyre and brake. This allows for the simultaneous measurement of tyre- and brake-related particles.

2 Measurement Setup

The measurement setup consists of a corner module test bench mounted on a chassis dyno is given. To measure the tire wear particles, an enclosure of the whole corner module is needed. Moreover, to isolate the brake related particles produced during a brake actuation, the whole brake system must be isolated from the external rim. For each enclosure, a dedicated dilution tunnel with blower is considered. The main role of the dilution tunnel is to slow down the laminar flow. A sampling probe with a calibrated inlet is inserted by a hole on the side wall. Whenever the brake is actuated, the constant airflow stemming from the brake enclosure transports all produced particles through the dilution tunnel to the sampling point. Likewise, the air flow from the corner module enclosure transport the tire wear particles through the second dilution tunnel to the sampling point. To eliminate/reduce the background particulate concentration, the incoming air is cleaned by a filter.

Figure 1: brake enclosure

Figure 2: tire cornersampling box

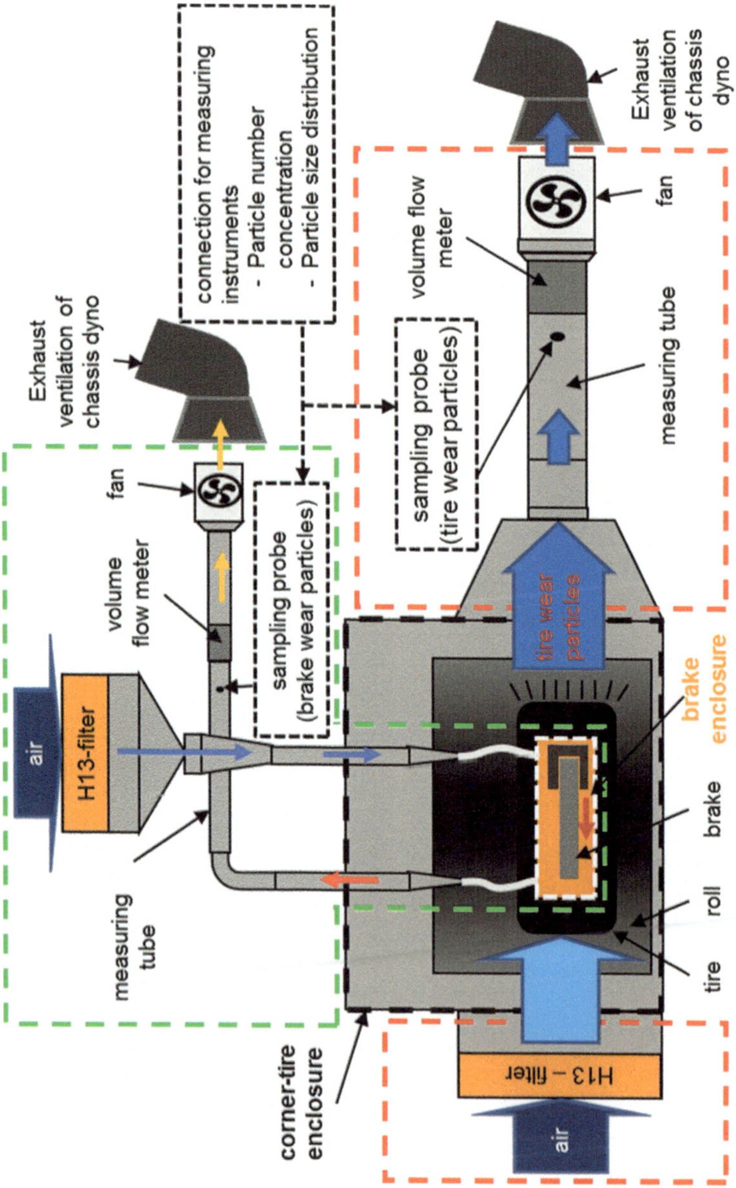

Figure 3: schematic measurement setup

3 Preliminary investigations

Preliminary investigations were carried out at the measurement setup and at the test specimen. For a high particle transport efficiency within the enclosure, the airflow is the deciding factor. Therefore, CFD simulations of the corner-tire enclosure were performed. The simulation shows a good flow around the tire towards the inlet of the measuring tube.

Tire contact area

The preliminary investigations on the test specimen are necessary to characterize the tire and the contact area. As an indicator for the classification of the tire with respect to the wear locations, the hardness profile of the tire tread should be used. Generally, it is known that the wear resistance increases with the hardness of a body. This is confirmed by studies in [1]. Depending on the manufacturer, the rubber compound can be variable across the tyre width to set a certain tyre characteristic [2]. Therefore, it is conceivable that it might show a different degree of wear pattern, which correlates with the local hardness of the rubber mixture. This is to be investigated by means of a hardness measurement at selected positions on the tyre tread strip. For this purpose, the shore hardness is measured at 5 points over the tyre cross-section and at 3 positions above the circumference. Furthermore, a local consideration of the extent of the profile depth reduction takes place as a function of the distance travelled. The aim is the complete measurement of the tyre tread pattern along both the tyre width and the circumferential direction. The tyre tread is measured with a laser line probe to identify local wear positions.

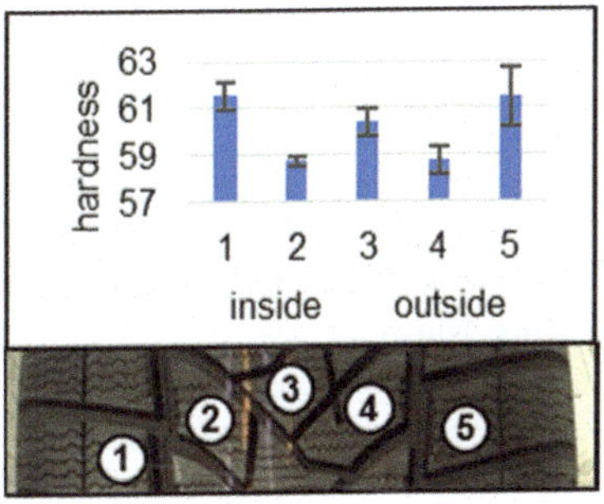

Figure 4: hardness grade of a winter tire

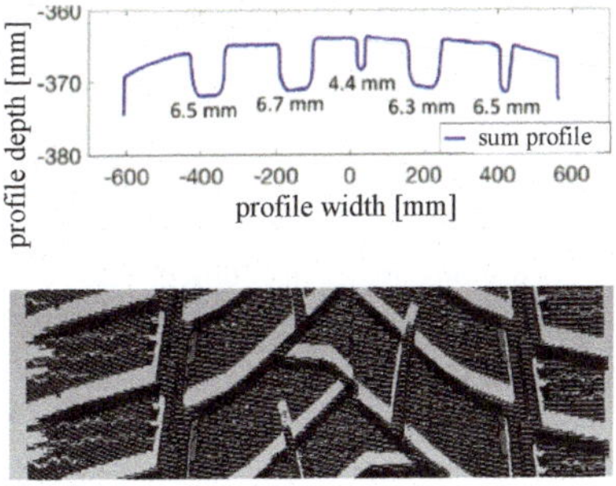

Figure 5: tire tread of a winter tire

Based on the consideration that the wear volume of a tribological system is proportional to the extent of the acting normal force, the tyre-road contact pressure distribution is calculated. To this effect, a pressure measuring mat is employed. The background to this is that the wear volume of a tribological system is approximately proportional to the normal force acting between the friction partners. Consequently, areas of the tire tread that are subjected to higher surface pressure are expected to wear our more easily and therefore produce more particulate. The pressure detection takes place here by piezoresistive sensors on a surface of (268 x 317) mm^2.

Tire temperature

According with preliminary tests, the tire temperature has a great influence on the particle emissions. The developed system measures the tire temperature during the test implementation at 5 Hz. The aim is to detect the tire temperature as a function of the space and time. For this purpose, three pyrometers were installed in the tire tread and distributed across the tire width. Both the temperature of the negative profile and the positive profile are recorded. The fourth infrared thermometer detects the temperature of the outer tire sidewall.

Background conditioning

To reduce the measurement noise and capture even the lowest particle emissions, a low background particle number concentration must be guaranteed. Thus, the supply air of the chassis dyno test is filtered in two stages. In addition, a HEPA 13 filter is installed at the air inlet of the corner-tire enclosure. To get the lowest background concentration the chassis dyno is operated in permanent climatic conditions. By means of these measures, a background concentration of 300 #/cm^3 is achieved

4 Quality of separation: brake – tire wear

In order to prevent the undesired mixing of the tire particle and brake particles, the brake system was encapsulated. Thanks to this design, the tire and the brake particle flows are physically decoupled. The volume flow for the brake cooling is generated by a fan. The enclosure consists of several components, which are sealed against each other. To examine the quality of the sealing, three braking operations were carried out in succession from 120 km/h to 80 km/h. Subsequently, the brakes housing was dismantled and the test series repeated. The particle number concentration was measured in the tire particle measuring section with the TSI CPC 3776 test. Figure 6 shows the significant dominance of the brake particle emissions. The sealing quality is reflected by the effective separation of the brake particles, as it can be seen in Fig. 6. Moreover, the high dynamics of the particle and the very short time delay between particle generation and measurement can be also noted.

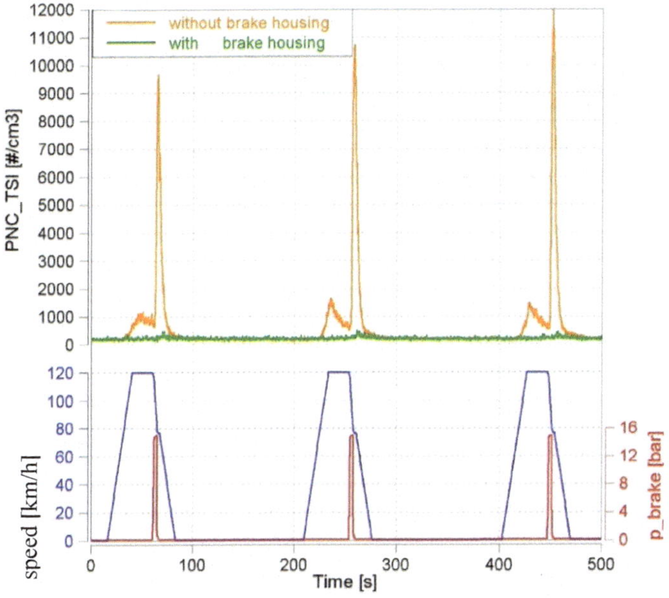

Figure 6: PNC with and without brake housing

5 Influence factors on tire wear

To identify influencing parameters on the particle emissions, a series of tests on a winter tire is carried out. The experiments are carried out once at an ambient temperature of 20 ° C ± 2 ° C, and a further time at 5 ° C ± 2 ° C. The initial tire temperature at the start of the test runs is 37 ° C ± 2 ° C for the measurements at an ambient temperature of 20 ° C. The tire is heated accordingly by means of a conditioning procedure. The initial tire temperature in the low-temperature tests is set at 17 ° C ± 2 ° C. The relative humidity is set to 25% ± 5% during the tests using the air conditioner. The standard tire inflation pressure is 2.6 bar. This corresponds to the recommended air pressure for light load with a wheel load of 600 kg. In addition, the tire inflation pressure influence on the tire wear emission at 450 kg wheel load and a tire inflation pressure of 1.5 bar.

Test cycle

With regard to the vehicle speed, three speed levels are examined in static observation, the distance between them being 75 km / h each. With the 50 km / h level, the inner-urban traffic and the moderate overland travel to be mapped. The 125 km / h cover an average of faster overland trips and moderate motorway speed. The 200 km / h level is for high speed research. The acceleration and deceleration is constant at 2.78 m / s^2 through the chassis dynamometer. These parameters guide to a distance of 1.7 km, 4.6 km and 7.8 km for the respective speed profile.

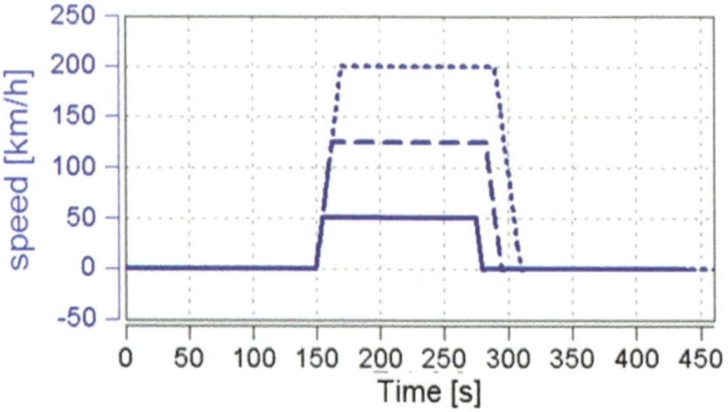

Figure 7: test cycle / speed profile

Each speed profile is repeated 5 times. The evaluation of the particle number concentration takes place during the 120 s constant speed section and only for the last three speed profiles. In this case, the evaluation corridor is corrected on the basis of the delay time of the measurement tube. This is on average three seconds. Furthermore, the background particle load prevailing during the experiment must be taken into account. For this purpose, the mean value of the present background concentration of the last 100 seconds before the start of the speed profile is determined for each individual test run. This value is subtracted from the average particle number concentration present during the constant speed plateau of the test cycle to ultimately conclude the effective tire particle emission. This difference is given in the following statistical evaluations.

Results

First, several series of experiments were evaluated with regard to the reproducibility of the measurement results. This shows a standard deviation within a test cycle of a maximum of 20 % relative to the mean of the three evaluated speed profiles. In most cases, the deviation is less than 10% of the mean value.

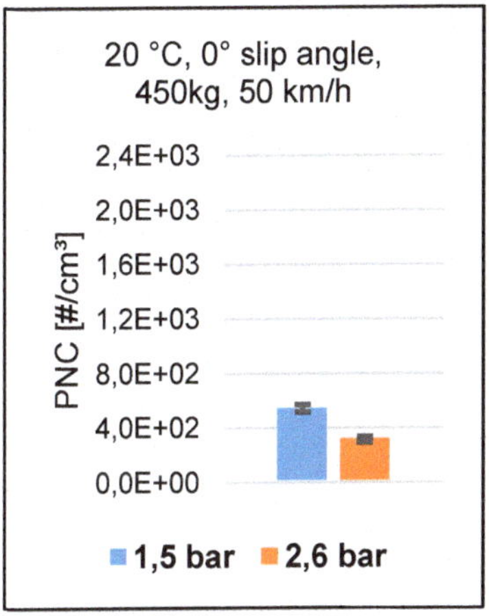

Figure 8: influence of tire inflation pressure

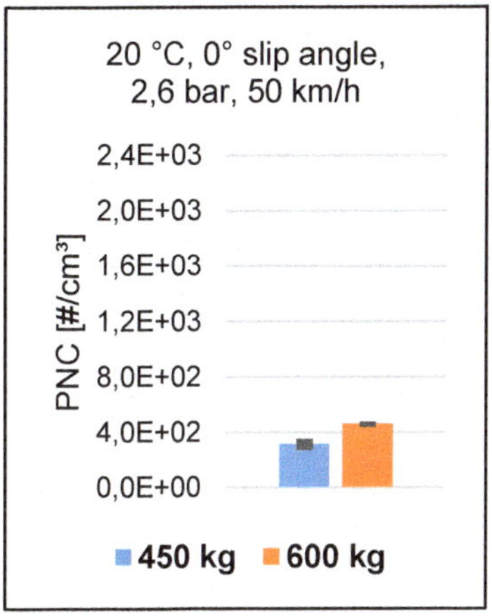

Figure 9: influence of wheel load

A low tire inflation pressure leads to higher particle number concentration. Likewise, the particle emission increases with increasing wheel load.

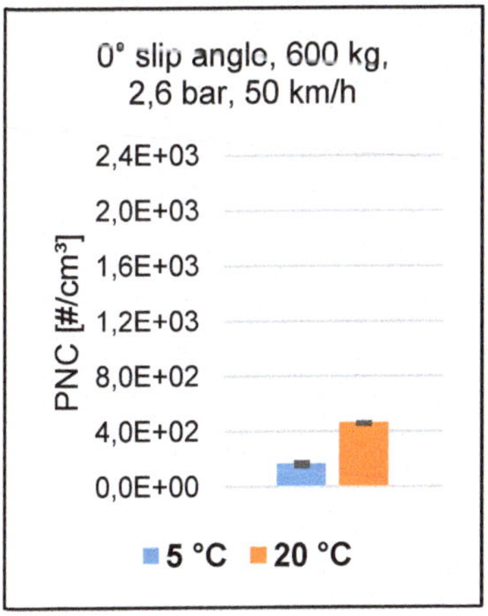

Figure 9: influence of ambient air temperature

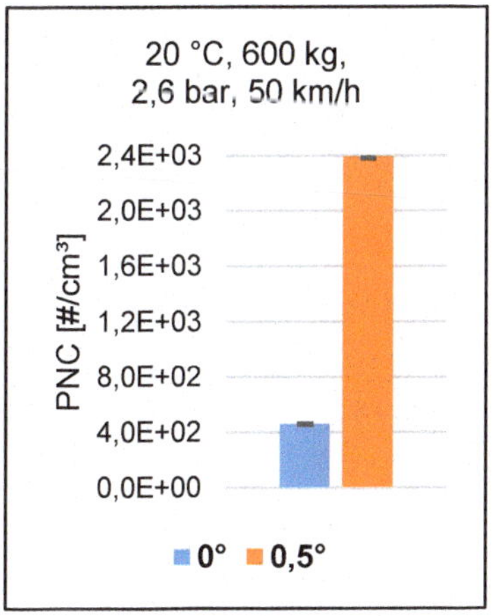

Figure 10: influence of slip angle

The ambient temperature and the slip angle have a greater influence. The graph in Figure 11 shows the speed as the largest parameter of influence on the particle number concentration.

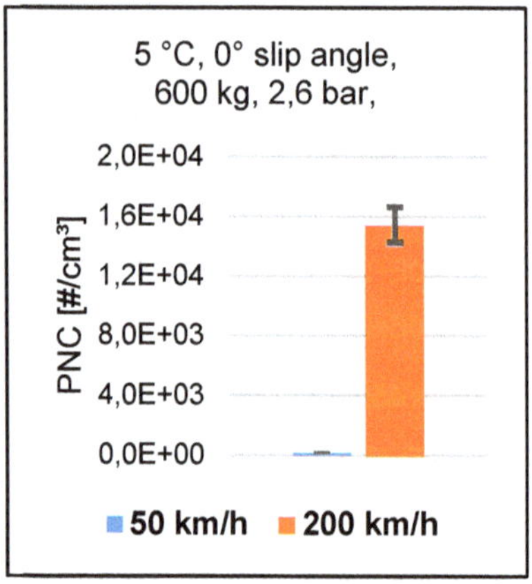

Figure 11: influence of vehicle speed

In addition, the investigation was carried out with a dynamic speed cycle. The WLTC was selected for this. The slip angle was investigated as variation parameter. Figure 12 shows the speed profile and particle number concentrations for 0° slip angle (blue) and 0.5° slip angle. It shows comparable results to the stationary tests.

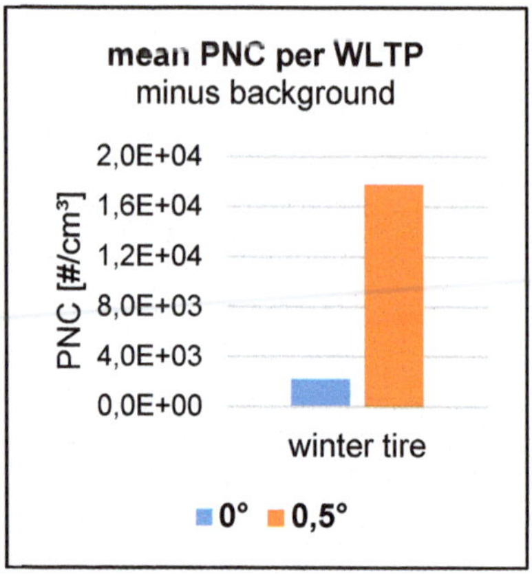

Figure 12: influence of slip angle over WLTC

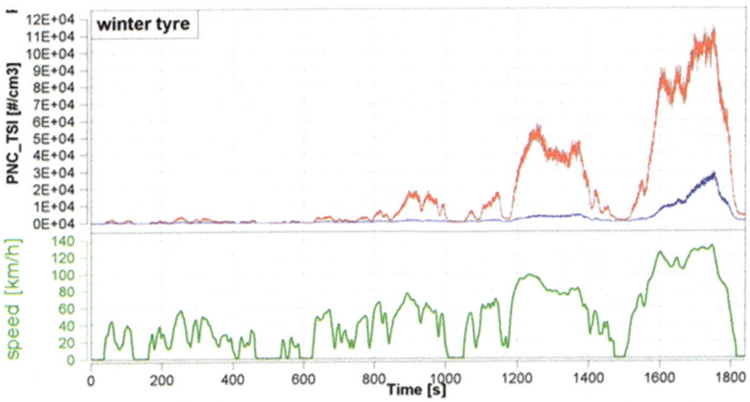

Figure 13: particle number concentration over a WLTC

To become information about the particle size distribution, a DEKATI ELPI + was connected to the tire particle measuring tube. The time courses for the particle number concentration measured at the individual stages show no rash, although an increase in the particle number concentration is detected by the TSI CPC 3776. Therefore, an initial estimation of the particle size distribution is based on the particle mass concentration measured with the TSI Dust Trak II. The investigations are carried out at various operating points using the impact filters of the DustTrak II TSI. With the filters, particles larger than 10 µm, 2.5 µm or 1.0 µm can be deposited, so that the aerosol can be classified with regard to the components PM10, PM2.5 and PM1.0. The results are summarized in Figure 14.

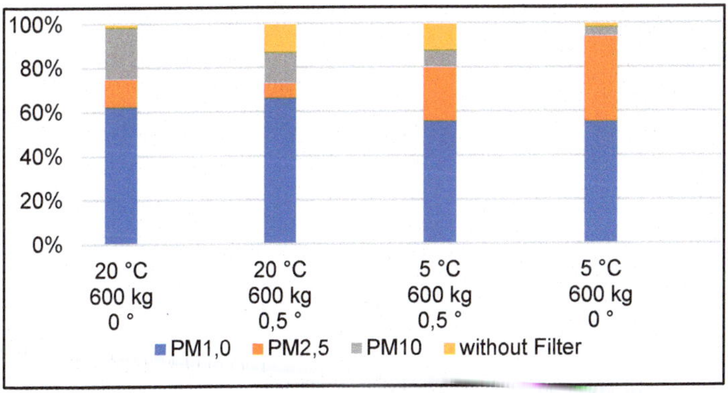

Figure 14: influence of vehicle speed

The evaluation of the particle size distribution happened on the tests with a speed profile of 200 km / h, since the measured particle mass concentrations are very low and are also below 10 µg / m³. For all test runs, a PM1.0 content of 50 to 60 % can be determined. With a larger slip angle, the proportion of larger particles increases. At lower ambient temperatures, the smaller particles tend to increase in the range of PM2.5. Although the measured absolute values of the emitted tire particle mass are very low, the evaluation in Figure 14 proves the existence of tire particles in the fine dust range.

During the tire particle measurements, tire abrasion particles are deposited within the corner-tire enclosure. Especially many particles accumulate behind the tire outlet at the transition to the tire measuring tube in the area of the funnel. These particles are collected after completion of the experiment and taken for further analysis. The evaluation of the particle size distribution is done with a CAMSIZER X2. The sample is dispersed by means of compressed air and the particles are visually evaluated. Particles with a size of 100 µm to 4 mm were detected. However, only the particle number up to a particle size of 1 mm is sufficient for a statistical evaluation. Therefore, only the particles up to 1 mm are evaluated for the particle size distribution. The maximum is about 160 microns. The sum distribution indicates that about 50% of the fines of the sample have a particle size of less than 300 µm.

Figure 15: collected large particles

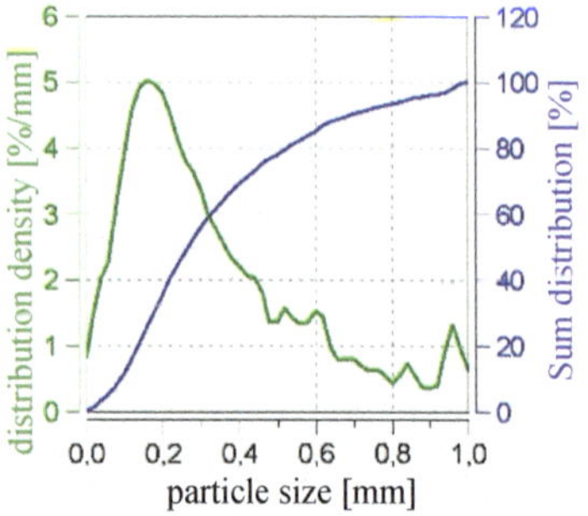

Figure 16: particle size ditribution

6 Conclusion

The authors introduce a measurement setup for tire wear particles. This considers the layout of the testing environment, which is represent by a corner-tire module mounted on a chassis dynamometer, air conditioning system, brake- / corner-tire-enclosure and measuring section. Additionally, the used measurement devices to characterize the particle number concentration and the particle mass are introduced. A method for separating the brake and tire wear are developed. The measurement setup were intensive optimized to realize a low background particle number concentration of 300 #/cm³.

As a next step, intensive preliminary investigations to characterize the tire were carried out. This considers the hardness grade, the tire tread and the tire contact area.

Furthermore a fist simple test cycle for stationary investigations and to identify influence factors is given. With this procedure an evaluation method for tire wear particles is defined. The influence factors with the highest impact is the vehicle speed and the slip angle. As dynamic cycle, the WLTC is used.

Using the measurement approach, the influence factors on tire wear particle emissions on a winter tire are presented. The investigation of the particle size distribution revealed particle size from 1 µm to 3 mm.

Further investigations are necessary on the run in behaviour respectively the bedding procedure and the characterisation of different tires. For more realistic test cycles the transferring of dynamic wheel settings from test drives to the dynamic actuators on corner-tire-module are planned.

7 References

[1] Czichos, H., Habig, K.-H.: Tribologie-Handbuch (3. Auflage). Wiesbaden: Vieweg + Teubner Verlag, 2010

[2] Leister, G.: Fahrzeugräder – Fahrzeugreifen (2. Auflage). Wiesbaden: Springer Vieweg Verlag, 2015

Printed by Printforce, the Netherlands